Geography and Drug Addiction

Yonette F. Thomas · Douglas Richardson ·
Ivan Cheung
Editors

Geography and Drug Addiction

Foreword by Douglas Richardson

Springer

Editors
Yonette F. Thomas
National Institute on Drug Abuse
National Institutes of Health
Epidemiology Research Branch
Division of Epidemiology,
Services and Prevention Research
Bethesda, MD
USA

Ivan Cheung
Insurance Institute for Highway Safety
1005 N. Glebe Road
Arlington VA 22201
Suite 800
USA

Douglas Richardson
Association of American
Geographers
1710 16th St., Nw
Washingotn DC 20009
USA
drichardson@aag.org

Library of Congress Control Number: 2008925135

Printed on acid-free paper

9 8 7 6 5 4 3 2 1

springer.com

Foreword

Making Connections: Geography and Drug Addiction

Geography involves making connections – connections in our world among people and places, cultures, human activities, and natural processes. It involves understanding the relationships and 'connections' between seemingly disparate or unrelated ideas and between what is and what might be.

Geography also involves connecting with people. When I first encountered an extraordinarily vibrant, intelligent, and socially engaged scientist at a private dinner several years ago, I was immediately captivated by the intensity of her passion to understand how and why people become addicted to drugs, and what could be done to treat or prevent drug addiction. Fortunately, she was willing to think beyond the bounds of her own discipline in her search for answers. Our conversation that evening, which began with her research on fundamental biochemical processes of drug addiction in the human body, evolved inevitably to an exploration of the ways in which research on the geographical context of drug addiction might contribute to the better understanding of etiology of addiction, its diffusion, its interaction with geographically variable environmental, social, and economic factors, and the strategies for its treatment and prevention.

This fascinating woman, I soon learned, was Nora Volkow, the Director of the National Institute on Drug Abuse as well as the granddaughter of Leon Trotsky. Our chance encounter that evening led to further wide-ranging discussions during several subsequent months on the interactions between geography and drug addiction, resulting ultimately in an agreement between the Association of American Geographers and the National Institutes of Health's (NIH) National Institute on Drug Abuse (NIDA) to jointly sponsor a special symposium on research topics related to geography and drug addiction.

This special AAG/NIDA Symposium eventually took place in March 8, 2006 in conjunction with the 2006 Annual Meeting of the AAG in Chicago, Illinois. We invited interested geographers, neuroscientists, GIScientists, medical researchers, epidemiologists, geneticists, and others with expertise in geographical dimensions of drug addiction and abuse to apply to participate in the symposium. Themes addressed included:

- Spatial patterns of drug use and addiction
- Linking spatial models with neuroscience and genetics in drug abuse research

- Interaction of social and environmental factors with biochemical processes of addiction
- Geographic analysis linking demographic and genetic characteristics related to drug addiction and treatment
- Locational analyses of drug addiction treatment and service delivery facilities
- Neighborhood scale studies of geographic factors (including the built environment) and their interaction with drug addiction, treatment, or prevention
- Use of Geographic Information Systems to better understand and respond to drug addiction
- Spatial diffusion modeling of addictive drug usage and its changing characteristics, including predictive modeling
- Interaction of other spatially dependent variables with drug addiction, or with prevention and treatment strategies
- Other geographic research relevant to better understanding the etiology of drug use and addiction

Attendance at the Geography and Drug Abuse Symposium was open to all and generated wide-ranging discussion and many new ideas for research and collaboration. Results of the symposium and subsequent conversations among the participants appear in this book, which we hope will help guide the development of future research agendas within geography and GIScience, and within NIDA and more broadly at NIH.

There has not been a great deal of past research on the connections between geography and drug addiction. Thus, it is important to note that the purpose of this book is to explore the relatively new terrain of an embryonic field of research. As such, this book represents an initial attempt to identify research ideas, connections, and research pathways which point to some promising avenues for future work in this area.

It is our hope that our initial explorations of research pathways and agendas in this book will generate far greater interest in and significant funding for this important new field. If we are successful in this goal, we look forward to publishing subsequent volumes reporting on what we believe will soon be a rapidly growing and mature field of research, essential to understanding and treating drug addiction. I would like to thank our publisher, Springer, whose editors were quick to appreciate the significance of this new field of research and encouraged our early efforts by publishing not only this first volume exploring these linkages and research needs, but by also by initiating a new series of books on this theme, with this book as the initial volume in the series.

We also hope that the 'connections' forged between the topics of geography and drug addiction – and between the AAG and NIDA – will provide geographic context and analysis to support NIH's ongoing efforts to understand the complex processes of drug addiction. I believe this book and these connections have the potential to create an extraordinarily fertile new field for geographic research, one which has significant potential for real-world benefit through better understanding and treatment of the scourge which is drug addiction.

I would like to thank Nora Volkow, who helped 'brainstorm' this collaborative process and who delivered the symposium's keynote address, and our distinguished NIDA colleagues Yonette Thomas and Wilson Compton for their sustained support and friendship, as well as the many colleagues and contributors from the worlds of geography and medical sciences who made the symposium and this book possible.

Douglas Richardson
Executive Director
Association of American Geographers

Contents

List of Figures

List of Tables

Chapter 1
Placing Substance Abuse

Geographical Perspectives on Substance Use and Addiction

Sara McLafferty

Abstract This chapter discusses how the concept of place and related geographic methodologies has been used in understanding health and health care, and it suggests avenues of exploration for research on substance abuse. 'Place' provides a foundation for understanding how substance abuse behaviors, prevention, and treatment relate to place environments from the personal to the global scale. I describe three main areas of contribution: exploring geographic inequalities in health; understanding the associations between place environments and health; and analyzing disparities in health care access and location. Each of these topics is discussed, in turn, emphasizing important themes and recent developments in health/medical geography literature and their applicability to research on substance use and abuse.

Introduction

Substance abuse is one of the most pressing health issues in the US. Defined as a harmful pattern of use of substances, such as drugs or alcohol, substance abuse has captured the attention of public health researchers and policy-makers in the recent decades. Although much substance abuse research has focused on biomedical pathways, increasingly researchers are considering how peoples' everyday environments and the political and cultural contexts in which they live influence the prevalence and consequences of substance abuse. Exploring how place environments relate to health is the very essence of health or medical geography. Until recently, however, the linkages between health geography and substance abuse were relatively unexplored. In this chapter, I argue that geographers can contribute to research on substance use and addiction by teasing out the connections between place environments and health

S. McLafferty
Department of Geography, University of Illinois at Urbana-Champaign, IL, USA

Y.F. Thomas et al. (eds.), *Geography and Drug Addiction*,

and by utilizing geographic tools and methodologies to explore these associations. I draw upon recent work in health geography to sketch out current and future directions for substance abuse research.

In health geography, 'environment' is used in a very broad and all-encompassing way to refer to the 'lived space' outside the body. It includes nature, the built environment, human social networks and interactions, local services, amenities, and institutions. It exists in multiple and overlapping scales from the global and regional scales to the personal spaces of everyday life. Place, the concept that underpins geographical research on health, builds on but moves beyond the concept of environment (Kearns 1993). Places are lived environments. They are environments infused with the meanings that people bring to them. Agnew (1987) describes places as 'meaningful environments'. The notion of place links the spaces of everyday life with peoples' experiences, perceptions, and well-being. Medical geographers have used the lens of place to understand a wide range of health issues from women's experiences of multiple sclerosis (Dyck 1995) to mental health care provision (Foley and Platzer 2006) to neighborhood quality and ill-health (Ellaway and MacIntyre 1998).

Place is not just an expression of social differentiation, but an integral part of it. Social interactions and activities are structured in place. The opportunities and hazards that exist in place shape people's lives and livelihoods. Through local institution – schools, shops, recreational, and employment opportunities – places provide the material basis for everyday life. In addition, places are sites for environmental hazards, such as air and noise pollution, and social hazards, such as crime and racial or ethnic discrimination. Social interactions within places affect peoples' senses of belonging and exclusion. People respond differently to places and have different susceptibilities to place characteristics. Responses and susceptibilities vary with age, gender, race, ethnicity, and class, and are often framed by local cultures and economies that are rooted in place. The intersection between place exposures and socially defined susceptibilities creates the geography of risk (Jerrett and Finkelstein 2005).

At the same time, people and social institutions engage in processes of place-making, shaping, and re-shaping the built environment and the social interactions within it. "People create places and places create people" (MacIntyre and Ellaway 2003). Places are contested. Place landscapes are imprinted with signs of powerful and not-so-powerful interests attempting to craft the environment to accommodate their interests. This politics of place has important effects on health and health care inequalities.

The concept of place provides a foundation for situating substance abuse research – for understanding how substance abuse behaviors, prevention, and treatment relate to place environments from the personal to the global scale. One can think of substance abuse as a specific health outcome that is shaped by the place context of daily life, including access to resources and services, the local availability and cost of substances, social networks and cultural norms, and specific settings in the local environment that either enhance or reduce the likelihood of substance abuse. In addition, the quality, quantity, and effectiveness of treatment and prevention programs vary from place to place, reflecting political, social, and economic processes that intersect around a politics of place. Thus, 'placing' substance abuse research involves looking beyond individuals to examine substance abuse and treatment in context.

Alongside the increasing interest in place and health has been the rapid development of geographic tools and methods centered around geographic information systems (GIS), but encompassing a much wider array of tools for spatial and social analysis. In addition, vast quantities of geocoded environmental and social data are available, and the supply of geospatial data from sources ranging from government agencies to satellites to GPS-tracking devices is rapidly increasing. These data, and inventive new tools for analyzing such data, provide a strong foundation for innovative place-based health research.

This chapter discusses how place concepts and GIS-based tools have been used in understanding health and health care, and it suggests avenues of exploration for research on substance abuse. I describe three main areas of contribution: exploring geographic inequalities in health; understanding the associations between place environments and health; and analyzing disparities in health care access and location. Each of these topics is discussed, in turn, emphasizing important themes and recent developments as represented in health geographic literature. This is not meant to be an exhaustive review of the literature but rather a series of signposts to point the interested reader to newer directions.

Exploring Geographic Inequalities in Health

An important area where geographers can contribute to substance abuse research is by using visualization and spatial analysis methods to explore geographic inequalities. Maps have long been essential tools for understanding health issues. From John Snow's well-known map of cholera in 1850's London to today's electronic atlases of disease distribution, maps can shed light on the etiology of health concerns, such as substance abuse, and provide a geographical foundation for health policy-making. Maps reveal the sharp geographic contours of health inequalities and the associations between health and environmental/place characteristics. While mapping continues to be important in health research, advances in GIS have shifted the emphasis from creation of static maps to a more exploratory process of spatial data visualization and analysis. Gatrell and Bailey (1996) describe three classes of spatial analysis and GIS tasks, which can enrich public health research and policy formulation: visualization, data exploration, and modeling.

Visualization refers to the creation of graphical and map displays of health data. Maps are a commonly used visual tool for exploring health data, and there are many interesting recent examples of mapping of substance abuse and related health issues. For example, a map of hepatitis C in Connecticut created by Trooskin et al. (2005) shows a highly uneven pattern with concentrations in the major urban areas of the state and in places where injection drug use is concentrated. A series of maps of disaster-related stress in New Orleans following Hurricane Katrina (Curtis, Mills, and Leitner 2007) depicts the complex dimensions of vulnerability to natural disasters and the strong ties between poverty and vulnerability. Mapping has also been used to assist field research methodologies, such as capture–recapture methods for estimating hard-to-find populations (Kruse et al. 2003). Interactive, web-based mapping is becoming increasingly popular. Online, interactive health atlases, such as the Atlas of Cancer Mortality in the US (Bell et al. 2006), make health data accessible to a wider audience and communicate geographic variation. These online atlases are designed to facilitate data querying and exploration, so that the user controls, to some extent, the map output. Visualization also includes innovative forms of display, such as cartograms (Dorling, Barford, and Newman 2007), animated map sequences (Goovaerts 2006), and traffic light maps (red-green-yellow) for identifying priority areas for health care intervention (Boulos and Phillips 2004).

Figure 1.1 maps the uneven spatial distribution of recorded drug use among pregnant women in Brooklyn, NY, based on the mothers' residential location. The data come from vital statistics birth records of all women who gave birth in 1990. Use of drugs, such as cocaine, heroin, and marijuana, was recorded at or near the time of delivery. Although it is likely that a significant fraction of drug use is not recorded, if we assume that the rate of recording error is relatively uniform across the borough, then the map provides a general sense of the residential neighborhoods in which drug use among pregnant women is concentrated. Concentrations of high numbers of pregnant women with recorded drug use appear in northern and central Brooklyn. Roughly 60% of the cases occur in high-poverty neighborhoods, indicating a strong association between drug use and neighborhood disadvantage, as observed in a previous research (Galea 2004).

Spatial analysis methods can be used to 'add value' to mapped information by making patterns clearer and easier to identify. For example, one challenge in mapping substance use at a detailed geographical scale, such as census tract, is that the numbers of events may be quite small leading to high random variability. This is known as the 'small numbers problem'. In Fig. 1.1, the small numbers problem is evident in the patchy map pattern, with zero and non-zero values side-by-side. One way of handling this problem is to 'borrow strength' by incorporating data from neighboring areas via a smoothing process. Smoothing involves moving a small 'window' across the map and calculating the incidence or prevalence of health events within the window to provide an estimate of incidence within a small, localized area. There are many different methods for spatial smoothing (Langford 1994). One important method is kernel density estimation (Bailey and Gatrell 1995). This method has been employed in several health geographic studies, including estimating geographic variation in environmental risk (James, Matthews, and Nix 2004), visualizing spatial clusters of disease (Kingham, Gatrell, and Rowlingson 1996), and evaluating immigrants' spatial access to prenatal care (McLafferty and Grady 2005). For the drug use data of Brooklyn, Fig. 1.2 displays a contour map, created via kernel estimation, of the density of mothers (women per square mile) with recorded substance use in 1990. Areas of high concentration are clearly visible as 'peaks' on the map, and the irregular pattern, as evident in Fig. 1.1, has been smoothed to reveal the overarching geographic trend.

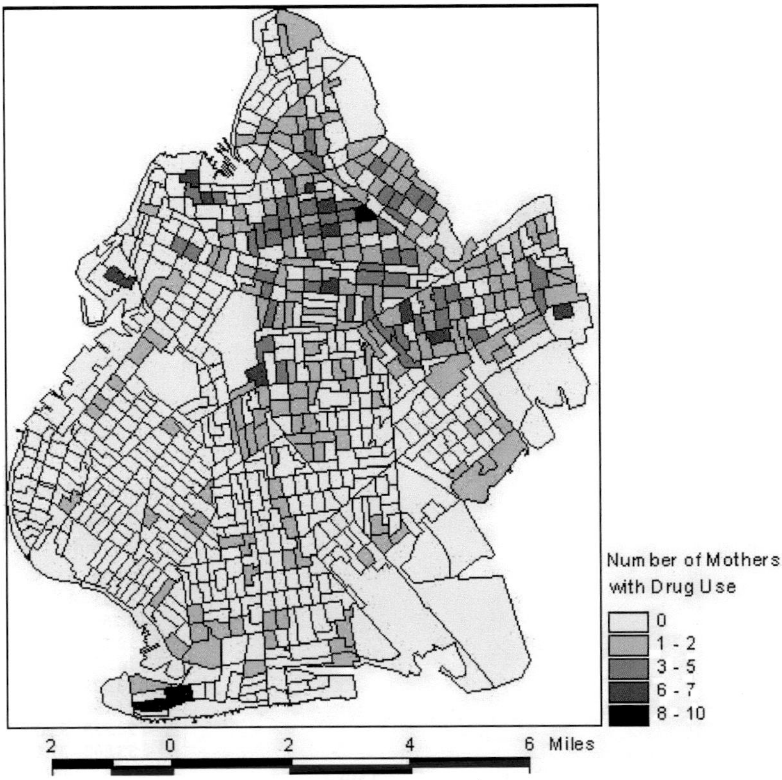

Fig. 1.1 Spatial distribution of recorded drug use among pregnant women in Brooklyn, NY, 1990

Spatial analysis tools also facilitate data exploration, helping us see patterns in data by detecting unusual clusters of health events. Spatial clusters are localized areas that contain an unusually high or low concentration of health events. Since the early 1990s, methods have been developed to scan maps searching for clusters of cases that are unusual geographically and statistically (Anselin 1995; Kulldorf 1997). From breast cancer (Sheehan et al. 2004) to birth defects (Rushton and Lolonis 1996) to hepatitis C (Trooskin et al. 2005), these methods have been widely used in exploring geographic variation in health, although they have not been widely used in examining substance abuse (an exception is Latkin, Glass, and Duncan 1998). More recent research extends these methods to address issues such as analyzing clustering in time and space (Avruskin et al. 2004), controlling for individual-level factors, such as age and socio-economic status, in evaluating clusters (Sheehan et al. 2004) and detecting clusters that are irregular in shape, such as those that might occur along roads or waterways (Aldstadt and Getis 2006; Yamada and Thill 2007). Recent studies compare the relative performance and strengths and weaknesses of different methods so that analysts can decide which method best fits a particular research or policy question (Kulldorf et al. 2006).

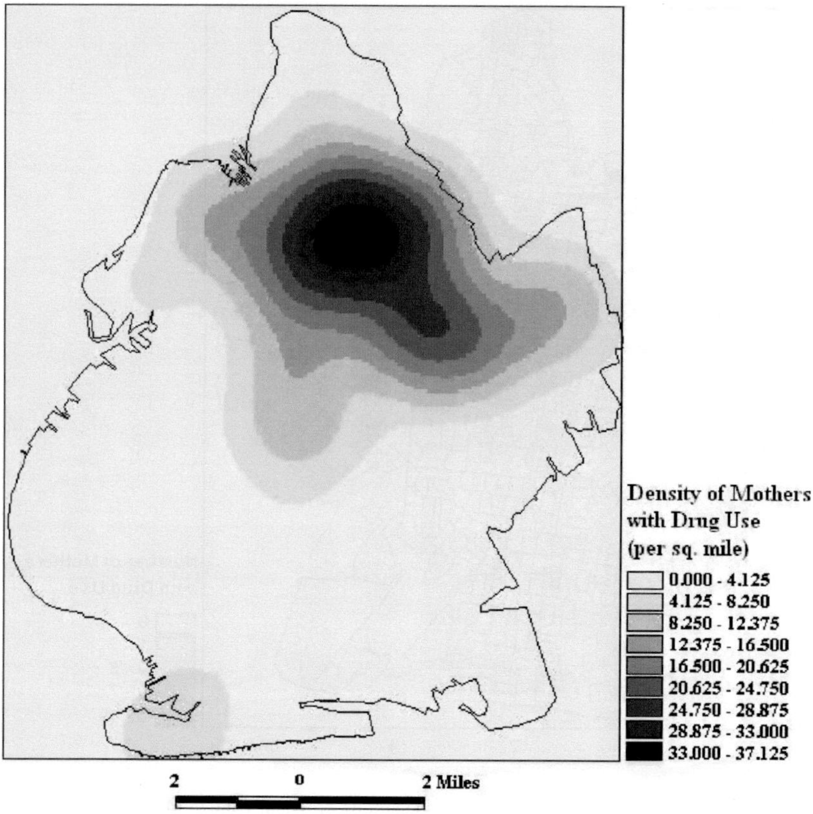

Fig. 1.2 Density map of mothers with drug use created using kernel estimation. Density refers to the number of mothers per square mile

Exploratory spatial analysis is also important for highlighting spatial processes that incorporate the time element – e.g., space–time geography. Spatial diffusion – movements of diseases and other phenomena through space and time – has long been an important topic in medical geography research. Studies on infectious diseases like measles and avian flu reveal the increasingly global exchanges of infectious agents as well as local flows and intensification (Cliff and Haggett 2004). Animated map sequences are highly effective in depicting the spread and retreat of infectious diseases. Although substance abuse is not an infectious disease in the strict sense of the word, many substance use issues undergo spatial diffusion processes as they shift from place to place across the landscape. An early study by Hunt and Chambers (1976) tracked the spatial diffusion of peak heroin use in the US as it moved over time from coastal to inland cities and hierarchically from large cities to smaller ones. In the UK, animated maps and GIS-based visualization were used to develop a system for forecasting the spatial diffusion of drug misuse (Ditton and Frischer 2001; Field et al. 2001). Analyzing changes in the spatial

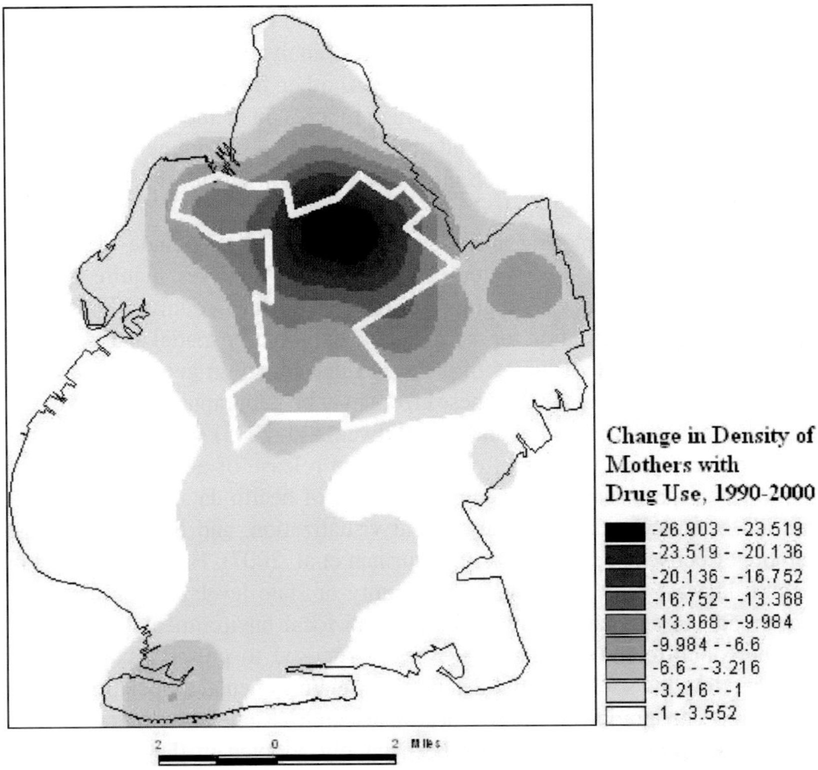

Fig. 1.3 Change in density of mothers with drug use, 1990–2000. The catchment area for the Healthy Start program is shown on the density map

patterns of drug activity over time is also valuable for studying the effects of policy changes on substance use. Fig. 1.3 illustrates the change in drug use among pregnant women in Brooklyn from 1990 to 2000. The map shows a substantial decline in drug use, especially in the areas that had the highest density of use in 1990. Superimposed on the change map is the eligibility area for the Healthy Start program, a program that provides prenatal services and education for needy pregnant women. The Healthy Start target area corresponds closely with the areas of drug use decline, suggesting that this geographically targeted program may have had a positive impact.

Exploring Geographic Inequalities: Software and Challenges

Developments in spatial analysis methods have been accompanied by great increases in the availability of computer software, some of which are free for download. Several of these have features that make them highly useful for exploring geographic data on health issues. GEODA, developed by Luc Anselin, provides a wealth of spatial analysis tools, including linking and brushing of map and tabular data, calculating spatial disease incidence rates, analyzing spatial clustering, and spatial regression modeling (Anselin 2003). Another freely available software is SatScan, developed by Martin Kulldorf, which focuses on methods for analyzing the clustering of disease cases in space and time. There are also commercially available systems, such as Clusterseer, which include a variety of cluster detection methods and tools for spatiotemporal analysis.

Critical issues in exploring and visualizing health data include the accuracy and scale of data being represented. These kinds of analyses require geocoded health data, which includes a geographical reference (state, county, address). Such geocodes provide a foundation for mapping and exploratory spatial analysis in GIS. However, there are many sources of error and inaccuracy in geocoding processes (Cromley and McLafferty 2002). These errors may be geographically biased, which in turn bias the results of spatial analyses (Oliver et al. 2005). Furthermore, privacy and confidentiality restrictions often prevent the release of geocoded health data at a fine geographic scale. The geographic scale of health data limits the level of detail and accuracy in spatial analysis and visualization, and finer scale data enable more effective spatial analysis (Schuurman et al. 2007). However, many kinds of health data are only available at the county or state levels, which are not detailed enough to see patterns of clustering and local environmental associations. A number of different strategies have been proposed to preserve confidentiality while providing the sort of spatial detail and/or environmental associations that are necessary for geographically based health research (Kamel Boulos et al. 2005); however, these approaches have not been widely adopted in the health research community.

Geocoded data on substance use present some additional challenges. Privacy and confidentiality issues are paramount for substance use, and so data by residential address, block, or census tract are rarely made available to researchers. In addition, populations of substance users are often highly mobile and transient; some are homeless. Not only are geocoded data at a detailed scale difficult to come by, but also the residential address has little meaning or accuracy for those who are homeless or highly mobile. Field data collection about residential locations and movements through time and space is essential for understanding such populations (e.g., Walker, Mason, and Cheung 2006). Despite these concerns, geocoded information is essential for exploring geographic inequalities in substance use, and strategies that facilitate effective spatial analysis of both primary and secondary health data can enrich and enhance such geographically based research.

Understanding the Place Contexts of Substance Use

Geographers can also make important contributions toward understanding substance use and its impacts. There is growing recognition in substance use literature about the significance of contextual/environmental factors – how social, cultural, and place characteristics affect both the incidence and intensity of substance use and its effects on individuals, families, and neighborhoods. In a recent article, Galea, Nandi, and Vlahov (2004) advocate a 'social epidemiology' framework that emphasizes social context – social networks and interactions – as a determinant of substance use. Focusing on injection drug use, Rhodes et al. (2005) look even more broadly at the 'social production' of HIV risk. They call for research investigating risk environments, "the social situations, structure and places in which risk is produced" (p. 1027). These innovative new directions intersect with the intellectual traditions in health geography emphasizing place and health. Themes within this broad area include analyzing how place environments at varying scales affect health; how individuals and groups negotiate place environments; and how place environments are constructed and re-constructed in influencing health.

The ties between place environments and health have been a central theme in medical/health geography since its inception. The classic emphasis on disease ecology has given way to a more critically informed perspective that seeks to understand the intricate ties between local, national, and global processes of health and well-being. This work acknowledges both the positive impacts of place on health and healing, especially the notion of places as therapeutic landscapes (Gesler 1992; Williams 1998), and the negative impacts associated with environmental hazards and material deprivation (MacIntyre, Ellaway, and Cummins 2002; Wakefield et al. 2001). An important theme in current research is to identify the pathways through which contextual/place factors affect health. This requires delineating salient attributes of places and determining the linkages with health for diverse populations. Research on obesity, for example, looks at features such as neighborhood walkability and the local availability and prices of healthy foods (Moon et al. 2007); studies of mental health show that neighborhood deprivation and access to health care are key determinants of hospitalization for psychiatric disorders (Almog et al. 2004). The multifactorial nature of most health issues and the varying individual responses to place make this a challenging but important area for research in health geography.

Current research highlights the importance of geographic scale in studies of health and place. Contextual factors operate at varying scales from the household and neighborhood scales of daily life to the national and global scales of political and economic processes. Health geographers have used multilevel modeling to explore the statistical associations between place characteristics at different scales and health outcomes (Duncan, Jones, and Moon 1998). Research on topics ranging from limiting long-term illness (Gould and Jones 1996) to low birthweight (Grady 2006) suggests that, after controlling for individual risk factors, place characteristics have significant associations with health risks and behaviors. Despite these important contributions, much of the research in this area has glossed over issues of scale linked to model specification. Multilevel studies often focus on the local or neighborhood scale as representing contextual effects. In defining contextual variables to represent the local scale, researchers typically utilize predefined areal units such as census tracts, zip codes, and states, which may have little relationship to health-related exposures and interactions. Such areal units "are particularly sacred once they have become established, even though they later may become serious obstacles to solution of contemporary problems" (Abler, Adams, and Gould 1971).

More relevant for health research than pre-defined areal units are activity spaces – the areas in which people interact and conduct everyday activities. Figure 1.4 shows the daily activity space for a hypothetical single mother in a large city. The mother's activity space links home, day care, work and shopping, and extends well beyond the bounds of her home census tract. Individual activity spaces are complex and vary by age, gender, race/ethnicity, and a host of individual characteristics, thus complicating the efforts to define appropriate areal units for multilevel analysis (Gesler and Meade 1988; O'Campo 2005). Even low-income populations have been found to have highly complex and relatively extensive activity spaces that do not conform well to pre-defined census boundaries (Matthews et al. 2006). Similar findings have been uncovered for homeless individuals. A study of homeless individuals in the Skid Row area of Los Angeles documented frequent trips outside the area for the purposes of maintaining friendship and family ties and obtaining social services (Wolch, Rahimian, and Koegel 1993).

Researchers have proposed several strategies for defining local areal units that better represent activity patterns and social interactions for multilevel studies. One is to 'build' socially homogeneous zones that approximate socially defined neighborhood areas (Cockings and Martin 2005). Zones can be created based on social and geographical criteria, and the impacts on research findings of alternative zone definitions can be easily investigated. Others have advocated using GIS to represent activity spaces based on travel patterns between home, work, and other activities. Sherman et al. (2005) discuss several GIS-based methods for characterizing activity spaces. Time is also important in people's daily activity patterns. A time–space prism is a three-dimensional representation of people's movement patterns through space and time. Sophisticated methods are being developed for visualizing such patterns with the use of aquarium diagrams and for analyzing the sorts of time–space constraints that emerge from these complex activity patterns (Kwan 1999). Finally, GPS technologies make it possible to monitor individual movement patterns directly through time and space (Elgethun et al. 2003). Such monitoring raises thorny questions about privacy and confidentiality, however, and analyzing the vast quantities of real-time data poses significant challenges.

Research by Mason, Cheung, and Walker (2004) highlights the importance of place-based activity spaces and social networks for substance abuse research. For a sample of adolescents in Washington DC, the authors collected data on daily activity locations and perceptions of safety and risk in everyday environments. These data were entered into GIS and used to characterize the balance of risk and protective factors in individual respondents' local environments. A highly individualized 'risk profile' was generated providing key understandings for development of individualized substance use prevention programs and policy-making. This research is important not only in the detailed characterization of activity spaces but also in its efforts to describe and model risky and protective attributes of those spaces in relation to individual perceptions, meanings, and needs.

Place and Substance Use: Networks, Interactions, and Adjustments

Although characteristics of local activity spaces are clearly significant for health, the social interactions that occur in and beyond such spaces are equally important. Social capital – the advantages that accrue to individuals from social interactions – and resources are a central focus of health-related research in the social sciences (Diez-Roux 2001, Putnam 2000). Historically, social networks were strongly place-based, rooted in local activity spaces; however increasingly, these networks extend across regional and national borders, facilitated by advances in telecommunications technology. Social networks play a central role in substance abuse, and recent studies focus on understanding the connections between place environments, social networks, and substance abuse. Wylie, Shah, and Jolly (2007) describe how local meeting places can facilitate social interactions that increase the risk of substance abuse. Rothenberg et al. (2005) find that social networks for persons at risk of HIV infection are often tightly clustered in space. Although these studies highlight the importance of localized, place-based social interactions, some kinds of substance use networks that extend via phone and Internet may also be relevant.

Much research on place effects privileges the local scale, but the influences on health often extend to scales beyond the local neighborhood. Regional economies and cultures affect access to jobs, services, and social support with implications for health status. Health policies are framed at the state and national scales; access to health care is often constrained by health maintenance organizations and insurers that have complex geographical webs of influence. These linkages are often considered in studies of health care, but their effects on individual health outcomes are less well understood.

It is also important to conceptualize the relationships between place and health over longer time scales, e.g., by looking at changes over time in individual life histories, and at changes in place characteristics and individual responses to them. The effects of migration on health have been studied in a variety of contexts and in relation to the process of acculturation (Elliott and Gillie 1995). Migration imposes physical and emotional stresses, and place characteristics of host communities affect how people adjust to the displacement of migration. Substance users often circulate between residential neighborhoods and treatment facilities, and the physical and social distances between these sites have implications for treatment success. Understanding how clients negotiate safety and risk in these linked place environments is an important topic for substance abuse research.

Other kinds of place adjustment are also relevant for research on substance abuse. Illness affects how people negotiate space, and people in turn modify their place environments in their efforts to cope with the experience of illness (Dyck 1995). A qualitative study of people diagnosed with HIV shows how the spaces of daily life changed with the progression of the disease (Wilton 1996). Rather than becoming more diminished as the disease progressed, people's worlds went through cycles of

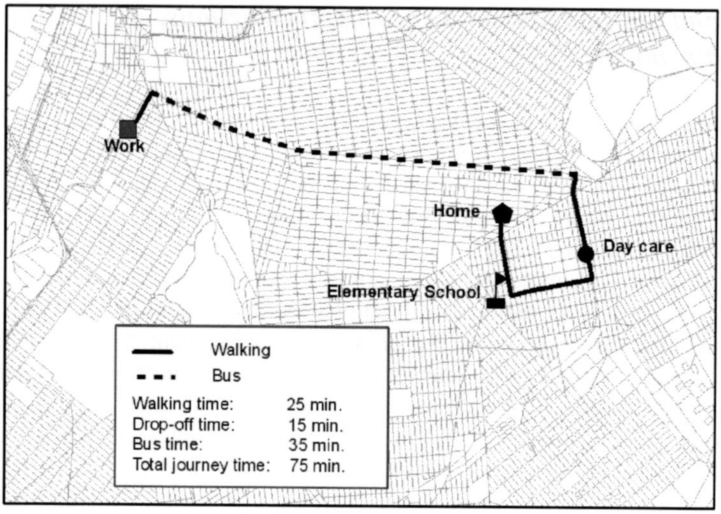

Fig. 1.4 Daily activity space for a single mother involves dropping children off at school and day care and taking the bus to work

expansion and contraction in response to illness. Health issues including addiction impact people's lives and the lives of people around them, thus altering the spaces of everyday life.

Understanding the dynamic relationships between place and health calls for innovative research methods that integrate understandings of place characteristics, social networks, human experiences and behaviors, and health outcomes. Mixed methodologies – methods that combine both qualitative and quantitative data collection and analysis – are especially important. A promising strategy is to link qualitative data, such as photographs, diaries, narratives, and oral histories, with more traditional maps and GIS data to paint a rich picture of spatial and social variation (Curtis, Mills, and Leitner 2007; Matthews, Detwiler, and Burton 2006).

In summary, GIS and other more qualitative geographic methods are greatly advancing how place characteristics are measured and how their associations with health are assessed. However, many challenges remain. One is to determine the appropriate metrics for measuring socio-environmental characteristics, metrics that need to represent environmental features and qualities that are meaningful for health. Another is to incorporate varying individual responses to those characteristics to reflect differences in vulnerability related to age, class, gender, and so on. Finally, perhaps the greatest challenge is to tease out the pathways through space and time by which places affect health and vice versa. This requires integrating geographical, social, and biological forms of understanding – a daunting but exciting challenge that will occupy health research well into the next century.

Analyzing Disparities in Health Care

Analyzing geographic disparities in health care location and access is another area where geographers can contribute to substance abuse research. Analysis of health care provision has been an important theme in health geography for several decades, and increasingly researchers are drawing upon place concepts to understand inequalities in the availability of health care and people's access to such care. The vast majority of research in this area has focused on formal health care provided by public and private institutions, although informal health care, provided by individuals, families, and friends, has begun to attract research attention. A consistent finding in geographic research on formal health care is the uneven and unequal geographic distribution of health services. Both the quantity and quality of services vary from place to place at scales ranging from the local to the global. In many countries and regions, the spatial distribution of health services follows an inverse care pattern in which the availability and quality of services is inversely related to the need for services (Joseph and Phillips 1984). Ironically, places where the need for health care is greatest are often less likely to have access to high quality care. Such inequalities differ according to the scale of analysis A recent study of mental health services in London identified a relatively equitable distribution of services at the borough scale, but significant inequalities at the local level (Foley and Platzer 2007).

Geographic analyses of health care emphasize the effects of distance on health care access and use. The choice of health service providers and the frequency of service use are strongly influenced by distance: people are less likely to use services located far from home. This distance decay effect arises because of the time and cost of traveling long distances and people's lack of knowledge about and familiarity with services located far from home. Researchers have documented distance decay for diverse health services in a wide range of settings (Joseph and Phillips 1984). For substance abuse services, distance and travel time pose significant barriers to service utilization, recidivism, and aftercare (Schmitt, Phibbs, and Piette 2003). The frictional effect of distance varies with service characteristics, such as the size and quality of services offered, and with individual and household characteristics, such as age, income, and access to transportation (Haynes et al. 1999). People whose mobility is constrained by low income, disability, age or lack of transportation are typically more reliant than others on services close to home (Allard, Tolman, and Rosen 2003; Ricketts et al. 2001). In exploring geographical access, we need to take into account the full range of barriers that vulnerable populations face in obtaining health care, including lack of social support and economic, cultural, and time-space constraints (Young 1999).

Some health services, such as those for substance abusers, carry a stigma that confounds traditional notions of distance decay. Clients may be reluctant to use services in their own neighborhoods fearing public recognition and stigmatization (Parr 1997). Yet using services outside the neighborhood involves added travel time and cost, which discourage service utilization. Exploring the interactions between proximity, stigma, service utilization, and treatment effectiveness is an important topic for substance abuse research.

The role of distance provides a foundation for investigating geographic inequalities in access to health care and the match between service needs and resources. The gravity model, which posits that patients choose health service facilities based on a trade-off between distance and service attractiveness, continues to be important in these efforts (Yang, Goerge and Mullner 2006). Gravity models have been used to predict catchment areas for new health facilities and to estimate the impacts on travel patterns when existing health facilities close their doors. Recent work on the association between alcohol outlets and problem drinking cites gravity-like attraction as a mechanism linking problem drinkers with specific alcohol outlets (Gruenewald 2007). In the past decade, gravity models have been enhanced in a number of ways, including the use of more accurate, network measures of distance or travel time; refinements in specifying 'attraction' factors for health services; and developments in model calibration. An interesting gravity-based approach to modeling spatial access to health care is the two-step floating catchment method (Wang and Luo). A floating 'window' is moved across the map, and the ratio between service needs and the local availability of services is computed within the window providing a local indicator of service access (Fig. 1.5).

Social, economic, and political processes that affect the uneven spatial distribution of health services from the global to the local scales have also attracted attention from health geographers. These studies point out the importance of national health care policies and modes of provision in influencing the locations of health care providers and the quantity and quality of services offered. National policies such as the neoliberal 'reterritorialization' of health care in New Zealand – an effort by the central government to reassert the importance and power of local citizens in local health care decision-making, alter the balance between local and national control over health care (Prince, Kearns, and Craig 2006). At the local scale, economic and political forces embedded in local places shape the changing delivery of health care. The economic viability of health care providers is closely tied to the economic and demographic health of the communities in which they are located. Major providers such as large hospitals are a significant economic force in many communities, wielding political and economic power that extends beyond their role in providing health care. At the same time, voluntary organizations, health insurance companies, and regulatory agencies exert control over health institutions, leading to shifting and complex webs of interaction. Community groups have become more vocal in challenging decisions made by health care institutions and policy-makers, adding another voice to policy debate. These processes play out differently in different places, and place landscapes are crucial for health care evolution and policy development (Mohan 2002).

Thus, for substance abuse services, the history and politics of place are critically important in affecting geographic variation. The stigma associated with substance abuse creates a distinct place politics similar to that observed for other types of unwanted services, such as mental health facilities and hazardous environmental sites. NIMBY (Not In My Back Yard) politics lead communities to oppose location of facilities in their neighborhoods (Takahashi 1998), and there are hierarchies and levels of community opposition that reflect local cultural constructions of

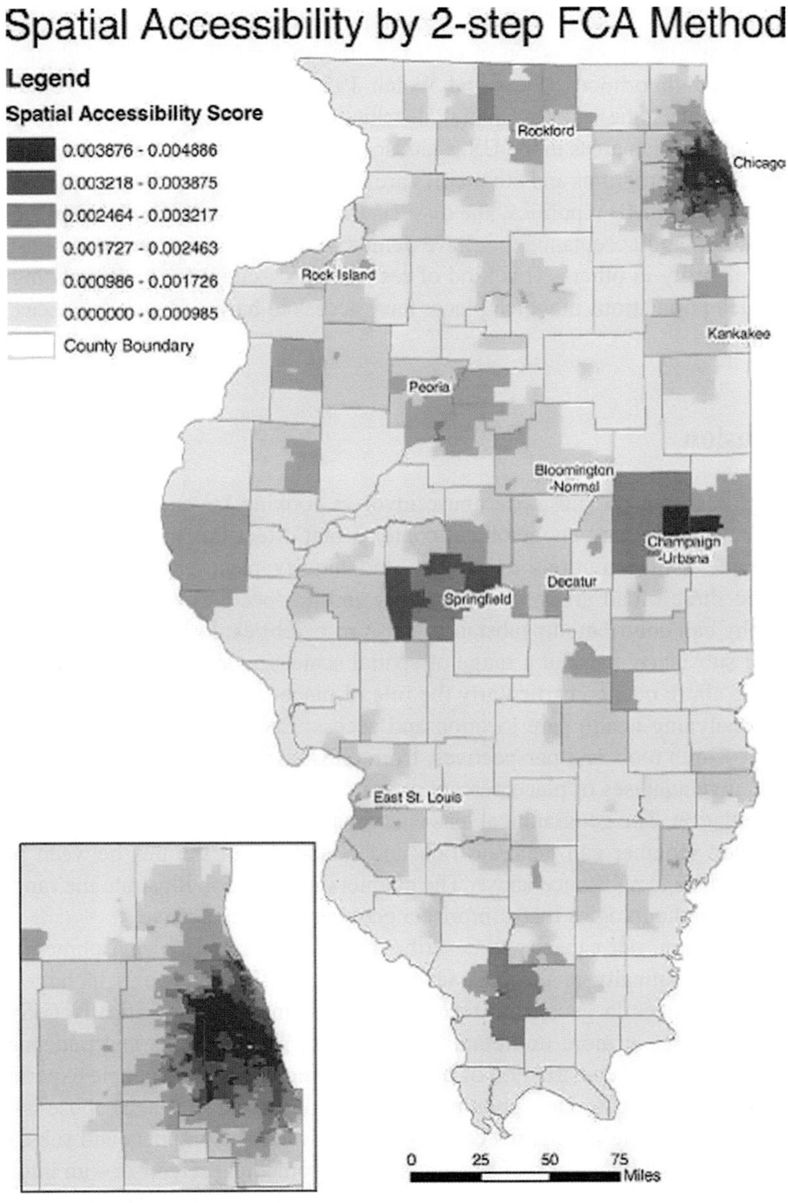

Fig. 1.5 Spatial accessibility to health care in Illinois based on the two-step floating catchment area method. Reprinted from Wang and Luo (2005)

difference (Wilton 2000). In many cases, inequalities in political power are manifest in the spatial concentration of such facilities in 'service-dependent ghettos' in inner city neighborhoods (Dear and Wolch 1987). In a recent article, Tempalski (2007) describes the uneven spatial distribution of syringe exchange programs across metropolitan areas in the US, a landscape where some large cities have no syringe exchange programs and other cities are relatively well served. From the War on Drugs to local NIMBY politics, she dissects the national, state, and local processes of opposition and acceptance that have denied services in some cities and ensured their availability in others. This kind of research is essential for understanding not only which populations in which places have access to harm reduction services, but also why.

Conclusion

'Placing' research on substance abuse involves looking beyond the individual to consider the social and geographical contexts that affect this important issue, and how people and institutions experience and modify place environments. I have identified three broad areas where insights and perspectives from health/medical geography can contribute to substance abuse research: exploring the uneven geography of substance abuse at a range of spatial scales; understanding how and why substance abuse occurs, particularly the role of place-based contextual factors; and finally, analyzing health care location and access. Each of these topics requires a mix of research tools and perspectives, from GIS-based mapping and visualization to qualitative analyses of place perceptions and experiences and the politics of policy formulation. The geographical lens encompasses diverse perspectives that when triangulated together can generate rich understandings of the ties between place environments and substance abuse. The chapters in this book illustrate the range of topics and methodologies that comprise a geographical perspective.

In conclusion, our capacity to describe and understand place environments has increased dramatically in the past several decades. This change is in large part linked to advances in geospatial technologies and methods, specifically GIS and spatial analysis, but more importantly, it is tied to new concepts and understandings about the social construction of place environments and how people experience them. These approaches have great potential for addressing substance abuse issues. Achieving this potential requires collaboration between geographers and substance abuse researchers, between social scientists and biomedical scientists – an intellectual marriage that will facilitate and enhance place-based understandings of substance abuse and its broader impacts.

Chapter 2
Integrating Geography and Social Epidemiology in Drug Abuse Research

Yonette Thomas, Douglas Richardson and Ivan Cheung

Abstract This chapter discusses connections between place, environment, and health, and how geographic tools and methods can help better understand the nature of these associations. Ongoing collaborations between the National Institute on Drug Abuse (NIDA) and the Association of American Geographers (AAG) exemplify this integrated perspective. NIDA's commitment to understanding the social epidemiology of drug abuse and the AAG's emphasis on integrating geography and geographic methodology into the public health equation have provided opportunities for unique and innovative approaches to understanding drug abuse, drug addiction, and human immunodeficiency virus (HIV)/AIDS. This chapter discusses the recent NIDA/AAG Geography and Drug Addiction Symposium, from which this book originated, and provides examples of how researchers included in this book are beginning to use geographic methods to explore aspects of the social environment, which are central to drug abuse research. A brief précis of the chapters of this book is also included.

Introduction

Utilizing the methods and insights of geography and geographical information systems (GIS) to enhance understanding of the social epidemiology of drug abuse and human immunodeficiency virus (HIV)/AIDS is a central part of the mission of the National Institute on Drug Abuse (NIDA). Achieving this mission requires a focus on improving the nation's public health, and NIDA does this, in large part, by promoting integrated approaches to understanding the interactions between individuals who abuse or are at risk of abusing drugs and their environments; by assessing the continuum of problems and causes related to drug abuse; and by fostering and providing guidance for new research. The collaboration between the NIDA and the

Y. Thomas
National Institute on Drug Abuse (NIDA), National Institutes of Health, Epidemiology Research Branch, Division of Epidemiology, Services, and Prevention Research, Bethesda, MD, USA

Y.F. Thomas et al. (eds.), *Geography and Drug Addiction*,

Association of American Geographers (AAG) exemplifies this integrated perspective. Geographers have long used geographic methods to explore how place and environments are linked to health and disease. However, the link between geography and drug abuse and drug addiction has been relatively unexplored.

The NIDA's commitment to epidemiology and the AAG's emphasis on integrating geography and geographic methodology into the public health equation have provided opportunities for unique and innovative approaches to understanding drug abuse, drug addiction, and HIV/AIDS. NIDA has effectively used epidemiological approaches to better understand drug abuse and addiction through research tools such as cross-sectional and longitudinal research designs, epidemiologic surveillance and modeling, and clinical epidemiologic experiments. In addition, the Institute also employs a combination of ethnography, medical sociology, demography, and medical anthropology to explore how multilevel methodological approaches can help identify the diverse pathways linking individuals to their environments and to health and disease, and to advance our knowledge of the determinants and correlates of drug abuse and its consequences, specifically HIV/AIDS, as well as other infectious diseases.

On March 8, 2006, NIDA and the AAG sponsored a joint symposium on "Geography and Drug Addiction" at the AAG's annual meeting in Chicago, Illinois. This unique collaboration provided a new and productive forum for drug abuse researchers and geographers to discuss connections between place, the environment, and health, and how geographic tools and methods can help better understand the nature of these associations. As McLafferty (Chapter 1) argues, geographers can contribute to drug abuse and addiction research by studying the linkages between place, environments, and health, and by integrating geographic tools and methodologies, such as GIS, to explore such associations.

Participants in the symposium were asked to explore the dynamic and integral link between geography and drug abuse and addiction, including ways in which GIS and spatial modeling might help to better understand and respond to drug abuse and related health and medical consequences. The symposium presentations and discussions engaged a wide range of potentially productive research areas and topics, including spatial analysis of drug use, abuse, and addiction patterns; spatial diffusion modeling of the spread of drug use, abuse, and addiction (including predictive modeling); employing locational analyses of drug treatment and service delivery facilities and their service areas; neighborhood studies of geographic relationships between drug abuse and built environment; and GIS locational analysis and integration of very large-scale social, economic, demographic, medical, and environmental databases with patterns and trends of drug abuse, addiction, treatment, and prevention. Also discussed were relationships and interactions of spatially dependent variables and HIV/AIDS and drug abuse, including related prevention and treatment strategies; geographic research and models relevant to better understanding the etiology and epidemiology of drug use, abuse, and addiction, including interrelationships among the transmission pathways of HIV/AIDS, other sexually transmitted infections, and hepatitis; and other health consequences of drug abuse and addiction.

In the process of these investigations, scientists were particularly encouraged to consider broader factors such as health disparities, socioeconomic factors at the neighborhood and community levels, family engagement, dysfunction, and disruption, and other related aspects of the social/environmental landscape. This book is an outgrowth of the NIDA/AAG symposium and subsequent scientific discussions. The chapters that follow emphasize innovative research that addresses the importance of studying drug abuse as the behavior of individuals in the context of place and environment.

Background and Significance of this Book

NIDA's social epidemiology of drug abuse focus encompasses research directed at examining drug abuse trends in the contemporary world where a variety of substances are available to increasingly diverse populations and in an expanding range of settings, geographic areas, neighborhoods, and communities. Since behavioral patterns and social trends related to drug use and abuse vary considerably between countries, regions, and communities, an important way to assess the status of drug abuse and the factors associated with its incidence, prevalence, and spread is to focus on geographically derived data. For example, how do neighborhood contextual factors influence the prevalence of cocaine users? How are overdose mortalities related to potential sources of environmental exposure and the location of treatment resources? Are rates of methamphetamine use elevated around production outlets? These are fundamentally geographic inquiries. However, the role of geographic methods and technologies in public health management and research related to drug abuse and addiction has been very limited to date. This book points out that geography's core emphasis on research of human/environment interactions, on place-based studies, and on inter-disciplinary approaches and integrative science is well-suited to helping to understand complex issues and relationships between drug abuse and its environmental context (Richardson and Solis 2004). In addition, these traditional research strengths are further enhanced today by powerful new geographic technologies, such as GIS and interactive GPS/GIS mapping and modeling systems, which enable geographers, medical researchers, and public health experts to collect, integrate, and analyze location-based information on multiple characteristics of a population, its drug use and health, and the environment in which people live, within a consistent geographic framework at variable spatial and temporal scales (Richardson 2006).

Application of geographic technologies, such as GIS, is an important step toward better understanding of drug abuse issues and their inherent complexities. When used alongside more traditional epidemiological techniques and geographic methods, GIS provides epidemiologists the ability to formulate new questions and refine or enhance existing analyses. The ability to evaluate and visualize geographic information provides a unique perspective on public health issues such as emerging and shifting epidemics, the utilization of treatment services, and rapid assessment of the impact of incidents, ranging from health impacts of natural disasters to bioterrorism.

In the following sections, we discuss the value of integrating geography and GIS in the social epidemiological approach in drug abuse research, drawing on some of the research and methods presented in this book to illustrate this integration. The remaining chapters of this book elaborate on these initial observations in many related contexts.

A social epidemiological approach is predicated on the notion that the social environment influences health and that population health is not simply an aggregate of individual health. In other words, we cannot understand the health outcomes of particular populations without considering the context within which these populations are living. Diez-Roux (2001) and Schwartz and Diez-Roux (2001) point to the inescapable role of the social environment and that individual autonomy and choice are constrained by social position and physical environment. Further, according to Barnett and Casper (2001:465), "components of the social environment include built infrastructure; industrial and occupational structure; labor markets; social and economic processes; wealth; social, human, and health services; power relations; government; race relations; social inequality; cultural practices; the arts; religious institutions and practices; and beliefs about place and community."

In drug abuse research, the social environment is considered a fundamental actor in drug-abusing behavior and its consequences. Just as constitutional differences account for differential vulnerability to substance abuse among individuals, there are characteristics of the social environment that make communities particularly vulnerable to elevated rates of substance abuse. These include social cohesion and integration, material deprivation, norms and attitudes about drugs and drug use, psycho-social hazards, the built environment, and the availability of drugs. Not only do each of these factors directly contribute to rates of drug use, but they may also exacerbate propensities to use or abuse drugs among certain individuals or groups of individuals. A multi-level research perspective stresses the need to parameterize and measure the social environment just as thoroughly and rigorously as similar undertakings at the molecular level. The continuum between an individual and his/her environment encompasses a complex myriad of functional and geographical units. GIS methods and tools present an effective approach to collect and organize this geographically referenced data.

Many researchers in this book explore the use of geographic methods, spatial statistics, and GIS to organize large arrays of social, economic, and health data as well as to describe and visualize spatial patterns of drug use and their relationship with explanatory factors. Others illustrate more advanced spatial analytical techniques. For example, Gopal, Adam, and Vanelli (Chapter 25) demonstrate the use of Local Indicator of Spatial Association (LISA) statistics while DiMaggio, Bucciarelli, Tardiff, Vlahov, and Galea (Chapter 27) illustrate the potential use of Bayesian hierarchical modeling techniques.

Below we briefly describe how researchers are beginning to use geographic methods and GIS to parameterize and measure several aspects of the social environment, which are central to drug abuse research. We highlight, as examples, three broad areas of focus. These are socioeconomic factors (e.g., poverty and income, race and ethnicity, and unemployment and educational attainment); social interactions; and the built environment.

Socioeconomic Characteristics

Community-level or neighborhood-level demographic data (typically drawn from a national or local census) are often used in characterizing the social environment (including its resources and vulnerabilities), within which individuals live. For example, Galea, Ahern, and Karpati (2005) show that neighborhoods characterized by social and economic vulnerability displayed substantial variability in mortality rates (such as homicide or HIV/AIDS rates) associated with causes that may be sensitive to social conditions. Numerous authors use GIS to organize and compare socioeconomic information at various geographic scales (ranging from state, district, census tract to block group). These data are typically tabulated from either individual- or household-level observations. The aggregation process is often constrained by arbitrary administrative boundaries, such as census tract, block group, and district. Although spatial statistics and GIS may provide efficient tools for integrating large numbers of these socioeconomic variables, researchers must pay particular attention to the geographical scale at which these variables influence behavioral and health outcomes, such as drug use and abuse. For example, residential segregation and income inequality may contribute to increased drug abuse prevalence in different ways at different geographic scales.

Poverty and Income

The potential association between poverty level and drug abuse has received a great deal of research attention (e.g., Braveman et al. 2005; Galobardes, Shaw, Lawlor, Lynch, and Smith 2006; Gordon 1995; Nandi et al. 2006). In this volume, most researchers rely on the use of third party data sources such as national census income data to parameterize poverty measures. For example, median household and monthly income is used by Banerjee, LaScala, Guenewald, Freisthler, and Treno (Chapter 7), Lee and Pang (Chapter 8), and Lu and Burnum (Chapter 12). Similarly, Brouwer, Weeks, Lozarda, and Strathdee (Chapter 3) and Snedker, Herting, and Walton (Chapter 4) discuss the use of percentage residents receiving less than minimum wage or living below federal poverty level, while Archibald (Chapter 22) and Yang (Chapter 19) use a poverty rate.

Race and Ethnicity

Community-level racial and ethnic composition is often considered a potential explanatory factor in drug abuse research (e.g., Acevedo-Garcia 2000; Jayakody et al. 2006; Martinez, Lee, and Nielsen 2004; Subramanian et al. 2005; Williams and Collins 2001; Williams, Neighbors, and Jackson 1993). In this volume, Banerjee et al. (Chapter 7) examine racial and ethnic composition by analyzing numbers of Hispanic, Black, and White persons, whereas Lu and Burnum (Chapter 12) discuss the use of percentage minority as explanatory factors. Valdez and Cepeda (Chapter 10) illustrate how historical immigrant settlement and creation of Mexican ethnic enclaves may have long-term influence on the prevalence of drug use in San

Antonio, Texas. Conversely, internal rural-to-urban migration within China is examined in Yang's (Chapter 19) study.

Unemployment and Education Level

Research has shown that there is an association between drug use (particularly opiates, cocaine, and crack cocaine) and the risk of unemployment at the individual level (e.g., Crew and Davis 2003; MacDonald and Pudney 2000, 2001). In this book, Archibald (Chapter 22), Banerjee et al. (Chapter 7), Brouwer et al. (Chapter 3), and Stahler et al. (Chapter 21) explore the relationship between drug abuse and unemployment at the community level. In all cases, unemployment is only one of many factors. Perhaps, from a social environmental context, employment opportunity in the community may be a far more critical factor than unemployment. Similarly, Brouwer et al. (Chapter 3) and Lee and Pang (Chapter 8) discuss the association between education level and drug abuse. However, it is difficult to determine if educational attainment at the community level, however defined, influences prevalence of drug abuse.

Social Interactions

In studying social interactions and drug-related behavior within a geographic framework, many researchers focus on "distance" as a geographical constraint upon which interactions operate. Waldo Tobler's "first law of geography," in fact, famously states that "everything is related to everything else, but near things are more related to each other" (Tobler 1970). Schilling, Monterroso, Fontdevila, Fernando, and El-Bassel (2004) assessed the association between proximity to a needle exchange program (NEP) and lower levels of HIV-related risk behavior among injection drug users (IDUs) in East Harlem, New York. GIS analysis and routing algorithms allow for easy computation of Euclidean distance between geo-referenced points, such as the centroids of a geographical unit of analysis (e.g., a neighborhood boundary or census tract) or to establish service area buffers or travel time contours around locations of interest, such as a treatment facility. In this volume, many researchers illustrate the value of GIS analysis in assessing the links between social interactions, physical distance, and drug using behavior. For example, Archibald (Chapter 22), Cahill and LaVigne (Chapter 6), Green and Pope (Chapter 23), Lee and Pang (Chapter 8), and Romig and Feidler (Chapter 18) illustrate how distance and accessibility to substance abuse and mental health service facilities may affect the level of drug use and abuse in a community. As well, Hunt, Kennedy, Summer, Frabutt, and Scholten (Chapter 24) evaluate proximity to local drug markets, Cahill and LaVigne (Chapter 6) and Valdez and Cepeda (Chapter 10) examine trafficking routes and drug-related behaviors. Malm and Tita (2006) use GIS to evaluate the effectiveness of specialized policing teams that target domestic marijuana production in British Columbia, Canada. Hunt et al. (Chapter 24) illustrate the applications of geographic technologies in analyzing drug market locations and curtailing drug-related violence.

Increasingly, researchers are using geographic research methods and models to measure spatial accessibility to heath care and drug treatment services. For example, Luo and Wang (2003) illustrate the method of establishing "a floating catchment area" whereas Guagliardo et al. (2004) used a gravity model-based kernel density estimation method to analyze accessibility to treatment services. More complex representations of the distance parameter may include the use of digitized street networks and other transportation modes. By integrating topological relationships, such as connectivity (of street segments and transportation routes) and adjacency (of neighboring and/or interacting neighborhoods and communities), geographic models and new technologies can help social scientists, medical researchers, epidemiologists and public health service providers to better understand complicated dimensions of social interactions. For example, Mason, Cheung, and Walker (2004a, b) discuss the utility of GIS analysis in measuring street network travel distances when evaluating positive and negative environmental influences on teenage drug users.

Built Environment

The built environment or the physical human-built environment is an important component of the milieu in which the individual lives. Weich et al. (2001) define the built environment as the "housing form, roads and footpaths, transport[ation] networks, shops, markets, parks and other public amenities, and the disposition of public space." Geographers (see for example, Dijst and Kwan 2005; McCray et al. 2005; Weber and Kwan 2002) have long focused on the physical structure of cities, communities, and neighborhoods as operational factors in accessibility and human activity and well-being. More recently, epidemiologists have begun more directly to assess the role of the built or physical environment in the health and well-being of populations. Consequently, recent studies (see for example, Diez-Roux 2003; Evans 2003; Gorden-Larsen et al. 2006; Klitzman et al. 2006; Northridge et al. 2003; Savitch 2003) have shown a strong link between health and the built environment. Similarly, studies linking drug using behaviors to the built environment are emerging. For example, Hembree et al. (2005) used multilevel analyses to demonstrate the association between neighborhood-built-environment and the likelihood of overdose death in New York City. There are many opportunities in which the integration of geography and social epidemiology may further our understanding of the degree to which the built environment contributes to drug use, abuse, and addiction.

Though not directly studying drug use, Galea, Vlahov, Ahern, Rudenstine, and Wallace (2005) use geographical analyses to evaluate the linkage between poor quality built environment and likelihood of depression. Other examples can be found in this book. Brouwer et al. (Chapter 3) speculate that the use of remote sensing and satellite imagery may enable better characterization of road condition and travel environments. They also discuss factors such as availability of electricity and indoor plumbing at the household levels. Banerjee et al. (Chapter 7) include location-based/place characteristics, such as the amount of vacant housing in their model. Median age of the structures, often used as a proxy in characterizing built environment in sociological research, is factored into Lu and Burnum's discussion in Chapter 12. These are some examples of how researchers may incorporate the characterization of the built environment into the integrative framework when studying the social epidemiology of drug abuse.

Précis of Chapters in this Book

In the context of geography and drug addiction, the chapters in this book expand the collective bounds of drug abuse epidemiology and geography, specifically health geography. From these chapters, we learn that understanding drug abuse and its adverse outcomes in individuals and population groups requires us to consider place and related geographic methodologies. Part I (Chapters 1 and 2) provides a rationale for this collaboration and the link between drug abuse epidemiology and geography. Part II (Chapters 3–14) presents geo-epidemiology in drug abuse research. Part III (Chapters 15–19) explores the geography of IDUs and HIV. Part IV (Chapters 20–24) examines the geographic dimensions of drug treatment and prevention. Part V (Chapters 25–27) presents emerging research directions.

In Part I, McLafferty (Chapter 1) considers how the concept of place provides a foundation for understanding how substance abuse behaviors, prevention, and treatment relate to place environments from the personal to the global scales. This chapter provides the rationale and background for linking geography and the social epidemiology of drug abuse and the role that NIDA and the AAG play in this connection.

In Part II, Brouwer et al. (Chapter 3) describe spatial and environmental issues such as migration, neighborhood characteristics, and proximity to services, which may affect drug use behaviors and risk. In this instance, the confluence of drug trafficking routes, migration, and income inequalities form unique environmental influences and drug use scenes at the US/Mexico border. Snedker et al. (Chapter 4) use geographic methods to explore features of the neighborhood environment and spatial patterns of alcohol use among adolescents. Specifically, the relationship between neighborhood alcohol availability, disadvantage, and crime as factors in alcohol use is analyzed, while controlling for individual, family, and peer characteristics. Lankenau et al. (Chapter 5) examine the mobility of homeless youth, specifically young IDUs, across broad geographic regions as a factor in the spread of bloodborne diseases. Cahill and LaVigne (Chapter 6) examine the spatial concentrations and mobility of drug abusers released from prison and how those patterns may differ from those of the general parolee population. Banerjee et al. (Chapter 7) examine the social ecological processes that link social disorganization and community disorder, legal markets for alcohol, and illegal drug market activities to violence. The chapter provides one approach to statistically modeling these social ecological relationships over time to help identify the contributions of these structural determinants to violence in community settings. Lee and Pang (Chapter 8) use a GIS-based approach to integrate data from the Central Registry of Drug Abuse, Census, HIV reports/seroprevalence studies, and methadone clinic service statistics to assess trends in heroin addiction and HIV risk in Hong Kong. Cooper et al. (Chapter 9) analyze the relationships of two dimensions of racial residential segregation to prevalence of IDU among Black residents of 93 large US metropolitan statistical areas. Valdez and Cepeda (Chapter 10) explore the relationship between ecological containment and heroin practices – i.e., the consequences of socio-historical and ecological processes for contemporary heroin use and addiction among Mexican Americans in San Antonio, Texas. Dasgupta et al. (Chapter 11) describe spatial patterns in opioid overdose mortality in metropolitan and non-metropolitan areas. Lu and Burnam (Chapter 12) examine spatial patterns of clandestine methamphetamine labs in Colorado Springs, Colorado and the roles that contextual socioeconomic characteristics play in their distribution. Romig and Feidler (Chapter 13) explore distinct spatial patterns that exist when examining methamphetamine crimes of production and consumption, and this geographic duality in the landscape of abuse. Specifically, they discuss the problems faced by the State of North Dakota in limiting methamphetamine production and consumption and the importance of applying a more holistic approach to understanding drug abuse. Chandra and Swoboda (Chapter 13) explore the role of spatial aspects of drug consumption in the broader analysis of the economics of drug consumption and addiction in India.

In Part III, Beyrer (Chapter 15) demonstrates how the spread of the HIV is associated with heroin trafficking routes. Specifically, he examines the routes leading from two primary regions for illicit opium poppy cultivation and heroin manufacture: the Golden Triangle of South-East Asia and the Golden Crescent of Central Asia. Friedman et al. (Chapter 16) describe the interrelationship of metropolitan area characteristics, injection drug use, and HIV incidence among injectors. Shedlin (Chapter 17) examines patterns of drug use, regional differences, and the relationship between drug use and HIV transmission in two Nicaraguan cities. Moran (Chapter 18) examines risk environments for drug use and HIV/AIDS in Post-Soviet Russia. Yang (Chapter 19) examines the impact of residence and residential mobility on substance abuse and HIV in China.

In Part IV, Tempalski (Chapter 20) presents a framework for considering place-based processes through which syringe exchange availability may be understood – i.e., that the geographic distribution of syringe exchange programs (SEPs) in the United States is linked to the social and political conditions of particular localities. Stahler et al. (Chapter 21) utilize GIS to investigate the effects of individual, neighborhood, and program factors on substance abuse treatment participation, compliance, and relapse. Archibald (Chapter 22) explores the reciprocal effects of substance abuse treatment provision and area substance abuse. Green and Pope (Chapter 23) utilize a GIS framework to assess hurricane recovery needs of substance abuse center clients in Katrina and Rita-affected areas. Hunt et al. (Chapter 24) utilize GIS to identify drug markets, understand the elements of drug markets, and develop a strategy for implementing a focused deterrence model.

In Part V, Gopal et al. (Chapter 25) show how spatial analysis and modeling might be used to improve the identification and treatment of substance use disorders (SUDs) and co-occurring mental disorders in clinical practice. The study is based on the work of a unique collaboration between a practicing psychiatrist and a geographer. DiMaggio et al. (Chapter 27) explore spatial analytic approaches to explaining the trends and patterns of drug overdose deaths.

Kwan and colleagues (Chapter 26) propose a new conceptualization of socio-geographical context for analyzing the potentially complex relationships between contextual risk factors and drug use, abuse, and addiction. In other words, in examining neighborhoods as social contexts for illicit drug use and related behaviors, future research must progress beyond the narrow, static, resident-based understanding of the role of neighborhoods that pervades current research, toward a more dynamic view.

Such thinking proffers an operational definition of neighborhood that is comprised of the patterns of movement of neighborhood residents and non-residents across time and space and how such movement influences individual drug abuse behaviors. Specifically, where individuals actually spend time while engaged in daily activities should be a key point of observation and measurement. These new insights provide exciting opportunities and challenges for future research in geography and drug addiction.

Summary and Future Challenges

The role of geographic methods and technologies in drug abuse research continues to evolve. Employing long-standing geographical research methods focused on human/environment interactions, and applying transformational new geographic research technologies, such as GIS, are important steps toward better understanding drug abuse issues and their inherent complexities. When used alongside more traditional epidemiological techniques, geographic research methods, models, and technologies enable epidemiologists to formulate new questions, and to refine or enhance existing analyses. The ability to evaluate, visualize, and model geographic information provides unique insight into public health issues such as emerging and shifting epidemics, the utilization of treatment services, the prediction of drug abuse diffusion trends and proactive intervention, and the assessment of context and the environment in drug abuse and addiction research.

Chapter 3
Integrating GIS into the Study of Contextual Factors Affecting Injection Drug Use Along the Mexico/US Border

Kimberly C. Brouwer, John R. Weeks, Remedios Lozada and Steffanie A. Strathdee

> *"When [drugs] wreak their devastation, they respect no boundaries of income, race, occupation or geography" – Kofi Annan, Secretary General (United Nations 2006)*

Abstract While the relationship between individual-level factors and drug use or associated risk behaviors has been investigated in-depth, comparatively little is known about the influence of contextual determinants. Geographic information systems (GIS) enable users to simultaneously display a number of environmental data layers and use spatial statistics to explore the relationships between contextual and individual-level variables. This makes GIS a potentially powerful tool in substance use research. One area where geography has a striking influence on health is at the US/Mexico border, where the confluence of drug trafficking routes, migration, and income inequalities form unique environmental influences on drug use scenes. In this chapter, we describe some of the spatial and environmental issues, such as migration, neighborhood characteristics, and proximity to services, which may affect drug use behaviors and risks. A number of practical methods to measure geographical indicators are also described, focusing on a case study of injection drug use in the Mexico/US border city of Tijuana, Mexico.

Introduction

While much research has been done to characterize individual-level risk factors for drug use and risk behaviors, comparatively little is known about contextual determinants. In recent years, public health and substance use researchers have come to appreciate how macro-level or structural factors may contribute to the

K.C. Brouwer
Division of International Health and Cross-Cultural Medicine, Department of Family and Preventive Medicine, University of California School of Medicine, San Diego, CA, USA

Y.F. Thomas et al. (eds.), *Geography and Drug Addiction*,

"risk environment" (Diez Roux 2001; Rhodes, Singer, Bourgois, Friedman, and Strathdee 2005; Rhodes et al. 1999), especially in regards to transmission of the human immunodeficiency virus (HIV) and other infectious diseases (Gleghorn, Jones, Doherty, Celentano, and Vlahov 1995; Harrison, Vlahov, Jones, Charron, and Clements 1995; Holmberg et al. 1995; Latkin, Mandell, Vlahov, Oziemkowska, and Celentano 1996; Manoff et al. 1996). These determinants include social, geographic, economic, and political factors, encompassed in an ecosocial approach where multiple categories of determinants are simultaneously considered (Poundstone, Strathdee, and Celentano 2004). Social determinants of health present in day-to-day life can exert influence on health above and beyond individual characteristics. A compilation of data from studies of injection drug users (IDUs) globally showed that HIV prevention interventions that focus solely on individual behavior change result in only partial reduction of risk of transmission, ranging from 25% to 40% (Rhodes et al. 2005). Developing "structural" interventions which act on the social and physical environment represents an avenue for addressing risk not affected by individual-level interventions and may eventually help to reduce health disparities (Burris et al. 2004; Diez Roux et al. 2001).

Most studies of injection drug use have largely described the geographic boundaries and characteristics of environments without analyzing what constitutes a risk environment. Geographic information system (GIS) models and the application of spatial statistics within a GIS provide a palette of tools that can help to improve understanding of the relationship between the environment and individual-level characteristics. The ability to simultaneously present a number of environmental data layers and explore their relationship with drug use variables through spatial statistics makes GIS a powerful tool in substance use research.

The illegality of drug use and the often mobile nature of illicit drug users makes mapping in substance use studies more complicated than in a typical health study. A further complication comes when working in countries where the digital mapping infrastructure tends to be underdeveloped. Both complications mean that researchers are generally required to create their own digital boundary files, or at least adapt those produced by others. For these reasons, this chapter will primarily draw upon examples from an ongoing bilateral study (Project *El Cuete*) of injection drug use in the Mexico/US border city of Tijuana, Mexico.

Tijuana is the northwestern-most border city in Mexico, with the highest prevalence of drug use in Mexico (SSA 1998). The US/Mexico border area is unique in that geography has a striking effect on health. Prevalence of infectious diseases in the 100 km area directly north and south of the international border are unusually high for both Mexico and the United States (Brouwer et al. 2006; Doyle and Bryan 2000; United States–Mexico Border Health Commission 2005). A number of cities along the border also lie along the major drug trafficking routes, further complicating the health situation. This chapter will briefly describe some of the spatial/environmental issues, such as migration, neighborhood characteristics, and proximity to services, which may have an influence on injection drug use behaviors and risks. Some practical issues involved in measuring geographical indicators are also covered, focusing on the case study of our Mexico/US border field project.

GIS As a Tool for Studying Drug Use

In moving the field of drug addiction research forward, innovative methods are needed to measure macro-level risk factors in an effort to characterize the complex interactions between factors operating at the level of the individual, network, and environment, thereby setting the stage for the development of structural interventions. Since epidemiology is an inherently spatial science, GIS applications have taken off rapidly in public health as new geospatial techniques have become available (Cromley and McLafferty 2002; Jenks and Malecki 2004; Mullner, Chung, Croke, and Mensah 2004).

There are as yet, however, few published reports of the application of GIS to the study of injection drug use, let alone substance use in general. Ecologic associations have been shown to be important in a number of studies of alcohol abuse, where, for example, it was found that higher density of alcohol outlets in Los Angeles was associated with higher rates of assault, independent of unemployment, age, income, female-headed households, or household size. In fact, alcohol outlet density explained 7% of the variance in the rate of assault (Scribner, MacKinnon, and Dwyer 1995). In the US city of Baltimore, Maryland, GIS was used to assess patterns of drug use. Type and frequency of drug use were associated with specific geographic areas, independent of neighborhood characteristics (Latkin, Glass, and Duncan 1998). Recently, Trooskin et al. used a GIS model to explore clustering of hepatitis C virus cases in the US state of Connecticut (Trooskin, Hadler, St Louis, and Navarro 2005). Most clusters occurred in known injection drug using areas, where the state was already providing limited needle exchange services; however, a new cluster in an area without such services was also identified, suggesting the need to rethink the distribution of service areas. GIS has also been used to model social networks of urban youth who were or were not substance users (Mason, Cheung, and Walker 2004). This analysis compared distances of homes to risky or safe places identified by young people. The above studies begin to illustrate the power of GIS to inform formation and optimization of public health interventions and increase understanding of at-risk drug using populations.

Background of Tijuana, Mexico and Project *El Cuete*

The more than 2,000 mile border separating Mexico and the United States forms a unique environment encompassing over 12 million inhabitants in the 100 km area directly north and south of the international boundary (United States–Mexico Border Health Commission 2005). It is the most extensive land frontier separating a developed and developing country, and the income gap is the largest between any two contiguous countries (INEGI 2000b). Located on the far northwestern edge of the border region, Tijuana, Mexico, a city of 1.3 million, sits just south of twin-city San Diego (INEGI 2000b).

Situated along the Tijuana/San Diego border area is a major drug trafficking route through which heroin, cocaine, and methamphetamine are smuggled to the United States (Bucardo et al. 2005). "Spillover" from these shipments has created a robust local drug consumption market (Bellis 2003; Magis-Rodríguez, Marques, and Touze 2002; Medina-Mora et al. 2003; SSA 1998). In fact, Tijuana has one of the fastest growing IDU populations in Mexico and the highest prevalence of consumption of illicit drugs in the country (Magis-Rodríguez, Marques et al. 2002; SSA 1998). There are estimated to be 10,000 IDUs and more than 200 shooting galleries in Tijuana (Morales, Lozada, Magis, and Saavedra 2004), where *people who do not necessarily know each other get together to inject drugs*.

HIV prevalence among IDUs and other high risk populations in Mexico has thus far remained low (Bastos, Strathdee, Derrico, and Pina 1999; Güereña-Burgueño, Benenson, and Sepulveda-Amor 1991; Magis-Rodríguez, Marques et al. 2002). Of the estimated 160,000 people living with HIV, approximately 2–6% are believed to be IDUs (Magis-Rodríguez, Rivera Reyes, and Bravo-García 2002; Noriega-Minichiello, Magis, Uribe, Anaya, and Bertozzi 2002; UNAIDS 2004). However, recent research by Strathdee and her associates in Tijuana has indicated that risky injection behaviors are rampant (Strathdee et al. 2005). Further, studies suggest that HIV prevalence is increasing dramatically in sentinel populations in Tijuana, such as pregnant women who have used drugs (Viani et al. 2006). A recent modeling exercise estimated that 2–7% of Tijuana IDUs were HIV-infected and comprised the at-risk group with the second largest number of infected persons (just behind men who have sex with men) (Brouwer et al. 2006). Although Mexico is considered a country of low HIV/AIDS prevalence with a concentrated epidemic, the window of opportunity for prevention may be closing rapidly since IDU-associated HIV epidemics are often explosive and can quickly become generalized (Rhodes et al. 2002; UNAIDS/WHO 2003). Studying the environment in which drug use occurs may help to identify risks associated with transmission of blood-borne viruses in this city and aid in the development of effective intervention strategies.

To illustrate some of the key subjects and practical issues that can be explored using GIS to study IDUs and expand upon unique aspects of border areas, we present here a description of Project *El Cuete*, an ongoing collaborative research study to assess infection prevalence, risk behaviors, and possible interventions in injection drug using populations along the Mexico/US border (*El Cuete* is a slang term commonly used among border IDUs to refer to a syringe). Project *El Cuete* is a three-phase research project funded primarily by the National Institute on Drug Abuse that is being undertaken by the University of California San Diego along with governmental agencies in Mexico [Centro Nacional para la Prevención y el Control del VIH/SIDA (CENSIDA), Instituto Nacional de Salud Pública] and several Mexican non-governmental organizations (Patronato ProCOMUSIDA, A.C., CIRAD, A.C., and Programa Compañeros, A.C.). Institutional review boards of the Tijuana General Hospital and University of California, San Diego approved the study's protocols.

Phase I of Project *El Cuete*, conducted in 2004, consisted of in-depth qualitative interviews administered to 20 IDUs in Tijuana and 24 in Ciudad Juarez (just south of El Paso, Texas, USA) to gather exploratory information on drug use and injection, and sexual behaviors. From February–April 2005, Phase II consisted of a cross-sectional study using respondent-driven sampling (RDS) methods to collect quantitative information on HIV risk behaviors among 222 IDUs in Tijuana and 206 in Ciudad Juarez and conduct antibody testing for HIV, hepatitis C, and syphilis. RDS is a chain referral method whereby a group of "seeds" were selected based on diversity of gender, location, and drug preferences, and given three uniquely coded coupons to refer IDUs in their social network. It is increasingly being used to recruit subjects from "hidden populations," such as IDUs (Heckathorn 1997, 2002). Referral chains continued until approximately 200 were recruited at each site. IDUs who tested positive for any of these infections received counseling and referral to treatment. Phase III, a longitudinal cohort study of infection incidence, will follow 1,000 IDUs recruited through RDS in Tijuana. Eligibility criteria for project *El Cuete* include: having injected illicit drugs within the past month, aged 18 years or older, able to understand and speak Spanish or English, and willing and able to provide informed consent. Examples presented here will focus primarily on phase II results from Tijuana. The main reasons for incorporating GIS into Project *El Cuete* were to determine social and environmental factors potentially affecting the drug use scene and any barriers to accessing public health resources.

Practical Mapping Issues in Tijuana

The city of Tijuana is approximately 24 kilometers wide (from the Pacific Ocean on the west to the city's eastern edge) and covers a total area of 1,727 square kilometers. As is the case with all urban areas, Tijuana is divided into a large number of neighborhoods (more than 600), which are known in Mexico as colonias. Whereas new, sparsely populated colonias are largely created by administrative or commercial boundaries, colonias in the most populated areas of the city stem from neighborhoods with distinct historical traditions and characters, often informal in their origins. Tijuana colonias have an average of approximately 1,900 residents and are one of the smallest units for which public census data is available, which is an appropriate unit for analyses of neighborhood effects (Diez Roux, 2001).

Obtaining Base Maps

Working in countries without a long tradition of digital mapping often means constructing digital maps from scratch, by venturing out on foot to map with global positioning system (GPS) machines or by digitizing paper maps, satellite, or aerial photos. We were fortunate to be able to obtain from the Instituto Nacional de Estadística, Geografía e Informática (INEGI) – National Institute of Statistics, Geography, and Informatics – of Mexico computer aided design files of the Tijuana municipal area demarcating colonias and city streets. These were then geo-referenced by the researchers using ArcGIS 9.0 (ESRI corp., Redlands, CA, USA). Additional geographic layers were obtained from local sources, such as the Tijuana Instituto Municipal de Planeación (Municipal Planning Agency). Even with these resources, the rapid growth rate of the city (roughly 5% per year) (INEGI 2000b) presents a challenge in obtaining digital boundary and street data for new neighborhoods.

Mapping Individuals

Perhaps one of the most challenging aspects of mapping in any substance-use study is obtaining individual-level positional data. Over half of the *El Cuete* phase II sample population (55%) was homeless. Even for those with a fixed address, there is no regular system of addresses/street names. For example, houses on a given street may not be numbered sequentially. For this reason, creating a digital street address database, such as Streetmap USA (ESRI corp., Redlands, CA, USA), and performing computerized geocoding for Tijuana is infeasible.

Our strategy in responding to these challenges initially began with asking participants for the name of the neighborhoods where they live or inject drugs. While colonia names are not standard and personal concepts of colonia boundaries may vary from official demarcations, these data have allowed us to generally assign participants to one or another section of the city. A weakness of such data is that it is not very powerful in comparing trends between colonias if the number of participants assessed per colonia differs greatly, which is usually the case with respondent-driven sampling and the nature of the drug use scene in Tijuana. A strategy to obtain more precise data was to ask for the cross-streets nearest to where participants spend most of their time. However, in phase II, only 34% listed more than one nearby street. Further, except for a small percentage of cases, obtaining GPS waypoints of where participants live or inject was largely ruled out due to concerns for the safety of the researchers trying to obtain those data.

Most of the participants in our study travel around the city on foot and have very good mental maps of their immediate neighborhoods. In Phase III, we have begun to tap into this knowledge by asking participants to identify the areas they live, use drugs, earn money, or buy drugs by pointing to specific locations on paper maps of each colonia of the city. This strategy is quite successful after a few landmarks or main roads are pointed out. As many of our field staff are former IDUs or work as health promotoras (health promotion field workers) in neighborhoods where drug use is common, they are comfortable in assisting with mapping. We have obtained data using similar methods in other settings, such as showing aerial photos of neighborhoods to study participants.

Many of the homeless in Tijuana live in one of the few parts of the city where they are able to find shelter – the Tijuana river bed. For participants living in the canal, we have taken advantage of the numbering system of portals (pipe drains) along the canal's length (Fig. 3.1A). Participants usually know the portal number closest to where they live, which enables us to link this information to GPS waypoints for these portals.

A

B

Fig. 3.1 Images portraying aspects of the urban environment in Tijuana. The numbering system of portals (pipe drains) along the Tijuana River canal facilitates mapping of homeless project *El Cuete* participants who live in the canal (Panel A). In colonia Zona Norte, it is common to observe injection drug users openly injecting near by the fence that separates Mexico from the United States (Panel B).

A number of ethical issues arise when collecting detailed location data on marginalized populations for whom we are also collecting sensitive health and behavior data. Although participants provide information willingly and risk of accidental release of data is fully explained in the informed consent process, to safeguard such information, no personal identifiers are used on paper nor digitized maps. All mapping results are aggregated or the scale decreased so that individual positions can not be determined, electronic files are password protected, and results of our study are shared with healthcare personnel, non-governmental organizations, and other health researchers only for the purpose of improving healthcare access and increasing understanding the risk environments faced by participants.

Characteristics of Neighborhoods

Neighborhoods are potentially important determinants of health for reasons that have to do both with their demographic composition and with the specific environmental context (Diez Roux 2001; Mitchell, Dorling, and Shaw 2002; Oakes 2004; Sampson 2003). Composition refers to the characteristics of people inhabiting a neighborhood. To the extent that neighborhood residents are poor and have poor health, for example, their behavior may negatively affect the health behavior of other residents, even beyond their own characteristics. Context refers to environmental characteristics that are exogenous to the demographics of residents. Proximity to polluted water or polluted air and inadequate health infrastructure would be examples of this. In Tijuana, we can note that living in a drainage canal represents the kind of environmental context that is conducive to poor health outcomes. Strategies for obtaining neighborhood-level data are discussed in the following sections.

Neighborhood Disadvantage

Neighborhood disadvantage is a demographic compositional measure of social disadvantage within the confines of a traditional neighborhood or amongst residents within a defined census tract; it is usually designed to reflect the constructs of area income and wealth, education, occupation and employment, and socioenvironmental characteristics related to area crowding, stability, and housing (Diez Roux 2001, 2004; Diez-Roux et al. 2001; Mitchell et al. 2002). A study by Galea et al. found neighborhood disadvantage to be a significant predictor of frequency of injection drug use, even after controlling for individual-level risk factors (Galea, Ahern, and Vlahov 2003). Although Galea's study did not explore causation, it is believed that injection drug use may be a way of coping with the psychosocial stress of living in highly disadvantaged neighborhoods. Conversely, it is theorized that concentration of IDUs may be higher in disadvantaged neighborhoods due to migration into the area by IDUs from wealthier neighborhoods in order to escape from social controls (Galea et al. 2003).

A recent study comparing IDUs from economically advantaged and disadvantaged neighborhoods found that the former were more likely to get syringes from a single source and also more likely to inject at home, rather than in a social setting, such as a shooting gallery (Buchanan, Shaw, Teng, Hiser, and Singer 2003). Interestingly, the same study showed a disadvantage for residents in wealthier neighborhoods in that they were much less likely to come into contact with outreach workers (Buchanan et al. 2003).

In beginning to build a GIS model of the risk environment for Project *El Cuete*, census data were obtained from INEGI that provide information on a number of colonia-level economic and social indicators. A full census is conducted every 10 years in Mexico so characteristics of colonias undergoing rapid migration or changes in infrastructure may not fully reflect current conditions. However, certain census variables, such as data on available health services, are updated annually and Mexico also conducts a mid-decade mini-census. INEGI census data have been previously used to calculate a "social well-being index," similar to neighborhood disadvantage but at the municipal (county) level, which was subsequently compared to morbidity levels for certain diseases (Ochoa-Diaz Lopez, Sanchez-Perez, and Martinez-Guzman 1996). The latest colonia-level Tijuana census in 2000 collected 227 variables on neighborhood characteristics such as percent unemployed or underemployed, average number of occupants per room, education level, percent receiving less than minimum wage, and number of houses with electricity or indoor plumbing. We are defining neighborhood social disadvantage based on a combination of these census-derived variables, referencing recent work by Diez-Roux (Diez Roux 2001; Diez-Roux et al. 2001). Additional data [e.g., road conditions, "broken window" index etc. (Cohen et al. 2000; Cohen, Farley, and Mason 2003)] are being garnered by taking a number of field measurements in the colonias in which most of our participants are clustered, and by comparing remotely sensed imagery of these neighborhoods. One of the coauthors has performed similar analyses of remotely sensed imagery of Accra, Ghana in order to characterize the built environment by developing a "slum index" for that city (Weeks, Hill, Getis, and Stow 2006).

Mobility and Injection Drug Use

A major challenge to the study of drug using populations is their mobility. When it comes to applying GIS tools to such a population, the difficulties of assigning a fixed "place" become even more evident. Exploring and understanding mobility, however, can lead to a better understanding of the dynamics of drug use trends and spread of disease, and guide the targeting of public health measures.

Local Mobility and Social Networks

Social networks and cultural factors, including knowledge, beliefs, and customs, may affect acceptability of drug use, injection practices, chance of encountering HIV-positive persons, and utilization of drug abuse treatment (Kottiri, Friedman, Neaigus, Curtis, and Des Jarlais 2002). By using data regarding who recruited whom, garnered from our respondent-driven sampling technique, we are able to use GIS to compare the spatial distribution of "seeds" and their referrals. One hypothesis to be tested is whether those who recruit a greater distance from their neighborhood will have a higher prevalence and incidence of blood-borne infections, based on their likely role as a bridge between social networks.

Only 53% of the *El Cuete* Phase II participants lived and injected in the same neighborhood. This is an important consideration in the study of the risk environment of an individual since, while one may live in a relatively safe area close to services, a subject may be spending most of his/her time in a riskier area. Homelessness and local mobility also means that participants will likely move during the course of longitudinal studies. For this reason, in Phase III of the study, we are collecting locator information every 3 months during the 18 months of follow-up. Not only does this assist in tracking down participants due for follow-up visits but it also enables us to explore temporal changes in the risk environment.

National Mobility

Mobility is not just local. In the United States, at-risk youth are known to travel throughout the country and IDUs, in particular, are known for their migratory nature (Jones, Davidson, Bisset, and Brettle 1988; Perlis, Torrico, and Settembrino 2002). The case is no different here. In fact, Tijuana, as a rapidly growing border town, attracts migrants from all parts of Mexico and Latin America. Proximity to the United States has created economic opportunities that attract migrants from other areas of Mexico, as reflected in the fact that over half of Tijuana's population in 2000 was born outside of the state of Baja California (where Tijuana is located) (INEGI 2000b). In a survey of homosexuals/bisexuals, prostitutes, and prisoners in Tijuana in 1991, only 22%, 5%, and 21%, respectively, originated in the Baja California region (Güereña-Burgueño, Benenson, Bucardo Amaya, Caudillo Carreno, and Curiel Figueroa 1992). In the case of *El Cuete* participants, although 76% had resided in Tijuana for at least 5 years, 70% were born outside of Baja California. Of these, 18% had come to Tijuana intending to cross to the United States and 20% ended up in Tijuana after deportation from the United States. This has important implications for health as Mexican migrants who have tried unsuccessfully to cross the border into the United States often feel of a lack of identity or attachment; many harbor a distrust of local officials, thus discouraging healthcare or treatment-seeking (Montiel-Hernandez, Muniz, Baez-Villasenor, and del Rio 1996).

In characterizing drug use trends in Mexico in general, spatial relationships to drug trafficking routes may also put certain neighborhoods and cities more at risk than others. This phenomenon has been documented in Brazil along major highways used for drug trafficking (Bastos et al. 1999), in overland heroin routes in Southeast (Beyrer et al. 2000) and Central Asia (Parfitt 2003), and in parts of Nigeria which have become transit points for heroin trafficked by air (Adelekan and Stimson 1997; Stimson, Des Jarlais, Ball, and Organization 1998). Similar patterns are seen in Mexico, where, for instance, drug use prevalence is two to three times the national average in border cities located along drug trafficking routes to the United States (SSA 1998). The state of Baja California also has the highest cumulative national AIDS incidence, following only Mexico City (CONASIDA 2004). Figure 3.2 shows the relationship between drug of primary impact for drug treatment admission and major methamphetamine producing states. The concentration of methamphetamine production in western Mexico is reflected in higher methamphetamine use in the West.

Cross-Border Mobility

International mobility and living in a border region between nations with disparate laws, customs, and healthcare should also be considered when mapping "place" in drug use studies. Migration has been linked to lower socioeconomic status, power inequalities, social and cultural alienation, a breakdown of family units and fear of deportation and violence (Brockerhoff and Biddlecom 1999; Massey, Arango,

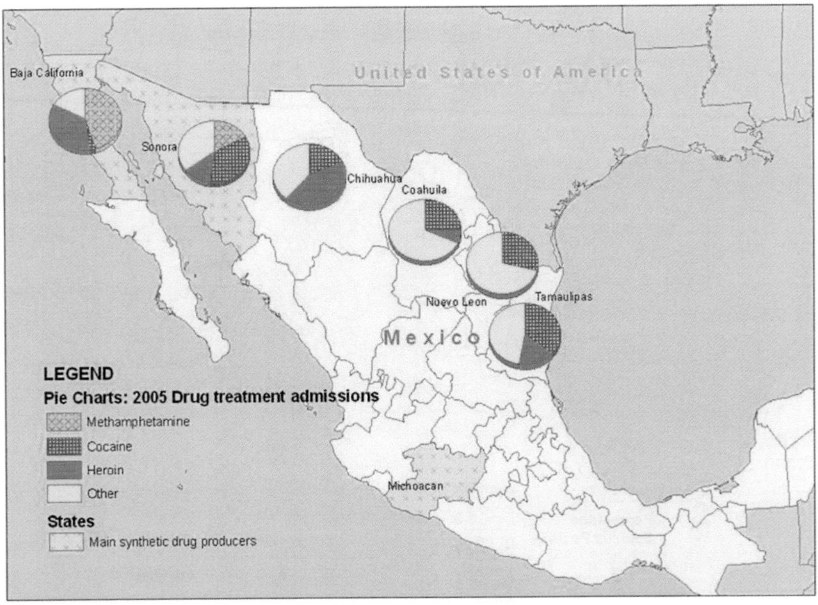

Fig. 3.2 Map of 2005 drug treatment admissions in Mexican states bordering the United States. Pie charts above each of the six Mexican states bordering the United States show the percentage of drug treatment admissions by primary drug used. The "Other" category includes marijuana, inhalants, alcohol, tobacco, and a variety of veterinary products. Major synthetic drug producing states are shaded
Source: Treatment data were compiled from the Mexican Addiction Epidemiologic Surveillance System or SISVEA (Maxwell, Cravioto, Galvan, & Cortes, 2005; SSA, 2002). Data on main synthetic drug producing states were derived from the Sistema Estadistico Uniforme para el Control de Drogas (SEUDC) (PGR, 2000). (See also Plate 1 in the Colour Plate Section)

Hugo, Kouaouci, and Pellegrino 1994; Peterson 1958; Rachlis et al. 2007). Often, drug trafficking takes place in regions where there are porous borders and previous research suggests that border regions can be magnets that heighten HIV susceptibility through social disruption and the coming together of vulnerable populations including IDUs and commercial sex workers (Lyttleton 2002; Organista 2004; Rhodes 2005).

The San Ysidro border station between Tijuana and San Diego is the busiest point of entry along the US/Mexico border (indeed, in the world) with up to 50,000 vehicles and 25,000 pedestrians crossing the border at this point each day (INEGI 2000b; US General Services Administration 2006). In effect, the border is an ambiguous line, failing to fully separate populations that have so many social, economic, and cultural ties. IDUs are known to congregate in the neighborhoods near the international boundary, which are more difficult for police to patrol due to structural barriers such as high walls and busy roadways (Brouwer, Firestone, Lozada, Magis-Rodríguez, and Strathdee 2005) (Fig. 3.1B).

In applying GIS to this unique drug scene, we are exploring whether frequency of border crossings and the distance between a drug user's residence and the

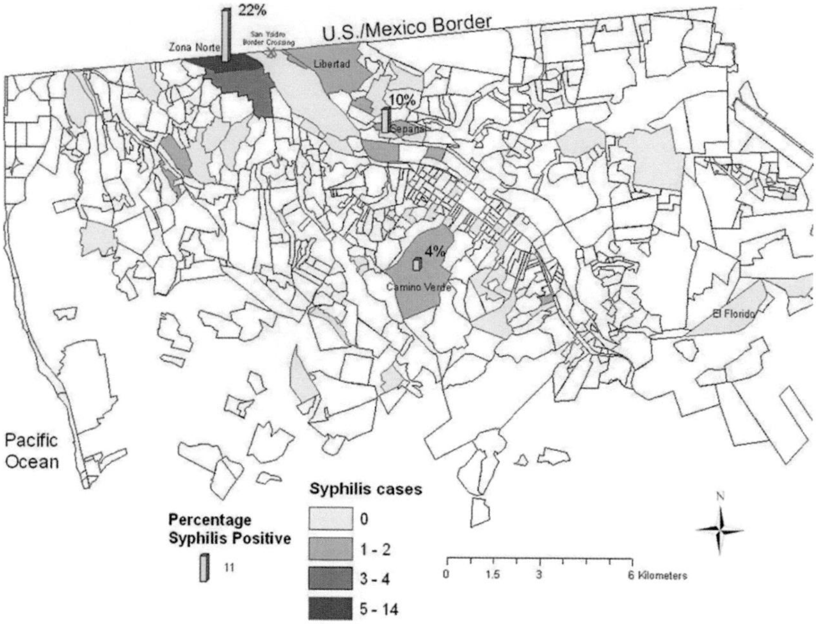

Fig. 3.3 Distribution of syphilis-positive project *El Cuete* participants, 2005 ($n = 220$). Colonias of residence of syphilis-positive project *El Cuete* participants are shaded according to how many cases live in that colonia (Note: values are not normalized by number interviewed per colonia). Height of bars represents percentage of participants positive for syphilis at each of three interview sites. Syphilis was higher nearer the border (χ^2 for trend, $P = 0.002$) even when adjusting for age, gender, and prevalence of commercial sex work (Frost et al, 2006)

Mexico/US border is related to type of drug used and prevalence of blood-borne infections. In preliminary analyses, we have already seen an increased prevalence of syphilis in our interview site closest to the border, while it was lowest furthest from the border (Fig. 3.3). However, it is not possible to fully distinguish between spatial distances and neighborhood effects. The border area is the oldest part of Tijuana and one of the most run down; it also has a lot of dead-end streets and a busy highway which may be among the reasons why it is not developed and is a haven for shooting galleries. Therefore, any findings based on spatial distance from the border will have to be considered in light of the characteristics of neighborhoods near and far from the border.

Mapping Infection Distribution and Proximity to Safe/Risky Places

Proximity to risky or protective areas has been shown in a number of studies to be related to risk behaviors and treatment seeking (Mason et al. 2004; Scribner et al. 1995). We detail below a few environmental factors whose distribution can be mapped and compared with IDU behaviors and infection prevalence.

Distribution of Shooting Galleries

The IDU social environment in Tijuana is highly influenced by attendance at shooting galleries [picaderos], where injection norms, such as renting and/or buying used needles and injecting on the premises, are often determined. The Mexican government reportedly closed 1,400 picaderos in Tijuana in 2002 (Oficina de la presidencia de la republica 2002), but over 200 are believed to currently exist in the city. During phase II of project *El Cuete*, we found that 69% of participants had injected in a shooting gallery in the past 6 months, and 57% claimed it was the place they most often injected (n = 222). Shooting galleries are uncommon in western US cities, but are widespread in the eastern US and in Puerto Rico, which have experienced severe IDU-associated HIV epidemics (Latkin et al. 1996; Noriega-Minichiello et al. 2002; UNAIDS 2004).

By virtue of their clandestine nature and territorial issues in drug trafficking, mapping of shooting galleries or "picaderos" can be dangerous. However, participatory mapping method with IDUs, especially those who frequent or manage picaderos, is one strategy by which to gain information on the density of these venues in neighborhoods. Inquiring through interviews about the estimated number of picaderos in the neighborhood where a participant injects can provide estimates of picadero distribution. This information will allow us to explore whether colonia shooting gallery density is related to the risk of acquiring a blood-borne disease.

Harm Reduction Resources

Risky injection behaviors, such as needle and syringe sharing, are likely enhanced when supportive services, such as needle exchange programs (NEPs) or pharmacies, are lacking (Rhodes et al. 2003). To reduce drug-related harm, services must not only be available, but also be accessible. Studies in Vancouver and New York City found that IDUs who lived further from NEPs were less likely to use them or properly dispose of syringes (Rockwell, Des Jarlais, Friedman, Perlis, and Paone 1999; Wood et al. 2004). In a similar situation, IDUs in Baltimore who had to travel longer distances were less likely to enter drug treatment programs (Strathdee et al. 2006).

As in most Latin American countries, few non-governmental organizations in Mexico appear to be involved in prevention activities aimed at drug users (Magis-Rodríguez, Marques et al. 2002). Harm reduction programs, especially NEPs, have often been met with opposition within the Mexican government. To our knowledge, there is only one documented NEP in all of Mexico, which is operated in Ciudad Juárez (Ramos 2000). Although no formal NEPs currently operate in Tijuana, there are more than 1,600 registered pharmacies where IDUs can theoretically purchase syringes legally over the counter. The purchase of needles and syringes in Mexico does not require a prescription. However, in areas of high drug activity some pharmacists limit sales of needles and syringes to those who appear to be drug users, by either saying that they have "run out" of the type of needles popular with drug users or by artificially raising prices (Strathdee et al. 2005). Qualitative studies in other settings have identified pharmacists' lack of knowledge regarding the laws and regulations governing syringe sales as a barrier (Blumenthal, Springer, Jones, and Sterk 2002; Wolfe, Amelunxen, Torres, Jenison, and Churchill 2002). Store policies or commercial considerations may also be at play. Using the Tijuana municipal pharmacy registry, which is updated annually, we are able to map pharmacy density per colonia and explore whether it affects direct or indirect syringe sharing by IDUs.

Legal Controls

While legal controls do have their place in society, enforcement of laws restricting drug use can have unintended consequences and even worsen the risk environment for disease transmission. Studies in Hong Kong and Thailand found that increased enforcement of laws prohibiting the opium trade and prosecution of users in urban areas led to increased prices for opium, and subsequent transition from opium smoking to injection of heroin (Westermeyer 1976). Thus, a problem that was primarily composed of drug use itself was compounded with all the health threats of blood-borne pathogens through higher rates of parenteral drug use.

A recent study in Togliatti, Russia, which has seen an explosion of HIV among IDUs, found fear of police detainment to be associated with needle and syringe sharing (Rhodes et al. 2003). In various contexts, fear of police detainment or arrest can discourage IDUs from carrying needles, leading them to share needles at the point of sale or inject with rented needles in shooting galleries (Harvey et al. 1998; Koester 1994; Rhodes et al. 2003; Strathdee, Zafar, Brahmbhatt, Baksh, and ul Hassan 2003). This practice can promote disassortative mixing, which in turn can increase the risk of transmission of blood-borne infections. Prescription laws, such as those limiting access to clean needles, can also increase risk of blood-borne infections. Freidman et al. have shown HIV prevalence in the United States to be related to local prescription laws (Friedman, Perlis, and Des Jarlais 2001). Both internal and external legal controls have helped to shape the drug scene in Tijuana. Efforts to crack down on drug trafficking across the Mexico/US border following September 11, 2001 led to reported increases in heroin availability and decreases in prices in Mexican border towns (Chavira 2003; Medina-Mora and Rojas Guiot 2003). In-depth interviews from phase I of Project *El Cuete* revealed that policing practices negatively affected accessibility to sterile syringes and promoted use of shooting galleries as it was common for IDUs to be arrested for carrying used or sterile syringes (Miller et al., 2008).

Tijuana is divided into seven administrative delegations, with policing areas based on sub-delegations (INEGI 2000a). Using municipal data, we will map locations of police delegations and stations and overlay neighborhood crime statistics in order to examine the impact of these influences on injection behaviors and risk of blood-borne infection. With the aid our study partner, COMUSIDA, and data from the Centro de Integracion Juvenil [juvenile hall], we are able to map "high risk"

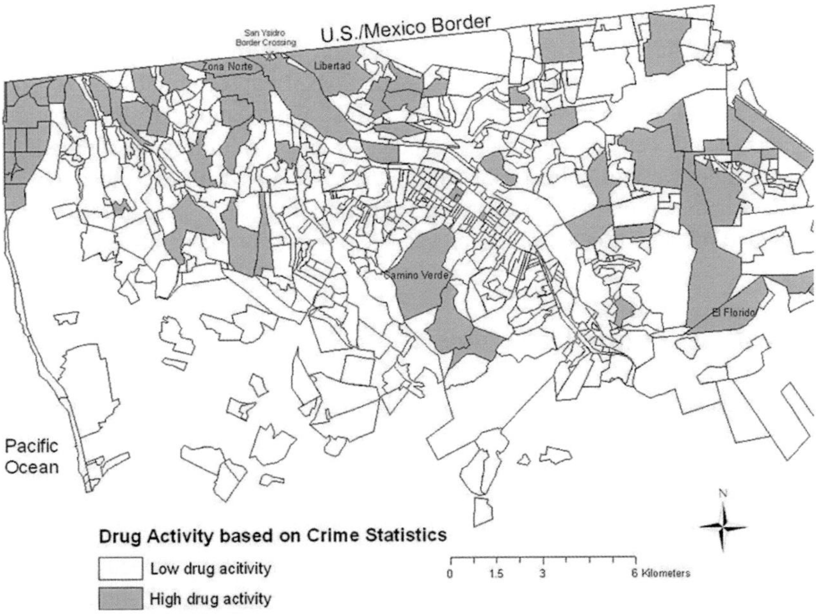

Fig. 3.4 Colonias with high drug activity, Tijuana, Mexico, 2005. Colonias identified as areas of high drug activity based on year 2005 local crime statistics for sale and consumption of illicit drugs Source: Mexican Department of Justice and COMUSIDA, A.C. Tijuana, Mexico

zones of the city according to neighborhood crime statistics for assaults, burglaries, vehicle thefts, and sale and consumption of drugs. Figure 3.4 shows a map containing estimated locations of high- and low-density injection drug use colonias based on these crime statistics.

Distribution of Drug Treatment Programs and Health Services

Lack of health and supportive resources has been repeatedly cited as a primary reason for not seeking drug treatment. For instance, an assessment of barriers to drug abuse treatment in Mexico and the United States found the principal obstacles to include lack of treatment "slots" and ignorance regarding facilities (Appel, Ellison, Jansky, and Oldak 2004).

Drug use in most areas of Tijuana occurs in an atmosphere of limited access to support services (Magis-Rodríguez, Marques et al. 2002). There are an estimated 20 residential drug treatment programs in Tijuana with a capacity to treat 3,500 persons per year (Trillo 2002). Taking into account estimates of the number of substance users in the city, however, the coverage of these programs is believed to be less than 20%. In light of this deficiency, some drug users report traveling to the United States to seek drug treatment that is considered unavailable in Mexico (Ferreira-Pinto and Ramos 1997; Ramos 1990).

The Mexican government is committed to providing treatment for all those with HIV/AIDS and has increasingly taken efforts to provide services to drug users, yet treatment often does not reach targeted populations. In applying GIS to Project *El Cuete*, we will use handheld GPS devices to map the locations of drug treatment programs, health centers, and IDUs' residences and injection neighborhoods. At the same time, we will collect other attributes associated with each location, and evaluate whether proximity to treatment centers is related to past drug treatment seeking and participation as well as prevalence of blood-borne pathogens.

Implications/Future Directions

In this chapter, we have endeavored to demonstrate the feasibility of applying GIS to the study of the IDU risk environment. A goal of our work in Tijuana, similar to health projects in other areas, is to eventually convert our GIS database into an internet-based tool to be available to those seeking drug abuse treatment and to facilitate decision making by policy makers and service providers trying to find the best way to allocate sparse resources. GIS mapping has already helped guide us to the neighborhoods which our mobile study clinic, the *Prevemovihl*, should visit. A limitation of ecological studies is that it is difficult to distinguish between the effects of the social/spatial context itself versus the characteristics of individuals making up the area under study. However, new techniques in multi-level analysis allow one to distinguish between the relative contribution of within- and between neighborhood effects, as well as estimating how much variability at the contextual level is due to individual factors (Diez Roux 2001, 2004; Oakes 2004). Although there are many limitations in trying to obtain accurate positional data in drug use studies, applying GIS to this field will substantially improve understanding of interactions between drug users and the risk environment over what is currently known.

Acknowledgment The authors gratefully acknowledge support from the National Institute on Drug Abuse (NIDA) through grants DA09225-S11 and DA019829, donor support for the Harold Simon Chair in International Health and Cross-Cultural Medicine, the UCSD Center for AIDS Research (P30 AI36214-06), and a career award supporting Dr. Brouwer (NIDA grant K01DA020364). We are extremely thankful for the staff of Proyecto *El Cuete*, CIRAD, Patronato ProCOMUSIDA and Programa Compañeros, and to Dr. Peter Hartsock of NIDA.

Chapter 4
The Spatial Context of Adolescent Alcohol Use*

Assessing the Role of Neighborhood Alcohol Availability and Disadvantage

Karen A. Snedker, Jerald R. Herting

Abstract The purpose of this chapter is to explore the spatial context of adolescent alcohol use. Using a regional sample of adolescents and geocoding individuals and matching the data to census tracts, we use geographic methods to explore spatial patterns of alcohol use among adolescents and features of the neighborhood environment. With hierarchical linear modeling, we examine the relationship between neighborhood alcohol availability, disadvantage, and crime on alcohol use while controlling for individual, family, and peer characteristics, including an individual's distance to sources of alcohol. Analyses revealed that neighborhood disadvantage has a direct negative effect inconsistent with typical neighborhood disorganization or environmental stress frameworks, while there is a weak positive effect of alcohol liquor law violations on adolescent use. Neighborhood disadvantage, crime, and alcohol availability also appear to moderate the effects of key individual risk and protective factors, specifically deviant peers and family resources. For example, living in areas with greater neighborhood disadvantage decreases the effect of deviant peers on adolescent substance use. In addition, there is evidence that living in high alcohol outlet density areas reduces the protective effect of family resources, but neighborhood crime increases the protective effects of family support. These moderating effects highlight the importance of assessing the influence of peer and family factors on adolescent behaviors within the broader neighborhood context. We discuss the implications of our findings for ongoing research on neighborhood spatial analysis and contextual effects and adolescent behaviors.

K.A. Snedker
University of Washington, Psychosocial and Community Health, Box #358732, Seattle, WA 98195-8732, USA
e-mail: snedker@u.washington.edu

*This research was partially funded by CDC (CCR015606), NIDA (DA10317; 5 T32 DAO7257-13; 1 R03 DA019622-01A2), and University of Washington Alcohol and Drug Abuse Institute. The authors would like to thank Peter Graham for geographical mapping assistance and the University of Washington Reconnecting Youth project team.

Y.F. Thomas et al. (eds.), *Geography and Drug Addiction*,

Introduction

There is increased interest in studies that assess how broader ecological spheres in which individuals are embedded influence behavior. Recent studies have applied spatial analysis and geographic methods to enhance the understanding of the role of geographical spaces on problematic behaviors at the individual and aggregate levels. Substance use and abuse is particularly promising research area in this regard. Guided by this broad spatial framework and using geospatial methodological approaches, we explore the impact of neighborhood context on adolescent alcohol use. Based on disorganization, stress, and opportunity frameworks that posit problematic contexts lead to problem behaviors, we examine the effects of neighborhood alcohol availability, disadvantage, and crime on alcohol usage among youth.

Research indicates that measures of alcohol accessibility and outlet density are linked to various problematic drinking-related behaviors. Alcohol establishments, both on- and off-premise outlets, are important access points for consuming and purchasing alcohol. Alcohol availability is important to include in a model of adolescent alcohol-use behaviors for several reasons. First, a higher concentration of alcohol outlets may directly increase alcohol consumption. This is in line with research linking alcohol outlet densities to adolescent drinking and driving, arguing that areas with greater numbers of alcohol establishments increase opportunities for youth to purchase alcohol (Treno et al. 2003). Second, alcohol outlets might create a riskier neighborhood context. Alcohol outlets provide spaces for people who are drinking to congregate; neighborhoods with high alcohol outlet density may signal an area with a normative structure supportive of alcohol consumption and perhaps lax social control. Moreover, this feature may coincide with other problematic neighborhood conditions (such as crime, disorder). More research needs to address the geospatial relationship between alcohol availability and alcohol-use behaviors as well as some of the limitations of previous research that focused almost exclusively on adults and used various units of analysis. This chapter is an important step as it addresses the gap in our understanding of the impact of neighborhood and individual factors on adolescent alcohol-use behaviors, with a specific emphasis on the role of alcohol availability at both the individual and neighborhood level on adolescents.

Previous work found that the neighborhood context has both direct and moderating effects on alcohol and marijuana use among youth (Snedker, Herting, and Walton; 2008). Specifically, neighborhood disadvantage has a negative direct effect on adolescent alcohol and marijuana use; adolescents living in economically disadvantaged neighborhoods have lower rates of use. Contrary to the social disorganization perspective, neighborhood disadvantage did not act as a risk factor increasing alcohol and marijuana use, but rather was associated significantly with lowered substance use among adolescents. While this finding appears counterintuitive to the general disorganization and stress frameworks and work primarily focusing on adult behaviors, it is consistent with research on neighborhood context and adolescent substance use (Chuang 2005; Hoffman 2002). These inconsistent findings for adolescents raise issues about the special nature of early initiation and use patterns of adolescents, and raise questions about whether and how spatial availability of alcohol may relate to adolescent alcohol-use behaviors.

The purpose of this chapter is to investigate the influence of neighborhood context on alcohol involvement among youth. Specifically, using data from Seattle, we explore spatial patterns of alcohol use among adolescents and features of the neighborhood. Our specific interest is in the role of alcohol availability on individual level on alcohol use, however, we include neighborhood disadvantage and crime (including alcohol-related violations) for theoretical reasons and to ensure if we find a direct or moderating effect of alcohol availability that the alcohol outlet measure is not simply a proxy for other community conditions not in the model. Not all research in this area includes other neighborhood factors when estimating a statistical model. If other key neighborhood factors are omitted, this raises questions about the impact of neighborhood alcohol outlet density on individual-level behaviors (Treno et al. 2001).

We assess if neighborhood measures of alcohol availability, disadvantage, or crime (including alcohol-related crime) influence adolescent alcohol use either directly (and indirectly) or as moderators on three known protective (personal and family resources) and risk (deviant peer networks) factors related to alcohol use. Adding indicators of alcohol outlets and availability at both the individual and the neighborhood level and crime measures will enhance our understanding of alcohol-use patterns among teenagers. Focusing on the possible additive and interactive effects of neighborhood conditions on youth alcohol use, while controlling for key individual factors, should help shed light on the possible mechanisms by which neighborhoods shape behavior. Moreover, getting underneath these relationships will yield a more complete understanding of how neighborhood context influences adolescent alcohol use ultimately leading to more focused prevention and intervention efforts.

Conceptual Model

In previous work, we developed a conceptual model (Snedker, Herting, and Walton; under review), which guides this research. The model is illustrated in Fig. 4.1; solid lines represent direct additive effects (of either individual factors or neighborhood context) while; dashed lines represent moderating effects of neighborhood context; and curved lines represent an association. In this chapter, we estimate Path C, which is the direct effect of neighborhood net of key individual-level factors (Paths A and B), and Path D, which represents the moderating effect of alcohol availability, disadvantage, and crime and on three specific individual factors. While we posit, in general, that at the individual level, baseline risk factors are related to levels of personal resources (e.g., personal control), social resources (e.g., level of family support), and peer associations (e.g., number of friends using drugs), and these protective and risk factors are directly related to a youth's level of alcohol use, we do not directly model these effects, rather we focus on how neighborhood context directly affects

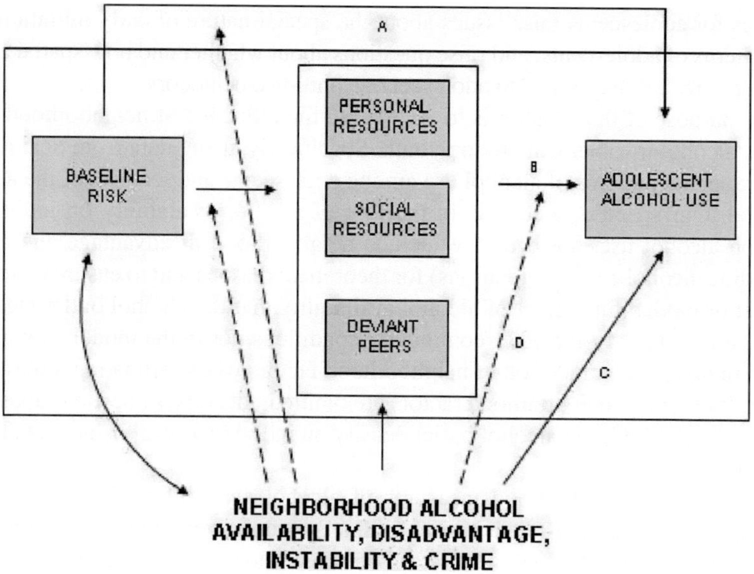

Fig. 4.1 Conceptual model of neighborhood effects on adolescent alcohol use. (The curved arrow between a baseline risk and neighborhood indicates the correlation between these two components.)

adolescent substance use (Path C) and how neighborhood context may moderate individual/social/peer resources (Path D). From this general perspective, one would expect positive personal and social resources to be attenuated, and effects of deviant peers to be greater in less advantaged areas. We expect that prior effects of variables of both the individual and family will remain significant.

Neighborhood Context

Neighborhood context is of broad interest to researchers studying substance use and other problematic behaviors. Below we outline a theoretical framework to capture a general neighborhood social control and social resource view of contextual effects coupled with a general opportunity/availability perspective of space.

Disadvantage and Crime

There is a general set of ecological features, among them neighborhood disadvantage is associated with problematic behaviors (Massey 1996; Shaw and McKay 1942; Wilson 1996). Theoretical perspectives, such as collective socialization, contagion, and institutional/resources (Jenks and Meyer 1990), argue that poverty, frequent social change, and crime weaken the conventional values system and nurture criminal and delinquent behavior. Neighborhood disadvantage and crime are seen as environmental stressors that affect adolescent substance use directly or moderate risk and protective factors by exacerbating the former and inhibiting the latter. Disadvantaged, unstable, and crime-ridden neighborhoods have fewer resources, employment opportunities, formal and informal forms of social control and monitoring, and overall collective efficacy (Bursik and Webb 1982; Elliott, Wilson, Huizinga, Sampson, Elliott, and Rankin 1996; Sampson and Groves 1989; Sampson, Raudenbush, and Earls 1997), which may be related to the substance-use behaviors of its adolescent inhabitants. Britt and colleagues (2005) used alcohol outlet density as their measure of collective efficacy.

Studies on adolescent substance-use patterns prioritizing neighborhood effects in assessing adolescent substance-use behaviors show mixed results in the directionality of neighborhood disadvantage. Relying on perceptions of neighborhood disadvantage, Crum and colleagues (1996) reported an increase in adolescents' exposure to cocaine. Brook and colleagues (2002) found that violence in the environment and everyday stress increases the risk of future drug use in a study of Columbian adolescents. Comparing adolescents who moved to middle-class neighborhoods to those who stayed in low-income neighborhoods, Briggs (1997) found that adolescents who remained in disadvantaged neighborhoods exhibited more signs of problem drinking and marijuana use.

Using a national sample of adolescents and matching to census tracts, Chuang and colleagues (2005), however, reported mixed results for the effect of neighborhood disadvantage on adolescent alcohol use. They found that disadvantaged neighborhoods (low SES) were associated with high parental monitoring that decreased adolescent alcohol use, but low SES neighborhoods were also associated with increased peer drinking that was associated with increased adolescent alcohol use. In contrast, high SES neighborhoods were associated with increased parental drinking leading to higher adolescent alcohol use. Using a national longitudinal dataset, Hoffman (2002) found that living in neighborhoods with a greater proportion of unemployed men, measured at the zip code level, was associated with an increased risk of drug use among adolescents controlling for family context. Notably, Hoffman (2002) found that the effect of poverty had a *negative* impact on adolescent substance use controlling for male joblessness.

Alcohol Availability

Alcohol outlet density is expected to be associated with higher adolescent alcohol use, reflecting greater opportunities for consumption and problematic neighborhood environments. Much research reports significant effects of alcohol availability, primarily measured at the neighborhood level or higher levels of aggregation on individual-level alcohol-related behaviors or aggregate-level problems.

For adults, research links alcohol outlet density to alcohol consumption (Gruenewald and Ponicki 1993; Scriber, Cohen, and Fisher 2000), drinking norms (Scriber, Cohen, and Fisher 2000), alcohol-related crimes (Sitt and Giacopassi 1992), self-reported drinking and driving (Treno, Grube, and Martin 2003), and alcohol-related automobile crashes and fatalities (Escobedo and Ortiz 2002; Scribner, MacKinnon, and Dwyer 1994). In addition to alcohol-specific outcomes, alcohol availability and outlet density are positively related to violent crime (Britt, Carlin, Toomey, and Wagenaar 2005; Gorman, Speer, Gruenewald, and Labouvie 2001; Nielsen and Martinez 2003; Reed, Hughey, and Peterson 2003; Scriber, MacKinnon, and Dwyer 1995; Speer, Gorman, Labouvie, and Ontkush 1998), neighborhood drunkenness and property damage (Donnelly, Poynton, Weatherburn, Bamford, and Nottage 2006), gonorrhea rates (Cohen et al. 2006), self-reported injuries (Treno, Gruenewald, and Johnson 2001), and suicide rates (Escobedo and Ortiz 2002).

A majority of studies focus on adults, while few studies put adolescents at the center of their analysis. At the neighborhood level, Freisthler, Needell, and Gruenewald (2005) were interested in the relationship between alcohol outlet density and neighborhood rates of child abuse and neglect, reporting that a higher concentration of bars is positively associated with rates of child maltreatment. At the individual level, Kuntsche and Kuendig's (2005) study of Swiss adolescents found a negative interaction effect of alcohol outlet density and perceptions of adolescent drinking in public in predicting individual alcohol use among adolescents. Their measure of alcohol outlet density was not objective, however, as it was based on the school master's perception of having a lot of small shops and kiosks near school. A recent study on adolescents (15- to 20-year-olds) in California, using a combined measure of alcohol establishments, found an association with drinking and driving and riding with drinking drivers (Treno, Grube, and Martin 2003). In their assessment of the density of liquor stores and bars in urban US, Romley, Cohen, Ringel, and Sturm (2006) reported that non-White youth face neighborhoods with higher densities of liquor stores than White youths.

Studies that identify the moderating role of neighborhoods in adolescent development and behavior (Klebanov et al. 1994; Kupersmidt et al. 1995; Spencer et al. 1997) and alcohol availability (Treno et al. 2000) are also emerging. Of those studies that are directly interested in adolescents at the individual or neighborhood level, there has been no study to date that examines the role of alcohol availability at both the individual and neighborhood level within a rich multilevel model assessing both direct and moderating effects.

Exploring Contextual Effects on Adolescent Alcohol-Use Behaviors

The effect of individual psychosocial risk and protective factors on adolescent development has been well documented (e.g., Brown, Schulenberg, Bachman, O'Malley, and Johnston 2001; Hawkins, Catalano, and Miller 1992). Both theory and research on adolescent substance use have focused on the role of individual factors, peers, and family characteristics. In our focus on neighborhood context and spatial impact on alcohol use, it is important that we control the direct effects of these key individual and family/peer features while assessing the role of spatial variables. In addition, while research often emphasizes the direct effects of context, there is also interest in the possible interactions with neighborhood measures and these individual factors. Our analytic framework discussed previously and later moves our exploration of contextual and spatial features of alcohol availability, disadvantage, and crime in this direction.

Methods

Data

Data for this chapter come from Reconnecting Youth (RY) prevention research project of adolescents in the city of Seattle from 1998 to 2003 and maps census data to the adolescent's home address. The RY project consists of both survey and prevention intervention research studies, including a drug use prevention program for high-school youth designed to reduce drug involvement. The data reflect a stratified (by high risk of school dropout) random sample of high school-aged youth in Seattle. The RY program targets potential high-school dropouts, invites youth to participate in the study, and then randomly assigns them to a control group (school as usual) or RY intervention. There is also a comparison group of low- or non-risk youth.

Nine high schools in a Seattle school district participated in the interventions and/or surveys. The analyses use data from three separate RY study sources conducted between 1998 and 2003 – 'Preventing Drug Abuse: Parents and Youths with Schools' (NIDA), 'Reconnecting Youth: Replication of an Indicated Prevention Program in Multicultural Settings' (Department of Education, DoE), and 'Assessing Suicide Risk among Adolescents' (CDC). The first two studies included the implementation of a comprehensive substance use prevention program that targets high school-aged youth at risk of school dropout. In the third study, a concurrent, non-intervention sample of non-risk youth was added. Merging the datasets increases our sample size and subsequent power and is appropriate given the data sets are compatible in terms of: (1) the sampling frame and definition of high risk; (2) the content of the survey and format of survey administration; and (3) the region/schools and time period. All participating youth were assented and parents provided consent in accordance with approved University of Washington IRB protocols. The analysis in this chapter is based on data prior to the participants knowing which condition, within the randomized controlled trial, they had been assigned. In total, 1721 individual respondents were included in the combined dataset. We weighted the data to reflect the over-sampling of high-risk youth in the sample; the weighting procedure reflected the high-risk population across schools in the study of 35%. Results did not differ between the weighted and unweighted results, and so we present the unweighted sample results.

Following other studies, we use the census tract as the aggregate unit of context. Research on alcohol and liquor outlet densities primarily relies on census tracts (Cohen et al. 2006; Gorman, Zhu, and Horel 2005; Gymiah-Brempong 2006; Nielsen and Martinez 2003; Nielsen Martinez, and Lee 2005; Reid, Hughey, and Peterson 2003; Scribner, Cohen, and Fisher 2000) or some combination of census tracts and block groups (Britt, Carlin, Toomey, and Wagenaar 2005; Pollack, Cubbin, Ahn, and Winkleby 2005). Density measured at larger geographic units than census tracts may distort the relationship between geography and alcohol outlet locations (Parker and Woltz 1979).

At the neighborhood level, data are compiled from the 2000 US Census, Seattle Police Department (1997–2002), and Washington State Liquor Control Board. Measures of alcohol availability are provided by the Washington State Liquor Control Board (2004). These data included information on address and premise type (liquor store, grocery store, convenient stores, restaurants, bars) of establishments licensed to sell alcohol. In line with previous research, we distinguish between on-premise alcohol establishments (bars, restaurants), off-premise alcohol establishments (liquor store, grocery store, convenient stores), and total alcohol outlet establishments (Britt et al. 2005; Cohen et al. 2006; Freisthler et al. 2005; Gorman et al. 2001, 2005; Pollack et al. 2005; Scribner et al. 1995; Sitt and Giacopassi 1992; Speer et al. 1998; Treno et al. 2001).

Neighborhood-level data were matched to individual records, following procedures similar to existing research on neighborhood analysis (Billy and Moore 1992; Crane 1991; Ku, Sonenstein, and Pleck 1993). Neighborhood-level variables were linked to the sample respondents through geographical mapping software (ArcGIS 9.2). Student addresses from the individual-level dataset are geocoded and spatially linked to the appropriate census. All census tracts that contain fewer than five individuals from the sample are aggregated with contiguous census tracts. After completing the matching and aggregation process, there were 111 (of 127 possible) census tracts in the neighborhood-level dataset from the city of Seattle that contained at least five individuals in the sample.

Measures

All outcomes and individual-level independent variables come from the RY High School Questionnaire (HSQ), a detailed multiscale self-report questionnaire capturing a range of youth behaviors, including substance use, peer and family relations, and school behaviors. The HSQ is designed to use a minimal number of indicators to capture a broad range of risk and protective factors associated with a set of diverse risk behaviors.

Outcome. A measure of substance-use frequency is used as the dependent variable in this analysis. We focus on the frequency of *alcohol use* (beer, wine, and hard liquor) during the past month. All frequencies of use are coded on a 7-point scale with the following classification: 0 = Not at All; 1 = Once; 2 = Two or Three Times; 3 = About Once a Week; 4 = Several Times a Week; 5 = Almost Every Day; 6 = Every Day. The frequency of alcohol use is a mean calculated from more than one type of alcohol; therefore, the values of these outcomes fall anywhere between 0 and 6. The outcome is based on ordinal frequencies yet analyzed as continuous outcomes. In addition to this approach, we used a dichotomous outcome and a logistic approach to the analysis ('use' or 'no use') and categories using multinomial analysis within hierarchical linear models (HLM). Results do not differ from those presented here and are available upon request.

Individual-level explanatory variables. We include measures of individual alcohol availability, psychosocial risk, and protective factors associated with substance use, peer group characteristics, demographic characteristics, risk for school dropout, mobility, family structure, and parent's educational attainment.

We created two individual-specific alcohol availability measures. Comparable with Pollack and colleagues (2005), we included a measure reflecting the nearest distance to an alcohol outlet by premise type from a respondent's home. We also calculated the total number of alcohol establishments within 0.5 mile radius of the respondents' home, attempting to reflect walking distance within the immediate environment. This measure is similar to others used in the literature (Pollack et al. 2005; Treno et al. 2001).

There are three measures designed to capture the psychosocial risk and protective factors: *personal control*, *family support*, and *deviant peer bonding*. *Personal control* reflects a mean score based on five items tapping into personal agency and coping abilities (Cronbach's alpha = 0.81). *Family support* comprised of five items based on the extent of and satisfaction with communication and help provided by immediate family members (Cronbach's alpha = 0.89). *Deviant peer bonding* captures the amount of close friends involved in six different delinquent behaviors (Cronbach's alpha = 0.77). Specific details about the questions that comprise each construct are summarized in Appendix A.

Controls. We include several control variables. A dummy variable corresponding to *race* (White vs. non-White) is included in the analysis along with the respondent's *age* and *sex*. A dichotomous variable representing *high risk for school dropout* is based on attendance and grades as assessed from school records. To capture individual mobility, *high school change* is measured as the number of times the respondent changed high schools prior to the survey. Intact *family structure*, representing living with both natural parents, is controlled with a dummy variable. Finally, the *educational attainment* for the respondent's parent with the highest level is included in the analysis as a proxy for individual or family socioeconomic status.

Contextual explanatory variables. Following past research, we evaluated several neighborhood-level measures of alcohol concentration based on counts of alcohol outlets (Nielsen and Martinez 2003; Nielsen et al. 2005; Scribner et al. 1994) and density of alcohol outlets per population (Britt et al. 2005; Donnelly et al. 2006; Escobedo and Ortiz 2002; Freisthler et al. 2005; Gorman et al. 2001, 2005; Gruenewald et al. 1993; Gymiah-Brempong 2006; Reid et al. 2003; Scribner et al.

1995; Speer et al. 1998; Treno et al. 2001). We included a count of on-premise, off-premise, and total alcohol outlets within a census tract. In addition, we calculated alcohol outlet density measures separately for on-premise, off-premise as well as the total outlet density from the number of outlets per 1000 people in the census tract.

Additional measures of neighborhood characteristics are designed primarily to represent economic disadvantage and crime in the neighborhoods as represented by the census tract unit. A composite measure is used to represent *neighborhood disadvantage* based on a scale constructed by Sampson and collaborators (1997). This index is a mean of four indicators of economic disadvantage at the census tract level: percentages of residents below the federal poverty level, female-headed households, residents receiving public assistance, and residents aged 16 years or older that are unemployed (Cronbach's alpha = 0.82).

Unlike census data that reflects 2000 information, Seattle crime data were averaged over the years 1997 through 2002 to reflect the entire period of sample data. We rely on different end-points (1997–2002) than the individual-level data (1998–2003) to more accurately reflect the crime rate when respondents were sampled. For example, for those respondents who were surveyed in 1998, the crime statistics from 1997 were probably more characteristic of the conditions they experienced. Crime data are averaged over the six-year period to rely on the same measure for each respondent regardless of the year he or she was sampled. Analyses reveal that averaging the crime rates does reasonably capture the entire time range and that crime rates are generally quite stable with a clear decline in crime rates, reflecting the city-wide reduction in crime. We assess three variants of Part I crimes: violent crime (murder and non-negligent manslaughter, forcible rape, robbery, and aggravated assault), property crime (burglary, larceny-theft, motor vehicle theft, and arson), and total crime (summing violent and property crime). In addition, we included an example of an alcohol-related Part II crime: liquor law violations.

Analysis Plan

Our analysis plan includes descriptive and analytical statistics. First, using GIS mapping software, we geocoded individual student addresses onto 2000 census tract maps of Seattle. This allowed us to look at individual spatial distance to neighborhood qualities and explore spatial patterns of alcohol use and neighborhood conditions. Second, we used multilevel techniques, such as HLM, to assess the impact of neighborhood context (measured at the census tract level) on alcohol use. A hierarchical model explicitly incorporates variables at the individual level and at the aggregate level, and accounts for the clustering of individuals in an aggregate unit (Raudenbush and Bryk 2002; Snijders and Bosker 1999). This analytical procedure specifically yields more reliable estimates of neighborhood effects on adolescent alcohol use. We were primarily interested in neighborhood factors that explain individual variability. We explored both direct and moderating roles of neighborhood context on alcohol use among youth. Adequate variation across neighborhoods was found for the alcohol-use outcome. We explored cross-sectional associations among neighborhood characteristics and alcohol and estimated three different types of effects.

In the first model, we examined the gross effects of neighborhood; the alcohol use-dependent variable is regressed on the neighborhood-level variables to test for direct effects of neighborhood alcohol outlet counts and density, disadvantage, and measures of crime on the alcohol use, controlling for key socio-demographic variables. See the two equations below:

Level 1: Individual model: random intercept

$$Yij = \beta_{0j} + \beta_1(\text{AGE}) + \beta_2(\text{SEX}) + \beta_3(\text{RACE}) + \beta_4(\text{PARENT EDUCATION}) + r_{ij}$$

Level 2: Neighborhood model (Model 1)

$$\beta_{0j} = \gamma_{00} + \gamma_{01}(\text{DISADVANTAGE}) + \gamma_{02}(\text{CRIME}) + \gamma_{03}(\text{LIQUOR LAW VIOLATIONS}) + \gamma_{04}(\text{ALCOHOL OUTLET DENSITY}) + u_{0j}$$

In the second model, individual alcohol availability, individual and family risk/protective factors and peer factors, and basic control variables and a neighborhood-level intercept are included in the model. All continuous variables (alcohol availability, age, high school change, parent education, personal control, family support, and high risk peers) were centered around their respective grand means. The neighborhood-level model remained unchanged. This alters the level-1 equation:

Individual model (Model 2)

$$Yij = \beta_{0j} + \beta_1(\text{AGE}) + \beta_2(\text{SEX}) + \beta_3(\text{RACE}) + \beta_4(\text{PARENT EDUCATION}) + \beta_5(\text{RISK STATUS}) + \beta_6(\text{HIGH SCHOOL CHANGE}) + \beta_7(\text{FAMILY STRUCTURE}) + \beta_8(\text{ALCOHOL AVAILABILITY}) + \beta_9(\text{PERSONAL CONTROL}) + \beta_{10}(\text{FAMILY SUPPORT}) + \beta_{11}(\text{HIGH RISK PEERS}) + r_{ij}$$

$$\beta_{0j} = \gamma_{00} + \gamma_{01}(\text{DISADVANTAGE}) + \gamma_{02}(\text{CRIME}) + \gamma_{03}(\text{LIQUOR LAW VIOLATIONS}) + \gamma_{04}(\text{ALCOHOL OUTLET DENSITY}) + u_{0j}$$

In the third model, interactions between the neighborhood measures and the three individual-level constructs are included; the effects of personal, family, and peer factors were allowed to vary randomly, allowing for variation by the neighborhood measures. This model was designed to test if neighborhood context moderates the impact of personal control, family support, and deviant peers on substance use. Simple cross-level interactions between contextual and individual-level variables were tested as shown below:

Level 2: Neighborhood model: moderating effects (Model 3)

$$\beta_{0j} = \gamma_{00} + \gamma_{01}(\text{DISADVANTAGE}) + \gamma_{02}(\text{CRIME}) + \gamma_{03}(\text{LIQUOR LAW VIOLATIONS}) + \gamma_{04}(\text{ALCOHOL OUTLET DENSITY}) + u_{0j}$$

$$\beta_{10j} = \gamma_{100} + \gamma_{101}(\text{DISADVANTAGE}) + \gamma_{102}(\text{CRIME}) + \gamma_{103}(\text{LIQUOR LAW VIOLATIONS}) + \gamma_{104}(\text{ALCOHOL OUTLET DENSITY}) + u_{10j}$$

$$\beta_{11j} = \gamma_{110} + \gamma_{111}(\text{DISADVANTAGE}) + \gamma_{112}(\text{CRIME}) + \gamma_{113}(\text{LIQUOR LAW VIOLATIONS}) + \gamma_{114}(\text{ALCOHOL OUTLET DENSITY}) + u_{11j}$$

Results

Descriptive Statistics

Table 4.1 reports the means, standard deviations, and ranges for all variables considered in these analyses. For the 1721 students participated in the analysis, the ages range from 13 to 21 years with a mean of 15.7. It was a racially mixed sample with 38% White, 23% Black, 15% Asian, and 4% Hispanic ethnicity. Because of the over-sampling of high-risk youth, 68% were at risk for dropping out of school. Just less than half the students came from homes with a biological mother and a father present (45%). Measures of parent educational attainment revealed an average of 15.2 years of schooling. The average distance to the nearest alcohol outlet was 1134.8 ft. The response options for personal control and family support ranged

Table 4.1a Descriptive statistics of level-1 variables ($N = 1721$)

Explanatory variables	Mean	Std. Dev	Min.	Max.
White race	0.38	0.49	0.00	1.00
Age	15.70	1.01	13.00	21.00
Female sex	0.49	0.50	0.00	1.00
High risk for drop-out	0.68	0.47	0.00	1.00
Number of times change high school	0.27	0.71	0.00	5.00
Lives with both natural parents	0.45	0.50	0.00	1.00
Distance (in feet) to nearest alcohol outlet	1134	783	0.00	5036
Parent's highest education	15.25	2.87	9.00	22.50
Personal control	4.28	1.28	1.00	6.00
Family functioning and support	3.44	1.63	0.00	6.00
Peer high-risk behaviors	1.61	1.20	0.00	6.00
Dependent variable				
Alcohol use frequency	0.74	1.09	0.00	6.00

Table 4.1b Descriptive statistics of level-2 variables ($N = 111$)

Explanatory Variables	Mean	Std. Dev	Min.	Max.
Neighborhood disadvantage	5.82	3.70	2.13	20.35
Percentage below the poverty level	7.36	7.50	0.00	44.40
Percentage female-headed families	9.31	4.84	1.10	27.90
Percentage receiving public assistance	3.31	3.43	0.00	16.60
Percentage unemployed (over 16 years old)	3.28	1.46	0.50	9.90
Crime rate per 1000 population				
Violent crime	6.15	7.03	0.11	47.98
Property crime	66.71	51.87	0.03	384.28
Total crime	72.86	57.83	0.14	432.26
Liquor law violations	1.67	2.28	0.00	15.67
Alcohol outlet density				
On-premise	.85	1.00	0.00	9.31
Off-premise	1.89	2.71	0.00	17.28
Total	2.74	3.49	0.00	22.86

from 1 to 6, with the averages being 4.28 and 3.44, respectively. Most students reported low levels of deviant peer bonding ($\bar{X}$ = 1.61, SD = 1.20). Alcohol use was generally low, with the average frequency of use falling below 'once a week'.

There is substantial variability among the 111 census tracts used in our sample of youth in Seattle. The average number of total outlets per census tract was 11.2 and the average density of total outlets per 1000 population was 2.7, with 0.9 for on-premise and 1.9 for off-premise. The mean neighborhood disadvantage score for the census tracts in our sample was 5.8%, falling within a wide range of 2.1–20.4%. There was substantial variation in crime. Across Seattle census tracts, the total average crime rate ranged from 0.14 to 432 per 1000 people with an average violent crime of 6.2 and property crime of 66.7. There were less than two liquor law violations on average per census tract with a maximum of 16 violations.

Maps

We created several maps from the geocoded survey data to explore spatial patterns of alcohol use and neighborhood conditions. Each map includes the census tracts of Seattle as the backdrop. Then, features of both individual and neighborhood are added to the baseline map. The first map represents a baseline map showing the census tracts within the city of Seattle and the distribution of sample respondents. From Fig. 4.2a, it is clear that there is a high coverage of survey participants in most areas of the city with the exception of the downtown corridor. We would expect a low number of adolescent respondents from this area, given the high business concentration and low family residential population in Seattle.

In order to assess if neighborhoods with higher concentrations of alcohol outlets have other neighborhood risk factors, we compared three maps of several neighborhood-level features. We report a map of neighborhood disadvantage, total alcohol outlet density, and liquor law violations. Due to the strong correlation between off-, on-premise, and total alcohol outlet density ($P < 0.001$), the results across all maps are comparable.

When we compare Fig. 4.2b of the percentage disadvantaged in a census tract with Map 4.2c of the total alcohol outlet density and Map 4.2d of the mean number of liquor law violations, it is clear that neighborhood economic disadvantage is not synonymous with a higher density of alcohol outlets or liquor law violations. Although the correlations between these neighborhood features are positive, the lack of significance between both alcohol measures and neighborhood disadvantage are consistent with the lack of any visible spatial patterns in the maps. A pattern between alcohol outlet density and liquor law violations can be seen reflecting the moderate positive correlation ($r = 0.28$; $P < 0.01$).

Next, we overlay Fig. 4.2c and d with moderate to high frequency drinkers in the sample (those with a score of 2 or higher, indicating 2–3 drinks or greater per month). Figure 4.2e of alcohol outlet density and moderate to frequent drinkers does not seem to show a clear pattern. It does not show a positive association as expected, but rather illustrates a negative association between alcohol outlet density and more

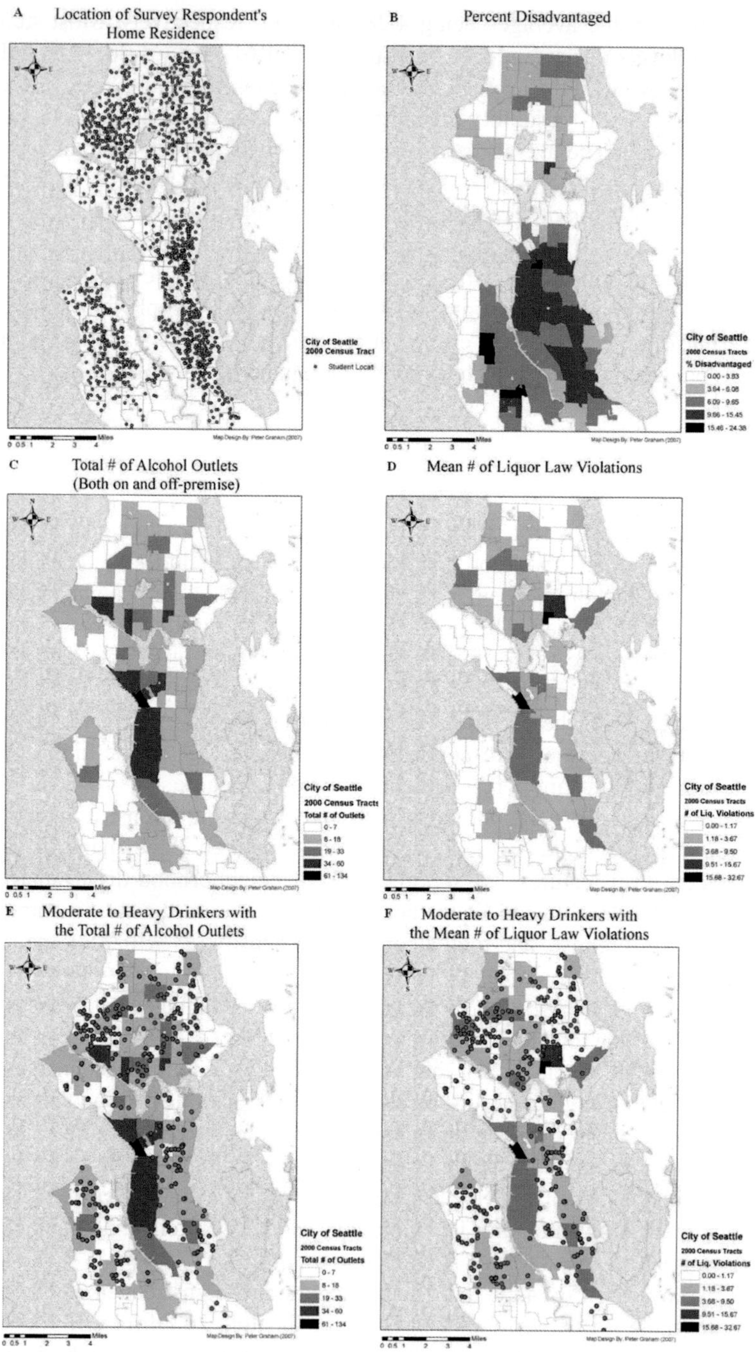

Fig. 4.2 (a) Location of survey respondent's home residence; (b) percent disadvantaged; (c) total number of alcohol outlets; (c) mean number of liquor law violations; (e) moderate to heavy drinkers with the total number of alcohol outlets; (f) moderate to heavy drinkers with the mean number of liquor law violations (See also Plate 2 in the Colour Plate Section)

frequent alcohol use among adolescents. Figure 4.2f of liquor law violations and moderate to frequent drinkers does not seem to show a clear pattern; however, it is suggestive of a positive relationship. Quantitative analysis was used to explore the direction and strength of the relationships observed in the maps.

Alcohol Use

Before adding individual-level covariates of substantive interest, we explored the direct relationship between neighborhood attributes and alcohol use while controlling for basic demographic variables (Model 1; see Table 4.2). We find that the direct relationship of neighborhood disadvantage on alcohol consumption is negative. That is, adolescents living in disadvantaged neighborhoods tend to report lower use of alcohol while controlling for individual demographics. We also find a positive relationship between liquor law violations and alcohol use. On average, census tracts with a greater number of liquor law violations have adolescents who report greater alcohol usage. There is no significant relationship for alcohol outlet density or crime. These direct effects remain present but with the addition of level-1 predictors in the following analyses.

In model 2, the individual-level model is added. At the individual level, none of the alcohol availability measures are significant; thus we only report the nearest distance to any alcohol outlet measure. We find that White, older, and high-risk youth reported higher levels of alcohol use. Interestingly, while girls report higher alcohol use, the coefficient is *not* significant. Parental education, as measured in years, is associated with less adolescent alcohol use. Two of the three psychosocial risk factors are significant. In line with previous research, positive family functioning and support is associated with less alcohol use, and having deviant peers is strongly associated with greater alcohol use. Model 2 also shows the consistent neg-

Table 4.2 Baseline effects of context on individual alcohol use net of basic individual demographic variables

Variable in Model	Unstandardized Regression Coefficient (S.E.)	P-value
Intercept	0.963(0.085)	0.000
Individual level		
Age	0.076(0.026)	0.004
Sex (female = 1; male = 0)	−0.092(0.048)	0.058
Race/Ethnicity (white=1 non-white=0)	0.211(0.061)	0.001
Years of education of parent	−0.027(0.010)	0.007
Contextual level		
Neighborhood disadvantage	−0.040(0.009)	0.000
Crime rate per 1000	0.001(0.001)	0.563
Liquor law violations	0.023(0.013)	0.085
Alcohol outlet density	−0.024(0.016)	0.147

Table 4.3 Effects of context on individual alcohol use net of key individual factors

Variable in Model	Unstandardized Regression Coefficient (S.E.)	P-value
Intercept	0.697(0.096)	0.000
Individual level		
Age	0.078(0.023)	0.001
Sex (female = 1; male = 0)	0.062(0.049)	0.202
Race/ethnicity (White = 1; non-White = 0)	0.278(0.056)	0.000
Years of education of parent	−0.023(0.008)	0.005
High risk status	0.193(0.051)	0.000
Number of school moves	0.051(0.039)	0.193
Household structure (dual parents of origin = 1; other = 0)	0.053(0.056)	0.345
Distance (in feet) to nearest alcohol outlet	−0.00001(0.00004)	0.973
Personal control	0.023(0.021)	0.290
Positive family functioning	−0.044(0.017)	0.011
Deviant peer behavior scale	0.326(0.027)	0.000
Contextual level		
Neighborhood disadvantage	−0.035(0.008)	0.000
Instability	0.010(0.008)	0.212
Crime rate per 1000	0.0001(0.001)	0.890
Liquor law violations	0.022(0.012)	0.071
Alcohol outlet density	−0.013(0.012)	0.293

ative effect of neighborhood disadvantage, which remains net of these additional individual variables. Liquor law violations also maintains its positive association with adolescent alcohol use.

Model 3 allows for an assessment of moderating effects. Consistent with the previous models, Model 3 reflects the same individual- and neighborhood-level results, but we only report the interaction effects. Of the three resource variables, only family and deviant peers showed significant variation across context. Modeling this variation produced three interaction terms that reached statistical significance. The interaction between alcohol outlet density and family functioning is positive. Family support is a protective factor for all youth's alcohol use, but living in high

Table 4.4 Unstandardized direct and moderating effects of contextual variables

	Direct Effect on Intercept	Moderating Effects on Family Functioning	Moderating Effects on Deviant Peer Behavior
Intercept		−0.030 (0.034)	0.456 (0.054)***
Neighborhood disadvantage	−0.037 (0.008)***	0.004 (0.003)	−0.020 (0.006)***
Crime rate per 1000	−0.0003 (0.001)	−0.001 (0.0004)**	0.001 (0.001)
Liquor law violations	0.020 (0.011)*	0.004 (0.009)	−0.013 (0.012)
Alcohol outlet density	−0.007 (0.012)	0.015 (0.007)**	−0.007 (0.014)

*$P < 0.10$; **$P < 0.05$; ***$P < 0.01$.

Model includes controls for all individual level variables in Table 4.3.

alcohol-outlet density areas reduces the protective effect of family resources. The interaction between crime rate and family functioning is negative. The protective effect of family support in reducing alcohol use is increased for youth in higher crime rate areas. The interaction between neighborhood disadvantage and deviant peers measure was also significant. This interaction result shows that, for adolescents living in neighborhoods with higher than average disadvantage, the influence of deviant peers on alcohol use is lowered. Alternatively, youth in advantaged neighborhoods react stronger to the presence of deviant peers.

Discussion and Conclusion

Our analysis revealed the expected statistical relations among individual-level factors and adolescent alcohol use (e.g., the positive effect of deviant peer association and alcohol use). Net of these individual factors, we also see significant relationships between alcohol use by adolescents and both neighborhood disadvantage and liquor law violations. We also report three significant moderating effects of neighborhood on adolescent alcohol use.

Consistent with our previous research on neighborhood context and adolescent substance use, we find that measures of neighborhood disadvantage are significant but in a negative direction. Interestingly, we find alcohol availability at the neighborhood level is not significant. Both effects are inconsistent with results from studies of adult behavior, but are not inconsistent with the literature emerging regarding substance use and adolescent behavior.

Furthermore, we report that individual measures of alcohol availability (i.e., distance a given person is from an alcohol outlet) did not show significant relations in our models, net of other individual characteristics. This finding is consistent with research by Scribner and colleagues (2000) where they reported the effect of alcohol outlet density (using off-sale measures) on alcohol-related outcome function through an effect at the neighborhood level and not at the individual level; they specifically report that at the neighborhood level, the mean distance to the closest alcohol outlet was negatively associated with drinking norms and alcohol consumption.

The number of liquor law violations by licensed establishments within an individual's neighborhood is positively related to adolescent use. This particular finding is of interest but not simply interpreted. It may reflect a normative view by business establishments that teen alcohol use is not a problem behavior, therefore representing a level of access (net of spatial closeness) not simply represented by closeness or density of outlets. It also may reflect the frequency of attempts by adolescents to obtain alcohol and reflect more of the level of normative drinking in the community (i.e., violations represent high levels of attempting). Areas with higher violations may actually be areas in which lax attitudes toward liquor laws create opportunities for youth drinking. This points out to an interesting avenue for further work.

Importantly, three interaction terms reached statistical significance – two with family resources and one with deviant peers. The decreased effect of deviant peers on youth's alcohol use in disadvantaged neighborhoods and the increased protective effect of family support for youth in higher crime rate areas are interesting. These results suggest that several risk factors at the individual and neighborhood levels might lower the impact of any one feature in increasing the problem behavior. In addition, it could be that residents of certain neighborhood become more resilient in responding to neighborhood risks. For example, although neighborhood crime may act as a stressor on residents, it might be the case that in response to neighborhood crime, families provide more resources to their children to counteract the criminal influence on adolescent behaviors. In either case, in this chapter, certain neighborhood risks were associated with both a decrease in the effects of risk factors and an increase in the effects of protective factors. Alternatively, the reduction in the protective effect of family support for youth living in high alcohol-outlet density areas suggests that certain neighborhood conditions can also reduce positive resources. These findings emphasize the value of assessing different social contexts on adolescent alcohol use, and that the influence of peer and family factors on adolescent behaviors is conditional upon the broader neighborhood context.

In general, neighborhood disadvantage and neighborhood alcohol outlet density do not work in the same direction for adolescents as they do for adults. The fact that neighborhood disadvantage is negatively related to alcohol use for adolescents, and alcohol outlet density is insignificant, while typically both are positively related for adults, is consistent with considering adolescent alcohol use as a different phenomenon, not just a precursor to adult problem behaviors.

The differences in neighborhood effects may be due to the nature of the phenomena being explored. Many neighborhood context studies focus on severe social problems and serious conditions, including crime, violence, and dangerous health behaviors. Adolescent alcohol-use behaviors may reflect initiation or experimenting behaviors and, therefore, indicate less serious, more common outcomes from what is typically analyzed in this literature. This distinction may argue for thinking differently about context and importantly adding other contexts, such as schools, to the typical neighborhood approach (i.e., institutional contexts as well as spatial context).

The lack of alcohol availability finding also raises questions about what access means for adolescent alcohol use. The alcohol outlet density measure was not significant and in the negative direction. It may be that for adolescents, neighborhood alcohol outlet is not the best context to capture access but rather access through their family or friendship networks. From the same study, we can compare the relationship between reported access to beer/wine and access to liquor to neighborhood outlet density. In doing so, we find that there is no correlation between access measure and neighborhood outlet density. This suggests those students who report easy access to alcohol are not measuring access by way of high alcohol availability in their neighborhood. However, there is a weak negative correlation ($r = -0.05$; $P < 0.05$) between access to both alcohol measures and individual-level distance measure to the closest alcohol outlet. However, in our study findings, we report no relationship once we control other individual-level measures. This suggests that research needs to continue to explore different access measures, including family, peer, and neighborhood contexts.

In part, results for adolescents and adults may also be a product of studies having differential ability to control key individual risk and protective factors. This study is unique in that it is able to control these factors and, therefore, make a strong contribution to the research on alcohol availability, neighborhood context, and adolescent substance use by examining context effects net of individual risk and protective factors. Without such controls, observed effects of context may be overestimated or simply spurious due to their correlation with other individual factors.

Despite the strengths of the current theoretical model and analysis, this study was not without limitations. First, given that this was a school-based sample and not a community-based one, we only accessed youth still in school and those who have dropped out of high school were not included in the sample. Allison and colleagues (1999) argue that youth not in the sample may be those with higher levels of drug involvement, which may alter the relationship between neighborhood indices and adolescent substance use. Another point to consider is the extent to which extremely disadvantaged neighborhoods drove previous results in the literature. Hogan and Kitagawa (1985) found a significant increase in teenage childbearing for lowest SES category but no difference for middle and high SES, and so it might not be that neighborhood disadvantage is the key feature but rather *extreme* neighborhood disadvantage. Examining school dropout, Crane (1991) found that there was only an effect for neighborhood in the extreme category, although in the direction consistent with a social disorganization framework. Our sample was drawn from the Seattle metropolitan area, of which there were relatively few extremely impoverished neighborhoods. However, a substantial portion of neighborhood studies rely on national data or single sites from medium-sized cities. Local samples generally have higher interrelations among neighborhood dimensions, thus limiting variation, but the strength of local studies is often in the quality of measures of deviant outcomes and more complete family and personal processes (Duncan and Aber 1997).

Finally, in applying contextual analysis to individual level processes, we also face some general limitations. First, while some of our measures are specific to spatial location (i.e., distance an individual is from nearest alcohol outlet), we are relegated to capture neighborhood by census tract boundary. Second, the context level derived from the census data is not specifically contemporaneous with the individual-level data. We have access to the 2000 census information, 2000 crime data, and 2004 alcohol and liquor establishment data, which span the timeframe during which the individual data were collected (1998–2003); the general stability of these contextual units makes this less problematic. Another issue is we are not in a position of controlling the level of exposure to the context in which individuals are placed. We do not know the length of residence at the given location, and by necessity we attribute current context as having an effect on behaviors. Future research that has measures of individual- and neighborhood-level measures over time can begin to address spatiotemporal relationships.

Finally, we are not in a position of directly accounting for selection of individuals into their spatial context. The role of family selection into neighborhood is an important issue. For example, family structure and resources inevitably influence residential decisions, household finances, and family experiences, while residential location influences peer selection. We make attempts to address the endogeneity problem by including controls for past movement by the family. Moreover, Seattle districts are open-enrollment schools so there is less direct movement by families to exist in neighborhoods that are tied to specific schools. While this is perhaps less of an issue for adolescents since they are not making their housing location decisions (i.e. they live with parents), it remains a significant concern for the general work in contextual analyses.

Despite the limitations of neighborhood contextual research, this study makes several contributions and addresses a number of problems faced in this work. Primarily, we were able to look for a broad array of contextual/spatial effects net of key individual factors related to alcohol use. In this sense, our finding that certain contextual factors matter should be reasonably stable given the array of controls. Furthermore, this study incorporates mechanisms known to impact adolescent behavior and possesses a more complex relationship between neighborhood characteristics and adolescent alcohol-use behaviors by exploring how context moderates the effects of known risk and protective factors. Our findings reflect both neighborhood direct effects and neighborhood moderating effects, and move us closer to understanding what types of neighborhood and individual processes matter for adolescent behavior.

Our findings continue to raise doubts about the general applicability of social disorganization perspectives to adolescent substance-use behavior, suggesting a need for more exploration and thought regarding what contexts are at play for adolescents and how they may differ from adult behavior or extreme problem behavior. Furthermore, our work suggests that porting the same spatial access and availability research applied to adults may have similar problems. If neighborhood disadvantage is negatively associated with adolescent alcohol use and alcohol outlet density is non-significant as reported in this chapter, this raises questions about policy choices to address adolescent substance use (e.g., reduction in alcohol outlet density may not lead to reduced alcohol use by youth). Gruenewald, Ponicki, and Holder (1993) argue that fewer outlets in a neighborhood may lead to a reduction in the purchase and consumption of alcohol and alcohol-related problems for adults but may also lead to an increase if distance traveled to alcohol outlet requires a car, thereby leading to an increase in drunk driving and alcohol-related traffic accidents/fatalities. For youth, the opportunity to use alcohol may result less from the density of alcohol outlets than from the failure to enforce drinking prohibitions against minors. If so, stronger enforcement measures against merchants and education about dangers of youth drinking may accomplish more than reducing supply through suppressing alcohol outlets. Regardless, applying geographical and spatial methods to explore neighborhood effects on adolescent substance use represents a much needed approach for effective policies and prevention programs.

Appendix A

Components of Psychosocial Risk/Protective Factor Scales (RY)

Personal control (mean of at least 3 of 5 items)
I feel confident that I can handle my personal problems.
When I try, I can make good things happen for me.
I can learn to adjust or cope with my problems.
No matter how bad I feel, I know that I will feel better eventually.
I feel capable and in control of my life.

Family support (mean of at least 3 of 5 items)
I feel satisfied with the way my family talks things over and shares problems with me.
I feel satisfied with the way my family and I share time together.
I am satisfied with the way my family expresses affection and responds to my emotions such as anger, sorrow, and love.
My parent(s) accept and support my activities.
I am satisfied that I can turn to my family for help when something is bothering me.

Deviant peer bonding (mean of at least 3 of 6 items)
How many of your close friends:
Skip school?
Have ever dropped out of school?
Often get into trouble at school?
Have gotten in trouble with the police/law?
Don't really care about school?
Have gotten into physical fights with other kids?

Chapter 5
Migration Patterns and Substance Use among Young Homeless Travelers

Stephen E. Lankenau, Bill Sanders, Jennifer Jackson Bloom, Dodi Hathazi, Erica Alarcon, Stephanie Tortu and Michael C. Clatts

Abstract Homeless youth are a diverse population of at risk adolescents and young adults who experience various negative health outcomes, including drug dependence, drug overdose, infectious diseases, and victimization. Previous studies have been directed toward understanding subgroups of homeless youth, such as injection drug users (IDUs) and young men who have sex with men (YMSM); yet limited research has focused on describing homeless "travelers," a migratory subgroup of homeless youth who move from city to city. Based upon a larger three site study of young IDUs recruited in Los Angeles, New Orleans, and New York, a total of 133 travelers were identified. A subsample of 56 travelers participated in follow-up interviews, and provided data points for mapping. Travelers in all sites had extensive histories of criminal justice involvement and injection drug use. Four common traveling routes within and across the United States were identified. Reasons for traveling often related to drug use, money-making opportunities, and law enforcement. Risk-reduction services, such as shelters, syringe exchanges, or HIV/HCV testing, were used infrequently or occasionally. Mapping data documents the mobility of young IDUs across both urban and rural areas, which suggests that migration among IDUs across broad geographic regions may be a factor in the spread of blood-borne viruses.

Introduction

Homeless youth are a diverse population of at risk adolescents and young adults estimated to be as large as 1.3 million (Washington Coalition 2004). Homeless youth comprise a variety of subgroups, such as injection drug users (IDUs) (Kipke et al. 1995; Lankenau et al. 2004; Milburn et al. 2006; Roy et al. 2002), young men who have sex with men (YMSM) (Clatts and Davis 1999; Gwadz et al. 2006; Lankenau et al. 2005; Whitbeck et al. 2004), and geographically mobile drug users (Des

S.E. Lankenau
University of Southern California, 6430 Sunset Boulevard, Suite 1500, Hollywood, CA 90028, USA. E-mail: slankenau@chla.usc.edu

Y.F. Thomas et al. (eds.), *Geography and Drug Addiction*,

Jarlais et al. 2005; Hahn et al. 2007; Hyde 2005; Lankenau et al. 2002). Based up on behaviors associated with these identities and practices, homeless youth experience various negative health outcomes, including drug dependence (Clatts and Davis 1999), drug overdose (Roy et al. 2004), infectious diseases (Roy et al. 2000), and victimization (Kipke et al. 1997; Whitbeck et al. 2004). While studies have been directed toward understanding homeless IDUs and homeless YMSM, limited research has focused on describing homeless mobile drug users, or "travelers," a migratory subgroup of homeless youth who frequently move from city to city over a period of months and years. In this manuscript, we describe common traveling pathways and modes of travel across North America, reasons for frequent movements from region to region, and the impact of traveling on accessing risk-reduction services among a sample of young IDUs recruited in Los Angeles, New Orleans, and New York.

Migration among homeless and disenfranchised youth is an enduring social phenomenon in the United States. During the great depression, for instance, an estimated 250,000 youth left home in search of work and adventure (Uys 1999). Leaving behind impoverished households, ruined farmlands, or devastated local economies, teenage boys and girls hopped freight trains or hitch-hiked rides in great numbers to reach new employment opportunities in faraway cities and states (Minehan 1934). Between 1933 and 1934, for instance, the Los Angeles Transient Intake Bureau, part of a federal effort to assist destitute persons of all ages, registered an average of 1000 migrant males under 21 per month, more than any other city in the United States (Uys 1999). While often romanticized, the life of a young "hobo" or "tramp" was frequently difficult and involved braving extreme temperatures, finding limited food, contending with police harassment and arrest, and avoiding other older homeless predators (Anderson 1923; Minehan 1934).

During the 1960s and 1970s, groups of disenfranchised young people also traveled the United States – though more often for reasons of disillusionment and exploration than poverty and desperation. Among these youth, who became known as "hippies" and numbered in the hundreds of thousand (Yablonsky 1968), traveling was associated with music festivals, anti-war protests, sexual freedom, and drug experimentation (Wolfe 1967). In particular, LSD and other psychedelic drugs were important ingredients in the development of new forms of music, alternative styles of living, and progressive modes of self-expression among young people (Lee and Shlain 1985). Common destinations included Haight-Ashbury in San Francisco, Venice Beach in Los Angeles, and the East Village in New York City, where "crash pads" were made available by fellow hippies for new homeless arrivals while others slept in public parks or at beaches (Yablonsky 1968). Converted school buses and Volkswagen microbuses became the preferred modes of traveling from city to city (Gaskin 1970).

The origins of "rave" – a culture of dancing, music and drug use – can be traced back to groups of traveling youth. In the mid to late 1980s, raves began to spring up in vacant fields, empty warehouses, beaches, and other unconventional locations in the United States and Great Britain (see Sanders 2006; Thornton 1995). Young people traveled distances to access these "secret spots," sometimes following a series of "map points" advertised by word of mouth hours prior to the beginning of the event.

Raves were clandestine due to their illegal features: loud, often outdoor, events attracting hundreds, sometimes thousands, of youth, where drugs such as marijuana, LSD, and ecstasy were common. Underground raves held outside conventional club environments, as well as large annual music festivals, such as Burning Man and the Detroit Electronic Music Festival, continue to take place in the United States (Sanders 2006). For both types of events, youth are required to travel, and for many, recreational drug use is a priority upon arrival.

Methods

Data in this manuscript are based upon a two-phased study of young IDUs recruited in New York, New Orleans, and Los Angeles as part of a larger project examining health risks associated with ketamine injection. Ketamine is a dissociative anesthetic that has emerged as a drug commonly used in the dance/rave scene (Jansen 2001) and among subgroups of young IDUs (Lankenau et al. 2007). Phase 1 comprised a cross-sectional, ethnographic survey of 213 IDUs recruited in New York, New Orleans, and Los Angeles. Phase 2 consisted of a 2-year longitudinal study of 101 young IDUs recruited in Los Angeles during Phase 1.

Phase 1

Data collection began with a community assessment process (CAP; Clatts et al. 1995) by trained ethnographers in each city to determine the locations of groups of young people who injected ketamine. Based upon the CAP, ethnographers recruited young ketamine injectors in each city – primarily from street and park settings – using a combination of chain referral sampling (Biernacki and Waldorf 1981; Penrod et al. 2003) and targeted sampling (Watters and Biernacki 1989). Both are non-random, yet effective methods for sampling hidden populations. Young IDUs were sampled in New York in 2004 during a 5-month period, in New Orleans between 2004 and 2006 during a 27-month period, and in Los Angeles between 2005 and 2006 during an 18-month period.

Young people were eligible for study enrollment if they were between the ages of 16 and 29, and had injected ketamine at least once within the past 2 years. These criteria were selected to enroll a sample of young IDUs who could describe recent ketamine injection events. A series of screening questions focusing on health behaviors, recent drug use, and history of homelessness were asked in order to hide the true enrollment criteria. Before beginning an interview, individuals signed informed consent documents approved by local Institutional Review Boards from each site. At the conclusion of each interview, which lasted approximately 1 hour, subjects received a $20 cash payment in New York and Los Angeles, a $20 drug store gift certificate in New Orleans, and referral information for local needle exchanges, health clinics, homeless shelters, and other service organizations for high-risk youth populations.

Phase 2

Ketamine injectors recruited in Los Angeles during Phase 1 were eligible for enrollment into the Phase 2 longitudinal study. The 101 IDUs enrolled during 2005 and 2006 in Los Angeles consented for participation in a series of seven follow-up interviews occurring approximately every 3 to 4 months. During the cross-sectional baseline interview in Los Angeles, locator information, such as telephone numbers and e-mail addresses, were collected from each participant. Additionally, ethnographers provided each participant with a toll-free telephone number that connected directly to the ethnographer's cell phone. Cash incentives increased for each interview by $5, so that participants earned $25 for the first follow-up, $30 for the second follow-up, and so on. Subjects were consented at each follow-up interview to detail any changes to the study design, and to remind them of their rights as human subjects.

Based upon early interviews conducted in Los Angeles (as well as New York and New Orleans), we discovered that our sampling methodology and enrollment criteria, while intended to capture young ketamine IDUs, were also enrolling high proportions of homeless IDUs and highly mobile homeless IDUs. We realized that homeless IDUs would be challenging to track for follow-up interviews, especially those leaving Los Angeles. In response, we modified the design to allow for follow-up interviews to be conducted over the telephone, and for respondent payments to be sent via Western Union (Lankenau et al. forthcoming).

Measures and Analyses

The Phase 1 interview guide, which was administered to all subjects recruited in New York, New Orleans, and Los Angeles, contained eight modules, and captured data on demographics, drug using histories, recent drug use, and risk behaviors. The Phase 2 interview guides, which were utilized only in Los Angeles, followed up on key risk behaviors, such as changes in homeless status, injection drug using behaviors, criminal justice involvement, and HIV and HCV serostatus. Additionally, new modules were included in each subsequent interview to probe important areas that emerged during earlier interviews, such as behaviors and attitudes associated with traveling. The bulk of data on traveling was captured during the first Phase 2 follow-up interview. Questions included: "Where are some of the places you've traveled to since we last spoke?"; "What do you like and/or dislike about traveling as a homeless person?"; "How does drug use fit into traveling?"; "How do you find clean syringes when you arrive in a new city?"; "How difficult is it to get tested for HIV or HCV upon arriving in a new city?" Some information on traveling, however, was also gleaned from Phase 1 interviews conducted in New York and New Orleans. Since many young IDUs recruited in those sites were also travelers, descriptions of traveling emerged during discussions of homelessness and drug use.

The Phase 1 and Phase 2 interview guides, which contained both structured, close-ended questions and probing, qualitative questions, were administered on laptop computers using Questionnaire Development Software, and interviews were recorded with digital recorders. Following the interview, digital recordings were transcribed into a text document and analyzed using ATLAS ti, a qualitative software program. Responses to closed-ended questions were analyzed using SPSS and SAS.

Identifying respondents as a traveler resulted from two procedures: coding specific quantitative questions within the interview guide; and analyzing qualitative data for indicators suggesting a mobile, nomadic existence. Subjects in all three sites were asked about current housing status, and were coded as either "homeless" or "housed" based upon their responses. During data analysis, transcripts and field notes were reviewed and each homeless youth's pattern of living over the past year was assessed. Youth who reported remaining primarily local or moved about mainly within the metropolitan areas of New York, New Orleans, or Los Angeles were coded as "homeless locals." Young people who did not remain in a city for longer than a few weeks or a month, and who reported regularly moving to different towns and cities outside of the recruitment locale were coded as "homeless travelers."

Additionally, details concerning travelers' appurtenances and associations with other travelers were recorded into ethnographers' field notes, which assisted in distinguishing homeless travelers from homeless locals. Travelers, who frequently associated with one another, tended to be visibly distinct from other homeless youth based upon certain observable identifiers, such as tattoos on the face, hands and neck, possessing dogs, carrying backpacks, wearing military or punk style clothing, or clothing with patches of punk band names and other swaths of clothing stitched onto them. These youth also often wore unique hair styles, which included a combination of shaved, dreadlocked, and/or hair dyed multicolored. Overall, the relatively unique style of travelers made them easily discernable from local homeless youth.

Follow-up Interviews and Mapping Methodology

Among the 101 IDUs enrolled in Los Angeles, 65 youth (68.3%) were identified as travelers. Among these travelers, 53 youth (81.5%) completed one or more follow-up interviews via telephone outside of the Los Angeles area. Additionally, two subjects recruited in New Orleans and one subject recruited in New York during Phase 1 were subsequently encountered by ethnographers during fieldwork in Los Angeles. These three subjects were enrolled in the Phase 2 cohort and have completed follow-up interviews via telephone. Collectively, 56 subjects completed a total of 290 follow-up telephone interviews, and the locations of these respondents during follow-up interviews were mapped in Fig. 5.1.

Data for plotting the location of subjects interviewed outside of Los Angeles were based upon information abstracted from Western Union receipts. After each follow-up telephone interview, subjects provided ethnographers with a Western

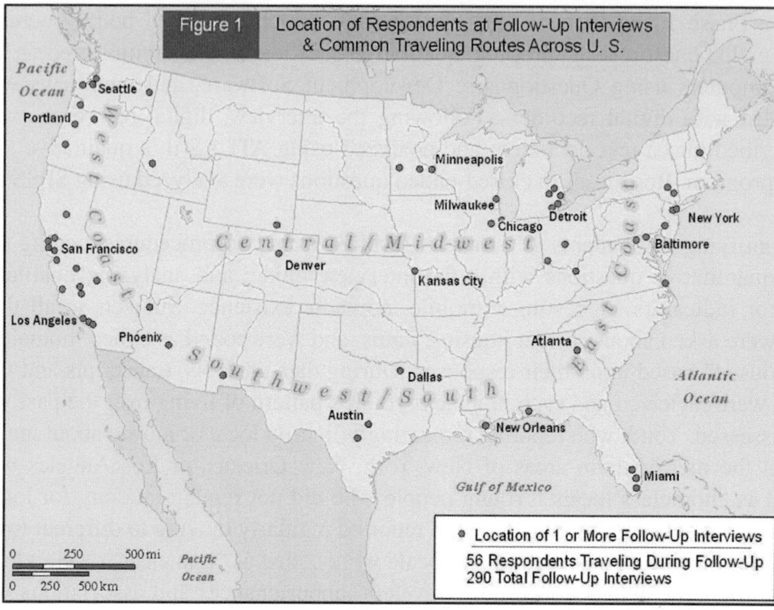

Fig. 5.1 Location of respondents at follow-up interviews and common traveling routes across the United States

Union address to wire an incentive payment. Following the interview, ethnographers recorded the zip code associated with the Western Union payment into a log. Zip codes were used to determine the 3-digit zip code tabulation area (ZCTA) for each location. Multiple interview events in one 3-digit ZCTA were aggregated to avoid ambiguity caused by subjects traveling in groups or settling for a time in one location. The locations of follow-up interviews were then mapped to the geographic center of the 3-digit ZCTA using cartographic boundary files obtained from the US Census. Common traveling routes, based upon respondent location data and qualitative data, were overlaid onto the map.

Results

Descriptive Characteristics

A total of 131 homeless travelers, or 61.2% of the three-city sample of 213 (see Lankenau et al. 2007 for a description of total sample) were identified across all sites. In Los Angeles, 65 youth (68.3% of the Los Angeles sample) were coded as travelers, 46 travelers (68.7%) were identified in New Orleans, and 20 travelers (40%) were identified in New York (see Table 5.1). Across all three sites, the sample is largely male, white, heterosexual, and in their early 20s. A majority graduated from high school, received a GED, and/or achieved a higher level of education,

Table 5.1 Demographic characteristics of travelers ($n = 131$)

	Travelers All Sites $n = 131$	Los Angeles Travelers $n = 65$	New Orleans Travelers $n = 46$	New York Travelers $n = 20$
Median age (years)	22	22	22	23
Male (%)	67.9	66.2	63.0	85.0
Race and ethnicity (%)				
White/Caucasian	83.2	84.6	84.8	75.0
Black/African American	–	–	–	–
Hispanic/Latino	5.3	3.1	10.9	–
Asian or Pacific Islander	0.8	1.5	–	–
Native American	–	–	–	–
Multiracial ancestry[a]	10.7	10.8	4.3	25.0
Sexual identity (%)				
Heterosexual	74.8	73.8	73.9	80.0
Gay/lesbian	1.5	1.5	2.2	–
Bisexual	19.8	21.5	17.4	20.0
Other/undecided	4.6	4.6	6.5	–
High school graduate or GED (%)	60.2	61.5	63.0	50.0
Employed full or part time (%)	19.8	6.2	34.8	30.0
History of drug treatment (%)	48.9	53.8	46.5	45.0
History of mental health care (%)	71.0	72.3	73.9	60.0
Ever arrested (%)	95.4	95.4	95.7	95.0
Ever in jail (%)	90.8	90.8	91.3	90.0
Ever in prison (%)	18.3	16.9	15.2	30.0
Tested for HIV (%)	94.7	95.4	93.5	95.0
HIV positive[b]	–	–	–	–
Tested for HCV (%)	82.4	86.2	73.9	90.0
HCV positive[b] (%)	22.1	26.2	21.7	10.0

[a]Of respondents reporting multiracial ancestry ($n = 14$): White/Caucasian 71.4%, Black/African American 7.1%, Hispanic/Latino 28.6%, Asian or Pacific Islander 7.1%, Native American 28.6%, Creole 7.1%.
[b]Self-reported.

such as college. Some worked part-jobs in the formal economy, such as day laborer, while most earned income through participation in the informal street economy, which included panhandling, sex work, and drug selling. Many had been to a drug treatment or detoxification facility. A majority received mental health treatment, which could have included therapy, psychiatric care, or a stay in a mental health facility. Almost all had histories of criminal justice involvement, such as an arrest or incarceration in a local jail or state prison. While rates of testing for HIV and HCV were high, none reported being HIV positive, but over one-fifth of those who had been tested for HCV reported a positive result. Travelers recruited in Los Angeles and New Orleans shared similar demographic characteristics with the exception of employment and HCV testing. In contrast, travelers enrolled in New York were more likely to be male, multiracial, and heterosexual, to have been in prison, and to self-report as HCV negative. Overall, Table 5.1 reveals a sample of high-risk youth with extensive experiences of criminal justice, drug treatment, and mental health care involvement, and high rates of HIV and HCV testing histories.

Drug using histories of travelers (see Table 5.2) were generally comparable regardless of recruitment sites with methamphetamine being the primary exception (see Lankenau et al. 2007 for more extensive drug history for entire sample). Travelers initiated marijuana, LSD, cocaine during late adolescence, heroin, ecstasy, and ketamine during mid to late teens. Methamphetamine was initiated the earliest among Los Angeles recruits and the latest among New York recruits. Heroin was typically the first drug injected across all three sites followed by methamphetamine in Los Angeles and cocaine in New Orleans. Notably, two-thirds of the total sample had histories of injecting heroin, cocaine, methamphetamine, and ketamine (all had injected ketamine due to the enrollment criteria). Marijuana was most commonly used in the past 30 days, whereas LSD and ecstasy were used least frequently. Recent use of heroin, cocaine, methamphetamine, and ketamine, while common, varied across the three sites with methamphetamine displaying the widest variability. Overall, travelers initiated most drugs by the age of 18, reported injection as a primary mode of administration, and demonstrated extensive past and current polydrug use (Sanders et al., 2008). With the exception of methamphetamine, which tends to be

Table 5.2 Past and current drug use among travelers ($n = 131$)

	Travelers All Sites $n = 131$	Los Angeles Travelers $n = 65$	New Orleans Travelers $n = 46$	New York Travelers $n = 20$
Mean age of initiation (years)				
Marijuana	12.4	12.0	12.8	12.5
LSD	15.2	15.2	15.1	15.7
Cocaine	15.6	15.5	15.5	15.9
Methamphetamine	16.2	15.7	16.4	18.0
Heroin	16.3	16.6	16.0	16.5
Ecstasy	17.0	17.0	16.6	17.7
Ketamine	17.7	17.5	17.6	18.4
First drug injected (%)				
Heroin	47.3	46.2	43.5	60.0
Methamphetamine	19.1	27.7	10.9	10.0
Cocaine	16.0	10.8	23.9	15.0
Ketamine	10.7	10.8	13.0	5.0
Other drug	6.9	4.6	8.7	10.0
Ever injected (%)				
Ketamine	100	100	100	100
Heroin	85.5	84.6	84.8	90.0
Cocaine	77.9	78.5	78.3	75.0
Methamphetamine	67.9	76.9	69.6	35.0
Any other drug	73.3	76.9	67.4	75.0
Drug use in past 30 days (%)				
Marijuana	82.4	87.7	76.1	80.0
Heroin	48.8	56.9	39.1	45.0
Cocaine	43.3	38.5	54.3	35.0
Methamphetamine	32.1	47.7	19.6	10.0
Ketamine	29.0	29.2	21.7	45.0
LSD	10.7	12.3	8.7	10.0
Ecstasy	7.6	7.7	8.7	5.0

more readily available in Western states, drug histories indicated a convergence of past exposure and current interest in a range of substances among travelers irrespective of recruitment locale.

Hitting the Road as a Homeless Traveler

Becoming a Homeless Traveler

Young people reported numerous circumstances as to how and why they initially became homeless. Travelers became homeless in fairly typical patterns: leaving home on their own; being asked or forced to leave by a parent; or running or fleeing from an abusive situation. The specific circumstances preceding homeless were numerous but often stemmed from problems within the family, such as neglectful or abusive parents, limited financial resources within the households, boredom, and drug problems:

> My parents kicked me out. I was doing bad in school. I was smoking a lot of pot. I went and dyed my hair black, and my Dad called me a faggot and a piece of shit. I told him to fuck off, and he hit me with his cane. I pushed him away and he kicked me out.

Upon hitting the streets, finding other similarly situated homeless young people was an important initial step toward surviving outside of the home, and the unique visible characteristics of the travelers assisted in connecting like-minded youth. For instance, other homeless youth provided skills and assistance toward living on the streets, such as making money and finding shelter:

> I got expelled from high school – boarding school – and I was in California. I was scared of going home because I had just lost a very large amount of tuition money, so I just went to Santa Monica Pier. It was the only place I knew of and I met people that taught me how to squat [find shelter in abandoned buildings].

Traveling out of one's home town and identifying with the lifestyle of a traveler was not typically the objective of a young person upon becoming homeless. Rather, deciding to travel often was often premised upon connecting with more experienced travelers:

> My parents kicked me out when I was 17, and I ran across a couple people and they said, "Hey, we're traveling, you know, this, that, and the other. You should travel with us." I started hitch-hiking. It wasn't until I was 20 or 21 that I started riding freight trains.

Becoming homeless and deciding to travel, however, was not always connected to a specific hardship. For some young people, traveling represented a rejection of middle class values, or even modeling their behavior after a parent's lifestyle from decades ago:

> I met all these people that were traveling – it seemed like a good idea. I didn't want to go to college, and I don't want to work. I don't mind working, but I just don't want to work to pay rent just so I have a place to live and shower. My mom always told me about her traveling stories. I guess she hitchhiked a bunch and rode around in her friends' buses and stuff. She did that for the whole hippie movement in San Francisco.

Modes of Travel

Travelers moved from state to state, and city to city via numerous modes of transportation, such as riding freight trains, hitch-hiking, getting rides in a friend's car, buses, and bicycle riding. Sometimes, travelers used multiple transportation modes during a trip:

> We hitched a ride up to Malibu, and then we ran across some home bum in Malibu who got us drunk and hooked up a ride on a van all the way to San Jose. But, the whole time the guy was drunk so we commandeered his van. We would go to different truck stops, panhandle, make money, get more beer, and then finally we got to San Jose, and we're like, "Later. Appreciate the ride." Then we got to Berkeley, hung out there for a couple weeks, had a good time, and caught a train out of Oakland.

While traveling by car or bus was most frequently reported, many youth revealed that they "train hopped" or "hopped freight" in compelling narratives. While train hopping is illegal, it provided a free and efficient means of traveling great distances. Additionally, many indicated that they enjoyed train hopping because of the adventure and opportunity to see different regions of the United States:

> Being on a train in the middle of nowhere and you got nothing on this side, nothing on that side, the only thing that's been touched is the railroad tracks you're on. And you look up in the sky and you're in the middle of nowhere and there's a million stars. I just like the scenery and the beauty of this country.

Even more so than hitch-hiking with strangers, travel itineraries via hopping freight, which were dictated by train routes, could be unpredictable and result in spending time in unanticipated places along the way:

> I went from southern California to Oregon, from Oregon to Idaho, and to Utah. Utah sucks. And, let's see, Wyoming, and then from Wyoming down here. Some of these places were just on the way. That's how the train line runs.

Train hopping, which often involved jumping on or off a fast moving train, carried certain bodily risks, such as injury, dismemberment, and death. Several travelers reported seeing or knowing others who had broken arms, severed limbs, or been killed hopping trains. One traveler enrolled in the study fell off a moving freight train and was killed. Due to the risks involved, train hopping attracted a particular subgroup of travelers who wore or carried specific identifiers, such as clothing, tattoos, or injury-related scars, and produced certain types of informal affiliations:

> We actually just had a gathering of about 70–80 train hopping kids who are like family. We all knew each other. We meet mostly by hopping trains, through random cities, or kinda meeting 'em on the street. Like "Oh dude! What's up! Went to CA, how's it going, just got into town".

In addition to bodily risks, train hopping also carried risks for arrest. Certain train yards, often in urban areas, were known for having more security patrolling the yards. The security of train yards has increased following the 9/11 terrorist attacks since trains are regarded as a vulnerable yet vital mode of transporting goods and passengers across the United States

Traveling Routes across North America

Travelers arrived in new towns and cities typically by bus, car, or train. Particular travel itineraries were often determined by a freight train's route or a driver who picked up hitchhikers. However, broader patterns of travel frequently emerged from the specific routes (see Fig. 5.1). Overall, these patterns generally followed railroad lines and major interstate highways. Primary destinations within these broader patterns included San Francisco, Portland, Tucson, New Orleans, and New York, which each contained a local youth scene and like-minded travelers.

Figure 5.1 plots the positions of 56 travelers based upon their locations during 290 follow-up telephone interviews, which occurred between 4 months and 16 months after baseline interview. Since all were initially interviewed in Los Angeles, many could be found in towns and cities in west coasts states, such as California, Oregon, and Washington. However, other travelers dispersed to locations in the Southwest, South, Midwest and East. Many of the locations of towns and cities fell along general traveling trajectories described during follow-up interviews. Based upon these locations and interview accounts, four common routes or travel pathways were identified:

West coast. One of the most frequent travel routes occurred along Interstate 5, which stretches the entire West Coast from Mexico to Canada. Route 101 and Highway 1 also extend for hundreds of miles along the California coastline and into Oregon. Between these three highways and primary north/south railroad lines, travelers left Los Angeles and headed north to Santa Barbara, San Francisco, Portland, Seattle, and sometimes into British Columbia. Likewise, travelers also headed south to San Diego and Tijuana.

Southwest/South. Another common route leaving Los Angeles was to head east along Interstate 10 or Interstate 40, which also included rail lines following parts of these same highways through Tucson, Albuquerque, Austin, and onto New Orleans. Both prior to and following Hurricane Katrina, travelers headed to New Orleans based upon its reputation for 24-hour partying and mild temperatures during the winter. Since Hurricane Katrina, the possibility of part-time employment in construction and other fields related to rebuilding the city have attracted new waves of travelers (Lankenau et al. forthcoming).

East coast. Interstate 95 spans the entire east coasts from Florida to Maine, and is also paralleled by various north/south railroad lines. New York City is a primary destination for travelers, particularly during spring and summer months. As with New Orleans, New York City's reputation for nightlife and status as a happening city attracted travelers from Atlanta, North Carolina, Washington, DC, Philadelphia, Boston, Toronto, and Montreal. One traveler's account, which demonstrates the amount of traveling and cities visited within a 3-month period, encompasses these three primary routes:

> After San Francisco, we went to Phoenix, Arizona and Tempe, and then El Paso, Texas and San Antonio and Austin, and then Tucson, Arizona, and then back to San Francisco. Then, down to Los Angeles overnight, and back to San Francisco, Eugene, Oregon, and now Portland. Originally, we were headed to New York and Philadelphia. So, we were just

> taking the southern route that way. Then, we decided to come back to Portland to drop off a friend's truck and hop freight east, but then we got separated, and I'm still here.

Central/Midwest. Traveling routes across the Midwest and Plain states also followed interstate and rail lines. Some young people traveled along freight train lines extending through the Midwest, often heading from the West coast to the East coast, or vice versa. Some traveled through the northern states of Wyoming and Wisconsin while others headed through more central states including Kansas and Missouri. These routes involved long trips requiring multiple stops, and therefore, travelers disembarked periodically to get supplies:

> The train stopped in Chicago, and I stayed there for a couple nights, and then I hopped out of Chicago, and I caught a ride into Cheyenne, WY, and I got off there because I ran out of supplies, so I got some more food and water, and then I came the rest of the way to Idaho.

Primary Features of Traveling: Drugs, Making Money, and Law Enforcement

Drugs

Drugs were an important part of these young people's lives, and the freedom of traveling was often associated with "getting high." Traveling was frequently a part of an explicit strategy to find particular drugs, better drugs, or cheaper drugs. For instance, several reported traveling to Mexico specifically to purchase ketamine, which could also be bought from pharmacies in some border towns, such as Nogales and Tijuana. Traveling to another city was sometimes initiated by a friend's recommendation of higher quality heroin:

> I remember I went to San Francisco about three years ago because the dope there was better. And then just this past month, that's just one of the reasons we went to San Francisco. "Oh yeah, the dope there is great". We went up there and it was the worst dope I've ever seen. I was doing a half gram and I wasn't even getting well. It was bad. So we left.

Upon arriving in a new or unfamiliar town, finding a drug source was important. Alcohol and marijuana, two drugs widely used while traveling, were often shared, traded, or purchased informally among travelers. Harder drugs, such as heroin or cocaine, usually required locating a drug seller:

> Like, if I first hit a town, if I haven't been there, one of the first things I do is try to find a drug dealer. Mostly, I just seek out other street kid looking people and ask them if they know where to cop [buy drugs]. Then start a conversation with them and proceed from there. A lot of it is through the drug culture community.

Other travelers, based upon years of moving around the country, knew where to find local drug sellers in many cities:

> I know where to get drugs in a lot of different places. I've been doing this a long time. If you named a city off of the top of your head that's a pretty major city, I could tell you where to go. I mean, to be honest, I can get heroin in every major city in the west coast and the east, so it's not a big deal for me.

Drug using practices, such as smoking, sniffing, and injecting, were influenced by the mode of transportation, the degree of privacy while traveling, and the persons traveling with. For example, a traveler's drug use could be restricted by those providing them with a ride:

> I've gotten rides from truckers and, I mean, they'll smoke speed, but they're not too cool... they don't trip. I've actually done a couple of shots [injections] in the back of a rig, and that ain't cool. But, he was like, 'You can do a shot, but I don't shoot.' He was like, 'Keep that on the DL [down low] and make sure you don't leave none of that shit in my rig because I'll get fucked over."

Ultimately, drug use while traveling was highly situational, given the numerous factors influencing drug availability, drug using practices, and traveling contingencies. For instance, some individuals ingested specific drugs to lessen the physical strain of travel:

> I use [drugs] sometimes on the road to keep myself awake, you know, if the coffee shops and the coffee don't really do it. Or, I drink [alcohol] or take sedatives and barbiturates to take the edge off to sleep.

Traveling to music festivals to buy, use or sell drugs was commonly reported. Given the large number of like-minded young people, festivals were an opportune time to experiment with new substances. For instance, several young people reported injecting ketamine for the first time at a music festival. Others reported consuming a variety of substances that were readily available at a festival:

> The end of the night, like I usually do 10 different drugs at festivals and shows. I was using nitrous oxide, probably LSD, probably Valium, probably coke. Everything goes there. I can get whatever I want pretty much when I'm at the shows.

Money Making and Work

Travelers frequently moved from city to city in search of new income opportunities. Travelers sometimes left for a new destination based upon rumor of work or offers of work from friends, and they frequently gained money from a variety of sources:

> I moved down to Long Beach for a little bit, and I started working for my friend. I was learning how to make or how to print t-shirts. But, he went off this summer to tour with a band, so I am traveling around again. I went up to San Francisco and then up to Portland for a couple days and then up to Seattle for a day then back to Portland for a couple days, and then back to San Francisco for like a month. And, I just left there about a week ago, and I am in Fresno, California as we speak. Since then, I do drug studies, I beg for money, I fly signs, and I sell drugs.

Seasonal work, such as picking fruit or vegetables, also prompted some travelers to move to a new locale. Though, since the work lasts for a few months, other types of income opportunities were necessary:

> I'm a hobo. I go to different towns to seek adventure, and income, such as jobs. Usually, I work harvests or work at a skate shop or something like that. Or, possible scams – receipt scams, sometimes government scams, see if I can scam a couple hundred dollars for a couple months.

In the main, most of the income generation described by traveler tended toward the illegal or semi-legal, such as selling drugs, stealing, scams, and panhandling. For these reasons, plus their unusual appearances and reputations for drug use, travelers had frequent contact with law enforcement.

Law Enforcement

Police and criminal justice involvement were major factors that impacted travelers' plans to stay in one town or move elsewhere. Police, or other security forces, were often inhospitable to street youth – particularly newcomers – so harassment was common. This practice was apparent in Hollywood, where local security guards, known colloquially as "Green Shirts," were responsible for displacing homeless youth from sidewalks (see Hyde 2005). These security guards, who carry handguns belted at their waist, are not police officers, and have no more power of arrest than an average citizen. Yet, on several occasions, ethnographers witnessed Green Shirts being heavy handed with homeless youth, using abrasive language and physically detaining such youth. Hence, avoiding the scrutiny or interrogation of the police was paramount while traveling or upon coming to a new city since some travelers had outstanding arrest warrants, or were frequently in possession of illegal substances or drug paraphernalia:

> When I'm hitchhiking around, I usually don't do anything but smoke pot just because I don't wanna get stopped by the cops and be drunk or like, have anything illegal on me. I usually try not to carry anything on me if I'm traveling.

In particular, upon arriving in a new city, travelers were vulnerable to being spotted by security or police officers on patrol since homeless youth frequently congregated or slept in visible public places, such as parks and along sidewalks:

> I was just walking down the street in Portland and a cop stopped me because he didn't recognize me. He works with street kids. Turns out I had a warrant for not going to court. I went to jail and got released 5 hours later and given a court date.

Traveling to another city or state to avoid facing criminal charges was commonly described, and sometimes served as an impetus to travel:

> I ended up getting a warrant in Missouri, and I can't go back there, so I went to California.

Typical charges, which often stemmed from being in public places, included sleeping on sidewalks, open container violations, and possessing drug paraphernalia. More serious offences included assault, robbery, or drug selling. After receiving a summons or being arrested, some failed to appear in court to face charges, which resulted in more serious charges. These subsequent arrest warrants increased a traveler's vulnerability should they be questioned or detained by police in the future in an unfamiliar city.

Impact of Traveling on Accessing Risk-Reduction Services

Shelter

Homeless travelers secured many different kinds of housing and shelter while on the road or upon arriving in a new city. None reported staying in a homeless shelter during their recent homeless periods. Many viewed homeless shelters as dangerous, unhygienic, inconvenient, controlling, or stigmatizing. As with finding drugs, locating shelter for travelers was facilitated by connecting with local youth with knowledge or access to shelter:

> Every time I hop a train or hitch into a city, I just start talking around and find the local punk scene, the anarchist kids, or anything like that. I just find a place through them – normally squats, punk houses, or anarchy houses. Sometimes, I'll sleep in the park if I don't feel like staying in a house.

Travelers were skilled at adapting to a variety of places, and often sleeping outside if the climate permitted. For some, finding shelter or being flexible as to where one stayed was based upon years of living a nomadic existence. Periods of being housed often follow periods of homelessness:

> Pretty much up until a couple weeks ago I was homeless. I just got back in Portland the first day of last April. So, before then I had just been staying – like, me and a friend had been traveling, and so we were staying just wherever we happened to be. A lot of times I stayed in the back of his truck or wherever. Since I've been back in Portland, I was staying mostly under bridges and then in a friend's apartment.

Syringe Exchanges

Since all of the travelers in the sample were active IDUs, finding sterile injection paraphernalia, such as syringes, cookers, cottons, and rinse water, was a concern for some. Like finding work, drugs, or shelter, locating a syringe exchange was frequently initiated by asking others on the street:

> I ask people where the drop-in centers are. I ask people that are walking around where there's a needle exchange. Sometimes you can go to pharmacies and buy them, but not in a lot of towns. Like, you can in Baltimore, but you can't do that in New Orleans. Even in Berkeley, I had to ask others for 3 days before I found some. I think I just wasn't looking hard enough, but there are some areas, like in New York, you're walking down the street and people are like, "Hey, do you need clean needles?" And you're like, "Ok."

Despite the availability of syringe exchanges in some cities, access was limited by hours of operation of the exchange, which may be open only at certain times on specific days. Additionally, since laws and regulations surrounding access to syringes from either exchanges or pharmacies varied from state to state, and sometimes from city to city within the same state, travelers encountered locations where syringes were scarce – particularly towns and smaller metropolitan areas:

> I had trouble finding cleans [syringes] in Hagerstown, Maryland, in New Hall, Pennsylvania, in Harrisburg, Pennsylvania, and in Syracuse, New York. I couldn't find shit.

Awareness of the risks associated with sharing syringes was evident in this sample of young IDUs since few reported sharing syringes during recent injection events. However, reusing the same syringe also carries certain health risks, such as vein damage, abscesses, and soft tissue infection. Service providers who offered clean syringes and other forms of assistance for homeless youth were a welcome haven for travelers arriving in a new or unfamiliar city:

> There was the needle exchange in SF that definitely helped allowing me to use clean needles all the time. I mean, I don't share needles anyway, but I was having to re-use them quite a few times in other towns. Up here in Portland, there's a drop-in for traveler kids, and that's cool because they have food and you can wash your clothes there, take a shower.

HIV/HCV Testing

As reported earlier, a high proportion of the sample had histories of HIV and HCV testing. Given the high-risk nature of the population, however, regular testing as advised by health care providers would require accessing testing sites in cities and towns along traveling routes. Despite the emphasis for regular testing, HIV and HCV testing was typically not a main concern:

> I don't really look for them [HIV testing services]. They just kind of pop up. More often it was available, but it wasn't really a top priority.

Rather than actively seeking out testing, as compared to clean syringes, travelers were often presented with testing opportunities and later decided whether tests were needed. For instance, street outreach providers, drop-in centers, homeless shelters, and research studies advertized testing services to travelers. Given the frequency of moving from one city to the next, however, travelers did not always remain in one place long enough to learn their test results:

> It's not very hard. You can call information, "Hook me up with a clinic that can help, test me for HIV". And they'll give you a big old long list. I've never had trouble finding them, just really lazy. If we weren't lazy, we weren't staying in town long enough to get the results.

While testing services were often free, access was limited by scheduling and opening availability on testing days. Also, unlike syringe exchanges, testing may be restricted to specific age groups if the organization or study was targeting young people. A youth-friendly staff was crucial for making travelers, who are often service resistant, feel comfortable about receiving testing.

> I went to a drop-in center in New Orleans, and they were really nice there, but there was this big guy who was like, "Spread 'em." And I was like, "Maybe I'll do it some other time." But, I can always rely on Planned Parenthood for being nice and having all the STD tests – especially because I am a youth. It's a lot easier because there are so many more youth-based groups than older people-based groups.

Discussion

A potential limitation to this analysis is that travelers were not recruited into this study from explicit enrollment criteria linked to behaviors and lifestyles associated with traveling. Rather, a sample of travelers was indirectly recruited based upon enrollment criteria for a study of ketamine injectors combined with a sampling methodology that targeted young people congregating in public places. Hence, this sample is biased toward IDUs, and those with a history of ketamine injection in particular. Consequently, the current sample is a high-risk subgroup, perhaps more so than travelers who may be less drug-involved. For instance, while not enrolled in the current study, reports were offered of travelers who identified as "straight edge" – those who abstained from alcohol and illegal drugs (Haenfler 2006). Additionally, we noted at the beginning of this manuscript that while some research had been focused on homeless IDUs and homeless YMSM, limited research had examined homeless travelers. However, these categories or identities are not mutually exclusive, and multiple types of risk exist. Travelers are particularly a high-risk group, given that all were IDUs, and one-quarter identified as lesbian, gay, bisexual or other.

Travelers comprised the majority (61.2%) of the three-city sample. The demographic characteristics and drug using histories among travelers across sites were largely similar: heterosexual white males in their early 20s with histories of drug treatment, mental health care, and criminal justice involvement. Initiation of significant "street" and "club" drugs occurred in a similar order, while injection drug using histories and recent drug use were comparable. These similarities across sites suggest that the forces causing or sustaining travelers, such as dysfunctional families, criminal justice involvement, limited participation in the formal economy, substance abuse, are not local in nature, but rather are regional or national in scope.

A primary difference between travelers recruited in each site was past and current use of methamphetamine. The earlier age of methamphetamine initiation, a common methamphetamine injection history, and recent methamphetamine use among travelers recruited in Los Angeles indicates greater access to this particular drug. This could be explained by travelers recruited in Los Angeles spending more time traveling the West Coast route, where methamphetamine has a long history of use and is more readily available (Sanders 2006). While methamphetamine production and use is now found in all regions of the United States, travelers recruited in New York and New Orleans injected and used methamphetamine less frequently.

Many of the locations of travelers during follow-up interviews were situated along one of four common traveling routes, which were often determined by interstate highways and railroads. Patterns of travel were influenced by numerous factors, including seasonal variations, festivals, disasters, e.g., Hurricane Katrina, as well as the pursuit of drugs and work, or avoiding law enforcement. Several hub cities were identified, including New York, New Orleans, Los Angeles, San Francisco, and Seattle. The availability and utilization of risk-reduction services, such as shelters, syringe exchange, and HIV/HCV testing, varied along these routes and may impact short- and long-term well-being of travelers.

Social networks among travelers were a key issue linking many aspects of the phenomena of traveling. In several of the domains described, such as becoming a traveler, determining a mode of travel, or finding drugs, work, shelter, or syringes, connecting with other travelers was a primary part of initiating a behavior or accomplishing an objective. Most of these connections originated in face-to-face interactions in public or semi-public spaces, such as streets, parks, or on trains. The distinctive, and often unique appearances among travelers, such as clothing, hair style, and tattoos, assisted travelers recognizing each other in unfamiliar cities and places. While travelers did report using the Internet and cell phones, most did not have the resources or regular access to make them a reliable form of communication. Rather, face-to-face interaction often suited traveler's immediate needs and formed the basis for associations and relationships with other travelers.

The findings suggest that competency as a traveler, along with developing social networks, was built upon accruing various forms of knowledge or capital over the course of months and years of traveling. Accomplishing key tasks associated with traveling, such as hopping the right freight train, locating shelter during inclement weather, finding a drug dealer in a new city, avoiding police while sleeping outside, or finding a syringe exchange, was facilitated by a knowledge base developed through experience. Elsewhere, we have discussed "street capital" among homeless youth (Lankenau et al. 2005), which refers to survival knowledge informally acquired prior to becoming homeless that later serves youth upon hitting the streets. Many of the travelers accrued street capital due to the hardships suffered prior to becoming homeless. Moreover, additional forms of street capital were acquired through traveling.

A salient feature of travelers' everyday existence was moving between formal structures, such as law enforcement, service providers, or railroads, and informal entities, such as networks among travelers, local drug markets, or earning money under-the-table. Formal structures are objective, bureaucratic, and hierarchical, whereas informal entities are subjective, fluid, and marginal (Laguerre 1994). Travelers' survival on the road or in new towns or cities was contingent upon avoiding or exploiting formal structures where possible, and largely participating in informal spheres. Attributes that made travelers more visible or problematic to formal entities, such as punk rocker style or IDU, often represented forms of capital that were valued among travelers in informal domains, including the street economy (Lankenau et al. 2004).

Lastly, incorporating newer spatial analysis and mapping techniques into this longitudinal study of drug users enhanced both the methodological rigor and the understanding of findings. First, plotting the location of respondents during follow-up interviews allowed for the emergence of traveling or migration patterns, e.g., West coast route, which would not have been observable from narrative descriptions alone. These data points also served as a general validity check on the respondents' particular descriptions of traveling patterns, e.g., Los Angeles to San Francisco to Seattle. Second, the mapping data demonstrates that patterns of migration among young IDUs were national or international in scope, and not limited to movement within local metropolitan areas. Furthermore, the mapping data documents the mobility of young IDUs across both urban and rural areas, which supports the proposition that migration among IDUs may be a factor in the spread of blood-borne viruses, such as HIV and HCV, across geographic regions (Rachlis et al. 2007). While most drug users are not as mobile as those described in this analysis, future studies of drug users – particularly longitudinal studies – should attempt to include spatial analyses to map the movement of drug types, drug using practices, and/or risk behaviors among diverse populations of users across different geographic regions. Such studies are likely to contribute to the growing understanding of the important linkages between patterns of drug use and geography.

Conclusion

These findings suggest that travelers are an amalgamation of characteristics of past generations of young people who traveled the country: often rooted in family hardships or economic difficulties; searching for adventure and freedom; an experimentalist orientation toward substance use; community defined through face-to-face interactions; and a resourcefulness toward providing necessities, including addictive substances. While sharing certain similarities with past generations, however, these travelers represent a particularly vulnerable population, given their drug using histories, loose commitment to risk-reduction services, and ease of identification for those wanting to harass or control. Compared to other high-risk youth, travelers are a unique group that may require specific intervention strategies.

Acknowledgment Support for this research was provided by the National Institute on Drug Abuse (DA015631; DA015631-S1).

Chapter 6
Residential Mobility and Drug Use Among Parolees in San Diego, California and Implications for Policy

Meagan Cahill and Nancy LaVigne

Abstract This research examines the spatial concentrations and mobility of released drug abusers and explores how those patterns may differ from those of the general releasee population. Using data on the drug use and residential histories of parolees in San Diego County, California, the findings reveal high rates of drug use (57% had known problems with drugs) and mobility (52% moved) across different types of parolees. Moves were also relatively far—on average, over four miles for the first move; parolees are likely changing their residences and their neighborhoods. However, parolees are living in areas that are far more disadvantaged than the average neighborhood in San Diego County, and when they move, they go to areas that are similar or worse-off. The research thus underscores the need for housing and other assistance for parolees that begins before release, and providers should work with parolees to ensure service continuity after a move. Understanding when, where, and how far parolees move is both essential to knowing where specific types of services should be located, and critical to maintaining accountability among parolees.

Introduction

Residential mobility has persisted for several decades as a principal topic of research among social scientists including economists, geographers, mental health researchers, and epidemiologists (Clark 1984; Clark and Ledwith 2006; Lamont et al. 2000; Lee, Oropesa, and Kanan 1994; Lix et al. 2007; Lu 1998; Quillian 1999). In more recent years, both researchers and local jurisdictions have increasingly engaged in the mapping and analysis of neighborhood-level incarceration and reentry data (see Cadora 2002; Harries 1999, 2002; La Vigne, Cowan, and Brazzell 2006; Rose and Clear 1998; Smith and Dickey 1999). There is little research, however, on the overlap of these two research agendas–the residential mobility of formerly

M. Cahill
The Urban Institute, 2100 M Street NW, Washington, DC 20037, USA

Y.F. Thomas et al. (eds.), *Geography and Drug Addiction*,

incarcerated persons. Of particular interest for the current research are the mobility patterns of those with histories of substance abuse, who may have greater needs for health services and treatment than their non-substance abusing counterparts. Policy-wise, understanding the residential mobility of formerly incarcerated persons with histories of substance abuse is key to developing better means for both assisting and monitoring this population.

This research uses data on drug use and residential histories of all parolees under supervision[1] in San Diego County, CA to explore these inter-related themes of residential mobility, prisoner reentry, and drug use. The objective of this chapter is to develop an understanding of the spatial concentrations and mobility of released drug abusers and to explore how those patterns may differ from those of the general releasee population. We are also interested in understanding the role of drug abuse in residential mobility among those parolees in our dataset. Knowing where released prisoners reside, how often they move, and how far they move has implications for post-release supervision as well as for the spatial allocation of housing, treatment, and other social services, especially crucial for the drug abusing population. A review of literature on the topics of residential mobility in general, the spatial distribution and mobility of released prisoners specifically, and the role that drug use and treatment play in both residential mobility and reentry success, provides a framework for the present research inquiry.

Background

Residential Mobility

Clark and colleagues' (1984) classic work on residential mobility confirmed the influence of age, home ownership, and life changes, such as marriage or the birth of a child, on mobility. That work focused mainly on the changing space requirements of families as they proceed through their life course. Clark and Ledwith's (2006) more recent work draws from data collected as part of the Los Angeles Family and Neighborhood Survey (LAFANS), which, after weighting, is representative of the population of Los Angeles. The authors found trends in the LAFANS sample to mirror national mobility trends. The data reveal that Native Americans have the highest mobility rates, closely followed by African Americans. Latinos have the lowest mobility rates, which can be partially explained by lower income levels and larger families among that population, both of which constrain housing choice. Lower income levels, however, were also prevalent among African Americans, who had the second-highest mobility rates. While this exception to the positive mobility-income relationship was noted by the authors, the LAFANS data precluded further investigation of the phenomenon. In addition, among all racial/ethnic groups, individuals under 30 years of age had the highest mobility rates, followed by those between 30 and 44 years of age. Because the population profiles of Los Angeles and San Diego are very similar, findings from the LAFANS data have important implications for the current study of released prisoners in San Diego County. The San Diego data are expected to reveal lower rates of mobility among Latino releasees and higher rates among African American releasees, regardless of their drug or alcohol use. Also, younger releasees are expected to have higher mobility rates.

Housing status is an important aspect of residential mobility that is especially relevant for released prisoners and for those with drug and alcohol problems. Research into the relationship between housing status and psychological stress among those with drug or alcohol problems has highlighted the mental health issues that can accompany different living arrangements, including homelessness. Individuals with drug and alcohol problems have been found to experience lower levels of psychological stress when their living arrangements are more dependent, i.e., they are living with family or friends, than when living independently (Wong 2002). The same study also found that physical health was the number one predictor of psychological stress among this same population of individuals with drug and alcohol problems (Wong 2002). As the author notes, drug or alcohol addictions can lead to poor physical health, which can lead in turn to greater levels of psychological stress. Poor physical health combined with subpar living arrangements can further compromise an individual's mental well-being. The research, therefore, suggests that addressing any drug or alcohol problem in concert with housing issues is important to ensuring both the physical and mental well-being of these individuals. In addition, a network of social support appears key to the mental well-being of drug addicted individuals with unstable living arrangements. The type of housing is also important; for released prisoners with drug or alcohol problems, transitional facilities, group homes, or living with family or friends may provide better prospects for success than living independently.

Research on the relationship between neighborhood characteristics and mental health has consistently shown that living in a more distressed area has negative effects on mental health (Dupere 2007). Studies have also found that moving to areas with low poverty levels can reduce an individual's level of distress and depression and improve outcomes for children (Leventhal and Brooks-Gunn 2003). Parolees released to disadvantaged areas are therefore likely to experience more stress than those who return to areas with a higher quality of life. Experiencing such mental health issues can in turn affect a parolee's ability to successfully complete any parole period, and can compound the difficulty of recovery from substance abuse.

Residential mobility research has also focused on the role of neighborhood characteristics in housing choice. Lee and colleagues (1994) found little relationship between neighborhood characteristics, including age, economic status, tenure, density and the racial mix of an area, and mobility in their study of mobility in Nashville, Tennessee. The authors suggest caution, however, in interpreting that result because they found neighborhood characteristics to have a greater influence on thinking about moving. These findings indicate that individual factors will have a greater effect on actual mobility among releasees, but that desired mobility may be higher among those who are living in more distressed neighborhoods, with individual factors constraining their ability to move. Other studies have, however, shown links between structural characteristics of neighborhoods, such as age, economic status, and home ownership rates, and mobility outcomes, including actual mobility and future mobility (Lu 1998). Clark and Ledwith (2006) also explored the relationship between neighborhood characteristics and future or expected mobility with the LAFANS dataset. They found that perceiving future mobility was positively associated with dissatisfaction with one's current neighborhood and perceptions that the neighborhood lacks close ties among residents. Perceptions of the current neighborhood being unsafe, however, were not related to increased future mobility.

Other relevant research has focused on the differences among racial groups. St. John and colleagues (1995) found that African Americans have lower rates of intra-urban residential mobility and are less likely to improve their residential environments than other racial groups when they do move. Reisig et al.'s (2007) investigation of the effect of neighborhood characteristics on recidivism found that released African American prisoners who resided in areas with high levels of racial inequality (as measured by poverty and other measures of deprivation) had higher rates of reincarceration and reconviction. The authors recommend further examination of the characteristics of areas to which prisoners–particularly African Americans–are released in an effort to stem the rates of recidivism experienced among that reentry population.

Prisoner Reentry and Mobility

Research conducted by the Urban Institute has shown that, within cities such as Baltimore, Chicago, and Cleveland, residences of released prisoners are clustered, with a small number of communities accounting for a large share of releasees. In Baltimore, for example, 30% of prisoners returning to the city ended up in just 6 of 55 Baltimore communities.[2] In Chicago, 6 of 76 communities accounted for 34% of those returning there. Social disorganization theory supports these findings that returning prisoners will cluster in space (see Shaw and McKay 1942), and also suggests that releasees will be relatively mobile (see Morenoff, Sampson, and Raudenbush 2001). Early work conducted by Shaw and McKay (1942) and later by Harries (1974) demonstrated the relationship between population stability in an area and decreasing delinquency rates. From this early work, researchers identified three area-specific factors–high economic deprivation, high residential mobility or population turnover, and high racial or ethnic heterogeneity–as characterizing socially disorganized and distressed areas. The combination of these different characteristics in one area can lead to growing "illegitimate opportunity structures and dysfunctional lifestyles," including violence and crime (Elliott et al. 1996, p. 394). These places, in turn, become more attractive to many released prisoners. Rose and Clear (1998) considered the effects of reentry on neighborhoods from a social disorganization perspective, arguing that when released prisoners cluster in specific neighborhoods, the social networks that exist in those neighborhoods are weakened, reducing the level of informal social control in an area. Because the social ties in these areas tend to be weak or non-existent, they encourage a greater level of mobility among residents, including released prisoners. Research on the reentry population has shown that parolees tend to cluster in areas with high levels of disadvantage, including poverty, female-headed households, and unemployment (La Vigne and Thomson 2003; La Vigne, Visher, and Castro 2004).

While spatial concentrations of formerly incarcerated persons are now well-documented, the question of residential mobility of this population is under-explored, although a few notable exceptions to the scant research on this topic exist. One such exception is La Vigne and Parthasarathy's (2005) study exploring the residential mobility of recently released prisoners in Chicago. The study, based on surveys conducted with approximately 145 released prisoners at three points during the post-release period, found that at 1–3 months after release from prison, only 12% of the parolees had changed residences. By 1–2 years after release, that share grew to just 28%, with only 10% of the sample moving more than once over the study period. In addition, mobility was found in many cases to be an effort on the part of the parolee to improve his or her living arrangements or become more independent, financially or otherwise. Finally, the research found that movers and stayers were not significantly different in terms of their illegal drug use behaviors. These counterintuitive results, limited by a small sample size, highlight the need for further investigation of parolee mobility.

Another study of parolees in Georgia considered the role of residential mobility in risk for recidivism (Meredith, Spier, Johnson, and Hull 2003). That study found that, with every change of residence for a parolee, the risk for recidivating increased by 25%. In other words, the risk for recidivating doubled with only three moves, or four total residences. It is vital, then, that parolees find stable housing with supportive services nearby. Unfortunately, many cities lag in providing services for releasees, housing assistance included. The authors also noted their difficulty in culling this information from the Georgia data, suggesting that, because mobility has such a large impact on risk for recidivism, corrections departments pay closer attention to their data collection methods and systems.

Drug Use Among Prisoners

Drug use among current and formerly incarcerated prisoners is prevalent. The Substance Abuse and Mental Health Services Administration (2006) found that more than one-quarter (26.3%) of adults on parole or supervised release from prison in 2005 reported current illicit drug use. This figure reveals a considerably higher level of drug use among parolees than among adults not on parole or under supervised release, for whom the rate of illicit drug use was found to be only 7.7%. Drug use among inmates of state prisons is especially common, and can compound the difficulties experienced by parolees upon release. Mumola and Karberg's (2006) analysis of 2004 Bureau of Justice Statistics (BJS) survey of State inmates revealed that 83% had used drugs at some time in the past, with more than half (56%) having used drugs within a month of the offense of which they were convicted. Using standard criteria,[3] the BJS survey results indicated that 53% of State prisoners were considered drug dependent or drug abusing. The BJS survey also found that drug dependent and abusing prisoners had experienced homelessness during the year prior to their admission to prison at a rate twice as high as that of other prisoners; 14% of the drug dependent and abusing population reported such homelessness (Mumola and Karberg 2006). The drug dependent and abusing population was also more likely than other prisoners to report growing up in single parent homes, having lived in foster homes, and having a parent incarcerated. The BJS statistics provide a portrait of drug dependent and abusing prisoners as having weaker family structures on which to rely after their release and being less likely to have a home to return to after their release from prison. These factors suggest increased mobility among the drug using population upon release, possibly associated with an increased likelihood of recidivism.

Mitchell and colleagues (2006) report, however, that both recidivism and drug use are reduced with pre-and post-release drug treatment. Winterfield and Castro's (2005) analysis of drug treatment during and after incarceration concluded that screening for and assessing drug abuse problems at the point of intake and then providing services during and after incarceration to treat drug abuse problems can help reduce the likelihood of post-release drug use and increase an individual's chances of success on parole. Their findings, based on surveys of prisoners in Illinois and Ohio, also found that over a third of prisoners who reported having a drug abuse problem at admission to prison did not receive any treatment for that problem, either during or after incarceration. The authors characterized this "mismatched" delivery of substance abuse treatment services in prison—whereby many of those in need do not receive treatment and a significant share with little or no need do—as problematic (Winterfield and Castro 2006).

While that study was confined to a sample of prisoners in Illinois and Ohio, it is likely that other state prison systems are experiencing similar problems with treatment delivery. Burdon and colleagues (2004) investigated outcomes for released prisoners in California who had participated in prison-based therapeutic community (TC) treatment programs. Previous research has shown the effectiveness of these specific programs in California, but focused solely on the prison-based portion of treatment; Burdon et al.'s work focuses on aftercare and participation in TCs after release from prison in California. This work is especially relevant for the current research because of its focus on programs in the state of California, its consideration of length of stay in a TC, which would affect mobility immediately after release from prison, and the effects of this type of treatment on recidivism. Their work considered 12 month return-to-custody rates among those who participated to some degree in a prison-based TC program. Among that population, the authors noted higher return-to-custody rates for those in suburban and rural areas compared to those in urban areas; this may hinge on the accessibility of aftercare programs, a function of such factors as prevalence of treatment programs, funding for such programs, and transportation. Higher return-to-custody rates were also found for African Americans compared to other racial and ethnic groups in the study; while Hispanics had lower rates of participation in aftercare programs, the authors speculate that they rely more heavily on family and social networks than on aftercare and are subsequently more successful after release from prison. This finding echoes Wong's (2002) conclusion on the benefits of dependent living for individuals with drug and alcohol problems. Not surprisingly, in Burdon et al.'s (2004) study, those who participated in some form of aftercare, after release from prison, had lower return-to-custody rates than those who did not during the first year after release. From these and other findings, the authors noted the importance of beginning substance abuse treatment while in prison and following it with an extended period of aftercare.

The above review of research on residential mobility patterns among the general population as well as those of returning prisoners indicates that we can expect that mobility levels will be relatively high for those reentering the community from prison, and particularly for those with histories of substance abuse. Such high levels of mobility after release, if confirmed by this research, are likely to compound the challenges of delivering critical drug treatment to populations that are in most need of such services. Thus, understanding the parolee population's mobility and drug use patterns is imperative to providing adequate drug treatment, housing, employment services, and other support to those in need.

Research Questions

Given the importance of understanding where released prisoners are returning and the impact of drug use behaviors on the success of parolees, we have posed a number of research questions. These questions begin by exploring the spatial patterns of all released prisoners, whether drug-using or not, followed by more specific questions related to former prisoners with histories of substance abuse. We therefore first seek to explore what the spatial distribution of parolees released to San Diego County, CA is, and how transient returning prisoners as a whole are. We then turn specifically to the subpopulation of prisoners identified as having substance abuse histories to determine whether and how spatial clusters and mobility patterns differ for this group and to explore the role of drug use in residential mobility. Finally, we investigate relationships in the data that could influence the level of mobility experienced by released prisoners. While the data preclude us from conducting any predictive modeling, they do allow us to suggest the appropriate design of predictive modeling for future research in this vein.

Data

The data used in this research were obtained from the San Diego District Attorney's Office and were collected from the State of California Department of Corrections and Rehabilitation. The data represent a snapshot of all individuals on parole in

San Diego County, California on August 30, 2004 ($N = 8,536$). The data include the date of release to parole, each reported address, and the effective date of each address. The parolees are split almost equally between those who moved at least once during their parole period ("movers") and those who did not move at all during their parole period ("stayers"), with 52% moving and 48% staying. The San Diego data also include demographic information, conditions of parole supervision, (e.g., drug testing requirements, curfews), and known problems with drug use, identified by the parolee's self-report and/or by information known to a corrections official.[4] The racial information collected by the County included several mutually exclusive categories including White, African American, and Hispanic.[5] Because the categories are mutually exclusive, we assume that all race categories other than Hispanic exclude anyone that is Hispanic. Of the parolees in the data set, 37.5% were White, 29.1% African American, and 27.8% Hispanic. Just under 12% were female.

This research identifies drug problems using three indicators of drug use based on the parole conditions and known problems fields: required registration as a narcotics offender; required anti-narcotics testing as a condition of parole; and the parolee's known problems with drug use (self-reported or as noted by officials). The majority of narcotics offenses requiring registration are "felony possession" of certain controlled substances not including marijuana,[6] and "felony possession with intent to sell" offenses[7] involving but not limited to methamphetamine, marijuana, and controlled narcotic substances such as cocaine and heroin. These offenders are required to register with the local police department or sheriff in their new residence within 30 days of moving. The data do not indicate whether a parolee actually did register with the local law enforcement agency, only his or her legal requirement to do so. Approximately 39% were required to register as drug offenders. Note that this figure is limited to felony drug offenders, as those are the only offenders required to register as narcotics offenders. According to BJS, in 2004 just over 21% of all state prisoners nationally were convicted of drug offenses (Harrison and Beck 2005); as evidenced by the number of parolees in San Diego required to register for felony narcotics offenses, the levels there are considerably higher than the national average.

Nearly 90% of these narcotics offenders were required to submit to drug testing as a condition of parole. Drug testing is a standard requirement for parolees upon release and is not necessarily based on prior known drug use. Furthermore, the data do not include the results of any drug tests that were performed, only the stipulation that drug testing was a requirement. Due to the large number of parolees subject to drug testing, the variable was not used further in this research to distinguish between drug using and non-drug using parolees.

The known drug problems data specify, where possible, the types of drugs used by the parolee. These data are based on self-report by the parolee or report by the parole officer regarding a parolee's known drug use problems. The field in the data set allowed officers to enter specific drugs a parolee was known to use or to simply confirm the existence of a drug problem with no information regarding specific drugs used. In most cases, specific drugs were identified and included in the dataset; in only 2.6% of the cases where drug use was noted was a specific drug not mentioned.

For the purposes of this analysis, we refer to those who have a reported problem with narcotics as "drug users." As so defined, drug users constituted 57% of parolees in the dataset, a considerably higher proportion than those required to register as drug offenders (39%). While our classification of parolees as drug users is not confined to any specific drug, as part of our analysis we conducted a more in-depth examination of the characteristics of those parolees with reported problems with specific drugs identified. Table 6.1 demonstrates that among all State prisoners nationwide, approximately 21.4% admitted to cocaine use during the month prior to the offense for which they were imprisoned, 40% used marijuana during the month prior to their offense, and nearly 11% used methamphetamine during the month prior to their offense, identifying these drugs as the most prevalent among State prisoners (Mumola and Karberg 2006). The same three drugs were also the most prevalent among the San Diego releasees, although in different proportions. Of all parolees in the San Diego data set, nearly one-fifth reported using cocaine (19.33%), nearly one-quarter reported using marijuana (24.63%), and more than one-third reported using methamphetamine (35.85%).

The 2004 BJS survey also found that cocaine use among State inmates had dropped since 1997 and that methamphetamine use had increased over the same period (Mumola and Karberg 2006); the data from San Diego seem to confirm this trend. The percentage of cocaine users in San Diego is similar to that among State prisoners nationwide, but the levels of marijuana use appear much lower in San Diego than the national average and the levels of methamphetamine use appear much higher. The low numbers for marijuana may be due to the fact that the variable used to measure drug use identifies known problems with drugs; the widespread use of marijuana and the often frequent attitude that it is not a serious drug may suggest that it is not always specified as a problem drug. The higher proportion of methamphetamine users in the San Diego data set reflects regional preferences for the drug: California ranked third among all states for methamphetamine treatment admissions in 2003 and western states have long had high methamphetamine use rates (Hunt, Kuck, and Truitt 2006).

Several caveats to the data should be noted. First, the data reflect only known addresses; it is possible that a parolee may not report all moves and that unreported moves are unknown to the parole officer. This creates a bias towards those parolees who are more likely to report moves. It is unknown, however, whether those more likely to report moves are more or less likely to either move or to use drugs. Second, even if parolees are reporting changes of address, there is no way of

Table 6.1 Prevalence of Specific Drugs and Drug Offenders as a Percent of State Prisoners Nationally and of San Diego Parolees

Variable	State Prisoners, 2004[8] (%)	San Diego Sample, 2004 (%)
Used cocaine	21.4	19.3
Used marijuana	40.0	24.6
Used methamphetamine	11.0	35.9
Convicted of drug offense[9]	21.0	39.0

determining whether they are actually residing there. The address could represent where the parolee receives mail, yet he or she may spend most nights at an intimate partner's house or at some other location entirely. Thus, there may be much more transience and mobility among this population than can be identified through these data.

Another important caveat is the fact that the data do not include information on recidivism. Therefore, a comparison of mobility and drug use behaviors of those who successfully complete parole and those who re-offend cannot be made. Also, it is unknown how completely the California Department of Corrections and Rehabilitation fills in the data for any given parolee; it is likely that some information may not be reliably updated for parolees.

Finally, not all addresses could be geocoded, or electronically matched to a map location. Nine percent of the addresses were not usable, meaning they were incomplete or missing, and 2.5% of the given addresses could not be geocoded. A small number of parolees (3.4%) provided addresses at some point during their parole supervision that indicated they were transient or homeless. Only 1% of parolees in the data set were found to have reported drug use problems and to have been identified at some point during their parole as transient. The low level of transience identified among parolees in San Diego is likely due to the data collection methods. Addresses available in the data set are those provided by parolees; if a parolee is transient or homeless, it is likely that he or she will not report that information to a parole officer. We did not conduct any additional analysis on the transient parolee population due to the small number of parolees reporting this information and the resulting likelihood of error in the data on this subject.

An additional consideration is that this research focuses on drug use among parolees, and those who have admitted drug use problems are more likely to seek drug treatment upon release from prison, whether being required to as a condition of parole or doing so voluntarily. These individuals may move less frequently immediately after release because of a stay in a residential treatment facility, halfway house, or group home, and moves may increase after 30 or 60 days on parole, when the individual leaves the facility. Unfortunately, explicit data on whether a parolee is released directly to a residential treatment program or any type of group home is not collected. The only way to determine whether a parolee stayed in such a facility is to examine the parolee's provided addresses and examine whether the parolee has been listed in "care of" a facility such as a halfway house, shelter, group home, transitional housing, or treatment center ("transitional facilities").

The data were examined and transitional facilities identified using the criterion that multiple parolees provided the facility name in the "care of" information. This resulted in the identification of 101 transitional facilities. Previous research has suggested that at least 10% of parolees spend time in homeless shelters within 2 years of their release (Metraux and Culhane 2004). Approximately 9% of the parolees in the San Diego data set listed one of the identified facilities as their first address upon release. However, approximately 19% stayed at one of the identified facilities at some point during their parole, whether immediately after release or during a subsequent period. This is higher than previous estimates, but in San Diego, the data on transitional facilities includes homeless shelters–the focus of other estimates–and other types of residential settings, so the higher numbers in San Diego are not surprising. A small number of parolees (3%) stayed at more than one of the facilities identified; some of those stays may represent moves from a treatment center to a halfway house or other group home. Interestingly, only 21% of the parolees who were reported to have a problem with drug use stayed at any of the transitional facilities identified, echoing Winterfield and Castro's (2005) and Burdon and colleagues' (2004) findings that a large number of inmates and parolees needing substance abuse treatment fail to receive it after release. This figure may be lower than the actual percent receiving treatment if not all parolees reported their stay at one of the facilities identified or because our method may not identify all of the transitional facilities used by parolees in the data set. The locations of these facilities are mapped and discussed below, and the implications of a stay at such a facility for parolee mobility are further explored.

Despite these limitations, the data remain extremely valuable; parolee data that track addresses beyond the first address given at parole are rarely available, and data that also include information on drug use habits are doubly rare. Important insights can thus be drawn from the existing data set. Our analysis of the data is divided into four sections: (1) a portrait of the average parolee, including characteristics of drug use and mobility; (2) an examination of the spatial distribution of parolees, with a special focus on those who use drugs; (3) an examination of the characteristics of places where parolees live; and (4) an examination of predictive modeling possibilities. The chapter ends with a discussion of the policy implications of the findings about drug use and residential mobility among parolees.

Average Parolee

To better characterize the demographics and mobility behaviors of parolees in the San Diego data set, average values for a number of different measures are provided in Table 6.2. The values are provided for five main groups of parolees: all parolees, movers, drug users, African Americans, and Hispanics. Only the African American and Hispanic groups are mutually exclusive with each other; those two groups and the other three groups demonstrate some level of overlap in their membership. Difference of means tests were conducted to compare values for the Hispanic group and the African American group where feasible, but such tests were not appropriate for other comparisons where parolees could be a member of two groups at the same time (e.g., a drug user and Hispanic or a mover and a drug user).

The median age for all parolees was 37 years. The average age shows a small amount of variation across all groups of parolees described in Table 6.2. Hispanics were the youngest group, with a median age of only 33, and African Americans were the oldest at 39. While not shown in the table, a t-test was performed to compare the average ages of African American and Hispanic parolees. The results revealed that Hispanic parolees were significantly younger than African American parolees.

Table 6.2 Average Parolee Demographic and Mobility Characteristics by Group

	All Parolees	Mover	Drug User	African American	Hispanic
Number of parolees	8,536	4,422	4,848	2,482	2,373
Percent of all parolees	100	51.8	56.8	29.1	27.8
Median age	37	38	36	39	33
Average time on parole (years)	1.9	2.1	1.9	2.0*	1.8*
Percent drug users	56.8	62.3	100	52.8**	55.5**
Percent using cocaine	19.3	21.6	34.0	33.4*	13.7*
Percent using marijuana	24.6	26.1	43.4	26.1	24.5
Percent using methamphetamine	35.9	39.1	63.1	15.5*	37.8*
Mobility rate (%)	51.8	100	56.8	56.7*	39.7*
Average places lived 1 year***	1.7	2.3	1.8	1.8*	1.4*
Med. time before 1st move (days)***	117.5	117.5	123	133	137
Med. distance moved, 1st move (mi.)***	4.3	4.3	4.5	3.9	4.2
Percent released to transitional facility as first residence	8.8	10.6	8.5	10.8*	6.1*
Percent residing at transitional facility at any time during parole	19.5	31.2	21.2	22.2*	13.4*

*Hispanic and African American averages significantly different at the 0.01 level
**Hispanic and African American averages significantly different at the 0.1 level
***Individuals who listed a transitional facility as their first address are excluded from these figures

In addition, movers were slightly older than the average parolee population while drug users were slightly younger. Higher mobility levels among older parolees may be the result of a longer period spent in prison, increasing the stress on social ties with family and friends outside of prison. As La Vigne et al. (2005) found, older released prisoners reported significantly lower levels of tangible family support (food, clothing, housing, and money) compared to their younger counterparts. Therefore, older parolees may not have the same kind of support when they are released from prison as younger parolees, increasing the likelihood that they will have to move more frequently.

We also considered the average time on parole: at the time the sample was taken, all parolees had spent an average of 2 years on parole. While the average time on parole varied little across the five groups, the difference between African Americans (2 years) and Hispanics (1.8 years) was statistically significant, indicating either that African Americans have longer sentences on average or that Hispanics are generally less successful on parole and are thus returned to prison earlier than their African American counterparts. Unfortunately we do not have data on either sentence length or on recidivism, so we can neither confirm nor reject either of these possibilities.

Next, Table 6.2 provides information on drug use behaviors among the five groups of parolees. Nearly 57% of all parolees had a known problem with drug use. That number was slightly higher among movers at 62.3%. Finally, a significantly smaller percentage of African Americans had known drug problems (52.8%) than did Hispanics (55.5%). Recalling the top three drugs among all parolees–cocaine, marijuana, and methamphetamine–Table 6.2 provides information on the prevalence of their use among the five groups of parolees. Methamphetamine has the highest percentages of users for all groups except African Americans, for whom methamphetamine use is the least common of the three drugs. Only 15.5% of African Americans noted a problem with that drug. While not shown in the table, the percent of White parolees in our sample with a known problem with methamphetamine use is just over 50%—the highest of all racial groups in the dataset. The average age of methamphetamine users in the data set is 36, the same as for all drug users. Previous research has shown methamphetamine users to be largely White and in their 20s and 30s; evidence from multiple data sets also shows low prevalence of methamphetamine use among African Americans (Hunt et al. 2006). The San Diego data follow these trends. The use of cocaine follows the opposite pattern; it is the least common among all groups except African Americans, for whom it is the most common drug. Just over 33% of African Americans had a known problem with cocaine. The differences between African Americans and Hispanics in terms of methamphetamine and cocaine use are statistically significant. The differences in prevalence of marijuana use, however, are small and not statistically significant. Again, it is likely that marijuana use among this population of parolees is underreported due to data collection methods discussed earlier.

Finally, Table 6.2 outlines the mobility behaviors of the five parolee groups. The mobility rates, or percent that moved, show wide variation across the three subgroups of parolees. While just over half of all parolees moved at some point during their parole period, that number is higher for both drug users and African Americans, with just under 57% of each of those groups moving. Hispanics have a drastically lower mobility rate, however, with only 39.7% moving at some point during the parole period. This finding is not surprising as previous research has shown that the Hispanic population in general tends to have lower mobility levels than other ethnic or racial groups (Clark and Ledwith 2006).

Three measures describe the timing and distance of mobility among the parolee groups: the average number of places lived in the first year after release, the median time in days before the first move, and the median distance moved for the first move. Parolees who listed a transitional facility as their first address ($N = 748$) are excluded from all three of these measures.[10] Not surprisingly, the average number of places lived in the first year on parole is highest among movers. The averages for drug users and African Americans were similar to that for all parolees, but the averages were much lower for Hispanics, at only 1.4 places lived during the first year. The median time before the first move is less than 5 months for each group and shows little variation across groups, with Hispanics waiting the longest before their first move at 137 days.

While the median distance moved also varied little across the groups, the distances were much higher than expected. Even African Americans, who moved the shortest median distance, moved nearly four miles, and drug users moved four and a half miles. These lengths indicate that parolees are not simply moving around within a familiar, small neighborhood but that they are making significant moves to new neighborhoods. Part of the distance moved is a function of the nature of San Diego–the city is less dense and more car-reliant than other major cities with similar population sizes, like Chicago. The distance moved may also represent parolees' attempt to improve their situations by moving to less disadvantaged areas, or they may be moving to other similar areas as the result of wearing out their welcome in their first residence. While considering motivation behind the moves is beyond the scope of this study, the types of places between which parolees move are considered below.

Table 6.2 also provides statistics on stays at transitional facilities among parolees, revealing that 10.8% of African Americans were released to a transitional facility while only 6.1% of Hispanics were, a statistically significant difference. This finding echoes Burdon et al.'s (2004) finding that Hispanics rely more heavily on familial and social networks than on treatment centers after release from prison. An interesting pattern that emerges in the data on transitional facilities is the greater likelihood that movers will stay at a transitional facility than drug users. This pattern is likely explained by the inclusion of multiple types of transitional facilities, including treatment centers, group homes, and shelters, in this measure. Because of the method we used to identify transitional facilities and parolees who stayed at such facilities, isolating drug treatment facilities from other residential facilities was not feasible. Parolees who stay at such a facility, then, are more likely to move, and these figures are not necessarily a good comparison of what proportion of different parolee groups received drug or alcohol treatment. Finally, Hispanics were the least likely to stay a transitional facility at any point during parole, while movers had the highest prevalence of all groups at 31.2%.

Parolees' Mobility Behavior

This section provides a more in-depth look at the mobility behaviors of drug users. Figure 6.1 displays the average number of places lived up to 1 year after release on parole for both drug users and non-drug users. Excluded from the drug use and non-drug use categories are parolees released to a transitional facility; those parolees are displayed separately. Drug users display a higher level of mobility with more places lived at each time period than non-drug users, although both groups have a similar rate of increase for the first 6 months. After 6 months, drug users have an increasing rate of mobility. Using t-tests, we determined that for every time period, the difference in average places lived is significant. This finding is not surprising as drug users generally have more difficulty maintaining employment and tend to

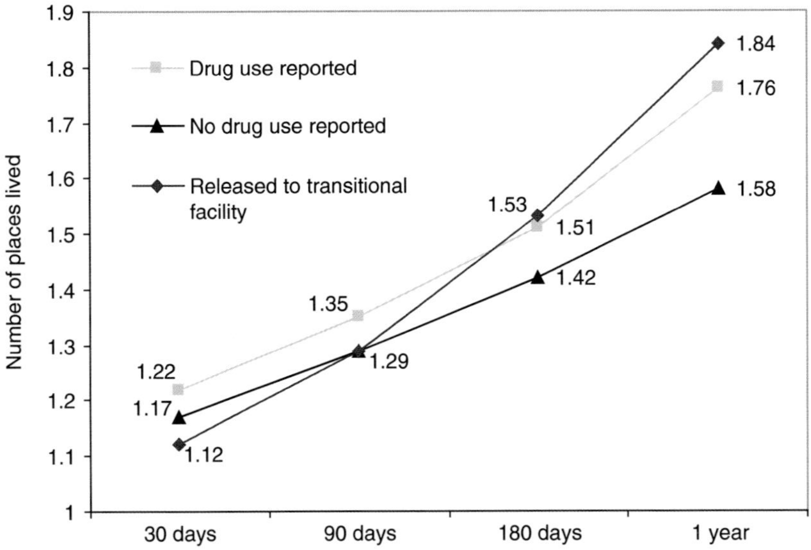

Fig. 6.1 Average number of places lived 30 days to 1 year after parole by reported drug use and release to transitional facilities (See also Plate 3 in the Colour Plate Section)

have weaker support structures, making maintaining a stable residence more difficult. They both choose and are forced to move more frequently than parolees who do not have problems with drugs. Also not surprising is that, while not shown in Fig. 6.1, drug users have a significantly higher total number of moves than non-drug users, confirming that drug users are the more mobile group over the entire parole period.

The role of stays at transitional facilities in the mobility of parolees was raised earlier and is pertinent to this finding. If a parolee spends a considerable amount of time in a transitional facility, be it a treatment center or halfway house, does that parolee have fewer moves immediately after release but have more moves after leaving the transitional facility? To investigate this possibility, we compared the average places lived during the first year on parole for those parolees who stayed at a transitional facility to the drug user and non-drug user groups. Figure 6.1 reveals that after 30 days, parolees released to a transitional facility have the lowest average of places lived. By 3 months out, however, those parolees have the same average places lived as non-drug users, and by the sixth month have surpassed the places lived of drug users. By 1 year out, those who were released to a transitional facility had lived in a significantly higher number of places than those who did not, regardless of drug use behavior. While low immediately after release, the mobility rate increases greatly within the 1-year period for this group of parolees.

The findings suggest that those released to a treatment facility or transitional housing have more difficulty in locating a stable residence while on parole than those who did not enter treatment and that housing services specifically aimed at those exiting transitional facilities may aid in their success on parole. The population released to such facilities faces more challenges than parolees who have family, friends, or even an independent but stable living situation upon release. Those released to treatment facilities face drug or alcohol addictions, and the struggle to stay sober compounds the already difficult situation facing parolees at the most basic levels, such as finding employers willing to hire ex-convicts and locating housing that they can afford and are eligible to rent. Those in our sample released to group homes or shelters also face greater difficulties; rampant drug use is common at many shelters and some research has shown that parolees who stay at homeless shelters are more likely to recidivate (Metraux and Culhane 2004). This group of parolees, then, faces significant challenges above and beyond the average parolee and addressing those challenges for this small, at-risk group released to transitional facilities may help to reduce recidivism significantly.

We also considered the median time between moves for drug users and non-drug users, again assessing those who were released to a transitional facility separately. The difference in time between moves for all three groups was very small and the median time before the first move for all groups was less than 6 months.[11] As expected, the median time before the first move was greatest for those released to a transitional facility, at 171 days. After the first move, however, that group had the shortest periods between moves, or the most frequent moves, reiterating the finding above that stays in a transitional facility are associated with higher levels of mobility.

The median time between moves drops dramatically after the first move for both drug users and non-drug users, but while the time between moves continues to get shorter with each subsequent move, the decrease in length is not as dramatic as between the first move and the second. One explanation for this pattern of a long first stay and shorter subsequent stays at each residence is that a parolee may be able to find a residence immediately following release on parole, often staying with friends or family. If parolees wear out their welcome and are eventually forced to move from that residence, they may have trouble finding suitable alternative housing, leading to more frequent moves between different friends, family members, or low cost housing facilities, with ever-shorter stays at each residence.

Finally, the median distance of each move for the three groups was considered, but did not reveal a discernible pattern. The first move is longer for drug users (4.52 miles) than for non-drug users (3.99 miles); this remains true for each subsequent move. The differences in length for subsequent moves, however, are very small for both groups. For drug users, the median lengths between moves hover around 4.5 miles while for non-drug users, the median lengths are steady around 4 miles. We expected those leaving a transitional facility to move longer on average for their first move in order to return to the neighborhood where family or friends live. In other words, parolees will first go where the services are being offered regardless of location, then move to their preferred area of residence. Parolees in the dataset, however, did not follow this pattern. Instead, those released to a transitional facility had a shorter median length between moves for the first move (4.3 miles) than drug users, and for subsequent moves the variation in lengths followed no distinct pattern.

While length between moves has implications for service provision and surveillance of parolees that is discussed below, there is no obvious relationship between frequent or numerous moves and length of move.

Spatial Distribution of Parolees

Figure 6.2 displays the distribution of all parolees released to San Diego County, California based on the first address provided by each parolee.[12] This map is provided as a point of comparison for the spatial distribution of drug offenders and drug users, discussed below. The map shows the rate of parolees per thousand persons by census tract based on the first address after release, and displays only the central-southwestern part of San Diego County, the area where the number of parolees is highest. San Diego is a very large county, a high percentage of which is rural and parolee numbers are low in those areas. The area with the highest rate of parolee residences is in downtown San Diego, on the western edge of the county along the water, across San Diego bay from Coronado. Other clusters of returning parolees also appear east of downtown in the Spring Valley area, northeast of downtown in the El Cajon area, and south towards Mexico. This pattern of parolee clustering is not surprising as the clusters are in some of the areas of highest population in the county and in areas where essential services for parolees, such as drug treatment services or employment opportunities, are likely to be located.

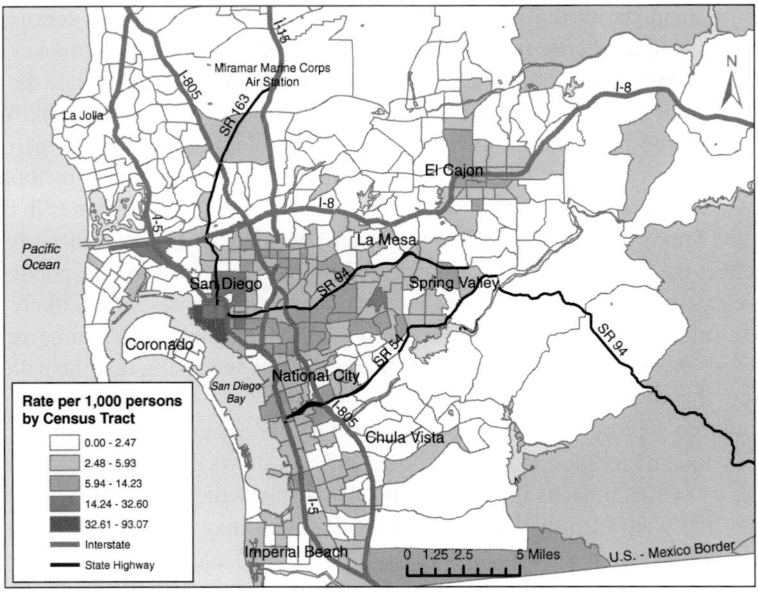

Fig. 6.2 First address by census tract for all parolees, San Diego County (See also Plate 4 in the Colour Plate Section)

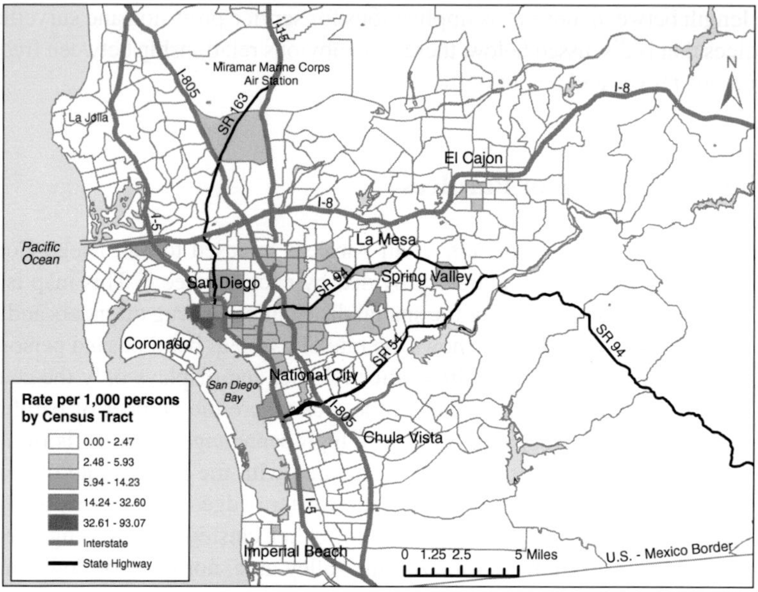

Fig. 6.3 First address after parole by census tract for drug offenders, San Diego County (See also Plate 5 in the Colour Plate Section)

Figure 6.3 displays the spatial distribution of drug offender parolees in the county, as determined using the "required to register as narcotics offender" variable. Again, the map shows the rate of parolees per thousand persons by census tract based on first address after release. The data are displayed using the same key as in Fig. 6.2 for all parolees to facilitate comparison. The data used to calculate the rates by census tract included those individuals released to transitional facilities. While the rates of drug parolees are lower than for all parolees, the pattern of narcotics registrants is very similar to that of all parolees. The highest rates are found in the downtown area of San Diego, with other pockets of high parolee rates in the El Cajon and Spring Valley areas and south towards Mexico. Again, this cluster pattern may reflect the location of services accessed by parolees; for this sub-population, drug treatment facilities, transitional housing, and group homes would likely have the most influence on the spatial pattern. In addition, because such a large percent of parolees in San Diego were drug offenders, it is not surprising that the pattern is similar to that of all parolees.

In order to explore whether parolee clusters are indeed correlated with the location of treatment facilities and group homes, Fig. 6.4 shows the rate of drug offender parolees by census tract, as in Fig. 6.3, and also displays the location of transitional facilities that were identified in the data set. Parolees may have visited additional facilities not shown in the current study if the facilities' addresses were not identified by name in the "care of" information or if only one parolee listed the facility. Figure 6.4 shows a smaller part of San Diego County, focusing on the areas where the largest number of facilities was identified. A large cluster of transitional facilities

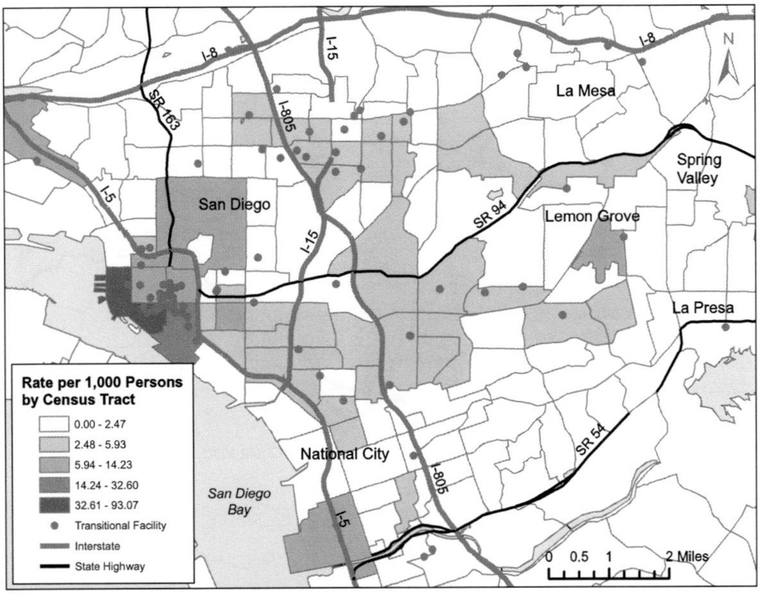

Fig. 6.4 Transitional facilities and first address after parole by census tract for drug offenders, San Diego County (See also Plate 6 in the Colour Plate Section)

is located in the downtown San Diego area, the area with the highest rate of parolee residences. This pattern confirms the expectations discussed above that the location of transitional facilities would be correlated with high rates of parolee residences. The similarity between the two patterns is likely circular: parolees go to areas where transitional facilities are located, and new transitional facilities locate where the demand is greatest, or where parolees are more prevalent. Some notable clusters of returning parolees, however, have few, if any, transitional facilities nearby. This is especially true in the Lemon Grove area and south of National City. Adding services in these areas may improve the chances of success for parolees returning to those areas. Until additional facilities locate in these areas, parolees' chances for success can be improved with efforts to connect parolees with appropriate treatment facilities elsewhere and ensuring the availability of transportation, whether public or otherwise, to essential services.

Figures 6.5–6.7 display the spatial patterns of parolees' first addresses based on the known drug problems indicator, and are broken down by the three main drugs of interest: cocaine, marijuana, and methamphetamine. The maps of parolee drug use are density maps, representing the concentration of parolees per unit of area instead of rates per thousand persons. Density maps were chosen for this portion of the analysis because the number of parolees using specific drugs was relatively small and calculating population-based rates would have resulted in very low rates with little variation over space. The maps of parolees by drug use echo the pattern for all drug offenders' first addresses; parolees using all three types of drugs cluster in

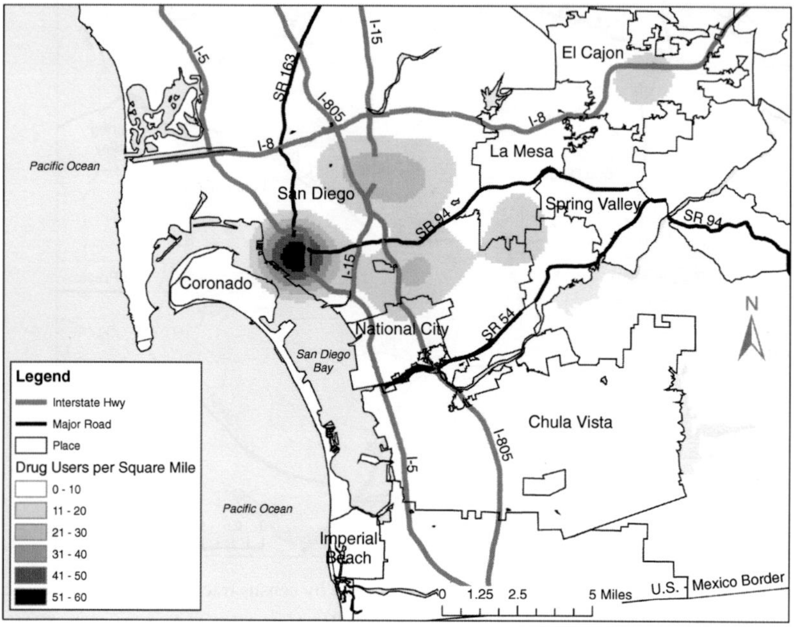

Fig. 6.5 First address after parole, cocaine users, San Diego County (See also Plate 7 in the Colour Plate Section)

downtown San Diego and east from the coast to El Cajon. Figure 6.5 reveals that the density of cocaine users is the most concentrated of the three drugs, with the highest densities in the downtown San Diego area. Figure 6.6 shows that marijuana users are also densest in the downtown San Diego area, but additional high concentrations are found further south toward the Mexican border. The concentration of cocaine users may reflect the availability of the drug in those areas. The density of marijuana users is assumed to be underestimated in this data set, and is likely more widespread across the area than Fig. 6.6 suggests.

Figure 6.7 shows the same downtown San Diego concentrations as the cocaine and marijuana densities but illustrates a stronger and broader concentration of methamphetamine users south toward Mexico than was found for cocaine and marijuana. Mexico is a large supplier of methamphetamines to U.S. markets (Hunt et al. 2006), so it is not surprising that Fig. 6.7 shows a greater density of methamphetamine users along major transportation networks leading to Mexico. While domestic production of methamphetamine includes small local labs that are mainly found in rural areas, most of the domestic production comes from so-called "superlabs" that can produce at least 10 pounds of methamphetamine. The only domestic superlabs are located in California, with at least one located in San Diego County (Hunt et al. 2006). The influence of such superlabs on the distribution of methamphetamine users, however, is unknown and the current dataset precludes further investigation.

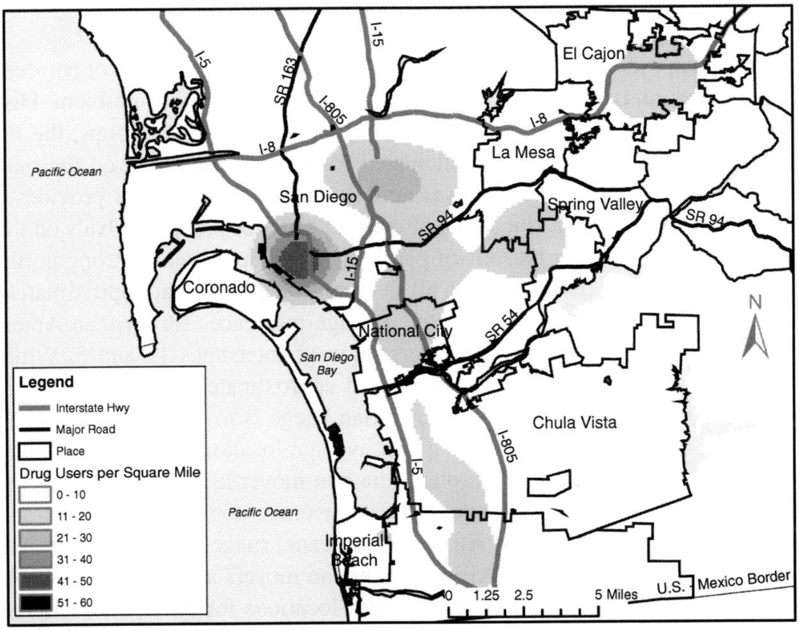

Fig. 6.6 First address after parole, marijuana users, San Diego County (See also Plate 8 in the Colour Plate Section)

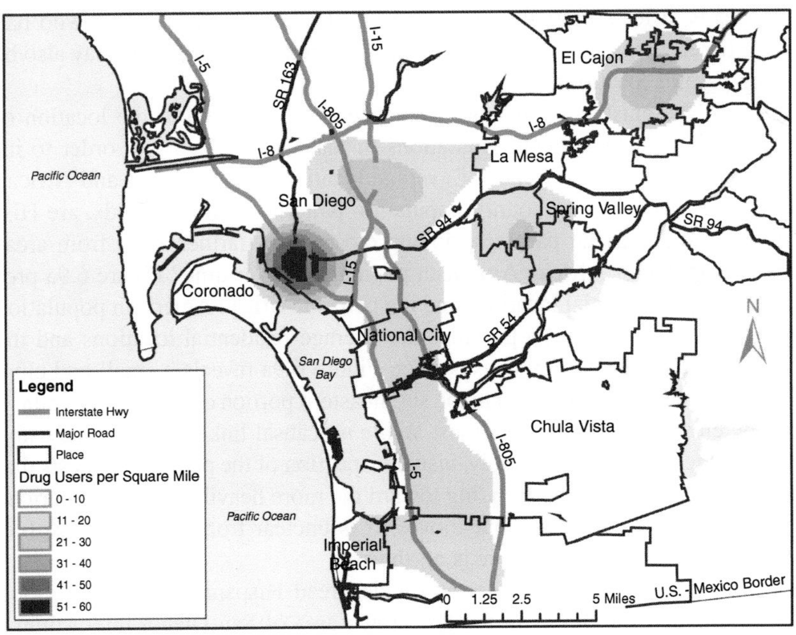

Fig. 6.7 First address after parole, methamphetamine users, San Diego County (See also Plate 9 in the Colour Plate Section)

Average Parolee's Mobility Experience

The following maps display the average locations for four different types of parolees introduced above: all parolees, parolees with some known drug use problem, Hispanic parolees, and African American parolees. For each of these groups, the location of the average residence was calculated for stayers (identified on the map as " St") and for the first five residences among movers.[13] Figure 6.8 provides a large-scale map of the average locations of the four types of parolees. Ovals on the map identify each of the four clusters of parolees. The map reveals a strong north-south pattern of average residences for all five parolee groups that approximately follows the route of Interstate 15, with the average residences for African Americans located the farthest south at the intersection of Interstates 15 and 8. While not shown in Fig. 6.8, these clusters are located approximately 7 miles northeast of downtown San Diego. Because downtown San Diego is in the southern portion of San Diego County, it is not surprising that average locations were pulled more northward than southward. For all parolees, the non-mover location is close to the cluster of mover locations; the distance between average non-mover location and first mover location is greatest for African Americans, suggesting that there may be significant differences between African American movers and their more stable counterparts in terms of location. The first three locations for African Americans and Hispanics are slightly more spread out than for drug users and all parolees, but are nonetheless separated by less than a mile. An interesting pattern to note is the movement of African Americans' fifth average residence northward, while the first four trend southward. While the distance of the fifth address from the initial cluster of residences is at least partially due to the smaller number of parolees who had moved five times, the trends indicate that some characteristics of places may also be influencing parolee locations.

In Fig. 6.9, the clusters of average residences are compared to the location of Hispanic and African American populations in San Diego County in order to investigate whether the spatial trends of average locations for Hispanics and African Americans are influenced by county population patterns. In other words, are Hispanic or African American parolees moving closer to or farther away from areas with larger Hispanic or African American populations over time? Figure 6.9a provides the average residential locations and the percent African American population by census tract while Fig. 6.9b provides the average residential locations and the percent Hispanic population by census tract. Figure 6.9a reveals a small and clustered African American population in the southeastern portion of San Diego County, south of the cluster of average locations. While no causal links can be made using the data presented in the current study, visual inspection of the patterns suggests that African American parolees are trending toward the more heavily African American areas of the county for the first four moves. It is unclear from the data available, however, why the fifth average move is northward.

Figure 6.9b reveals a much larger and widespread Hispanic population in San Diego, with the densest cluster in the downtown area of San Diego. Interestingly, the average locations of Hispanics tend to be trending northward and away from

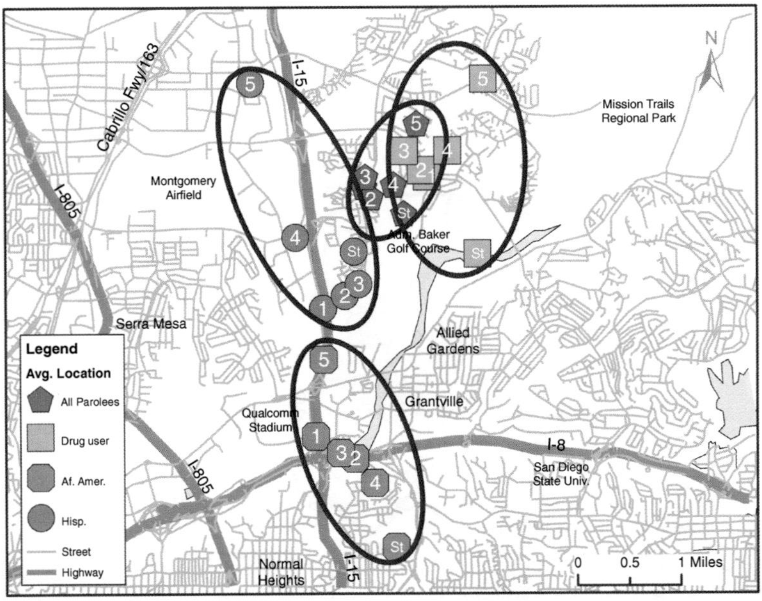

Fig. 6.8 Average residential mobility experiences of four types of parolees: All parolees, parolees with known drug use problems, and African American and Hispanic parolees (See also Plate 10 in the Colour Plate Section)

that dense cluster of Hispanic population. This is counter-intuitive as past research has indicated the reliance of Hispanic parolees on family and social networks. The expectation, then, is for Hispanic parolees to trend spatially toward areas with high concentrations of Hispanic persons. Because the data do not provide the level of detail needed to assess the motivation behind these moves, we can only provide speculation as to why this spatial pattern exists. It may be that among Hispanic parolees, those who lack strong family and social networks are movers and those who have the family and social support they require upon release from prison are stayers. Hispanic parolees that move may have different characteristics than those who do not move, and thus may have behavioral patterns not yet outlined in research on ethnicity and mobility that can affect their success on parole and the supportive services required upon release from prison.

Parolees and Neighborhood Characteristics

We assessed the characteristics of places that parolees chose with five standard measures collected at the census tract level: percent unemployed, percent below the poverty level, percent female-headed households, percent rental housing units, and percent vacant housing units. For each of the five parolee groups introduced above, we considered the average values for each place characteristic for the first and last residence in order to investigate how the characteristics of places may

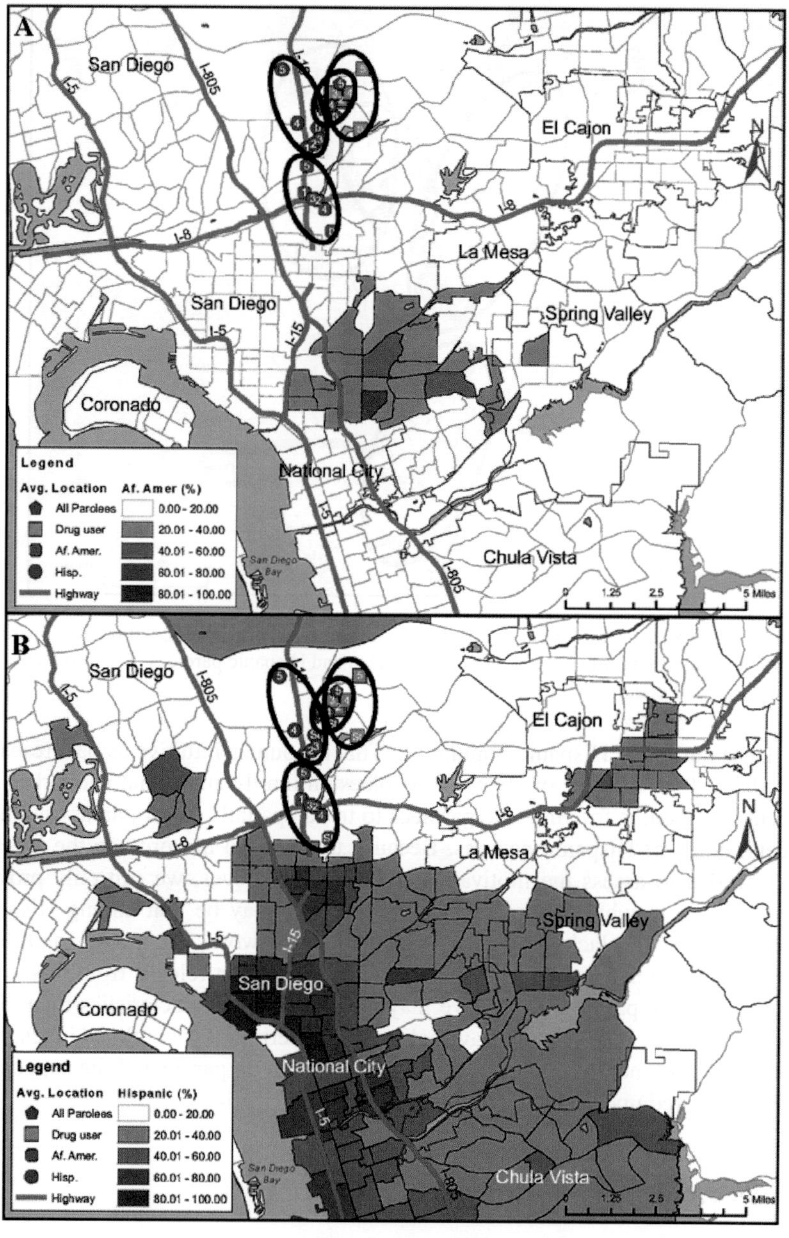

Fig. 6.9 Average residential mobility experiences of four types of parolees and percent (**a**) African American and (**b**) Hispanic by census tract (See also Plate 11 in the Colour Plate Section)

change over time. Parolees who were released to a transitional facility were excluded from the measures for the first residence under the assumption that neighborhood characteristics play a very small role in the choice of transitional facility. We also examined the characteristics of stayers' residences separately. Table 6.3 provides the average values of the five measures for San Diego County (for comparison purposes), stayers, and movers' first and last residences. Across all parolees—both stayers and movers—the places they resided were much worse-off than the average census tract in San Diego, with higher rates of unemployment, poverty, and female-headed households, and a greater proportion of rental and vacant housing units. The differences in place characteristics between residences of stayers versus movers, however, are less distinct. Stayers did not clearly choose less disadvantaged places than movers; in fact, the differences between stayers' and movers' first residences were very small on each measure, with stayers actually residing in areas with slightly higher poverty levels and greater proportions of female-headed households. Movers' last residences, however, were clearly more disadvantaged than movers' first residences, and while small, the differences between the two places were statistically significant for all measures except vacant housing units. This indicates that parolees are not improving their surroundings when they move, and suggests that moving can have serious impacts on their mental and physical well-being and affect their chances for success while on parole.

While not shown in Table 6.3, the differences between first and last residences for other groups of parolees described in above—drug users, African Americans, and Hispanics—followed similar patterns as for all movers. The differences were very small and in every case the last residence was in a more disadvantaged area than the first residence. Few of the differences were statistically significant, however, indicating no net change in disadvantage from first residence to last residence, and reiterating previous findings on Chicago parolees' neighborhood characteristics (La Vigne and Parthasarathy 2005). While we do not have any information on parolees' motivations to move, these findings do provide us with enough insight to speculate on the reason behind the moves. Because parolees are moving to worse-off places, it is less likely that they are moving for positive reasons, or reasons that will aid in their success on parole, such as moving to gain personal

Table 6.3 Comparison of San Diego County Characteristics with Stayers and Movers Residences

	Unemployed	Below Poverty Level	Female Headed Households	Rental Units	Vacant Units
San Diego County	3.50	11.00	6.79	41.77	6.52
Stayers	4.85	18.63	15.35	55.59	4.02
All movers, first residence	4.84	18.22*	13.81*	57.54*	4.47*
All movers, last residence	5.02**	19.04**	14.11**	59.88**	4.52

Values are expressed as percentages
*Statistically significant difference from stayers at $p<0.05$ level
**Statistically significant difference from first residence at $p<0.05$ level

independence, financially or otherwise. Instead, it is likely that parolees are moving for negative reasons, such as wearing out their welcome at family or friends' houses. In this case, parolees have limited choices on the timing and location of their moves and their next residences, restricting their ability to improve their situations.

Modeling Parolee Mobility

In order to further investigate the relationships between demographic and mobility characteristics of parolees, we explored the potential for modeling the number of moves parolees made. Based on the literature, we expected to find significant and strong relationships between demographic measures, drug use behaviors, and characteristics of moves that might increase or reduce the number of future moves. Specifically, Hispanic parolees were expected to have significantly lower numbers of places lived, while African American parolees were expected to have higher numbers of places lived. We also expected drug use—as indicated by the general drug use measure and variables identifying of specific drugs used—to be associated with a higher number of places lived. Stays at transitional facilities were expected to increase the number of places lived, along with certain move characteristics, including higher levels of disadvantage in the first neighborhood and greater distances moved. Before conducting any predictive modeling, correlations between these key variables and the two places lived measures (places lived at 1 year out and total places lived) were examined.

Table 6.4 provides all significant correlations for the two places lived measures. The same set of variables was significantly correlated with each places lived measure. Surprisingly, we found small correlation sizes across key measures in the data. While a number of measures that were expected to be correlated with the places lived measures were indeed significant, the sizes of the correlations were extremely small, indicating that the relationships were not as strong as expected. Table 6.4 reveals that only one measure, having stayed at a transitional facility at some point during the parole period, was correlated with a places lived measure above 0.25. With a correlation of -0.12, being Hispanic was weakly associated with fewer places lived. All other variables had correlations with the places lived measures that were smaller than ± 0.10.

The correlation coefficients suggested that any predictive modeling of places lived would not result in very high levels of explained variance, nor would it yield any models that provided great insight into the influences on mobility behaviors of parolees. Any model that was produced from these data would likely suffer from model misspecification, as the strongest predictors of places lived are likely not included in the present dataset. One such measure not available in our dataset but likely providing additional predictive power is information on the motivation behind moves. Knowing whether moves were made for positive reasons (gaining independence) or negative reasons (wearing out one's welcome) might allow a more

Table 6.4 Significant Correlations Between Parolee Measures and Places Lived Measures

	Significant Correlation Coefficients*	
	Places Lived, 1 year	Total Places Lived
Age	0.07	0.08
Hispanic	−0.12	−0.12
African American	0.04	0.05
White	0.09	0.09
Drug user	0.08	0.12
Used methamphetamine	0.05	0.07
Used cocaine	0.04	0.08
Used marijuana	0.03	0.05
Stayed at transitional facility	0.34	0.47
Female headed households, first neighborhood (%)	−0.06	−0.08
Rental units, first neighborhood (%)	0.03	0.04
Vacant units, first neighborhood (%)	0.03	0.05
Unemployed, last neighborhood (%)	0.03	0.02
Below poverty level, last neighborhood (%)	0.03	0.02
Female headed households, last neighborhood (%)	−0.07	−0.08
Rental units, last neighborhood (%)	0.08	0.07
Vacant units, last neighborhood (%)	0.04	0.06

*All correlations significant at $p < 0.05$ level

powerful prediction of future mobility. Another missing but relevant measure is more specific information on transitional facilities. Separating transitional facilities, into specific types, such as residential drug treatment, homeless shelters, or group homes, and including more details on the requirements of residence at such facilities would also shed more light on future mobility.

The findings also indicate that a path analysis that first predicts places lived and then predicts recidivism would likely be appropriate. Ultimately, our interest in parolee mobility stems from the search for an understanding of factors that influence each parolee's chances for success while on parole, defined at its most basic level as completing a parole period without recidivating. Thus, an appropriate model would consider what factors influence mobility behaviors and in turn how mobility affects recidivism or success on parole. Because the data lack detail on recidivism or re-arrest events, such a model cannot be specified. Future work in this vein, however, should consider the explanatory benefits that such a path model could provide.

Discussion

Our examination of drug use and mobility behaviors of parolees in San Diego revealed some interesting patterns and raised important issues for ensuring parolee success after release. First, the comparison between movers, drug users, African Americans, and Hispanics on several key measures highlighted distinct differences between the different types of parolees, especially in terms of drug use and mobility behavior. The findings revealed high rates of drug use across all parolees, with over 55% of all parolees classified as drug users. While overall, methamphetamine use among San Diego parolees was found to be much higher than among State prisoners nationwide, echoing regional trends in the use of that drug, African American parolees were much more likely to be cocaine users. All groups also displayed a high level of mobility, although Hispanic parolees were found to have dramatically lower mobility rates than the other parolee groups. In addition, Hispanics had the lowest rates of stays at transitional facilities, while African Americans had the highest rates. These findings confirm previous research that Hispanics in general have low mobility rates and tend to have a greater level of family and social network support than do other racial or ethnic groups (Burdon et al. 2004; Clark and Ledwith 2005). This is especially relevant for releasees who are likely to be more successful on parole with this type of support.

Our exploration of spatial patterns of parolees' first addresses revealed that parolees cluster in specific areas of the county after release. The patterns were similar for all parolees, drug offenders only, and drug users by specific types of drugs used—cocaine, marijuana, and methamphetamine. The map for cocaine users was the most concentrated, with the largest cluster centered on downtown San Diego. This may be due to the availability of the drug in the downtown area. When we considered the location of transitional facilities, including treatment centers, halfway houses, and group homes, we found that, not surprisingly, the clusters of parolees with known drug use problems corresponded closely to the locations of those transitional facilities. However, there were some areas with large numbers of returning parolees but few transitional facilities. These areas would likely benefit from the provision of additional drug treatment services for parolees.

More than half of the parolees moved at some point during their parole supervision; this is quite a different finding from LaVigne and Parthasarathy's (2005) study of released prisoners in Illinois, where a large majority of releasees did not move. Based solely on that finding, we consider mobility among San Diego parolees to be relatively high. Taking a more nuanced approach to the concept of mobility, however, we expected that parolees who moved would exhibit high rates of subsequent mobility due to instability in terms of jobs, family relationships, and housing, especially immediately after parole. However, we found that while a large proportion did move at some point while on parole, few moved very frequently. In fact, more than 90% of all parolees in the data set moved less than three times over their entire parole period, a finding that holds even after excluding those released to a transitional facility from the analysis. Parolees that moved, though, did tend to move early on in their parole period, with the average move occurring within the first 6 months on parole. Drug users and those who spent some time in a transitional facility experienced higher levels of mobility, moving more frequently within the first year and experiencing the highest mobility rates over the entire parole period. The higher mobility for these two subgroups of parolees was expected based on prior research (Burdon et al. 2004; Metraux and Culhane 2004).

We also found that moves occur more frequently with time for both drug users and non-drug users. This phenomenon may result if parolees initially stay with friends or family upon release. Once they wear out their welcome at that residence, they are forced to move and maintain a residence on their own, which may prove difficult whether or not the parolee is using drugs or has unstable employment. Another scenario is that parolees may accomplish the initial tasks of finding housing and employment upon release but fail to make rental payments and meet the ongoing demands and expectations of a job over time due to poor life skills, drug addiction, health issues, and other challenges that often accompany prisoner reentry. This inability to sustain positive transition goals over time may lead to greater residential mobility.

One surprising finding was that moves are relatively far—on average over four miles for the first move—indicating that parolees are not only changing residences, they are also changing their neighborhoods. Moving farther distances may make job stability, maintaining service delivery, and supervising parolees more difficult. On the other hand, parolees may be moving in order to be nearer to employment opportunities, supportive family or friends, or needed services; because we do not know the motivation behind the moves we cannot classify moves as positive or negative. In addition, though parolees may be motivated to improve their housing situation and have the desire to move to a better neighborhood, they may lack the knowledge or skills to do so—they may not know how or where to look for better housing, safer neighborhoods, or areas with more employment opportunities. This can result in moves to areas that are similar or worse-off.

What we do know, however, is that regardless of the motivation behind any moves, parolees are living in areas that are far more disadvantaged than the average neighborhood in San Diego County. This more detailed look at the types of places in which parolees reside highlights the fact that they are not living in areas that provide many chances for success; instead, they are living in disadvantaged areas with fewer employment opportunities and higher poverty. These places are likely to contribute to psychological distress, depression, and compromised physical health, and can make staying drug-free and complying with all parole conditions very difficult. Our findings also indicate that when parolees do move, it is to similar or worse-off areas. Prior research has shown that, especially for African American parolees, the type of places in which parolees reside can have negative consequences for them, increasing their risk of recidivism and reconviction (Reisig et al. 2007).

The research thus underscores the need for housing assistance for parolees that begins before release, so that stable housing can be located in an area where parolees can succeed. With assistance, those who have a desire to move to a better neighborhood are much more successful than those with the desire but lacking the skills and knowledge required to conduct a broad housing search (Bembry and Norris 2005). In addition, regardless of the types of places in which parolees reside or the motivation for any moves that take place, service providers should work with parolees to ensure service continuity after a move. This may include locating feasible transportation to employment locations and service providers, and connecting the parolee with new service providers in the destination. Here, it is clear that service providers can be more effective by working together to provide a network of comprehensive services. Local policies should reflect this need for comprehensive service provision across the County and could include requirements that service providers maintain a common database on clients, making it easier to determine what services each client requires.[14]

Such relatively far moves may also have implications for parole supervision. Some researchers have recommended that assigning parolees to parole officers geographically would lead to more efficient and effective caseload management (Karuppannan 2005; O'Connell and Fleury 1999). The evidence from San Diego shows that this might not be feasible, however. If parolees are moving more than four miles to new homes after release, parolees would likely be required to change parole officers as frequently as they change residences. Under the current method of supervision—where parolees are not assigned geographically—maintaining steady contact with parolees and ensuring their compliance with parole conditions is difficult at best; changing parole officers when moves are long and parolees move out of an officer's geographic coverage area would complicate the supervision task even further and could be a counter-productive method of supervision.

While this work has offered some important insights on parolee mobility, several additional data elements would improve our findings and our ability to interpret our findings. The most important data we could add to this research are prior convictions and prison terms as well as information on parolees specific offenses. Additional knowledge of prior offenses would allow us to compare the mobility patterns of repeat offenders with first time offenders. If such an analysis revealed that repeat offenders were found to be more mobile than their first-time offending counterparts, suggesting that mobility is predictive of recidivism, service providers could adapt their services to those who pose the greatest risk to community safety. Knowledge on recidivism and specific offenses would also allow us to determine whether length of time on parole affects mobility; i.e., does mobility increase after 2 or 3 years on parole, or do parolees have an initial period of adjustment characterized by higher mobility followed by greater residential stability in the long run? Finally, as suggested by Meredith and colleagues (2003), better records of releasees' residences would assist parole officers and corrections officials in serving and surveilling the releasee population. Understanding when, where, and how far parolees move is both essential to knowing where specific types of services should be located, and critical to maintain accountability among parolees.

Acknowledgment We would like to thank Julie Wartell at the San Diego County District Attorney's office for her help in obtaining and understanding the data and her input and feedback on this research.

Notes

1. These data represent a "snapshot" of all parolees as of August 30, 2004.
2. In Baltimore, communities were identified by the Baltimore City Planning Department and the Family League of Baltimore City and represent clusters of census tracts into 55 broad communities.

3. Drug dependence or abuse among prisoners was identified using criteria from the Diagnostic and Statistical Manual of Mental Disorders, fourth edition (DSM-IV) (Mumola and Karberg 2006).
4. The term "problem with drug use" is not well-defined by the data source, and likely leads to wide variation in the severity of drug use and abuse that is reported among this sample of parolees. It is, however, the most reliable source of parolees' drug use that is available in this dataset.
5. Other racial/ethnic categories included American Indian, Japanese, Chinese, Hawaiian, and Other.
6. No possession offenses involving marijuana are felonies.
7. All offenses that involve the intent to sell are felonies.
8. From Mumola and Karberg's (2006) analysis of BJS data.
9. In the San Diego data, this is the percent required to register as drug offenders, and thus includes only the felony drug offenders.
10. Stays at transitional facilities tend to have a prescribed length of residence or attendance and they are not necessarily in the same area to which the parolee would have moved if not residing at the facility. In other words, stays at transitional facilities restrict a parolee's choice of when and where to move. While not shown in the table, the differences on the time and distance measures between those who were released to a transitional facility and those who were not are statistically significant at the $p < 0.05$ level. Those who were released to a transitional facility lived in more places during their first year on parole, had a longer average time before the first move, and moved longer distances for their first move.
11. While we discuss the median time between moves in the text, we conducted difference of means tests for the three groups and found that the differences were small but statistically significant ($p < 0.05$) for the first three moves.
12. All figures are available in color at http://meagan.cahill.googlepages.com/reentryresmob.
13. While individual parolees in the sample had as many as 15 moves, we selected a cut-off of five moves for analysis purposes because it was the last move for which each group had at least 100 members.
14. Because of HIPAA restrictions, this proposal would likely be most feasible to implement within or across county government agencies. Bringing in non-governmental service providers, however, would likely prove quite challenging.

Chapter 7
Social Disorganization, Alcohol, and Drug Markets and Violence*

A Space–Time Model of Community Structure

Aniruddha Banerjee, Elizabeth LaScala, Paul J. Gruenewald, Bridget Freisthler, Andrew Treno and Lillian G. Remer

Abstract The United States remains one of the most violent countries in the developed world. We believe that at least three separate sets of factors have contributed to these high rates: (1) social disorganization and related community structural factors, (2) over-densities in the distribution of alcohol outlets, and (3) illegal drug markets. Data were collected for the city of Sacramento, CA, for the years 1997 through 2001 and geocoded to 304 Census block groups. The outcome measure for the study was numbers of Emergency Medical System (EMS, ambulance, and fire) calls for service related to assault injuries. Data on population and housing characteristics related to social disorganization were obtained from the US Census. California Alcohol Beverage Control data provided the locations of active alcohol outlets in the city separated into bars and taverns and off-premise establishments. Sacramento Police Department incident reports related to drug sales and transport were used to indicate areas of the city where drug market activities were most prevalent. Bayesian disease models were used to assess statistical relationships between EMS assault injuries and these independent measures. The index of drug market activity and numbers of alcohol outlets were positively related to the number of assault injuries. Areas with lesser White and greater African-American populations had more assaults. Lower education, greater amounts of vacant housing, and more unemployment were also related to greater levels of violence. However, contrary to the expectations, more owner-occupied housing and greater household incomes were positively related to the levels of assaults, and greater numbers of households below the poverty line were related to lower levels of assaults. A substantive effect was observed for the interaction of drug arrests and poverty, indicating lower levels of violence in poor areas with high degrees of drug activity (actually increasing the

---

A. Banerjee
Prevention Research Center, Berkeley, CA, USA

*This chapter summarizes research supported in part by National Institute on Alcohol Abuse and Alcoholism Center grant P60-AA06282 and Merit Award R37-AA12927.

size of the anomalous effect noted in the previous paragraph). Substantive positive interactions were observed with respect to sizes of minority populations, African-Americans, and Hispanics. Drug market activities and alcohol outlets have separable and substantive effects upon observed rates of violence. These environmental conditions serve as catalysts for violence among at-risk populations.

Introduction

The United States remains one of the most violent countries among all those in the developed world. In fact, despite the relative uniformity of crime rates across the world, international comparisons indicate that the US continues to have the highest rate of firearm homicides among all developed countries (United Nations Office on Drugs and Crime 2000) and, in general, violence-related mortality in the US surpasses that of other developed countries (World Health Organization 2002). An important question to ask with regard to these high rates of violence is how the US has come to be one of the most violent of these industrialized societies? How do we account for such rates of assaultive violence? We believe that at least three separate sets of factors have contributed to these high rates: (1) social disorganization and related community structural factors, (2) over-densities in the distribution of alcohol outlets, and (3) illegal drug markets. Although concentrated poverty and the social disadvantages that are associated with community disorder provide the primary social conditions from which violence may grow, the legal and illegal marketing of substances that enable violence provides catalysts to violence in communities throughout the US. The social ecological processes that link social disorganization and community disorder, legal markets for alcohol, and illegal drug market activities to violence are complex and poorly understood. This chapter provides one approach to statistically modeling these social ecological relationships over time in order to help identify the contributions of these structural determinants to violence in community settings.

Social Disorganization and Related Community Structural Factors

Standard references to 'social ecological' approaches in the social science literature generally refer to one of two distinct fields of research, the 'bioecological' perspective of Bronfenbrenner (1979, 2005) and several sociological theories that introduce the ecological conditions of social action into explanations of the sources of social problems. The bioecological perspective makes the broad claim that multiple levels of influence, from peers and families to community structures, affect developmental processes that lead to social problems. The explicit social mechanisms by which these effects take place are not made clear. On the other hand, sociological theories of the causes and consequences of neighborhood disorganization have successfully incorporated aspects of the ecological contexts of social action into explanations of the conditions under which problem behaviors grow and thrive. The latter include theories of social disorganization (Kelling and Coles 1996; Skogan 1990), social capital (Putnam 2000), collective efficacy (Sampson et al. 1997), routine social activities (Felson 1987), and social network or core group theories (Eck 1995; Johnson et al. 2007) that suggest explicit social processes that may affect rates of crime and other problem behaviors. These theories suggest explicit social mechanisms by which crime and other problem behaviors may arise in community settings and suggest how spatial interactions across neighborhoods shape geographic patterns of crime (crime potential theory; Brantingham and Brantingham 1993, 1995).

Underlying these different ecological models are several different social processes emphasized by researchers as explanations for different crime outcomes. For example, LaGrange (1999) applied concepts from routine activities theory to the study of spatial distribution of minor crimes like property damage and posited a mechanism by which greater incidence of these crimes may lead to greater problems over time. LaGrange suggests that minor crimes become more common in areas where social controls are weakened (for example, in areas with greater amounts of vacant housing). He asserts that an excess of minor crimes in a neighborhood mark environments that provide opportunities for other crimes (e.g., through lower levels of guardianship), attract potential offenders, and may lead to more serious crimes like drug marketing and violence. These more serious crimes further degrade the organization of neighborhoods by eroding social and institutional resources that might otherwise act as stabilizing influences, leading to a 'spiral of decay' in community organization. One consequence of these processes is that social capital (Coleman 1988), a resource that develops through social interactions and neighborly ties, may be diminished as families feel unsafe spending time together outdoors, have fewer neighbors and nearby friends upon whom they can rely, and available local social services become over-taxed by the problems at hand. Another consequence may be that social networks for illicit activities, such as drug sales, may strengthen as core groups form for the distribution, sale, and use of illegal drugs (Eck 1995; Johnson et al. 2007).

Further applications of these theoretical approaches suggest other reinforcing mechanisms by which social disorganization and declines in community recreational and social services lead to more troubled neighborhoods. For example, outlets for the sale of alcohol tend to increase in number throughout lower income impoverished and socially disorganized neighborhoods as smaller retail vendors fill niches left behind by larger retail operations. These 'mom and pop' stores often fill a vital role in impoverished communities, serving both as a place to purchase small items and, often, as social gathering places outside the home. These outlets also serve as one of the associated signs of social disorder that affect community norms and enable the expression of problem behaviors (Sampson and Raudenbush 2004), and they sometimes become places where illegal activities of several kinds may take place (e.g., youth alcohol sales, prostitution, illegal drug sales; Alaniz et al. 1998).

Alcohol outlets. Independent of the many other correlates of violence in community areas suggested by social disorganization theories, the association between alcohol outlets, drinking, and violence has been clearly established in the empirical research literature for some time. One line of such research has found consistent associations between the number of alcohol outlets and violent events occurring across local areas (Gruenewald et al. 2006; Lipton and Gruenewald 2002) and within local areas over time (Gruenewald and Remer 2006). Another line of research conducted using data routinely collected by emergency departments and trauma centers has established an association between a presenting patient appearing as a result of a violent act and elevated Blood Alcohol Concentration (BAC) (Cherpitel 1993; McDonald et al. 2005; Treno et al. 1994). Still another line of research has linked social psychological characteristics associated with violence to drinking patterns and the propensity to drink in relatively risky locations (Treno et al. 2008). Recently, a series of community intervention studies targeting risky drinking practices and the environments in which they are likely to occur have found their efforts producing decreases in violence (Holder et al. 2000; Treno et al. 2008). At the micro-level of the contexts in which much problematic consumption occurs, notably bars, the environmental characteristics of such establishments along with the characteristics of those who frequent them are also related to aggressive behavior (Haines and Graham 2005). Thus, the addition of many alcohol outlets in socially disorganized neighborhoods may prove a volatile mix producing excessive levels of violence (Gruenewald et al. 2006).

Illegal drug markets. While much violence appears to be linked to the sale and transport of illegal drugs, less systematic study of the relationships of these illegal markets to violence has been conducted across communities in the US. In fact, although at the individual level of the drug seller and user relationships to violence and other health problems are apparent, relatively little is known at the population level either about the structure of these markets (Caulkins and Pacula 2006) or their relationships to population outcomes of interest to the public health (e.g., population distributions of sexually transmitted diseases and violence). What is known is that, since drug markets are proscribed, there are no legal venues to regulate activities and enforce 'contractual' agreements, and as a result, violence is often used as a strategy for regulation and enforcement of market activities (Goldstein 1998). Just as with legitimate businesses, wholesale and retail activities sustain illegal drug markets as functioning parts of community systems. Since some neighborhoods of communities are more active in this regard than others and because these activities themselves vary across time and space, there are dramatic variations in the intensity of drug trade-related violence that is observed from neighborhood-to-neighborhood and potentially different structures of distribution based upon which substance is considered (Eck 1995).

Goals of this study. Clearly, the three factors that have been linked to violence do not provide mutually exclusive explanations of the distributions of these problems across community neighborhoods. Indeed, since excesses in the numbers of alcohol outlets tend to concentrate in neighborhoods characterized by low levels of social organization and monitoring, the same places where drug sales also concentrate, it is quite possible that these different factors act synergistically to support and maintain these problems in troubled neighborhoods. Thus, although it is quite clear from recent research that illegal drug markets are likely to thrive in the presence of social disorganization and related community conditions (Freisthler, Lascala, Gruenewald, and Treno 2005; Mazerolle et al. 1998; Saxe et al, 2001), the unique contribution of each of these factors and the extent to which they interact in shaping patterns of violence across community neighborhoods are not known. In order to begin to fill this gap in the empirical literature, this chapter examines changes in the spatial distribution of assaults across areas of one community relating measures of social disorganization, alcohol outlet densities, and an indirect measure of drug market activities to this problem outcome over space and time. Two hypotheses will be tested in this study:

Hypothesis #1: Assault rates will tend to be higher in neighborhoods characterized by concentrations of alcohol outlets, low levels of social organization, and greater retail drug activity.

Hypothesis #2: These three factors will interact in important ways to produce higher rates of assault than can be accounted for by their separable effects.

Methods

All geographic data were collected for the city of Sacramento, CA, for the years 1997 through 2001 and geocoded to the 304 Census block groups of the city using ArcMap 8.3 (ESRI 2003a) against StreetMapUSA base maps (ESRI 2003b). The primary outcome variable for the study was numbers of assault incident calls for service to the Sacramento Police Department (SPD) provided by the SPD reporting system. These were obtained for the years 1997 through 2001 and geocoded by the nearest street address. There were 18,658 assaults in the five-year period with 3731.6 mean number of assaults per year and a standard deviation of 129.29 assaults over the entire time period. More than 99% of all calls to the police referring to an address location were geocoded to either the exact address of the incident or appropriate corresponding block face.

Data on population characteristics related to social disorganization were obtained from the US Census 2000 (2001), which include summary description of individuals living in block group units. These measures included racial/ethnic composition (number of Hispanics, number of Blacks), population effects (total population, number of Whites) and concentrated disadvantage (number of high school graduates, number of unemployed, numbers below poverty, median income).

Data on the environmental characteristics of neighborhoods related to violence in the city were obtained from three sources (2000 Census, California Alcohol Beverage Control, and SPD). Census data provided information on residential composition (number of vacant housing, number of owner-occupied housing). California Alcohol Beverage Control data provided the locations of active alcohol outlets in the city separated into bars, taverns, and off-premise establishments. Sacramento Police Department calls for service data were used to indicate areas of the city where drug market activities were most prevalent. Specific to this measure, we included incidents for sales and transport of drugs, but not for drug possession. It is important to note that this does not directly measure the activities of drug markets, but rather the reactivity of calls to police in response to the evidence of the existence of these markets. The surveillance of these activities by local individuals and police is, inevitably partial and cannot fully represent the extent of these otherwise invisible activities in the city. More than 99% of all drug crime incidents were accurately geocoded each year.

Statistical analysis. Bayesian disease modeling approaches were used to model Emergency Medical System (EMS) assault rates across 304 block groups of Sacramento over five years, from 1997 through 2001. This provided 1520 space–time units over which change in assault rates and associated correlates could be examined. Given the relatively low incidence of assaults in any given year for most units in the study, the expected number of assaults per block group per year was modeled using a Poisson distribution. A Poisson space–time model (Waller et al. 1997) was applied to the analyses of change over time in predictors and outcomes. This model includes unit and time random effects that represent, corresponding, differences between rates for assaults between block groups and variation between units in specific (linear) time trends related to changes in assault rates within units. A virtue of these modeling strategies is that explicit statistical controls may be introduced for spatial heterogeneity (i.e., spatial autocorrelation), the tendency of assault rates between block groups to be correlated with one another. Spatial autocorrelation among geographic units is a primary cause of the failure of the assumption of unit independence that invalidates the application of traditional regression methods to spatial data (Waller and Gotway 2004). The assumption of the current model is that parameter estimates of independent random effects and spatially autocorrelated (heterogeneous) random effects can be represented by Gaussian distributions (Bernardinelli et al. 1995).

Using WinBUGS, a Bayesian modeling tool, prior information on the parameters was specified and updated with respect to the fit of the model to available data using iterative maximum likelihood methods guided by a Markov Chain Monte Carlo (MCMC) search procedure (effectively searching for best parameter estimates that satisfy the constraints of the Bayesian model; Spiegelhalter et al. 2003). Starting with uniform or low-precision priors for each parameter of the model, MCMC procedures attempt to maximize fit to the observed data and estimate the joint posterior distribution of all parameters in the model, providing estimates of the distributions of most likely parameter values. These procedures require a 'burn-in' period during which the search algorithm converges to likely best estimators (1000 iterations in the current case). Continued iterations using the MCMC algorithm then provides samples from the joint posterior distribution of parameter estimates that can be used to characterize the distribution of parameters in the model (40,000 iterations in the current case). In the current study, convergence of the model was assessed using continued iterations to determine whether or not continued improvements in estimates were observed (see Spiegelhalter et al. 2003, for details).

In technical detail, posterior distribution (parameters θ given number of assaults y) is defined as:

$$p(\theta|y) \propto L(y|\theta)\, g(\theta) = L(y|\theta) \prod_i g_i(\theta_i)$$

where $g(\theta)$ is the joint prior distribution of θ (priors).

Assaults are assumed to be Poisson distributed:

$$y_i = \text{Poisson}(e_i\theta_i)$$
$$\theta = \frac{y}{e}$$
$$\log\theta_i = \sum \alpha_j + u_i + v_i + \beta t_k + \delta_i{}^* t_k$$

where $i = 1, 2, 3, \ldots 304$ block groups; $k = 1, 2, 3, \ldots 5$ years; α_j = parameter vector (overall level of relative risk plus covariates); u_i = spatial heterogeneity (area random effect, spatial autocorrelation); v_i = uncorrelated heterogeneity (area random effect); βt_k = linear trend term in time t_k; δ_i = interaction random effect between space and time.

With these definitions in place, the resulting space–time model has the following form:

$$\begin{aligned}\log\theta_{ik} = {} & \alpha + \alpha 1^*\text{Drug sales and transport} + \alpha 2^*\text{Number of bars and pubs}\\ & + \alpha 3^*\text{Lag effects} + \alpha 4^*\text{Number of off-premise outlets} + \alpha 5^*\text{Lag effects}\\ & + \alpha 6^*\text{White population} + \alpha 7^*\text{African-American population}\\ & + \alpha 8^*\text{Hispanic population} + \alpha 9^*\text{High school graduates}\\ & + \alpha 10^*\text{Vacant housing} + \alpha 11^*\text{Unemployed (civilian workforce)}\\ & + \alpha 12^*\text{Median household income}\\ & + \alpha 13^*\text{quadratic term} + \alpha 14^*\text{Owner occupied housing}\\ & + \alpha 15^*\text{Percent population 15–29 years}\\ & + \alpha 16^*\text{Number of households below poverty line}\\ & + \alpha 17^*\text{Poverty}^*\text{Drug activity} + \alpha 18^*\text{Black population}^*\text{Drug activity}\\ & + \alpha 19^*\text{Hispanic population}^*\text{Drug activity}\\ & + \alpha 20^*\text{Number of bars}^*\text{Drug activity}\\ & + \alpha 21^*\text{Number of off-premise outlets}^*\text{Drug activity} + u_i + v_i + \beta t_k + \delta_i^* t_k\end{aligned}$$

Main effects were incorporated in all models reflecting drug market activity, numbers of alcohol outlets, and other person and place characteristics related to violence. In addition, we added model terms interacting levels of drug market activity with measures of poverty, number of African-Americans, number of Hispanics, number of bars, and number of off-premise establishments. This procedure enabled the determination of the joint effects of drug market activities in conjunction with these neighborhood and population characteristics. Finally, given the prior observations regarding cross-lagged effects of alcohol outlets upon the rates of violence between adjacent neighborhoods (Gruenewald et al. 2006), we also tested the impact of spatial lags of bars and off-premise establishments on assaults. Thus, the numbers of bars and off-premise establishments in adjacent areas were used as predictors of violence in target block groups.

Results

The results of the Bayesian space–time analyses of assault incidents are presented in Table 7.1. The table presents the names of variables entered into the analysis and simple descriptive information about the parameter estimates. Confidence bounds are *not* presented, but rather critical intervals that best represent parameter estimates from the Bayesian model. Thus, the lower 2.5% and upper 97.5% bounds of the critical intervals reflect most likely lower and upper values of each estimated parameter with the median representing the most likely value of each parameter from the joint posterior distribution of the model. The upper part of the table presents the main effects of all measures. The middle portion of the table presents interaction terms testing the relationships between drug market activities and various features of local populations as they relate to assault rates, and the lower part of the table presents space–time and spatial random effects. These latter parameter estimates suggest that all random components related to space and time were substantive (linear time trend, spatial random effects, and space–time interactions), as was spatial heterogeneity (or autocorrelation) between topographically adjacent units.

As expected from previous research, the measure of drug market activity, incident reports for sale, and transport of drugs was positively related to the number

Table 7.1 Bayesian space-time analyses of assault incidents

Variable Name	2.50%	Median	97.50%
Constant term	1.653	1.894	2.13
Drug sales and transport	0.004391	0.01231	0.02011
Number of bars and pubs	0.0525	0.09962	0.145
Lag effects	–0.04402	–0.01625	0.01136
Number of off-premise outlets	0.09062	0.12	0.1508
Lag effects	–2.8E-05	0.0115	0.023
White population	–0.00103	–0.00078	–0.00056
African-American population	3.58E-05	0.000384	0.00073
Hispanic population	–9.6E-05	0.000175	0.000437
High school graduates	–0.00102	–0.00074	–0.00047
Vacant housing	0.001253	0.002468	0.003657
Unemployed (civilian workforce)	3.85E-05	0.000822	0.001626
Median household income	–0.06319	–0.05406	–0.04504
Quadratic term	0.01976	0.0279	0.03545
Owner occupied housing	0.00035	0.000764	0.001184
Percent population 15–29 years	–0.00048	–0.0002	9.24E-05
Number of households below poverty line	–0.00109	–0.00085	–0.0006
Poverty * Drug activity	–0.01684	–0.0096	–0.00241
Black population * Drug activity	0.007932	0.01695	0.02484
Hispanic population * Drug activity	0.008692	0.01476	0.02131
Number of bars * Drug activity	–0.00045	0.000244	0.000958
Number of off-premise outlets * Drug activity	–0.00092	–0.00019	–0.000544
Space–time interaction	0.2289	0.282	0.345
Spatial heterogeneity	0.01396	0.1803	0.897
Spatial random effects	0.1919	0.4349	0.5372
Time trend	–0.04423	–0.02561	–0.00789

of assaults, as was the total number of bars or pubs and off-premise alcohol outlets. Areas with lesser White and greater African-American populations had more assault incidents. Spatial lag effects related to alcohol outlets, however, were not significantly related to assaults, contrary to prior research.

Measures reflecting socioeconomic status, social disorganization, and social capital were generally related to assaults in ways expected from the sociological literature: Lower education, greater amounts of vacant housing, and more unemployment were all related to greater levels of violence. However, contrary to the expectations, more owner-occupied housing and greater household incomes were positively related to levels of assaults (with the quadratic term indicating positive acceleration in violence beyond incomes of $10,000 per year). Moreover, greater numbers of households below the poverty line were related to lower levels of assaults. The latter two observations are particularly notable in the context of the current statistical analysis. Since this analysis controls the levels of drug market activities and assaults related to alcohol outlets, both more prevalent in less wealthy areas of communities, the effects of wealth and impoverishment are quite different than typically observed in the previous studies of assaults.

Finally, the interaction terms introduced in the analysis revealed particular mixes of drug market activity and demographic characteristics that may reflect which groups of individuals are most prone to violence related to drug activities. As shown in the middle portion of the table, a substantive effect was observed for the interaction of drug arrests and poverty indicating lower levels of violence in poor areas with high degrees of drug activity (actually increasing the size of the anomalous effect noted in the previous paragraph). Substantive positive interactions were observed with respect to the sizes of minority populations – African-Americans and Hispanics. In these neighborhoods, greater levels of drug market activity were related to greater levels of violence.

Discussion

The results of these analyses for the city of Sacramento demonstrate support for the two hypotheses that were tested in this study. Considering the first hypothesis, the results indicate that the measures of drug market activity and outlet densities were positively related to assault rates. In addition, measures related to socioeconomic status, social disorganization, and social capital, such as vacant housing and unemployment, were generally related to violence in predictable ways. With regard to the expected main effects in the study, however, there were three key unanticipated findings: Greater levels of impoverishment were related to lower levels of violence and greater household incomes, and more owner-occupied housing was positively related to violence. These observations are counter-intuitive, in large part because the sociological literature has consistently linked poverty to a host of social problems.

There are several reasons why these ostensibly counter-intuitive effects may have been observed in this study: one reason may be that this was the first study to examine these relationships at the Census block group level. Prior studies have relied upon cross-sectional data often collected across relatively large geographic areas (Census tracts, zip codes, and cities), units of analysis that may have obscured these local relationships. Second, the present study tests these relationships over time. No study to our knowledge has done so. Finally, the current analysis is also the first to include controls for levels of drug market activities and violence related to alcohol outlets. When these features are incorporated into the study design, it appears that the effects of relative wealth and impoverishment are quite different than characteristically reported in studies of violence across community areas. If impoverished neighborhoods are home to more drug market activity and increased density of alcohol outlets, both of which are associated with greater violence, then the total effect of poverty with respect to the levels of violence will be mediated by these features of the environment. Clearly, it is not impoverishment alone that accounts for such levels of violence across neighborhoods, but more likely other characteristics of neighborhoods that are common in impoverished communities. Once these characteristics are taken into account, poverty appears to be associated with decreasing levels of violence across neighborhoods.

It is more difficult to explain the findings that both higher median incomes and more owner-occupied housing is related to more violence. These observations appear to suggest that relatively stable areas of communities, net of effects controlling for drug and alcohol markets, may be more prone to violence. Several very speculative explanations may be offered with regard to these outcomes. First, it is possible that a relatively small proportion of home owners raise the mean income in some neighborhoods, and increased income inequalities in these neighborhoods lead to greater levels of violence. Income inequality has been identified in previous investigations as an important contributor to rates of crime. Second, it is possible that the current study is tapping into differentiated aspects of violence that require more well-separated and detailed study. The measure of assault includes simple assaults (often associated with alcohol outlets), 'assault and battery' and violent assaults (often associated with drug markets and violent crime), and assaults related to intimate partner violence and child abuse (associated with home and family life). To the extent that these differentiable aspects of assault go unmeasured, predictions from the current model, and other similar models of assault, remain uninformed.

Core Groups and Shifting Patterns of Neighborhood Violence

Considering the second hypothesis of the study, it is notable that greater levels of drug market activity were uniquely related to greater levels of violence in both Hispanic and African-American neighborhoods. Thus, neighborhoods with a greater density of minority residents would appear to be particularly at risk for violence associated with the sale and transport of illegal drugs. It is a common finding in criminology that areas characterized by proportionally larger minority populations have higher drug possession and trafficking arrest rates (Mosher 2001). It is an uncommon finding that these activities should be uniquely related to greater levels of violence in these settings.

The results of this study suggest unique effects of drug market activities on assaultive violence within minority communities. From the perspective of social network and core group theories (Eck 1995; Johnson et al., 2007), a reasonable explanation for these excessive risks related to drug markets might be that greater proportions of active drug users as well as those in recovery may reside in Hispanic and African-American communities. With these groups in place, minority neighborhoods may be more susceptible to risks for violence related to increases in drug market activities. The existence of such core groups in association with high densities of on-premise alcohol outlets has been identified as one ecological factor affecting the spread of sexually transmitted diseases (Johnson et al. 2007) and maintaining drug markets at the neighborhood level (Eck 1995).

This viewpoint suggests that some neighborhoods are, in fact, more violent than others and that patterns of drug market activities and related violence shift over time and across neighborhoods in predictable ways. Much of this perspective rests on the fundamentals of core group theory. A core group represents a subgroup of individuals within a social network that shares common values and activities. For the purposes of this study, the common linkage for core group formation is illegal drug use. Neighborhoods vary in their numbers of core groups related to drug use, but all neighborhoods have some number of core groups inside their boundaries. It is assumed that within core groups some individuals are active drug users, and others, for reasons of recovery or other, no longer use drugs. However, the members of core groups all remain connected by social networks involving families, friends, drug dealers, case workers, police and others representing various sectors of the community at large. At any given point in time, an increase in drug market activities (a newly formed alliance for successful transport of some illegal substance) acts to catalyze the susceptible population of drug users to active drug use. This is similar to disease epidemic models where communities with endemic rates of disease lapse into epidemic rates because of opportunistic disease networks. We assume that neighborhoods of each city contain core groups in approximately stable equilibrium ready to be disturbed. An increase in any drug activity within certain neighborhoods initiates a new cycle of rapid growth in local markets, and associated 'systemic' violence.

One visible consequence of shifting drug markets as they move across neighborhoods with different at-risk populations should be a continual shifting pattern of rates of violence associated with drug market activities over time. Indeed, that is the pattern seen in the current study. As shown in Fig. 7.1, predictions from the Bayesian space–time Poisson model show areas of the city that correspond to increasing and decreasing risks for violence associated with these changing markets. This figure plots the predicted rates of assault across the city relative to the city average (1.32 relative risk per year) over time. Populations in areas most at-risk respond with greater levels of violence over the 5 years of the study, with predicted rates that are different by more than an order of magnitude for some neighborhoods of the city over time.

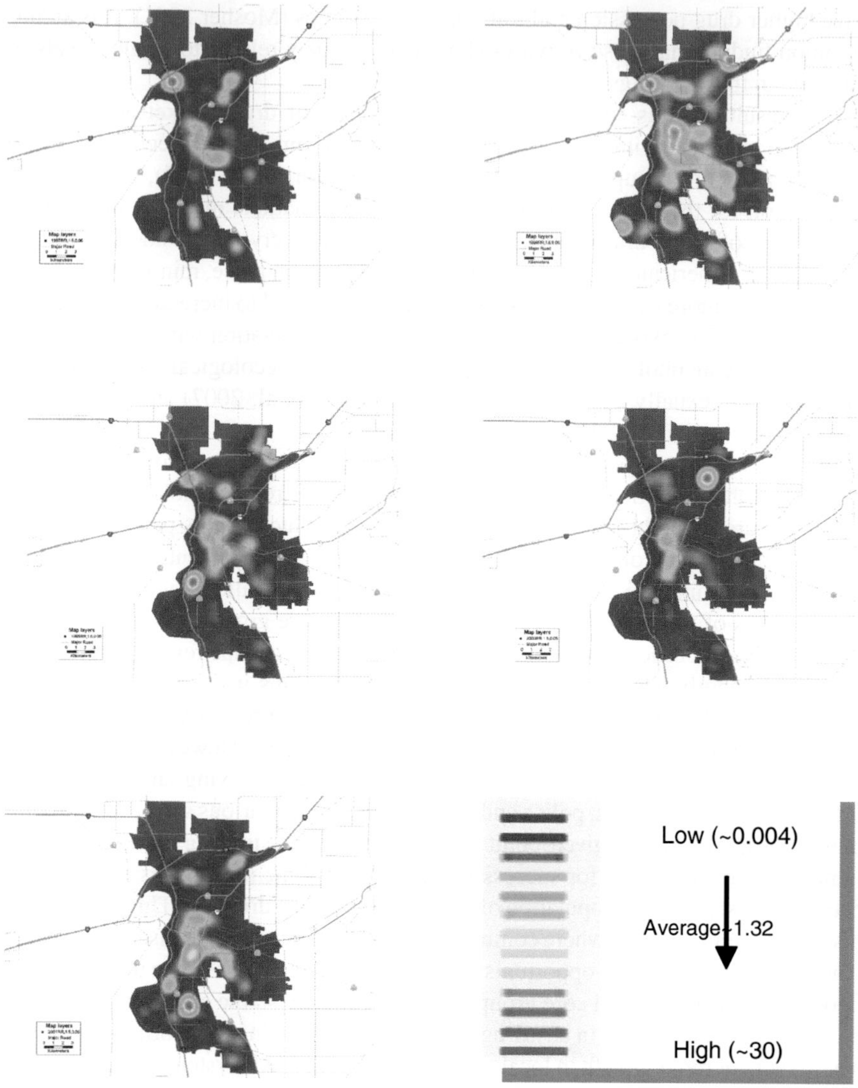

Fig. 7.1 Predictions from the Bayesian space–time Poisson model showing areas of the city that correspond to increasing and decreasing risks for violence associated with these changing markets. This figure plots the predicted rates of assault across the city relative to the city average (~1.32 per year) over time (See also Plate 12 in the Color Plate Section)

Future Research Directions and Study Limitations

The theoretical foundations of spatial ecological models for the explanation of patterns of drug-related violence across community areas are in their infancy. In the future, such models must, among other issues, identify and address the spatial interdependencies that exist between populations and characteristics of neighborhoods that support drug activities, explain the dynamic interactions of drug markets and enforcement reactions by police, and identify spatial equilibrium conditions that explain the spatial patterns of assaults seen in community studies. This is a very big theoretical task, but one that, when it is accomplished, will be of greatest benefit to communities attempting to reduce rates of violent crime in general, and rates of violent crime related to alcohol and drug markets in particular. As suggested by the current analysis, the predisposing characteristics that may lead persons to use drugs are enabled in different social and physical contexts, and these vary by area and change over time as communities develop new formal and informal social and institutional structures. Studies that employ space–time models to explore these relationships are critical to the advancement of the understanding necessary to provide a basis upon which to develop successful community prevention programs.

However, the theoretical models and empirical analyses presented in the current study are very crude and preliminary in nature and only suggest future possibilities for the development of successful ecological models in this area. From a theoretical perspective, little is known about the development of drug markets in community settings, the social conditions under which such markets thrive or fail, or the changing social dynamics that support such markets. From an empirical point of view, little is known about the best procedures by which to index the existence of drug markets, their level of activity, or their complete impacts upon neighborhood social conditions and problems. Rather obviously, taking the current study to task, reliance on police data for indexing drug-related activity may be very problematic. Although drug sales and trafficking incidents may track drug markets through a city, it is also very likely that these incidence patterns also reflect independent activities of police investigations. If police activities accurately track market activities, with enforcement efforts suppressing drug markets in one area than tracking them to another, incidence measures may accurately reflect change in drug markets across the landscape of a city (although with some temporal lag). If, on the other hand, incidence calls and police activities are weakly related to drug markets, and if these markets manage a degree of invisibility that makes them difficult to identify geographically, then the use of incidence data from police records may be a poor indicator indeed of drug market activity.

Given these observations, the current study provides an opportunity to make several important recommendations with regard to future research in the area. First, much more theoretical work should be directed at specifying the social mechanisms that underlie the observed patterns of illegal drug use and related crime. While the current study is suggestive of a few of these mechanisms, it constitutes only a first step toward their specification and provides limited guidance to the plausible interactions of these mechanisms in determining the development of drug markets. Second, empirical assessments of the development of drug markets will require extensive spatial longitudinal studies of drug market activities and related problems that extend well beyond the significant limitations of cross-sectional studies. Unless enforcement activities are fully ineffective, drug markets and their related social effects are never in equilibrium with regard to their social determinants. As such, the results of cross-sectional studies will not validly reflect the underlying social mechanisms that support these markets and lead to problem outcomes. Finally, studies of drug markets must eventually go beyond the examination of readily available archival data and incorporate survey data to show how users of illegal drugs access these substances, how sellers market these substances, and how these sales and marketing activities translate into the many social problems associated with drugs of abuse. With the successful development of these approaches, our new understandings of the extent to which different social mechanisms facilitate or impede the development of drug markets, drug use, and related problems can be turned toward the development of successful community programs for the prevention of drug problems.

Chapter 8
Integrated Assessment of Addiction Epidemiology in Hong Kong, 1996–2005

Shui Shan Lee and Phoebe TT Pang

Abstract In Hong Kong, the Central Registry of Drug Abuse (CRDA) is a register that forms the database for the study of drug addiction in the territory. Using a geographic information system-based approach, a pilot study was conducted, which integrated data from the CRDA, census, HIV (human immunodeficiency virus) reports/seroprevalence studies, and methadone clinics service statistics, to support the assessment of the trends of heroin addiction, HIV-related risks, and the interrelationship with sociodemographic attributes in heroin users and the general population. Apart from the visualization of spatial distribution of heroin users, the study had uncovered specific local patterns at district levels. It is noted that despite a general decline in heroin addiction in Hong Kong, some areas showed a rising pattern in selected subpopulations. The increasing use of multidrugs also gave a district-specific pattern. Against the background of a low HIV prevalence in heroin users, the temporospatial pattern of injection provides potentially useful clues to track the spread of HIV risk. The location and coverage of methadone clinics, an important HIV-prevention strategy, were assessed in context of their public health impacts. It is concluded that the extension of the project to a long-term system would be useful for the study of addiction epidemiology, so that lessons learned from Hong Kong can contribute to the global knowledgebase on the development of effective response to substance abuse.

Introduction

As a special administrative region (SAR) of China, Hong Kong has a long history of social and public health turmoils resulting from substance abuse, beginning with the Opium War that saw the unveiling of British rule in the middle

---

S.S. Lee
2/F School of Public Health, Chinese University of Hong Kong, Prince of Wales Hospital, Shatin, Hong Kong. Tel.: (852) 22528812; Fax: (852) 2635 4977
e-mail: sslee@cuhk.edu.hk

Y.F. Thomas et al. (eds.), *Geography and Drug Addiction*,

of the nineteenth century. Before 1945, the sale and use of opium were in fact legal. In the year of 1964, seizure of heroin in Hong Kong almost topped the world, second only to the much larger countries of United States and Thailand (Lee 2000). Thereafter, the territory's heroin addiction pattern has been shaped by the alteration of the social environment and interventions introduced by the Government, notably the methadone treatment program launched in the mid-1970s. For over thirty years, methadone is offered through a territory-wide program that has enabled over 60% of the heroin users to be treated (UN Regional Task Force on Drug Use and HIV Vulnerability 2002). Methadone maintenance or substitution treatment is now a main strategy for reducing human immunodeficiency virus (HIV) spread through the minimization of risk behaviors (Wong, Lee, Lim, and Low 2003). Remarkably, the HIV epidemic has spread through injection drug users in many countries in the region but left Hong Kong relatively untouched (Chan and Lee 2004). The situation is, understandably, a volatile one, as can be inferred from the intensive international travel and proximity to the Golden Triangle.

As a public health issue, heroin addiction carries a global dimension and a diversity of local implications. In this connection, the tracking of Hong Kong's heroin addiction situation is particularly meaningful, both for evaluating the local intervention programs and serving as a reference for other countries, including Vietnam, Malaysia, and neighboring Chinese cities, as methadone treatment began to be rolled out in the past years. Parallel with the launching of the methadone treatment program some thirty years ago, a Central Registry of Drug Abuse (CRDA) also came into being in the territory. The ongoing collection, collation, and analysis of drug addiction data through this register has opened up the study of addiction epidemiology. CRDA is a rich source of reference, and has become the core data source of addiction epidemiology in Hong Kong. Two open access statistical reports are now published by CRDA every year. However, unlike infectious disease epidemiology that looks at the prevalence and incidence of biological markers, addiction epidemiology is a more complex subject that is founded on the monitoring of behavioral markers, measurement of HIV risk, and trends prediction. Conventional means of constructing statistical analysis from a single data source seems to be inadequate to meet the aspiration of addiction epidemiology.

The advent of geographic information system (GIS) has opened up a new chapter for addiction epidemiology. In Hong Kong, heroin addiction has long been seen as a very local phenomenon. It is not uncommon to find more injection heroin users (IHUs) in one or some streets/areas but not the others. This was possibly the original rationale for the presentation of breakdown statistics according to districts in the CRDA report. The districts can be conveniently compared while being controlled for other common environmental and social characteristics. This chapter reports on a pilot study on the application of GIS in the enhancement of epidemiologic study on heroin addiction in Hong Kong, focusing on the trends during 1996 to 2005.

Capturing Data for a Pilot Study

The objectives of the pilot study were to: (1) determine the temporospatial trend of heroin addiction; (2) explore demographic and social factors associated with addiction; and (3) assess the risk of HIV and related infections. An integrative approach utilizing data from multiple sources, both on heroin users and the general population, was adopted. With a population of 6.8 million in an area of $1000\,\text{km}^2$ (Census and Statistics Department 2006b), the territory of Hong Kong is divided into 18 districts and 282 tertiary planning units (TPUs) (Fig. 8.1). The availability of district and TPU-level data on heroin addiction and the general population is invaluable in supporting the pilot GIS-based epidemiologic study.

In the study, heroin addiction data were obtained from the CRDA, a territory-wide register run on a voluntary basis and managed directly by the government. Over the years, changes have been introduced to improve the operation of the very system. The last major alteration was introduced in 1995, which included the redevelopment of the computer system, expansion of the reporting network from 34 to 67 agencies, and the application of revised questionnaires (Central Registry of Drug Abuse 2006). Sources of reports have included law enforcement departments,

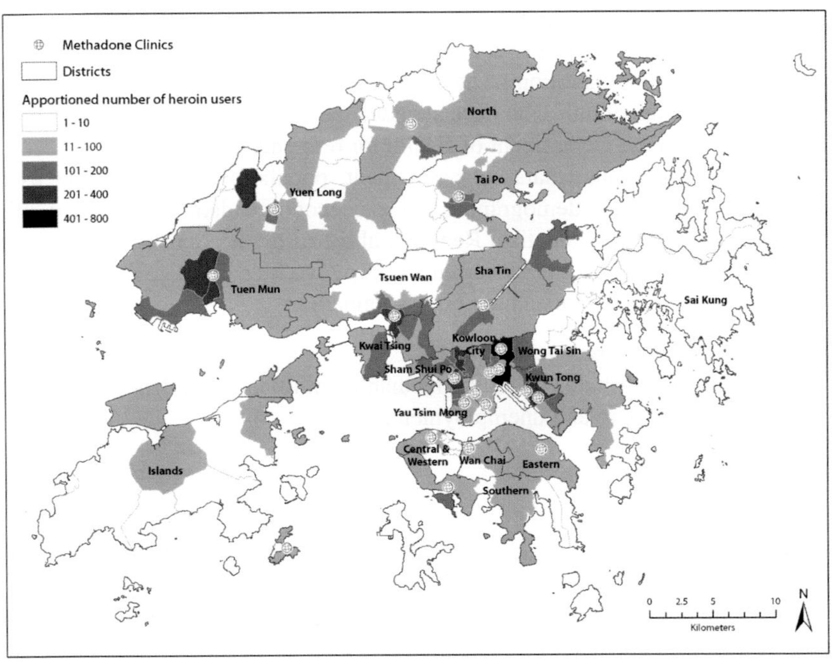

Fig. 8.1 Map showing the distribution of heroin users according to districts.
Number of heroin users 2005, acquired from the CRDA, is apportioned from districts into tertiary planning units. It is symbolized by graduated colours. The 20 methadone clinics are mapped according to their exact locations

treatment and welfare agencies, tertiary institutions, hospitals, and clinics. For consistency, we decided to focus on data collected between 1996 and 2005, so that the number of confounders can be minimized.

Relevant data were retrieved from the hard copies of the CRDA reports or website (www.nd.gov.hk/drugstatistics.htm), supplemented by aggregate data made available on request through the Narcotics Division of the Hong Kong Government. Each year between 10,000 and 16,000 heroin users were reported. Four major categories of yearly district-level data were collected from the CRDA for developing the pilot study. These were:

1. Prevalence and incidence of heroin users, as derived from the reported and newly reported numbers, and the practice of multiple drug (multidrug) use.
2. Demographics of heroin users comprising their sex and age.
3. Social characteristics of heroin users, namely, education level, housing type, conviction history, employment, and habit of using multidrugs.
4. Risk factors for blood-borne infection, using injection as the main surrogate.

In order to assess the risk of HIV infection, yearly statistics on HIV-infected drug users and seroprevalence studies in Hong Kong (without district breakdown) were extracted from the HIV surveillance reports published by the Department of Health of the Hong Kong Government (http://www.info,gov.hk/aids/english/index.htm). Knowingly, methadone treatment is a protective factor against HIV dissemination in drug users. There are now a total of 20 methadone clinics in the territory of Hong Kong. The effective registration of heroin users of each methadone clinic, defined as the number of individuals who have attended the clinic in the preceding 28 days, was accessed from the Department of Health for reference.

To provide background data for assessment, population statistics were used. Specifically, district-level demographic and socioeconomic data from 1996 to 2005 were extracted from the Statistical Reports published by the Census and Statistics Department of the Hong Kong Government (http://www.censtatd.gov.hk). Two categories of yearly statistical data have been obtained:

1. Population statistics on age and sex.
2. Socioeconomic attributes including median monthly income, dependency ratio, education attainment, and unemployment.

Building a GIS Framework for Addiction Epidemiology

Using ArcInfo 9.1, a study framework was designed to visualize the pattern of heroin addiction, analyze its correlations with demographic and socioeconomic factors at a district level, and determine the potential risk of HIV and related infections. A 1:5000 digital base map of 1996 and 2001 Tertiary Planning Unit and Street Block (TPU and SB) Boundaries was obtained from the Planning Department of the Hong Kong Government. The project was conducted beginning with conversion and standardization of data, through mapping and then an assessment of factors and impacts associated with heroin addiction.

The reported number of heroin users was converted into incidence and prevalence using district-level mid-year population as the denominator. Inconsistency of age classification between different data systems was adjusted on the notion that all individuals within an age interval were evenly distributed. Areal interpolation, using population-weighted method, was adopted to transfer data from districts to TPUs, a smaller geographic unit in Hong Kong. Areal interpolation refers to the apportionment of data values from source polygons to target polygons based on selected algorithms (Reibel and Bufalino 2005). In our study, a population-weighted method is adopted that assumes value of data in the target polygons is proportional to the number of population in the source polygons. The interactive selection of symbology for map layers was applied according to their categories or quantities to enhance flexibility in visualizing data. We explored the spatial relationships among variables by arbitrarily overlaying map layers or performing logical operations of data such as Boolean, adjacency, and areal interpolation. Finally, factors associated with heroin addiction and their impacts were visualized with maps, graphs, and tables. The approach has enabled the assessment of heroin addiction to be made in four dimensions: first, demographic characteristics of heroin addiction; second, the impacts of population patterns on heroin addiction; third, trend of socioeconomic factors associated with heroin addiction; fourth, public health impacts of heroin addiction, including HIV risk, abuse of multiple drugs, and crimes.

An Assessment of Addiction Epidemiology

Temporospatial Pattern of Heroin Users

Over the years, CRDA has provided a useful description of heroin addiction through the published reports. It is known that the number of reported heroin users, both total in a year and new cases, has been on a decline. Visualizing the numbers over districts gives a somewhat different pattern (Fig. 8.2). The overall decline is still observed, but the slope varies across districts. The prevalence (expressed as number per 100,000 population) is highest in two urban districts of Yau Tsim Mong and Sham Shui Po. Interestingly, no obvious decline is seen for female heroin users, but rather a paradoxical rise in prevalence in 5 of the 18 districts, which appears to be more marked after 2003. In male, the fluctuation is less marked. The incidence numbers show a similar pattern.

The district variation and time trends of the numbers of heroin users, both in terms of prevalence and incidence, are probably related to the demographic change of the background population. The Hong Kong population's age has been increased. The mean age of the general population has risen from 34.8 in 1996 (Census and Statistics Department 2002a) to 39 in 2005 (Census and Statistics Department 2005). Correspondingly, the age of heroin users has also been rising. The mean age

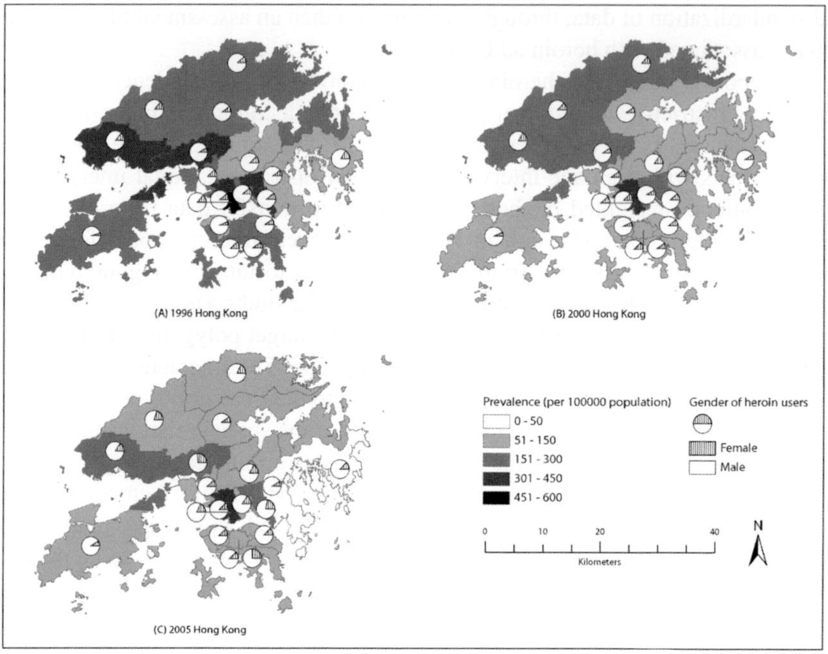

Fig. 8.2 Distribution and temporal change of the prevalence of male and female heroin users 1996, 2000, and 2005.
Prevalence of heroin users in (A) 1996, (B) 2000, (C) 2005 refers to the total number of heroin users per 100,000 population. Graduated coloring is adopted to visualize differences between districts. Gender of heroin users is shown as pie charts, with striped symbol for female and white symbol for male

of all heroin users was 40 (42 for male and 33 for female respectively) (CRDA 2006). The higher prevalence districts for heroin users correspond with those districts with a higher population of people above the age of 35. On the contrary, the absolute number of reported heroin users below the age of 21 has fallen from 2200 in 1996 to 77 in 2005 (CRDA 2006). The size of the population below the age of 21 has also fallen. Again there is a district variation of the trend, which is important in context of the development of corresponding intervention strategies (Fig. 8.3).

HIV Risk in IHUs

HIV is efficiently transmitted through needle sharing in IHUs. The risk of HIV spread in IHUs is a cause for concern in many countries, including China and South East Asia. In Hong Kong, the reported number of HIV-infected drug users has remained at a low level throughout the past 10 years, alongside the low seroprevalences determined in surveillance studies (Centre for Health Protection,

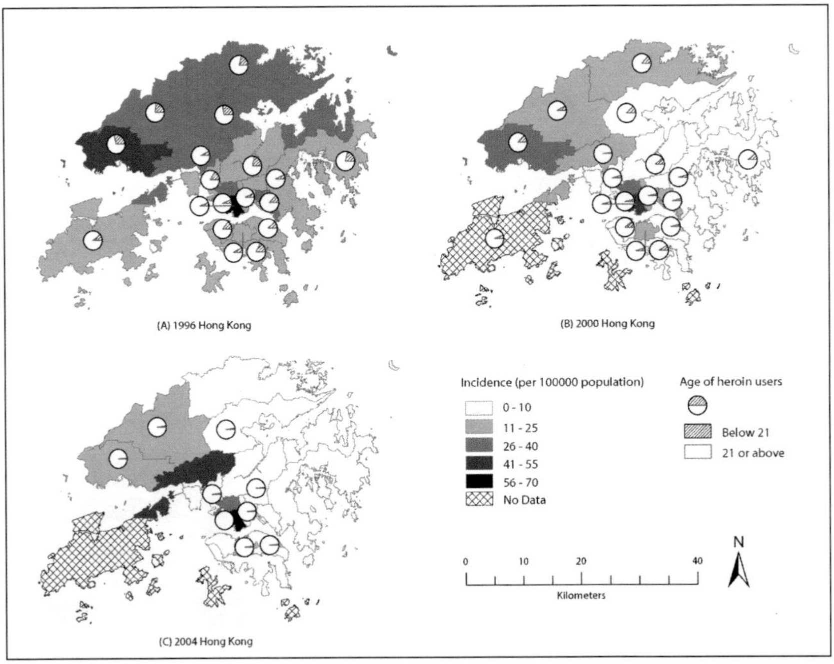

Fig. 8.3 Distribution and temporal change of the incidence of heroin users by age below 21 years versus 21 years or above – 1996, 2000, and 2004.
Incidence of heroin users, referring to newly reported number of heroin users per 100,000 population, is shown in graduated colors. Overlaid on incidence map are the pie charts that show the comparison between heroin users aged below 21 (striped symbol) and those aged 21 or above (white symbol). Absense of pie chart reflects the unavailability of required data

Department of Health 2006). Figure 8.4 shows the distribution of heroin injectors in Hong Kong, in 2005. The absolute number of injectors has been decreasing uniformly, in line with the fall in the number of heroin users. The percentage of those who have injected has varied from 40% to 65% in different districts. As these are prevalence figures reflecting the history of injection for the heroin users, this does not necessarily reflect the current pattern of risk behaviors. Overall, the picture suggests that there is no significant variation of HIV risk across districts.

Methadone treatment is one important strategy of reducing the population risk of HIV spread in IHUs. On a daily basis, about 8000 heroin users attend Hong Kong's methadone clinics. Dividing the effective registration on a selected day of year by the total number of heroin users for that year, we calculate the coverage of methadone treatment as 98.9% in 2005. District-based coverage cannot be directly determined from the available data and spatial location of methadone clinics. However, from the map in Fig. 8.1, it can be seen that the methadone clinics are closer to where the heroin users are geographically located.

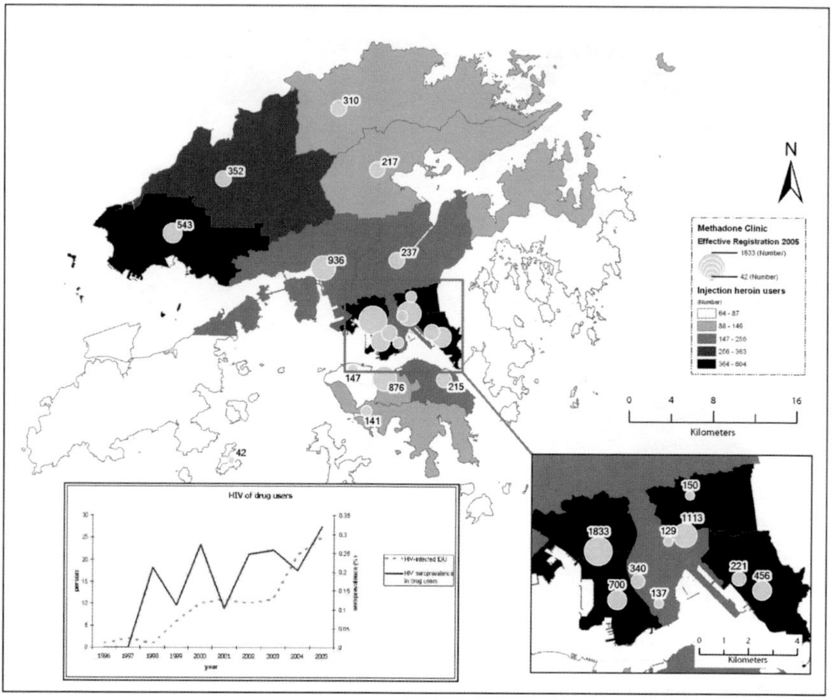

Fig. 8.4 Distribution of injection heroin users (IHUs), methadone clinics, and the relationship with HIV risk.

A choropleth map showing the total number of IHUs in 2005 by districts symbolized by graduated colors. The size of methadone clinics is proportional to the effective registration of each methadone clinic, the number highlighted in white, as of the end of December 2005. A chart showing HIV reports as well as the seroprevalence of heroin users of the years 1996 to 2005 is attached below the map

Relationship of Heroin Addiction with Social Environment

Heroin addiction is not an isolated social phenomenon. It is associated with one's socioeconomic background and also that of the society. In the year 2005, 63.5% of the reported heroin users were unemployed, 44.6% have received only up to lower secondary education, and 54.9% resided in aided housing (CRDA 2006). The geographic pattern of unemployment is shown in Fig. 8.5. The unemployment rate of the total population has risen from 2.2% in 1997 to 7.9% in 2003, which then fell to 5.6% in 2005 (Census and Statistics Department 2006a). The proportion of heroin users who were unemployed had not followed this trend. The absolute number of unemployed heroin users has remained largely unchanged, which implies a rising proportion between 1997 and 2001. In some districts, the proportion was much higher; for example, Tai Po, Yuen Long, and Sham Shui Po.

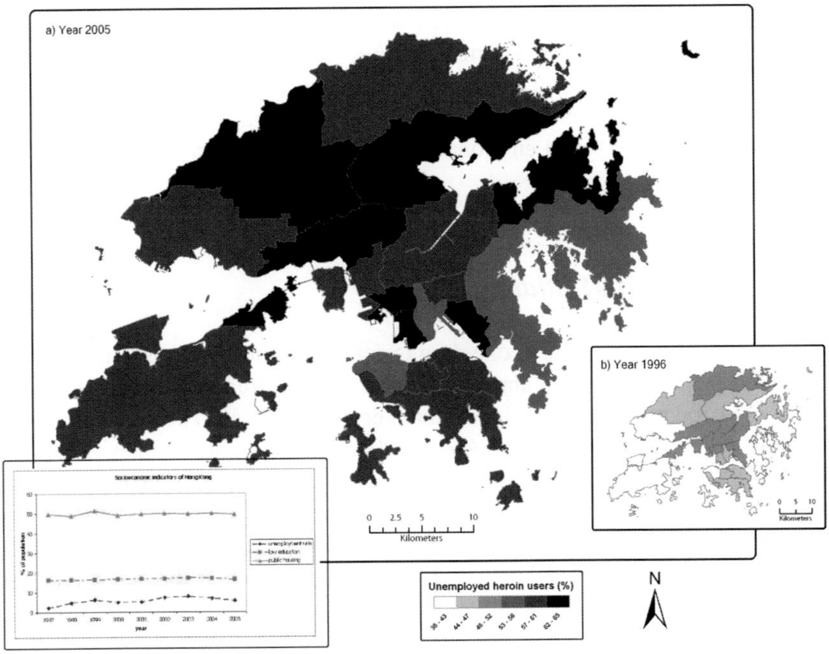

Fig. 8.5 Unemployed heroin users and socioeconomic indicators of the general population. The larger map exhibits the percentage of unemployed heroin users of all heroin users in 2005 by districts, contrasting with a smaller map showing the percentage in year 1996. Both choropleth maps used the same classification scheme with six equal intervals. The chart below shows the socioeconomic indicators of Hong Kong from 1997 to 2005

Low education is another presumptive vulnerability factor associated with addiction. The general education level in Hong Kong has risen. Between 2000 and 2005, the percentage of the general population who have only attained low education (lower secondary only, referring to that for people at and below the age of 15) has declined by an average of about 5% in all districts. For heroin users in general, the proportion of those attaining up to lower education level has remained the same at over 80%, which is much higher than that of the general population (39.7% in 2000 and 25.5% in 2005). The corresponding district-level figures for heroin users are, however, not available for analysis. Housing could be another means of assessing the socioeconomic status of heroin users. In Hong Kong, a very high proportion of the general population are residing in pubic housing (about 50%). Interpretation of the association with housing type as an attribute for socioeconomic status is difficult.

Criminal offense is one important marker of adverse outcome of heroin addiction. Over the 10-year period between 1996 and 2005, the crime rate in Hong Kong has stabilized at a relatively low level of between 1000 and 1300 per 100,000 population (Hong Kong Police 2006). The number of heroin users who had been convicted varied from one district to another. Interestingly, the four highest level districts (Kwun Tong, Yau Tsim Mong, Shum Shui Po, and Wong Tai Sin) in 1996 all showed a clear decline by 2005. All were densely populated districts with a higher number of heroin users reported. The remaining 14 gave a smaller and stable number over years. The association between heroin addiction and adverse social environment may go either way: heroin addiction leading to social problems on one hand, and adverse social environment predisposing to heroin addiction on the other.

The New Trend of Multidrug Use

The practice of using more than one drug is defined as multidrug use. In this study, we refer to heroin users who abuse at least one other drug on top of heroin. The latest CRDA report recorded a highest number of people (4037) abusing multidrug in 2005 (Central Registry of Drug Abuse 2006). The mapping of the situation has allowed us to examine the problem from a temporospatial perspective. First of all, not all districts began simultaneously to report the new pattern of multidrug use. We used the report of 20 persons as the threshold for multidrug use, as inconsistent fluctuation at lower level was likely. Eight out of 18 districts had reached this level

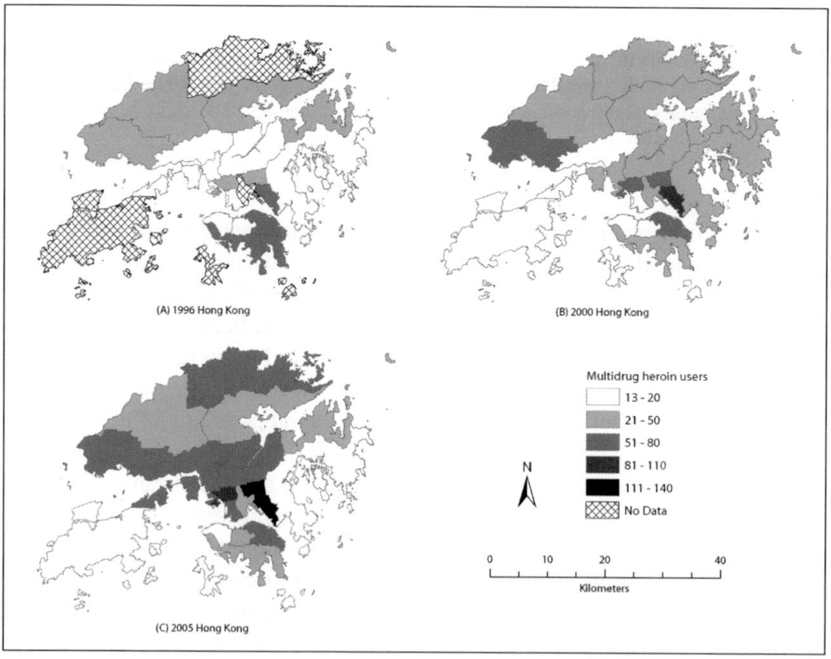

Fig. 8.6 Distribution and temporal change of the multidrug users 1996, 2000, and 2005. Number of heroin users consuming multidrugs in (A) 1996, (B) 2000, and (C) 2005 is displayed as graduated colors, with the darker ones representing a higher number of multidrug heroin users. Districts with no data appear in gray

in 1996 in the report of multidrug use in heroin users. The number has continued to grow and by 2005, all except two districts had reported multidrug use beyond the threshold level.

Geographically, there is a general trend of a rising proportion of heroin users abusing multiple drugs. In districts with a relatively early start of multidrug use, a plateau was reached the level of which had varied from one district to another. Kwun Tong, for example, had reached a plateau of about 108 multidrug abusing heroin users since the year 2000 (Fig. 8.6). For districts like Wan Chai, Islands, and Yau Tsim Mong, the percentage has continued to rise. The rising trend of multidrug use paralleled the decline in the overall number of heroin users, and the increase of the number of reports of soft drug abuse (Central Registry of Drug Abuse 2006). Triazolam and midazolam are the commonly used soft drugs in heroin users in Hong Kong. Unlike young people abusing multiple soft drugs, heroin users take triazolam/midazolam as substitute for expensive heroin.

Discussion

Epidemiology is the study of disease in context of the attributes described as a triad of "time, place, and person" (Lilienfeld and Stolley 1994). Spatial and temporal perspectives are crucial and are part and parcel of conventional epidemiology studies, but these may not have been explored to the fullest extent. Normally, epidemiologic study is conducted on the assumption that the causative agent of the disease is randomly distributed in the society. For infectious disease epidemiology, infectivity of the agent and immunity (size and duration) of the host are the key determinants of trends of spread of the infection. These premises are inadequate for the study of chronic conditions like addiction, which results from a complex interplay between human behaviors, environments, and is intricately linked with other forms of morbidity, notably HIV, and other blood-borne infections. A multilevel approach has been advocated for the study of HIV and sexually transmitted infection epidemiology (Aral, Padian, and Holmes 2005). The same principle actually applies to addiction epidemiology, though the definition of "transmission" is dissimilar – that there's the propagation of a behavior and social phenomenon, rather than an infective agent.

In our pilot study in addiction epidemiology, it is evident that a GIS-based approach can add value to the current mechanism of public health assessment with its capability in delivering information in an easy-to-understand visual medium (Lang 2000). Using the new framework, we are able to illustrate the pattern of heroin addiction and its multiple dimensions at district level in the territory of Hong Kong. The ecologic approach allows the incorporation of aggregate data from multiple data sources, which echoes the multilevel approach of epidemiology studies in other infections (Aral, Padian, and Holmes 2005). The study has enabled a number of new observations to be made, which would otherwise not be revealed through conventional mechanisms of assessment. First, despite the well-known declining trend of heroin use in Hong Kong, the pattern clearly varied from one district to another. The discovery of a higher proportion of heroin users in selected districts could prompt policy makers to prioritize their strategy on addiction. Second, behavioral practices were not identical among communities. Peer influence and social relatedness are the underlying reasons for the spatial variability. In recent years, we saw an increasing trend of multidrug use in Hong Kong. This has probably arisen from one or some districts and then "propagated" to other districts, a pattern that is otherwise inconspicuous if displayed on frequency tables and charts. Third, the interrelationship between addiction and social factors can be effectively visualized using maps instead of statistical tables. The influence of demographic changes (aging, for example), social deprivations, and education levels may vary from one locality to another, even within the same country. It is apparent that despite the general decline of heroin addiction, the prevalence has remained high in poor and densely populated districts. Again, there is the higher unemployment rates, lower education attainment and common phenomenon of association with crimes and therefore conviction that characterize the heroin-taking population. A GIS-based description would enable public health strategies to be developed to address social inequalities rather than just targeting respective individual behavior (Mackenbach and Bakker 2003).

HIV has become one important dimension of the complications of heroin addiction through the practice of injection and needle sharing. To this day, HIV has not taken root in the drug taking communities in Hong Kong, and a prevalence of less than 1% is consistently observed in surveillance studies (Chan and Lee 2004). The phenomenon can be clearly observed even without any enhancement of the current surveillance mechanism. Our pilot study has, however, enabled us to track the spatial variability of injections in different districts, watch for signals of changes that may predate HIV transmissions, and study factors which could be associated with heroin injection. With the incorporation of data on methadone maintenance, the same platform also serves to evaluate the effectiveness of public health interventions. A broad coverage of methadone maintenance is effective in protecting IHUs from the risk of HIV infection, an observation that has been made in Hong Kong. Coverage rate of methadone treatment can be used as an indicator of harm reduction on a population level (Chan and Lee 2004). Temporospatial assessment of coverage of methadone treatment, if conducted regularly over time, could become a surrogate of effective harm reduction.

The application of GIS in the epidemiological study of addiction is not without limitation. Available data sets are largely secondary data from multiple sources. These data are not uniform both in terms of spatial and temporal scale. With efforts, adjustments can be made to collect data, both in their conversion and standardizations, to support systematic assessment, calculations, and visualization. However, errors in the course of data engineering, although small, may have significant bearings on the interpretation of the final assessments. As currently geocoded data from individual heroin user is not available from the CRDA, nor is it available from methadone clinics, while district-level HIV statistics are likewise unavailable, the only applicable spatial referencing units available for addiction epidemiology are districts and TPU. Though the algorithm of areal interpolation used in this study is simple, which assumes that the number of population by district is the key factor in estimating the number of heroin users by TPU, the apportionment problem can only be eliminated when individual records georeferenced to point are available, which is unlikely to be practical (Simpson 2002). Population-weighted interpolation can be regarded as a pycnophylactic interpolation technique, on the assumption that the summation of estimated values in some target polygons is the same to the value of their corresponding source polygon (Reibel and Bufalino 2005). Outweighing the drawbacks of population-weighted interpolation, apportioned data can be mapped in a reasonable spatial resolution and can be generated into more informative maps after data interpolation, as shown in our study. Moreover, as areal unit is modifiable and subjective to one's definition, analytical result will be incomparable or biased. This is not a serious drawback knowing that the residential address of an infected individual may not carry significant meaning epidemiologically. Availability of geocoded data may, however, be useful for the planning of HIV services.

The Way Ahead

The pilot study has so far encompassed an assessment of addiction epidemiology in Hong Kong on a project basis, covering a period of 10 years. Useful results have been generated, leading us to conclude that such an integrated approach is feasible, and that this should be continued and further institutionalized to improve public health. In other words, an ongoing GIS-based system should be established, beginning with the standardization of fields, computerization of the databases, and the plan for a customized template for data entry. While it would be ideal eventually if geocoding of individual heroin user is introduced, apportionment shall continue to be useful to illustrate the distribution of heroin addicts. Apart from TPU, the 405 District Council Constituency Areas offers another means of apportioning heroin users. As each of such areas has approximately the same population size, comparison can be conveniently undertaken while controlling for population sizes. By integrating the various data sources, the pattern of heroin addiction can be presented in a user-friendly manner, with linkage to its association with HIV infection, HIV risk, social demographics, and the use of multiple drugs. The main advantage of an ongoing system is the commitment to the generation of addiction epidemiology results on a regular basis, an output similar to that of the production of surveillance reports in public health epidemiology.

Surveillance aside, the provision of an addiction epidemiology system would enable customized research to be developed. Hong Kong is uniquely positioned as this is one of the few places in the region with a low HIV prevalence in IHUs, against the background of an extensive substitution treatment program operated in the form of methadone clinics. While the reduction of risk behaviors in IHUs on an individual level is well known, the correlation between methadone treatment and population HIV risk has not been quantified with a validated scientific model. The GIS-based approach would allow us to establish practical models to determine the threshold of coverage that could create public health impacts. The configuration of social networks of IHU would be another dimension for the study of addiction epidemiology (Rothenberg et al. 1998). An integration of spatial epidemiology and social network analysis would be one important next step in the development of a model for studying the growth of the epidemic of addiction, and the effects of interventions. As for other public health issues, projection and prediction of patterns would be invaluable outputs to inform policy development.

Finally, Hong Kong has been an example of best practice in substitution treatment in the past decade (UN Regional Task Force on Drug Use and HIV Vulnerability 2002). The unique history of heroin addiction, low HIV prevalences in IHU, and the high coverage rate of methadone maintenance are hallmarks of a potential model the lessons from which have continued to emerge. The situation is quite different in Mainland China. Of the estimated 650,000 persons living with HIV in China, injection drug users accounted for 44.3% of the total (Sullivan, Metzger, Fudala, and Fiellin 2005). HIV prevalences of 50% or above have been reported in Yunnan Province, Xinjiang and Sichuan (Ministry of Health, UNAIDS, and WHO 2006). In the neighboring cities within the Pearl River Delta Region in Guangdong Province, the HIV prevalence has also reached 5% or above. An integrative study of addiction epidemiology between Hong Kong and neighboring cities would enable new lessons to be learned, which would be of useful reference to countries in or even outside the region. In the Pearl River Delta Region specifically, people sharing the same culture and heritage have been exposed to the same health condition. A GIS-based approach would allow innovative correlation studies to be undertaken on factors associated with the growth of the HIV/heroin addiction epidemic, and attributes of effective intervention. An enhancement of the existing CRDA, the core database of heroin addiction in Hong Kong, would be all that is required to advance addiction epidemiology in the region, beginning from the territory of Hong Kong.

Acknowledgment We thank Mr Peter Lee, ex-Commissioner of Narcotics, for his visionary approach in establishing the Central Registry for Drug Abuse, which was formally inaugurated in 1976. This chapter has been written to acknowledge the efforts of Mr Lee in commemoration of the thirtieth anniversary of the CRDA and the methadone clinics in Hong Kong. The authors are grateful to Miss Candy Wong and Professor Hui Lin for useful discussion in the course of planning the studies in the pilot project. The Department of Health, Hong Kong, is thanked for the provision of data on HIV reports and seroprevalence studies, and the registration statistics of the methadone clinics. Narcotics Division, Hong Kong, is thanked for the provision of supplementary district-based data from the CRDA. The project team is supported by the Research Fund for the Control of Infectious Diseases (RFCID) from the Health, Welfare and Food Bureau of the Hong Kong Special Administrative Region Government.

Chapter 9
Residential Segregation and the Prevalence of Injection Drug Use among Black Adult Residents of US Metropolitan Areas

Hannah L.F Cooper, Samuel R. Friedman, Barbara Tempalski and Risa Friedman

Abstract *Objectives*: We analyzed the relationships of two dimensions of racial residential segregation (isolation and concentration) in 1990 to the 1998 prevalence of injection drug use (IDU) among Black adult residents of 93 large US metropolitan statistical areas (MSAs).

Methods: We estimated IDU prevalence among Black adults in each MSA by analyzing three databases documenting injectors' encounters with the healthcare system. Multiple linear regression methods were used to investigate the relationships of isolation and concentration to the natural log of Black IDU prevalence, controlling for possible confounders.

Results: The median IDU prevalence was 1983/100,000 Black adults (interquartile range: 1422/100,000–2759/100,000). The median isolation index was 0.48 (range: 0.05–0.84): in half of the MSAs studied, the average Black resident inhabited a census tract where ≥48% of the residents were Black. The multiple regression model indicates that an increase of 0.50 in the isolation index was associated with a 23% increase in IDU prevalence among Black adults. Concentration was unrelated to the outcome.

Conclusions: Residential isolation is positively related to Black IDU prevalence in MSAs. Research into the pathways linking isolation to IDU is needed.

As recognized by the National Institutes of Health (National Institute on Drug Abuse, 2001), identifying the determinants of injection drug use (IDU) among Black adults holds importance for public health, given the substantial and persistent overrepresentation of Black Americans among people diagnosed with injection-related health problems, including HIV/AIDS and fatal illicit drug overdoses (Tardiff, Gross, and Wu, 1989; Substance Abuse and Mental Health Services Administration,

---

H.L.F Cooper
Assistant Professor, Rollins School of Public Health at Emory University, 1518 Clifton Road, Northeast Room 568, Atlanta, Georgia
e-mail: hannah.cooper@sph.emory.edu

Adapted with permission from an article published in the American Journal of Public Health titled, "Residential Segregation and the Prevalence of Injection Drug Use among Black Adult Residents of US Metropolitan Areas." Vol. 97, No. 02, ©2005 American Public Health Association.

2003; Harlow, 1990; Galea, Ahern, and Tardiff, 2003; Friedman et al., 1987; Friedman et al., 1988; Novick et al., 1989; Selik et al., 1989; Selik, Castro, and Pappaioanou, 1988; Centers for Disease Control and Prevention, 2001). Structural factors, including racial residential segregation, have been hypothesized to be potent determinants of drug use patterns among Black individuals and, in fact, some have suggested that such factors play a more important role in determining drug use among Blacks than Whites (Johnson and Mulffler, 1997; Roberts, 2000; Cooper et al., 2005; Friedman, 2002; Fullilove, 1993; Galea, Nandi, and Vlahov, 2004; Lillie-Blanton, Anthony, and Schuster, 1993; Taylor and Jackson, 1990; Wingo, 2001; King, 1997; Saunders-Phillips, 2002; Fuller et al., 2005). Few studies, however, have pursued related lines of inquiry (Lille-Blanton, Anthony, and Schuster, 1993; Fuller et al., 2005; Jones-Webb et al., 1997; Delva, Mathiesen, and Kamata, 2001). This omission is striking when placed within the broader context of public health, a discipline which has increasingly emphasized the structural determinants of health and health-related behaviors (Berkman and Kawachi, 2000; Diez-Roux, 2001, 2000, 1998). The micro-level focus of research on drug use patterns among Blacks is, however, consonant with the larger body of research into the etiology of licit and illicit drug use and dependence in the general population, a body of research which has tended to locate the causes of drug use and abuse within the individual, family, and peer group (Galea, Nandi, and Vlahov, 2004). This micro-level orientation is also evident in drug-related interventions: the two principle methods of addressing active drug use and addiction, drug treatment programs and, particularly for Black Americans, the criminal justice system, primarily target individuals (Harrison and Beck, 2003; Drucker, 2002, 1999).

The present analysis investigates the relationships of two dimensions of racial residential segregation, namely residential isolation and concentration, to the subsequent prevalence of IDU among Black adults residing in 93 large US metropolitan statistical areas (MSAs). Residential segregation has been found to adversely affect the physical and mental health status of Black populations across the lifecourse (Hart et al., 1998; Jackson et al., 2000; Polednak, 1996; Acevedo-Garcia, 2001; La Veist, 1989, 2002; Subramanian, Acevedo-Garcia, and Osypuk, 2005). While no research has yet investigated the association of segregation with IDU prevalence among Black adults, available studies allow us to trace a pathway linking these two phenomena. Black residents of segregated communities are at elevated risk of depression, anxiety, and general psychological distress (Aneshensel and Sucoff, 1996; Schultz et al., 2000). These mental health outcomes, in turn, create vulnerability to both engaging in IDU and using injectable drugs (Crofts et al., 1996; Irwin et al., 1996; Khantzian, 1997; Dunn and Laranjeira, 1999; Weinberg and Glantz, 1999; Swendsen and Merikangas, 2000; Fuller et al., 2001; Neaigus et al., 2001). Research regarding the relationship of structural factors to drug use in the general population also testifies to the salience of structural determinants: rates of unemployment, poverty, and arrest and neighborhood disorder have been found to be associated with patterns of IDU and heroin and cocaine use in various geographically defined communities (Friedman et al., 2004; Hunt and Chambers, 1976; Duncan, Duncan, and Strycker, 2002; Nurco, Shaffer, and Cisin, 1984; Bell,

Carlson, and Richard, 1998; Boardman et al., 2001). Since segregation concentrates and amplifies material deprivation in Black communities (Massey and Denton, 1988), this research collectively suggests that segregation may contribute to the prevalence of IDU in Black MSA populations.

Our examination of the relationship of each of two segregation dimensions to IDU prevalence among Black adults reflects emerging recognition in public health that residential segregation is a multidimensional construct. Until recently, public health research regarding segregation and health has almost exclusively conceptualized residential segregation as a uni-dimensional phenomenon consisting of unevenness (defined as the extent to which the racial composition of an MSA deviates from that of its constituent neighborhoods [Massey and Denton, 1988; Massey, White, and Phua, 1996]) and operationalized using the Dissimilarity Index (Subramanian, Acevedo-Garcia, and Osypuk, 2005; Collins and Williams, 1999; Acevedo-Garcia et al., 2003; Acevedo-Garcia, 2000; Acevedo-Garcia and Lochner, 2003). However, Massey and Denton's 1988 factor analysis of 20 segregation measures found that segregation is instead a highly complex phenomenon consisting of multiple dimensions, including but not limited to unevenness, isolation, and concentration, each signaling a particular spatial configuration (Massey and Denton, 1988). This complexity has historical roots: while the overarching origins of racial residential segregation lie largely in efforts to restore and maintain White supremacy in the wake of emancipation, its multidimensional nature is in part a product of variations across geographic areas in the specific methods employed to perpetuate this supremacy (Massey and Denton, 1993; Du Bois, 1962; Foner, 2002; Lemann, 1991). Acevedo-Garcia and others have thus recommended expanding inquiries into segregation and health beyond the current focus on unevenness to include these additional dimensions (Subramanian, Acevedo-Garcia, and Osypuk, 2005; Acevedo-Garcia et al., 2003; Acevedo-Garcia, 2000; Acevedo-Garcia and Lochner, 2003). The Index of Dissimilarity has itself been questioned because of concerns about its conceptual links to health and interpretability (Subramanian, Acevedo-Garcia, and Osypuk, 2005; Acevedo-Garcia et al., 2003; Acevedo-Garcia and Lochner, 2003; Lieberson and Carter, 1982). The following paragraphs review the definitions of the two segregation dimensions studied here, namely residential isolation and concentration, and trace their possible relationship to IDU.

Isolation refers to the extent of potential intra-racial contact for a group in its residential area and thus reflects a combination of the percent of the overall population constituted by that group in an MSA and its distribution across the MSA's neighborhoods (Massey and Denton, 1988; Massey, White, and Phua, 1996). Some have hypothesized that isolation is the segregation dimension that holds the most salience for health, and higher Black isolation has been associated with poorer self-reported health and higher mortality and homicide rates among Black Americans (Subramanian, Acevedo-Garcia, and Osypuk, 2005; Collins and Williams, 1999; Peterson and Krivo, 1993). MSAs with high levels of Black isolation were often produced through White violence and legal actions, including zoning laws and restrictive covenants, designed to exclude Black individuals and families from historically majority-White neighborhoods (Massey and Denton, 1993;

McMurry, 1998; Boyle, 2004; Massey and Mullen, 1984; National Association for the Advancement of Colored People, 2005). Ongoing discrimination by the real estate and banking industries against Black individuals seeking to buy or lease homes outside of majority-Black neighborhoods has perpetuated Black residential isolation (Massey and Denton, 1993; Bonilla-Silva, 2001). Drawing on past research indicates that isolated Black areas suffer high rates of unemployment, poverty, and violence (Denton, 1994; Galster and Mikelsons, 1995), each of which has been linked to IDU or the use of injectable drugs (Friedman et al., 2004; Hunt and Chambers, 1976; Duncan, Duncan, and Strycker, 2002; Nurco, Shaffer, and Cisin, 1984; Bell, Carlsom, and Richard, 1998; Boardman et al., 2001; Dee, 2001; Vermeiren et al., 2003), we posited that MSAs with elevated Black isolation would have a relatively high prevalence of IDU among Black adults.

Concentration refers to "the relative amount of physical space occupied by a minority group in the urban environment" (Massey and Denton, 1988). As Black migration to cities mounted between 1870 and 1970 (Massey and Denton, 1993; Lemann, 1991), Whites' refusal to permit Blacks to live outside strictly delineated areas created highly concentrated Black neighborhoods as existing housing units were divided and sub-divided to create new homes for the burgeoning Black population (Massey and Denton, 1993, Boyle, 2004). Mid-century urban renewal programs that relocated large numbers of Black households from "renewed" areas to majority-Black neighborhoods compounded this concentration (Massey and Denton, 1993, Lemann, 1991). As has been suggested previously (Jones-Webb et al., 1997; Wallace and Wallace, 1998), we posited that the overcrowded conditions characterizing concentrated Black areas would contribute to the prevalence of IDU, in part by creating intensely stressful living conditions. Further, the urban renewal programs that produced some concentrated Black communities might have disrupted social networks and institutions, both in the "renewed" community and the new host community, for a prolonged period (Fullilove, 2001). Such disruptions have been linked to increased IDU and injectable drug use (Wallace and Wallace, 1998, Rhodes et al., in press; Friedman et al., 1999; Sterk-Elifson and Elifson, 1992).

Methods

We tested the hypotheses that isolation and concentration would be positively related to Black IDU prevalence in a sample of 93 large US MSAs using a lagged cross-sectional design, a design commonly used in comparative research in which predictor variables precede the outcome variable (Mellor and Milyo, 2003; Lyson, Torres, and Welsh, 2001; Blakely, Lochner, and Kawachi, 2002), thus allowing the statistical model to mirror the conceptual model's temporal sequence. Defined by the US Census Bureau, MSAs are adjacent counties that include at least one central city home to 50,000 people or more that collectively form a single cohesive socioeconomic unit (Office of Management and Budget, 2000; U.S. Bureau of the Census, 1998). To be included in our sample, MSAs had to have been home to at least 500,000 residents in 1993. Ninety-six MSAs met this criterion. Three MSAs, however, lacked sufficient data on IDU among Black adults and were dropped from the sample. The boundaries of 50 MSAs changed between 1990 and 1998 (U.S. Bureau of the Census, 2005); all measures were operationalized using 1998 boundaries.

Measures

Segregation: Massey and colleagues have identified the isolation and relative concentration indexes (RCIs) as valid measures of their respective constructs (Massey, White, and Phua, 1996). The isolation index captures, for the average member of racial/ethnic group X in a MSA, the percent of individuals sharing his/her residential census tract who are also in group X (Table 9.1) (Lieberson and Carter, 1982; Lieberson and Carter, 1982; Bell, 1954). The RCI compares the surface area of census tracts occupied by one racial/ethnic group in an MSA to that occupied by another (Table 9.1) (Massey and Denton, 1988). As Massey and Denton note, this area-based measure also reflects tract population density: because tract boundaries are partially determined by population size, tracts with a smaller surface area are usually more concentrated than larger tracts (Massey and Denton, 1998). Where MSA boundaries remained constant between 1990 and 1998, index values were obtained from the 1990 Census. Otherwise, we calculated values using 1990 US Census STF1 data.

IDU prevalence among Black adults: It is difficult to estimate IDU prevalence in geographic areas because IDU is both illegal and heavily stigmatized (Larson, Stevens, and Wardlaw, 1994; Larson and Bammer, 1996; Hickman et al., 1999; Cox and Shipley, 1997). Our calculation method estimated the 1998 prevalence of IDU among Black adults in each of the 93 MSAs in a four-stage process: (1) estimating the proportion of injectors in each MSA who are Black; (2) calculating the number of injectors, regardless of race, in each MSA; (3) calculating the prevalence of IDU among Black adults using project data produced in Stages 1 and 2, combined with US Census data on the number of Black and White adults in each MSA in 1998; and (4) validating our IDU prevalence estimates. Project Stages 1 and 2 have been described in detail elsewhere (Cooper et al., 2005; Friedman et al., 2004).

Stage 1: We calculated the proportion of injectors who were Black in 1998 in each of the three databases that documented injectors' encounters with the health-care system and then averaged these database-specific percents to create a single estimate for each MSA (Cooper et al., 2005). The three databases analyzed were the Substance Abuse and Mental Health Administration's (drug) Treatment Entry Data Set (TEDS) and the Center for Disease Control's HIV Counseling and Testing Database (CTS) and AIDS Public Information Database (APID) (Table 9.2). Because the proportion of injectors who were Black in APID reflected racial patterns of both HIV seroprevalence and IDU, APID-based estimates were adjusted for the HIV seroprevalence among Black injectors in the MSA. We analyzed CTS, APID, and TEDS because each captures a slightly different segment of the underlying

Table 9.1 Construct definitions, operational definitions, and formula for calculating two dimensions of racial residential segregation in metropolitan statistical areas (MSAs) (Massey and Denton, 1988, 1998)

Construct and construct definition	Measure and operational definition	Measure formula, range, and interpretation
Isolation: Extent of potential contact among members of a single racial/ethnic group within their residential area.	*Isolation Index:* For the average member of racial/ethnic group X in a MSA, the percent of individuals sharing his/her residential census tract who are also in group X.	*Formula*: $\sum_{i=1}^{N} [x_i/X][x_i/t_i]$ where x_i = no. of members of group X in census tract *i* X = no. of members of group X in the MSA t_i = total population of census tract *i* *Range*: proportion of population in group X – 1.0 *Interpretation*: A value of 1.0 indicates total isolation.
Concentration: "The relative amount of physical space occupied by a minority group in the urban [and suburban] environment" (Massey and Denton, 1988 p. 289)	*Relative Concentration Index:* Ratio of the urban and suburban space occupied by one racial/ethnic group relative to that occupied by another in a MSA.	*Formula:* $\dfrac{\{[\sum_{i=1}^{n} (x_i a_i/X)]/[\sum_{i=1}^{n} (y_i a_i/Y)] - 1\}}{\{[\sum_{i=1}^{n_1} (t_i a_i/T_1)]/[\sum_{1=n_2}^{n} (t_i a_i/T_2)] - 1\}}$ where census tracts are ordered from smallest to largest in surface area and a_i = the area of census tract *i* n_1 = rank of tract where cumulative total population of tracts equals the total minority population of the MSA, summed from smallest tract up n_2 = rank of tract where cumulative population of tracts equals the majority population total from the largest tract down T_1 = total population of tracts from 1 to n_1 T_2 = total population of tracts from n_2 to n y_i = no. of members of group Y in census tract *i* Y = no. of members of group Y in the MSA X, x_i and t_i as defined above *Range*: no lower bound to 1.0 *Interpretation:* A value of 1.0 indicates that X's concentration exceeds Y's concentration to greatest extent possible.

injecting population in each MSA. Collectively, they should represent the racial demographics of this underlying population better than any single database could alone (Cooper et al., 2005).

Stage 2: To calculate the number of injectors in each MSA, we first adjusted the 1998 National Household Survey on Drug Abuse estimate of the number of past-year injectors nationwide to account for under-reporting of IDU and undercoverage

Table 9.2 Description of databases analyzed to calculate the prevalence of injection drug use among black adults in 93 Large US Metropolitan Statistical Areas in 1998

Database characteristics	Treatment episode data system (TEDS)	HIV counseling and testing service	AIDS public information database
Description	SAMHSA database recording admissions to public and private drug treatment facilities licensed by the state.	CDC database documenting HIV test incidents at 11,640 HIV counseling and testing sites. Participating sites include family planning and STD clinics, hospitals and private medical centers, drug treatment programs, correctional facilities, and freestanding counseling and testing clinics (Centers for Disease Control and Prevention, 1999).	CDC database describing newly-diagnosed cases of AIDS.
Coverage	SAMHSA estimates that the 1997 TEDS database described 87% of all admissions to facilities participating in TEDS and 67% of admissions to all treatment programs nationwide (Substance Abuse and Mental Health Services Administration, 2004).	No coverage estimates are available.	85% of all AIDS cases are eventually reported in most areas (U.S. Department of Health and Human Services, 2000).

of injectors (Friedman et al., 2004; Turner et al., 1998; Wright and Gfroerer, 1997). The adjusted nationwide figure was then apportioned to each of the 93 MSAs studied using data on national and MSA-specific patterns of utilization of IDU-related services and past MSA-specific IDU estimates (Friedman et al., 2004).

Stage 3: We calculated the number of Black injectors in each MSA by multiplying the proportion of injectors in the MSA who were Black (from Stage 1) by the estimated number of past-year injectors in that MSA (from Stage 2). Race-specific IDU prevalence estimates were then calculated by dividing the number of Black injectors in each MSA by the total number of Black adults aged 19–65 in that MSA in 1998 obtained from the US Census.

Stage 4: We investigated our estimates' validity by correlating them with two theoretically related variables, the prevalences of heroin and cocaine overdose fatalities (calculated using the CDC's Multiple Cause of Death database) and of injection-related AIDS among Black adults (calculated using APID). Because the prevalences of IDU, overdose mortality, and AIDS among Black adults were each highly correlated with the region of the country in which the MSA was located and the MSA population size and racial composition, we used partial correlation methods to validate our prevalence estimates that controlled these factors. Recognizing possible circularities inherent in analyzing the relationship between IDU prevalence (calculated using APID data) and the prevalence of IDU-related AIDS, we examined the relationship between the prevalence of injection-related AIDS and IDU prevalence calculated both with and without APID data.

Potential Confounders: Past literature suggests that MSA population size, racial/ethnic composition (percent Black and percent White), and geographic region might confound the relationship between segregation and Black IDU prevalence (Friedman et al., 2004; Massey and Denton, 1993). MSAs are nested within regions; dummy variables for these regions were included in the model as fixed effects that serve as proxies for all characteristics that vary across these regions. All of these variables were calculated using 1990 Census data.

Analysis

Multiple linear regression methods were used to test our hypotheses. To insure that our data met our model's assumptions, and to gain a comprehensive understanding of key variables' distributions and inter-relationships, we conducted extensive exploratory data analysis and regression diagnostics (Fox, 1991; Hartwig and Dearing, 1979). Because the distributions of Black IDU prevalence and population size were skewed, they were transformed using a natural log function. Observations that had undue influence in the multiple regression analysis, assessed using the DFFITS test, were re-weighted (Fox, 1991). An examination of variance inflation factors (VIF) in the multiple regression model indicated that the two segregation measures were not collinear and thus could be simultaneously included in the model (Fox, 1991). The percent of MSA residents who were Black was, however, collinear with the isolation index (VIF = 3.5) in this model, a relationship rooted in the index's incorporation of MSA racial composition (Table 9.1) (Massey and Denton. 1988). The variable denoting the percent of MSA residents who were Black was therefore dropped from the main analysis to increase the point estimates' precision (Fox, 1991). To investigate the extent to which the observed relationship between isolation and Black IDU prevalence was an artifact of MSA racial composition, we ran a second regression model that incorporated the variable percent Black and compared the magnitude of the relationship of isolation to IDU prevalence across the two regression models.

Results

Our IDU prevalence estimates indicate that, in half of the MSAs studied, there were 1983 injectors or more per 100,000 Black adults (Table 9.3). IDU prevalence among Black adults ranged considerably (interquartile range = 1422/100,000–2758/100,000; see Fig. 9.1). The validation analysis indicated that these IDU prevalence estimates were positively and significantly associated with the prevalences of overdose deaths ($R = 0.31$, $p = 0.003$) and injection-related AIDS among Black adults ($R = 0.49$, $p < 0.001$); the latter correlation persisted when IDU prevalence estimates were re-calculated without APID data ($R = 0.47$, $p < 0.0001$).

The median adult population size in the 93 MSAs studied was over 720,000 and the median percent of the total MSA population who self-identified as Black was 9% (Table 9.3). Parallel to other research on US MSAs (Massey and Denton, 1993), values were high on both segregation measures. In 47 of the 93 MSAs studied, the isolation index indicated that the average Black adult or child lived in a census tract in which at least 48% of the tract population was Black (Table 9.3; note that this is lower than published values of the isolation index for 1990 calculated using 1990, rather than 1998, MSA boundaries [Massey and Denton, 1993]). In 50% of the MSAs sampled, the RCI was $\geq$0.72, exceeding the 0.60 cutpoint. Massey and colleagues use to identify high levels of segregation in this dimension (Massey, White, and Phua, 1996).

Bivariate regression analyses indicated that the isolation index was not associated with the natural log of IDU prevalence among Black adults (Table 9.4). However, once we controlled for MSA sociodemographic characteristics and region, particularly the West where isolation was low and Black IDU prevalence high (Tardiff, Gross, and Wu, 1989),[11] a positive relationship between the isolation index and

Table 9.3 Sociodemographic characteristics, geographic distribution, and prevalence of Injection Drug Use (IDU) among black residents of 93 large US Metropolitan Statistical Area (MSA) residents

MSA characteristic	Descriptive statistics (N=93)
Adult population size, 1990 (median, range)	720,975 (256, 123 – 5, 684, 204)
Racial/Ethnic Composition, 1990 (median, range)	
% White, Non-Hispanic	79.00 (25.58 – 97.95)
% Black, Non-Hispanic	9.28 (0.90 – 40.59)
Region (number of MSAs in each region)	24
Northeast	21
South	21
Midwest	27
West	
Isolation index, 1990 (median, range)	0.48 (0.05 – 0.84)
Relative concentration index, 1990 (median, range)	0.72 (−1.02 – 0.94)
IDU Prevalence among Black adults age 19 through 65 (per 100,000), 1998 (median, interquartile range)	1,983.43 (1, 421.56 – 2, 758.59)

[11] Statistical tests indicate that region is a suppressor variable, that is, its inclusion in the model allowed the isolation index to explain more of the variance in the dependent variable.

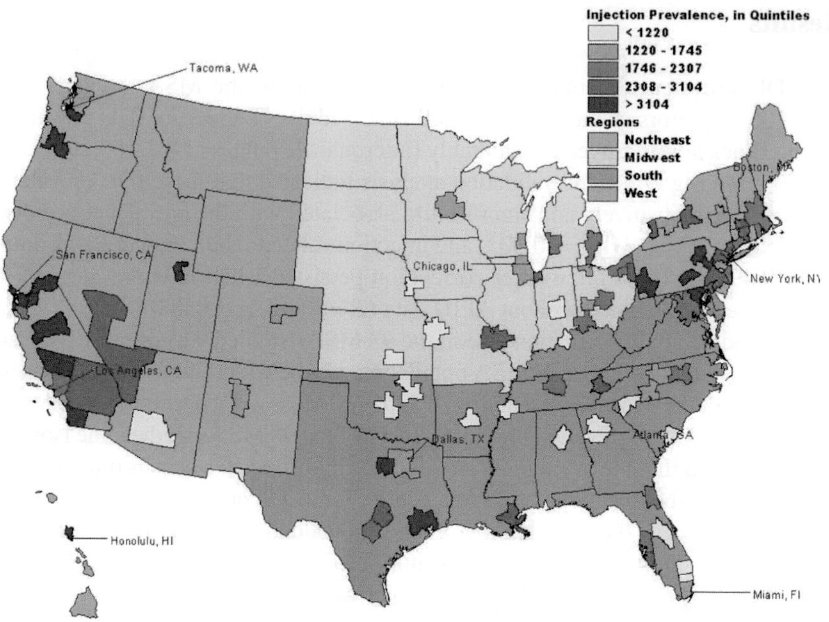

Fig. 9.1 Prevalence of injection drug use per 100,000 Black adult residents of large US metropolitan statistical areas in 1998 (See also Plate 13 in the Colour Plate Section)

the natural log of IDU prevalence among Black adults emerged (Table 9.4). By exponentiating the regression equation, we find that an increase of 0.50 in the isolation index was associated with a 23% increase in the (unlogged) IDU prevalence among Black adults. Adding percent Black to the model only slightly altered the magnitude of the relationship between isolation and the natural log of IDU

Table 9.4 Bivariate and multiple linear regression of two segregation dimensions on the natural log of the prevalence of injection drug use among black adult residents in 93 large US Metropolitan Statistical Areas (MSAs)

Covariates	Unadjusted coefficient (SE)	Adjusted coefficient (SE)
Intercept	N/A	3.72 (0.55)***
Natural log of the adult population size	0.08 (0.10)	−0.12 (0.09)
% Population Non-Hispanic White	−0.003 (0.002)	0.002 (0.002)
Region (reference category: Northeast)		
South	−0.35 (0.07)***	−0.33 (0.07)***
Midwest	−0.26 (0.07)**	−0.29 (0.07)***
West	0.03 (0.07)	0.15 (0.08)
Isolation index	−0.26 (0.15)	0.41 (0.20)*
Relative concentration index	0.05 (0.08)	0.07 (0.08)

*$p < 0.05$ **$p < 0.01$ ***$p < 0.0001$

prevalence, from $b = 0.41$ to $b = 0.43$, suggesting that the relationship between isolation and IDU prevalence was not an artifact of MSA racial composition. There was no relationship between the RCI and the log of Black IDU prevalence in bivariate or multivariate analyses. The model accounted for 31% of the outcome's variation in these MSAs.

Discussion

MSAs with higher levels of Black residential isolation in 1990 had a higher prevalence of IDU among Black adults in 1998 than other MSAs. It is noteworthy, however, that concentration was not associated with IDU prevalence. These divergent relationships testify to the distinct association each segregation dimension has with particular health outcomes and, more specifically, lend support to the proposition that residential isolation may be the segregation dimension that holds particular significance for health (Subramanian, Acevedo-Garcia, and Osypuk, 2005; Collins and Williams, 1999; Acevedo-Garcia et al., 2003; Acevedo-Garcia, 2000; Acevedo-Garica and Lochner, 2003).

There are multiple pathways through which Black residential isolation could be linked to IDU prevalence among Black adults in MSAs. In contrast to concentrated Black census tracts, which are primarily characterized by elevated poverty rates, isolated Black tracts tend to be associated with multiple indicators of social and material disadvantage, including poverty, unemployment, and violence (Peterson and Krivo, 1993; Denton, 1994; Galster and Mikelsons, 1995). Each of these characteristics, alone or in combination, could create vulnerability to IDU. Exposure to community violence creates a risk of illicit drug use (including heroin and cocaine use), perhaps because witnesses or victims of violence use drugs to manage subsequent depression, fear, and anxiety (Vermeiren, 2003). Likewise, poverty may promote the transition to (and continuation of) injecting among active drug users because IDU is a more efficient drug administration method than sniffing or snorting (Neaigus et al., 2001). Moreover, evidence suggests that injectable drugs are more overtly available in disadvantaged neighborhoods than they are elsewhere (Crum, Lillie-Blanton, and Anthony, 1996).

Another interpretation of our findings is also possible. Isolated Black census tracts may protect against IDU by creating a place where Black residents encounter, daily, organizations fostering a positive Black identity; endure little White-initiated interpersonal discrimination; and seek and offer solace and racial solidarity in the face of racial inequality and discrimination. The absence of these protections may create vulnerability to IDU among Black individuals living outside isolated Black census tracts (Yen et al., 1999; Martin, Tuch, and Roman, 2003; Bennet et al., 2005; Richman, Flaherty, and Rospenda, 1996; Landrine and Klonoff, 2000; Guthrie et al., 2002; Gibbons et al., 2004; Caldwell et al., 2004), though this risk may be offset by the greater access these individuals tend to have to socioeconomic resources and reduced exposure to violence (Jones-Webb et al., 1997).

Further research is needed to elucidate the pathways through which elevated isolation is associated with IDU prevalence and to determine, as we could not in this ecologic study of MSAs, whether the relationship is evident only among Black residents of isolated Black tracts, Black individuals residing outside such tracts, or both. Two additional avenues of research are also indicated. First, a positive relationship between residential isolation and IDU prevalence may exist in other racial/ethnic groups. Puerto Ricans, who appear to have the highest prevalence of cocaine use of all Latino groups (Amaro et al., 1990), are also the sole racial/ethnic group experiencing a level of isolation that approaches that of Black Americans (Massey and Denton, 1993). Research is needed to determine whether isolation promotes IDU among Puerto Ricans, and perhaps other racial/ethnic groups. Our findings also suggest an additional line of inquiry: studying the extent to which Black isolation shapes the distribution of IDU-related health problems across Black MSA populations by elevating the prevalence of IDU. This inquiry is particularly pressing given the high prevalence of overdose deaths and IDU-related AIDS borne by Black Americans (Tardiff, Gross, and Wu, 1989; Substance Abuse and Mental Health Services Administration, 2003; Harlow, 1990; Galea, Ahern, and Tardiff, 2003; Friedman et al., 1987; Friedman et al., 1988; Novick et al., 1989; Selik et al., 1989; Selik, Castro, and Pappaioanou, 1988; Centers for Disease Control and Prevention, 2001). Collectively, the results of this research could help injectors and their allies identify communities in need of drug-related health services, including drug treatment and syringe exchange programs.

Our findings must be understood in the light of their limitations, which lie principally in the study's ecological and cross-sectional design and measurement of IDU prevalence among Black adults. Because our unit of analysis was the MSA, we could not investigate the role of neighborhood- or individual-level factors, such as socioeconomic status, gender, and age, as confounders or modifiers of the relationship between MSA-level residential isolation (or concentration) and Black IDU prevalence (Diez-Roux, 2001, 1998). Multilevel research into these possibilities should address these limitations. Additionally, though we used a lagged cross-sectional design in which predictors pre-dated the outcome, the possibility of autocorrelation precludes assessing the causal direction of our findings. Our service-based method of calculating IDU prevalence leaves room for an alternative interpretation of our findings: while it is possible that residential isolation produces a higher prevalence of IDU, it is also possible that isolation results in a higher prevalence of health problems among injectors and thus greater use of drug-related services. Possibly, countering this bias, however, is the fact that isolated Black areas of segregated MSAs tend to be medically underserved (Institute of Medicine Committee on Understanding and Eliminating Racial and Ethnic Disparities in Health Care, 2003), and CTS and TEDS capture individuals accessing routine or non-emergency health services.

We place our findings in the context of past research regarding racial inequality and discrimination and health. Our analysis suggests that IDU should be added to the growing list of the adverse health behaviors and outcomes among Black Americans that may be generated by racial inequality and discrimination (La Veist,

2002; Polednak, 1996). Further and more specifically, this analysis also adds a new dimension to the body of research documenting the relationship between inequitable and discriminatory racial relations and licit and illicit drug use and abuse. Investigators have concluded that Black adolescents and adults who report higher levels of interpersonal or everyday discrimination are more likely to report lifetime smoking, smoking more frequently, and engaging in problem drinking than other Black individuals (Yen et al., 1999; Martin, Tuch, and Roman, 2003; Bennet et al., 2005; Richman, Flaherty, and Rospenda, 1996; Landrine and Klonoff, 2000; Guthrie et al., 2002; Gibbons et al., 2004). Likewise, Black adults reporting higher levels of internalized racism also report consuming more alcohol than other Black adults (Taylor and Jackson, 1990). Our findings extend this body of research by concluding that structural, in addition to intra-psychic and interpersonal, manifestations of racial inequality and discrimination may adversely shape IDU rates among Black adults.

If substantiated by additional research, our finding that a structural factor is related to Black IDU prevalence may also hold consequence for US domestic drug policy. As noted earlier, the criminal justice system, which locates cause within the individual and calls it culpability, plays a major role in the US government's response to illicit drug use among Black Americans that dwarfs its role in addressing White drug use (Harrison and Beck, 2003; Drucker, 2002, 1999). If, as our research suggests, the cause of IDU among Black adult MSA residents lies partially in isolation, related prevention, and intervention efforts may also benefit from altering social structures – and, more broadly, from eradicating racial inequality and discrimination in the USA – rather than from arresting and incarcerating large numbers of Black individuals.

Acknowledgment The authors would like to thank Drs. Greg Falkin and Holly Hagan for their invaluable comments on this paper. We would also like to thank the US Census Bureau for calculating the segregation indices and making them publicly available. While conducting this research, the first author was supported by a Behavioral Science Training in Drug Abuse Research postdoctoral fellowship sponsored by the Mental and Health Research Association of New York City, Inc. and National Development and Research Institutes with funding from the National Institute on Drug Abuse (5T32 DA07233). All other authors were supported by NIDA grant # R01 DA13336 ("Community Vulnerability and Response to IDU-Related HIV"). Points of view, opinions, and conclusions in this paper do not necessarily represent the official position of the US Government, Medical Health and Research Association, or National Development and Research Institutes, Inc.

Chapter 10
The Relationship of Ecological Containment and Heroin Practices

Avelardo Valdez and Alice Cepeda

Abstract In this chapter, we address how ecological and spatial factors contribute to behaviors that led to heroin use among Mexican-origin persons in the United States. Most of this contemporary drug using population is embedded in socio-economic environments that shape heroin use and other deviant behaviors. Using the experiences of San Antonio's Mexican American population, the chapter illustrates how the dynamics and consequences of heroin use among this group can best be understood when considering the intersection of context and culture from both a historical and current framework. Specifically, it addresses how the constellation of social-historical patterns of inequality, ecological and deviance containment, and proximity to vice districts in San Antonio sustained and exacerbated heroin use among this population. What emerges from this analysis is how these historical precedents together with ecological and urban spatial processes provide a theoretical framework that helps to explain Mexican American heroin practices.

Introduction

Participation in drug-related behavior among today's ethnic and racial minorities has been highly concentrated in mostly urbanized and socially disadvantaged minority communities. Mexican American drug users who, over the last six decades, have developed a clear preference for heroin and other opiates exemplify this social phenomenon (Bullington 1977; Casavantes 1976; Desmond and Maddux 1984; Moore 1978; National Institute of Justice 1996). The dynamics and consequences of heroin use among this group can best be understood when considering the intersection of context and culture from both a historical and current framework. Our argument is that decades of deviance containment, as illustrated by the location of vice districts in minority communities, shape perceptions, cultural patterns of learning, and opportunities. In this sense, heroin use and addiction is best explained by community level structural factors rather than individual characteristics.

A. Valdez
Graduate College of Social Work, University of Houston, Houston, TX, USA

Y.F. Thomas et al. (eds.), *Geography and Drug Addiction*,

The macro-social or community level explanation asks what it is about community structures and cultures that produce behaviors such as interpersonal violence, drug use, high-risk sex, etc. Is it the characteristics of the community that produce high rates of drug use or is it the characteristics of the persons living in these communities? This perspective goes back to the early work of Shaw and McKay's social disorganization model of delinquency in ethnic communities in Chicago in the 1930s and 1940s. In this model the goal of the research was to identify social-structural characteristics (low-income, ethnic heterogeneity and residential mobility) that lead to social disorganization and then to high rates of crime. This "leads to the rejection of individualistic explanations of deviant behavior focusing instead on the processes by which these behaviors were transmitted across generations in areas of social disorganization and weak social controls" (Sampson & Lauritsen 1994:45).

In this chapter the emphasis is on how ecological and spatial factors contribute to behaviors that led to heroin use among Mexican-origin persons in the United States. Most of this drug using population is embedded within a socio-economic environment that is an exposure risk, independent from the risks stemming from ascribed and achieved characteristics of individuals (Rankin & Quane 2000). The scarcity of meaningful employment and social opportunities within the community shapes the behaviors in these communities as well as associations with diverse cultural groups. Within this context, economic realities necessitate that family and peer networks create and maintain bonds and share resources. These behaviors are reinforced by ecological isolation of many of these communities where contact is mostly with others with similar social attributes (Bauder 2002). Communities characterized by high drug use and other risks behaviors, make it difficult for members not to be influenced by these behaviors. With limited societal resources or social capital outside the network, members are constrained and limited in opportunities to escape this environment (Rankin & Quane 2000). We discuss how this process is linked to the historical and current patterns of spatial context as it relates to this ethnic group's relationship to outsiders.

The History of Vice Districts and Heroin Use

The persistent presence of drug use among minorities during the last 30 years (1970–2000) is associated with the historical trend of segregating vice districts with their brothels, gambling joints, *narcotics*, and other stigmatized pleasures in or adjacent to segregated immigrant and minority ghettos and barrios. These "ethnic vice districts" have historically provided an ecological niche for immigrants and minorities in both legal and illegal entrepreneurial economic activities that otherwise would not have a chance of developing (Light 1977). During the first half of the 20th century, these areas tended to be located in low-rent urban districts that were easily accessible to the general population yet isolated from conventional businesses and residents. These vice or red light districts were found in most major cities like the Black Belt in Chicago, Storyville in New Orleans, and the Tenderloin District in New York City (Kornblum 1993). During prohibition (1920–1933), these were the areas where speakeasies and juke joints were located that offered alcohol and other vices primarily to whites. Also, during this period marijuana, cocaine, opiates (morphine and heroin), and other drugs began to be used throughout the U.S. Access to these drugs was primarily in these vice districts especially, after they became illegal substances. It was in this manner that these drugs were introduced and integrated into the leisure life of urban Hispanics, blacks, and other minorities. The diffusion of these drugs, especially heroin, was facilitated by minorities themselves becoming involved in the marketing of these illegal drugs within their communities and to outsiders as part of other informal economic activities.

As importantly, during the late 1930s, heroin use became more stigmatized compared to earlier periods. As has been documented, "opium use was seen as relatively innocuous, morphine more dangerous, and heroin especially dangerous" (O'Donnell and Jones 1970). As a result, opiates were met with mild disapproval during the late 1800s and early 1900s, but by the late 1940s, heroin was seen with a marked disapproval. Public opinion was reinforced by the perception that the intravenous heroin user was part of a deviant subculture that included criminality, sexual promiscuity, and other "deviant" behaviors. By the end of this period and in the mid 1960s, the use of heroin by whites began to decline and increase among blacks, Mexicans, and Puerto Ricans (Ball and Chambers 1970). Increasingly, heroin use and marketing became associated with minorities living in distinct geographic areas of major cities. The perception that heroin use was primarily a minority phenomenon persists to this day. However, we do recognize that there have been brief periods when heroin use spiked among whites like after the Vietnam War. This chapter delineates the consequences of these socio-historical and ecological processes for contemporary heroin use and addiction.

Contemporary Heroin Use and Trends Among Mexican Americans in San Antonio

The Mexican American drug using population has demonstrated a clear preference for heroin and other opiates and a nearly universal use of intravenous injection as the route of administration (Bullington 1977; National Institute of Justice 1996) Moreover, Mexican Americans have had a relatively high rate of heroin use compared to other groups (ADAM 1998; Desmond & Maddux 1984; DUF 1995, 1996). For instance, 4 of the 10 cities with at least 30% of Hispanic arrestees had the highest percent of arrestees testing positive for heroin (Rio Alba, NM, Manhattan, Albuquerque and San Antonio). Moreover, 28% of Hispanics were admitted for treatment episodes in the U.S. were admitted for heroin compared to 16% and 12% for Blacks and whites, respectively (Substance Abuse and Mental Health Services Administration 2003). Similarly, lifetime heroin use was 3.9% for Hispanics compared to 2.6% for both whites and Blacks (Morbidity and Mortality Weekly Report 2004).

San Antonio's high rates of heroin use are associated with impoverished neighborhoods on the West Side of the city in which crime, delinquent behavior, and gang activity are concentrated. The proximity of Mexico, a major source of heroin for this area of the United States, has made this as well as other illegal substances more accessible to users in San Antonio (and South Texas) compared to other regions of the U.S. For almost two decades, the Office of National Drug Control Policy (ONDCP), pursuant to the Anti-Drug Abuse Act, has designated the Southwest border as a High Intensity Drug Trafficking Area (HIDTA). The South Texas map indicates that this HIDTA includes all the counties adjacent to the U.S./Mexico border. This area is recognized as being a major staging area for bi-national narcotic drug trafficking operations entering the rest of the United States. The map also indicates the vast area Texas shares with Mexico as well as the proximity of the U.S./Mexico border to San Antonio and other cities. These factors result in Mexican Americans living near the U.S./Mexico border having more access to heroin sources and other drugs compared to others living far from the border. More importantly, such easy accessibility to these substances has resulted in the proliferation of Mexican American small-time drug dealers and large scale entrepreneurs that target Mexican American drug users.

High Intensity Drug Trafficking Area in South Texas

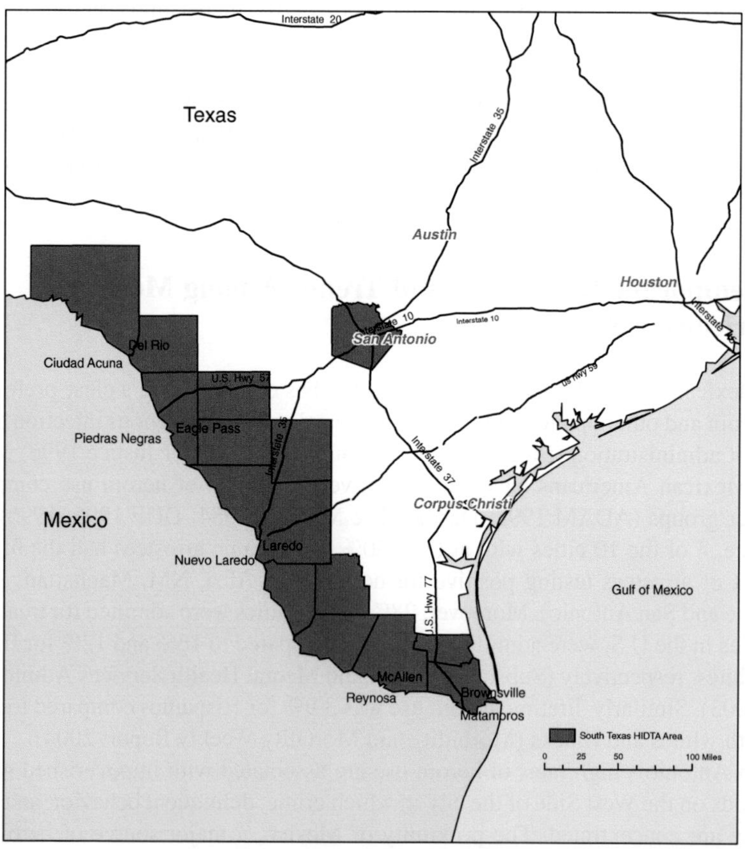

Although the most popular mode of heroin administration for Mexican Americans has been intravenously, an increasing number are reportedly using heroin intra-nasally and by other non-injecting methods (Maxwell 1999; Ramos 1995). Data on male heroin users entering publicly funded drug abuse treatment in the state of Texas in 1998 indicates that 9% reported intranasal use (Center for Epidemiological Work Group 1999). In an ethnographic study of San Antonio heroin users, Ramos (1995) found that a large number of adult respondents smoked heroin. The Texas Commission on Alcohol and Drug Abuse (TCADA) estimated that in 1999, the number of heroin users in San Antonio (Bexar County) was approximately 8,936 (Maxwell 1999). Authors of this chapter have also found high rates of non-injecting heroin use (NIU) in studies they have conducted on Mexican American-male gang members and female adolescents associated with these male gang members (Valdez 2005; Valdez, Kaplan, and Cepeda 2000; Yin, Valdez, Mata, and Kaplan 1996). This growth of NIUs has very important implications for addiction, the transmission of HIV and related social deviant behaviors, such as crime and high-risk sexual behavior.

Presented here is how the constellation of social-historical patterns of inequality, and vice district segregation sustains and exacerbates heroin use leading to addiction. We discuss how heroin was introduced and diffused into the Mexican American population by macro-social patterns of inequality, ecological containment of the truly disadvantaged, and proximity to urban vice districts. Specifically, it is argued that San Antonio's stigmatized, segregated characteristics, and historical patterns of vice market concentration have led to cultural adaptations that have sustained the use of heroin and addiction among Mexican Americans to the present day. Moreover, we will explain how new modes of heroin administration such as non-injecting practices is still influenced by these ecological and cultural processes. Within this framework, we address four specific factors that have contributed to these patterns of heroin use among Mexican Americans; the ecological containment and socio-patterns of inequality, changing market dynamics, accessibility, and increased heroin use and transition to injecting.

Methods

This research is based on data collected through a NIDA funded research project focused on determining the risks of transitioning to injecting and of acquiring HIV among Mexican American NIUs. The purpose of the study was to better understand the specific drug risk practices that NIUs engage in, and the influence of the drug market and social context on their drug and other risk behaviors. This study implemented a multi-methods design. Quantitative data collection consisted of a baseline questionnaire and two follow up interviews every 6 months among 300 non-injecting San Antonio heroin users who were 16 years and older, and were not in drug treatment. All had Mexican or Mexican American ethnic backgrounds.

Extensive qualitative data was collected through ethnographic interviews and field observations. Two outreach fieldworkers under the supervision of the authors conducted extensive fieldwork in the targeted community. This fieldwork resulted in the identification and observation of sites that were mapped based on spatial (neighborhoods) and temporal (time of day, day of the week) characteristics. In-depth, multiple ethnographic interviews were conducted with 20 subjects. Finally, field notes were collected by the outreach specialists based on their observations in the community while recruiting the targeted baseline sample. The ethnographic interviews were recorded and transcribed. The ethnographic interviews, field notes, and focus group notes were combined into an electronic qualitative database.

The Ecological Containment and Social Patterns of Inequality

San Antonio, Texas is located 140 miles from the U.S. – Mexico border. The population in 2000 was estimated to be 1.2 million, with approximately 60% of Mexican descent (U.S. Bureau of the Census 2000). The West Side community in San Antonio is comprised predominantly of Mexican-origin persons and is one of the poorest urban areas in the United States. According to the census data, the per capita income was $5,098, and the median household income was $14,352 for 22 census tracts that comprise this community. Fifty-five percent of the West Side families had children living in poverty, and only 23% of the families received public assistance (U.S. Bureau of the Census 2000, 2001, 2002). It is also an area that has a high concentration of crime, violence, substance use, and some of the highest rates of teenage pregnancy in Texas (second highest in the nation) (Alan Guttmacher Institute (AGI) 1995; Yin et al. 1996). A more relevant fact is that in these neighborhoods, there is a high concentration of heroin use.

We argue that the high rates of heroin use in this community has its origins in early settlement patterns associated with close proximity to vice districts. The West Side was the traditional settlement for Mexican immigrants beginning in the 1920s, when they arrived in large numbers escaping the political turmoil and poverty of Mexico. This migration steadily increased through the 1940s as San Antonio experienced an urbanization process. However, unlike other newcomers to the city, Mexicans were segregated into limited geographic areas that were primarily in neighborhoods on the near West Side adjacent to the central business district. These neighborhoods were characterized by inadequate housing, inferior infrastructure (water, sewers, electricity, etc), and limited public services (schools, police, clinics, hospitals). In the 1930s, San Antonio's barrios had some of the highest rates of tuberculosis, venereal diseases, and infant mortality in the United States (Garcia 1989).

The Mexican barrios of San Antonio were also where the city's vice was concentrated in close proximity to the Mexican businesses located immediately adjacent to downtown. In this sense, it is similar to what was found in other cities such as New York and Chicago, where vice districts were typically segregated in minority ghettos or barrios. According to one historian (Bowser 2003), in the first half of this century, this area of San Antonio had over 90 bordellos, saloons, gambling dens, and small shacks where prostitutes plied their trade. These establishments were sustained by

Anglo clients throughout the city. This area simultaneously was the entertainment center of the Mexican population where Spanish language theaters (vaudeville and film), cantinas, restaurants, outdoor markets, and dance halls were situated. It was the primary source of liquor during prohibition and illegal drugs in the late 1930s. This red light district operated under the approval of the city until 1941, when it was closed under the pressure of the military (Bowser 2003). Similarly, other Mexican American communities were located adjacent to vice districts like Los Angeles's Chavez Ravine, East Los Angeles and South Central that were located adjacent to that city's downtown vice districts (Normark 1999). U.S. border cities such as El Paso and Laredo were situated in close proximity from red light districts such Boys Town in Nuevo Laredo where prostitution was legal.

Identified in Fig. 10.1 is the location of San Antonio's vice district (1911–1941) superimposed on a current map of San Antonio's West Side and other Mexican American residential areas. Also located on this map is the enlarged and detailed representation of the vice district (1911–1941) and location of major theaters (Zaragoza, Nacional and La Alameda) and two retailers (Penners and Kleins) that catered to Mexican Americans. Included is the Plaza de Zacate, a major gathering place that featured Mexican musicians, other entertainers and outdoor restaurants (Broyles-Gonzalez 2001). The area was also a source of opium and heroin for white users including military personnel, medical professionals, ex-medical patients and

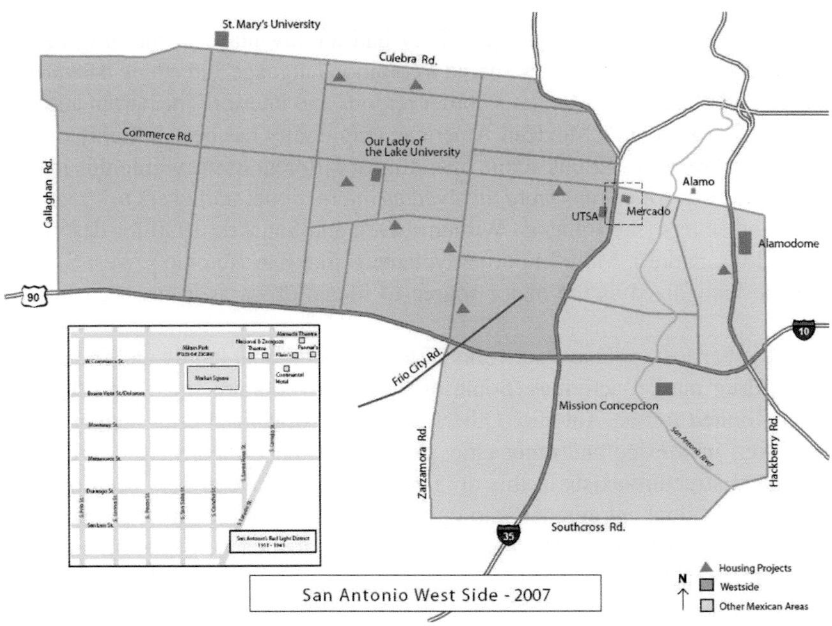

Fig. 10.1 San Antonio's red light district [note: includes two adjacent blocks east of Santa Rosa St. that were not recognized in David Bowser's (adapted from Bowser 2003) map of San Antonio's red light district.]

entertainers (i.e., musicians, etc). It is these persons we suspect may have introduced heroin to Mexican Americans.

Continual use of heroin during the following decades resulted in Mexican Americans developing a distinct heroin subculture in San Antonio (and other Southwestern cities and towns) among Mexican American users, or *tecatos. T*he term *tecato* tends to denote a chronic or career heroin user with a criminal orientation and repeated involvement with the criminal justice system (Quintero and Estrada 2000; Ramos 1995; Valdez et al. 2000). As a result, *tecatos* have developed a distinct street identity that revolves around a lifestyle characterized by heroin use, criminality, incarceration, unique style of dress, tattoos, and highly exclusive social networks. They have traditionally been stigmatized and socially isolated from the larger Mexican American community including from that of other drug users and street-oriented groups. For these and other reasons, investigators have described Mexican American heroin users as "clannish" (Casavantes 1976) even within their own neighborhoods and in the prison system.

Changing Market Dynamics

Parallel to the development of the *tecato* subculture, Mexican Americans became involved in the distribution and selling of opium, morphine, and heroin (Bullington 1977; Desmond and Maddux 1984). This period (just prior to World War II) corresponds with increased numbers of Mexicans using heroin in cities like San Antonio, El Paso, Houston, and Albuquerque. They had an advantage in the drug dealing business over others, in that they shared a common language and ethnic background with drug wholesalers in Mexico's border regions and interior. The heroin and drug market in San Antonio's Mexican American community has been a highly diversified marketplace with various actors operating at different levels within this market. That is, Mexican Americans were involved as international traffickers to street-level dealers and sellers (see Johnson, Williams, Dei, and Sanabria 1990 for description of U.S. drug market). More importantly, barrios like San Antonio's West Side began to be recognized as the major source of illegal drugs as it has been in other cities.

In these barrios, exclusive networks of multigenerational family and friends carried out drug market activities. Some of these networks are small, and their operations limited to San Antonio. Others are larger, organized drug networks with connections in Mexico and other cities throughout the United States. Although a hierarchical structure exists in this drug market, there have always been opportunities for individual entrepreneurs to operate given their more direct primary drug market sources in Mexico. This has resulted in an abundance of heroin and other popular drugs such as marijuana, cocaine, and prescription pills including tranquilizers, painkillers, and diet pills from a variety of resources (Valdez and Sifaneck 2004). The easy availability and abundance of illegal drugs in South Texas is distinct from other regions of the U.S. that have to go through various middle-men and over distances to research the market and consumers.

The West Side heroin market radically changed when Mexican American prison gangs entered into the marketplace in the early 1980s. This decade experienced a dramatic increase in incarceration rates in Texas and the United States, largely as a result of federal and state drug laws passed during the last three decades (Gray 1998). Thus, the American prison population increased to approximately 2 million during the 1990s. Young Hispanics and blacks disproportionately comprised the majority of this incarcerated population in the U.S. (Harrison and Beck 2003). Eventually, this incarcerated population returned to communities like the West Side that offered little opportunity, particularly for ex-felons. These ex-prisoners comprised the core of the three Chicago prison gangs that emerged in San Antonio. In these communities, convicted felons were absorbed into the everyday life of residents through family and friendship networks.

Pura Vida, one of these prison gangs, was the most visible in the West Side of San Antonio, home for many of the parolees. It was within this community, particularly in the city housing projects (see map), that they established their presence in the heroin marketplace. During the course of this study (1995–1999), *Pura Vida* gradually gained control of a large portion of the heroin market on the West Side. The takeover of the heroin market was accomplished through a highly regimented vertical organization using ex-felons recruited in the prisons and connections in Mexico. For instance, drug dealers were allowed to sell in these geographic areas, provided they pay a 10% surcharge, known as *el diez por ciento* to the gang. *Pura Vida* members enforce the surcharge through intimidation, physical threats, violence, and murder. It was in this manner that the heroin market transformed from one that was relatively open to one monopolized by the prison gang.

Changes in heroin use among West Side drug users in the early 90s were also the result of several other market factors. One of these was associated with consequences of federal drug policies that successfully increased interdiction of marijuana along the US/Mexico border in contrast to less bulk drugs such as heroin and cocaine. As a result, there was a drop in prices of the latter two drugs relative to marijuana. Also, there began to appear less black tar heroin and more brown powder heroin in the San Antonio drug market. Black Tar was a highly gummy form of heroin that needed to be dissolved by heating it in a "cooker" (usually a spoon or bottle cap) and drawn into a syringe using cotton as a filter. To use black tar non-intravenously, users had to dissolve it in the same manner but had to wait for it to dry. Once solidified, it was crushed into a powder form to snort it, smoke it, or "spray" it into a nasal cavity. Even though Black Tar heroin users could have snorted or used it in some other way, it was mostly injected. On the other hand, brown heroin was available in powder form that just needed to be heated to use intravenously or crushed in a finer powder and snorted. Brown heroin made it easier to use in more casual settings like powdered cocaine. *Brown* was being supplied to West Side users and sellers beginning in the late 90s (Valdez and Sifaneck 2004). Moreover, fieldwork data indicates that the purity (potency) of street-retail heroin has increased compared to earlier decades.

Accessibility

Mexican Americans living near the U.S./ Mexico border and involved in drug distribution have more direct access to primary illegal drug sources than others who must go through multiple layers of middle-men. The further away you are decreases profits and increases risks. Cultivating a drug connection in Mexico may also be easier for a Mexican American who may be more acceptable to a Mexican dealer since they share a common ethnic background. This type of access for persons in the Southwest makes it unnecessary for them to develop an organizational hierarchy to traffic in drugs as opposed to those residing in cities and towns located in other areas of the U.S. As a result, more drugs flowed through South Texas than any other region of the United States including heroin (Bucardo et al., 2005).

Brown heroin was marketed on the West Side just like any other retailer or wholesaler would market a more conventional product. Traffickers and dealers began to make the heroin more accessible by aggressive marketing strategies, including making the product more easily available, and lowering prices. Accessibility was facilitated by an increase of neighborhood-based user sellers and sellers at local bars, public and semi-public locations, and private residences.

Youth street gang members were one group who were recruited as independent sellers or as more formal associates of the gang. There were approximately 26 active youth street gangs in the geographical area identified as the West Side. There was a history of conflict between the street gangs and the prison gangs, often centered on control of drug markets. Nonetheless, many delinquent barrio youth, including gang members, have been eager to join adult prison gangs, giving the adult gang member's *status* within the street culture of the West Side of San Antonio. Also, many street gang members were co-opted by a prison gang when they were incarcerated (Valdez 2005). The "pinto" (prison veteran) was seen as upholding the highest of values of the "code of the streets" so admired by most young street gang members. In addition, many of these delinquent youth assume that by joining adult prison gangs, they would have access to more lucrative illegal enterprises and increased levels of protection from other street rivals.

Prison gangs like *Pura Vida* also made heroin more accessible by "fronting" drugs to potential sellers. Fronting is a technique that is commonly used by drug dealers. It is somewhat similar to giving credit to someone. These adult prison gangs will "front" some drugs to a gang member or other sellers and then they would be obligated to pay the debt after the drugs are sold. If the seller does not pay his full debt in the manner agreed upon, then he may be forced to continue to sell for the gang even after his debt is paid. Many times just the intimidation that *Pura Vida* and the other gangs wield in the neighborhood obligates them to sell for them. The positive side of fronting for the seller is that he now has the protection of the prison gang in the hood from rival youth and other adult criminals in the same area.

Accessibility to heroin was facilitated by lowering the prices and making it available in smaller quantities than in the past. For instance, it became common to have people selling nickel ($5) or dime ($10) bags of heroin. One young heroin user commented how easy it was to "score a $10 paper to party just to "kick back." You can get it anywhere in the neighborhood, from all kinds of people. Even young kids can get it for you." Another user made this observation about heroin availability:

> I live here in the west-side and it's know for *brown*. Where I live around my house there are a lot of connections. It's all over the place; it's all around me. It's easy for me to get. Even if I don't have money. I can just go and tell them, "hey I want some." More than likely they'll give me some, you know what I mean?

This type of accessibility makes heroin easily available for drug users on the West Side. Norma is a 22-year-old Mexican American that has been snorting heroin for about 3 years. She is very conscious that her NIU heroin could lead to more chronic use and possible addiction. But, it is hard to stay away from heroin when it is so pervasive and easily available. Here she describes an older neighbor who lives right across the street. She states:

> I went over to his house the other day. I was going to buy some weed from his nephew who I know. They are always partying over there like watching games, playing basketball in the street with my brothers. They are just neighbors. When I went over there, I met their uncle (about 35 years old). He had a balloon in his mouth. I ask, what is that a balloon? He said, "Do you know what it is for?" I said, for *chiva* (heroin). He said, "Do you want some?" I said, yes. So, he hooked me up with a couple of lines. He has been giving me some ever since then.

Norma describes this man as an injecting heroin user with large quantities of heroin.

> He sometimes comes over and asks me if I want any. Sometimes I go over there and he gives me some. When I get it, I bring it back here. Sometimes I share it with my sister or I just do by myself.
>
> When I asked, "Why do you think that he is giving it you?" She answered, "For sex, but he's not going to get any. I am not that way. Exchange sex for drugs, never."

Unfortunately, one of Norma's other source of drugs was her 23-year-old brother who also lives in the house with his wife and baby. Her other sources are her sister and "aunt" Cindy. Although, she continued to snort heroin for several months, she eventually transitioned to injecting.

Increased Heroin Use and Transition to Injecting

Along with the increase in NIUs and accessibility, there has been a process of transitioning to injecting heroin use, within this drug using community. As mentioned previously, during the last 50 years in Southwestern cities and towns Mexican American users, or *tecatos*, were viewed very disparagingly. *Tecatos* within these communities have a reputation of being untrustworthy and unreliable. Much of this is associated with the fact that heroin addicts spend most of their time scamming to get the resources for their next fix. One young poly drug user stated: "Nobody wanted to be a tecato. They'll steal from their own mother." As a consequence, most drug users tend to avoid engaging in injecting heroin for fear of being associated with this lifestyle (i.e., addiction, criminal involvement, incarceration, HIV infection, etc.) yet continued their non-injecting heroin use. However, as they continue to engage in NIU, many are unable to maintain their balance and transition to injecting.

Sara is an 18 year-old Mexican American female that has been snorting heroin for about 3 years. She has two young children, the first of which she had when she was 16 years old. Sara views sniffing brown as a way to minimize the risk of injecting and consequently addiction. She did heroin for the first time when she was 15 years old. "My friend and I bought a 'dime'. It made me feel sweaty, ugly and high. We were just hanging out at the Cassiano Courts with the guys and we decided to buy it. But it was a better buzz (than other drugs). A good high."

Sara, like many NIUs, expressed a concern about injecting and becoming addicted since she was familiar with the consequences on family and friends. She states:

> My *comadre* and my best friend are both addicted. I heard that my *comadre* is selling her body. She was into having a lot of sex when she was younger. Hanging out with gangs getting high all the time. Now she is really in bad shape. I feel sorry for her. My friend is the same way.

More importantly, many have personally experienced how heroin dependence and addiction destroyed the lives of their fathers, older brothers, or mothers. Sara said "Once my father started using heroin he didn't really care about the family."

Other NIUs avoided injecting because they were aware of the associated HIV/AIDS risks. This was the result of several community-based HIV/AIDS outreach educational programs that targeted injecting users for the last 10 years. Others became aware of these risks through health educational programs in the federal, state and local prisons, and jails. One female NIU expressed her reasons for avoiding injecting drug use:

> I'm scared, I'm scared of AIDS, HIV, Hepatitis, I'm scared of needles period. I don't like needles. That's the only reason I don't, I can't find myself shooting up. I think about it, but I don't do it, I don't think I'll ever do it. I've seen people who have shot up. I see them pretty, and a year later they're fucking ugly. Their teeth are falling out, they're arms are all ugly, they're skinny. I just won't do it.

Another NIU expressed his perception of the life trajectory of injectors:

> There's people that like it (injecting heroin) and wanna live on with that life. They start loosing everything, their kids, wife, house, clothes– start staying under a bridge. I don't want that. That's why I lay off (injecting) and I just snort.

Attitudes toward injecting change after NIUs transition to injecting. This transition is facilitated by increased dependence and association with heroin injectors. NIUs usually buy $10 bags of heroin. The content of these bags vary, but are not more than a 1/4 g. The powder is snorted or shabanged. Depending on the level of tolerance, a non-addicted user will get a high from this quantity. When a NIU is unable to get the effect (high) desired, the person may decide to inject. One male respondent illustrates the first time he injected:

> My friend Jeffrey was scoring for me and Shorty. We were waiting there at his house. He scores and goes to the restroom to shoot up. When we went in there he already had ours (heroin) in the rig. I was like I don't do it like that. I snort. And he said "You can shabang it (squirting the heroin into your nose with the syringe)." I was thinking about it. Shorty was already blasting. I thought about it. I was snorting it and it wasn't hitting me or it took too long (for the affect) All my friends were doing it, so I decided to blast it (inject).

Another NIU relates his first experience:

> I started shabanging and you know it wasn't like before. I wasn't getting as *browned* (high) out as I used to. They were always telling me that snorting was nothing compared to when you are blasting. And one time I asked my friend Rick to cook me up a dime bag of brown. He made thirty units out of it and he injected me. You know they were right. It felt way different. I injected three more dimes that day. I just loved it from the beginning.

Lastly, one NIU became frustrated that he was not getting the same high from sniffing heroin:

> I just woke up and I wanted to shoot up because it wasn't doing nothing to me no more. I told my grandmother's boyfriend to go buy some for me. I had enough to buy it... He fixed up the twenty for me. I got the cotton. He got the cooker and syringe. He cooked it for me. He then hit me up. That was the first time that I shot up.

One typical respondent indicated that he snorted heroin 6–7 times a week before deciding to inject. This transition happened even though in an interview the previous week, he had stated. "I woke up with a headache this morning. I felt like I wanted to do another line. I thought I better stop."

Discussion and Conclusion

Ecological segregation, social isolation, and vice containment is important in understanding how Mexican Americans have been susceptible over the years to heroin (and other drugs) and related deviant activities such as drug selling and other criminal activities. In this regard, context must be understood from both a historical and a current framework. The emergence of the *tecato* subculture in disadvantaged Mexican American communities is a consequence of this social process. This subculture is what has culturally sustained heroin use among this population within the structural context of socially disadvantaged communities. Moreover, the intensity and frequency of interaction reinforced by the relatively spatial isolation contributes to these behaviors.

Historically, the initiation to illegal drugs and other illicit activities for Mexican Americans in San Antonio was facilitated by the proximity of the city's vice district during the late 1930s and 1940s. We described how this area was located adjacent to the early residential communities of Mexican immigrants and the group's major commercial and entertainment centers. We surmise that introduction of opium and later heroin to Mexicans by white drug users occurred within this time, space and social setting that tolerated the mixing of racial and ethnic groups. This type of integration was not allowed in other spheres of life in South Texas which maintained a clear separation of these groups through an imposed de-facto Jim Crow social structure.

At this same time, Mexicans were introduced to intravenous injecting practices. This was the popular route of heroin administration. Prior to this period, whites used injected heroin subcutaneously and intramuscularly as opposed to intravenously. Also, during this time, there was a gradual decrease in supply that led to smaller doses and to the use of the hypodermic needle as a more effective route of administration. This was significant because intravenous injecting heroin leads to a more rapid trajectory toward addiction compared to other practices. As a result, public records beginning from this period document an increasing number of Mexican American heroin addicts in public drug treatment facilities (Chambers, Cuskey, and Moffett 1970).

During the decades that followed, Mexican American's minority status was solidified, especially in South Texas (Montejano 1987). This led to limited access to conventional opportunities and increased segregation into improvised neighborhoods like San Antonio's West Side. In response, many Mexican Americans began to engage in off- the book ventures, secondary labor activities like migratory farm work, cotton picking, pecan shelling and illegal activities, and involvement in the heroin and drug market. By the 1940s, Mexicans began to be associated with the drug market enterprises that extended throughout the United States (Redlinger and Michel 1970). It was in this manner that the West Side began to be a source of illegal drugs and other vices for people in and outside the community, especially after San Antonio's vice district was officially closed in 1942.

These activities continue to flourish to this day among groups of truly disadvantaged West Side residents who have been socially isolated and segregated from mainstream society (Bauder 2002). Moreover, given the physical isolation of these communities from larger society and the density of interaction with each other creates a social environment where it is difficult to internalize conventional norms, values, expectations and behaviors. Instead, they develop norms and values that are an adaptation to the social structural conditions of their communities that further increase their probability of engaging in behavior such as heroin use. The continual involvement of West Side residents in drug markets further stigmatized this population and further hindered their assimilation and integration similar to blacks in urban ghettos (Kornblum 1993).

The drug market created opportunities for poor minorities, but concomitantly made them highly vulnerable to arrest and incarceration. This has resulted in large numbers of Mexican American men being incarcerated during the last two decades. Upon being released as convicted felons, many return to communities such as the West Side of San Antonio that offer few opportunities. As a result, many have begun to take advantage of opportunities offered by drug markets centered in these barrios, often under the umbrella of the prison gangs. It was in this context that Mexican American adult prison gangs were able to dominate the heroin market and other street-based drug sellers and dealers such as Mexican American youth gangs. However, it shall be emphasized that the dominant illegal institutions (i.e., international cartels) of this drug industry continue to be located outside the barrio.

Within this context and cultural milieu, this drug using population becomes highly vulnerable to the vicissitudes of drug markets. As illustrated in this chapter, intravenous heroin use was historically highly stigmatized even among street-oriented poly-drug using populations. In contrast, sniffing heroin during the 1990s was socially acceptable, given that it was perceived as a way in which to avoid the consequences associated with the *tecato* lifestyle, including addiction. As a result, snorting heroin (or "sniffing") has slowly become an acceptable alternative among many drug users, particularly as a more powdered form of heroin (brown) became available in the West Side drug market primarily through the Mexican American prison gangs. They were able to create a demand for powdered heroin by encouraging the use of heroin through non-injecting methods, which was perceived as less of a risk than injecting. However, many non-injectors have eventually transitioned to injecting because it is a more effective route of administration and others because of their association with injectors. Other studies have found that social network characteristics, such as having friends or sex partners who are injecting drug users, are associated with the onset of injecting drug use and transitioning to injecting (Neaigus, Hagen, Friedman, Miller, and Des Jarlais 1998). This illustrates how ecological isolation, subcultures (i.e., *tecatos*) and deviance containment contribute to drug practices that lead to risk behaviors associated with addiction.

This chapter has begun to examine some of the contextual and cultural variations associated with heroin practices among Mexican Americans. Heroin use on the West Side of San Antonio is considered to be part of the cultural milieu for Mexican Americans involved in drug use and other street-oriented activities. This population has been exposed to a vice district that historically facilitated the social learning process associated with the consumption of heroin via family, friends and acquaintances, and the *tecato* subculture. More importantly, there is a legacy of segregation and marginalization that has contributed to the creation of vice market segmentation in which illicit markets and illegal economic institutions thrived. This parallels the history of ghetto containment of deviance experienced by African Americans and other U.S. minorities living in disadvantaged minority communities. What emerges from this analysis is how these historical precedents together with ecological and urban spatial processes provide a theoretical framework that helps explain Mexican American heroin practices.

Acknowledgment This research was supported by the National Institute of Health, National Institutes on Drug Abuse, Grant Number RØ1 DA 13560. We are grateful to the project's field staff, Richard Arcos and Ronnie Cardenas, for their contribution to this research.

Chapter 11
Comparing Unintentional Opioid Poisoning Mortality in Metropolitan and Non-Metropolitan Counties, United States, 1999–2003

Nabarun Dasgupta, Michele Jönsson Funk and John S. Brownstein

Abstract Numerous reports have documented the rise in medical and non-medical use of opioids in the United States since the early 1990s. Geographic variations in the medical and non-medical use of opioids have also been documented. In this chapter, vital statistics data from 1999 to 2003 were analyzed to describe spatial patterns in opioid overdose mortality. Deaths associated with prescription opioids, including methadone, showed large increases during the study period in both metropolitan and non-metropolitan areas. Despite previous research suggesting otherwise, metropolitan areas showed higher rates of accidental opioid overdose mortality than non-metropolitan areas. Between 1999 and 2003, the rate of fatal unintentional opioid overdose among 15- to 54-year-old residents of metropolitan counties increased from 3.85 to 5.45 per 100,000 per year, see Table 11.2. Among residents of non-metropolitan areas, the increase was from 1.65 to 5.01 per 100,000 per year. Cocaine toxicity was much more likely to be mentioned as a contributing cause of death in metropolitan areas, odds ratio (OR) = 2.60 (95% CI: 2.37, 2.84), as was alcohol, OR = 1.34 [95% confidence interval (CI): 1.23, 1.45]. Benzodiazepine toxicity was more likely in non-metropolitan areas, OR = 1.56 (95% CI: 1.44, 1.70), as was toxicity due to anti-depressants, OR = 1.42 (95% CI: 1.28, 1.57). Geographic variation in unintentional opioid mortality may be a function of drug availability and may be occurring in distinct populations in metropolitan and non-metropolitan counties. Reasons for these differences are discussed, including methodological considerations.

N. Dasgupta
Department of Epidemiology, School of Public Health, University of North Carolina at Chapel Hill, Chapel Hill, NC, USA

Y.F. Thomas et al. (eds.), *Geography and Drug Addiction*,

Introduction

In the United States, prescription opioids are the mainstay for treatment of moderate to severe pain (Noble 2007; Ventafridda, Saita, Ripamonti, and De Conno 1985; Wiffen 2003); the non-medical use of these medicines and fear of reflexive governmental action jeopardize the benefits that opioids provide to those who need pain relief (Rannazzisi 2007). Increases in the prescriptive use of opioids since the early 1990s have been paralleled by an increase in medical consequences caused by their abuse, as exhibited by emergency department admissions (Dasgupta et al. 2006) and treatment seeking behavior (Hopfer, Mikulich, and Crowley 2000). During the same time period, there has been a rise in the purity of heroin available in the United States, fuelled by increases in the planting of opium poppy in central and south America (International Narcotics Control Board 2006). Taken together, the use of opioids in the United States, both medical and non-medical, is likely to be higher now than ever before.

Respiratory depression can occur as a consequence of excessive or inappropriate use of prescription opioids (McNicol et al. 2003) and heroin (Warner-Smith, Darke, Lynskey, and Hall 2001). Respiratory depression can be followed by hypoxia and death if not treated with an opioid receptor antagonist or rescue/resuscitative breathing (White and Irvine 1999); pulmonary edema has also been reported as a consequence of opioid poisoning. Deaths due to unintentional drug poisonings in the United States increased by 218% between 1990 and 2002, with most of the increase since 1999 due to "narcotics and psychodysleptics" or "other and unspecified drugs" (Paulozzi, Budnitz, and Xi 2006).

Despite different mechanisms of production and potency, prescription opioids and heroin have similar pharmacological action at the mu-opioid receptor. Mu-opioid receptor agonists are central nervous system depressants and overdose occurs through respiratory depression and hypoxia due to diminished sensitivity to changes in blood oxygen saturation; pulmonary edema and asphyxiation are also contributing causes of death (White and Irvine 1999).

The most important risk factors for accidental opioid poisoning are recent release from prison or abstinence-based treatment program, polydrug use, and street-level variations in purity and contaminants (Oliver and Keen 2003; Warner-Smith, Darke, Lynskey, and Hall 2001). Since mid-2005, the United States has been confronting a multi-state "outbreak" of opioid overdose deaths due to heroin laced with fentanyl (CDC 2005).

Differences in the availability, use and misuse of prescription opioids have been observed between persons living in urban and rural areas (Cicero, Inciardi, and Munoz 2005; Green, Ndao-Brumblay, West, and Washington 2005; Havens, Walker, and Leukefeld 2006; Hodgson, Landsberg, Lehning, and Kleban 2006; Hughes and Crawford 1974). Conceptually, geographic location could exert influence on opioid overdose mortality through availability: illicit drugs are less likely to be available in areas of low population where drug distribution networks cannot make a large enough profit to justify the expenses and risks associated with trafficking. In these places, access to opioids for misuse would be postulated to be achieved mainly through diversion of prescription medications. However, the prescriptive use of opioids, as measured in terms of quantity, is the highest in metropolitan areas. In these markets, the availability of heroin, especially of high purity, may attenuate, or at least offer an alternative to, the non-medical use of prescription opioids. Therefore, there may be a geographic balance in non-medical use of various opioids.

In Michigan, prescription opioids were less likely to be available for medical use in pharmacies in minority communities, which were predominantly in metropolitan areas (Green, Ndao-Brumblay, West, and Washington 2005). A rural preference for non-medical use of opioids has been observed in cross-sectional studies of prison and community samples, as well as from Medicaid databases (Havens, Walker, and Leukefeld 2006; Leukefeld et al. 2002; Schoeneberger, Leukefeld, Hiller, and Godlaski 2006). However, prescription opioid misuse also occurs in metropolitan areas (Paulozzi 2006; Peters, Williams, Ross, Atkinson, and Yacoubian 2007). Variations in the form and purity of heroin (e.g., "black tar" in the western United States versus "China white" in the northeast) can be seen regionally in the United States and is thought to influence HIV risk and incidence of overdose (Ciccarone and Bourgois 2003; Sperry 1988).

Surveillance for accidental opioid poisoning deaths has traditionally been conducted through state medical examiners' offices; reports of drug overdose deaths have been published on a smaller scale based on county-level mortality data (CDC 1983, 2000; Gill and Graham 2002; Mikolaenko, Robinson, and Davis 2002) and on a state or national scale by the Centers for Disease Control and Prevention (CDC) and other health authorities (Ballesteros et al. 2003; CDC 2004; Mueller, Shah, and Landen 2006; Paulozzi, Budnitz, and Xi 2006; Sorg and Greenwald 2002). Multiple prescription opioid surveillance systems catering to industry exist (Akbik et al. 2006; Cicero et al. 2007), as do prescription monitoring programs in many states and surveillance of poison center data (Hughes, Bogdan, and Dart 2007; Sims, Snow, and Porucznik 2006); however, the ability of these systems to predict opioid poisoning mortality has not been conclusively determined (Hoppe-Roberts, Lloyd, and Chyka 2000). Federal vital statistics data show high concordance with state-level medical examiner databases (Landen et al. 2003), although there may be some discrepancy due to access to primary sources of information at the level of the state medical examiner office. National vital statistics records can be used for surveillance, placing opioid mortality in context, and informing policy decisions (Minino, Anderson, Fingerhut, Boudreault, and Warner 2006); however, national vital statistics reports on injury do not focus solely on poisonings and are not intended to explore questions regarding the influence of geography.

Studies using vital statistics data for injury (including poisonings) mortality are subject to a set of specific biases, as illustrated in Fig. 11.1. Non-fatal injuries are not reflected in vital statistic reporting systems, necessitating supplementary injury surveillance systems (Mack 2004; Vyrostek, Annest, and Ryan 2004). Variations in toxicological assessment and definitions of poisoning can be a major source of bias (Poulin, Stein, and Butt 1998). Case detection and classification can result in under-reporting among elderly and female decedents (Dijkhuis, Zwerling, Parrish, Bennett, and Kemper 1994). Determining intent is also fraught with difficulty and

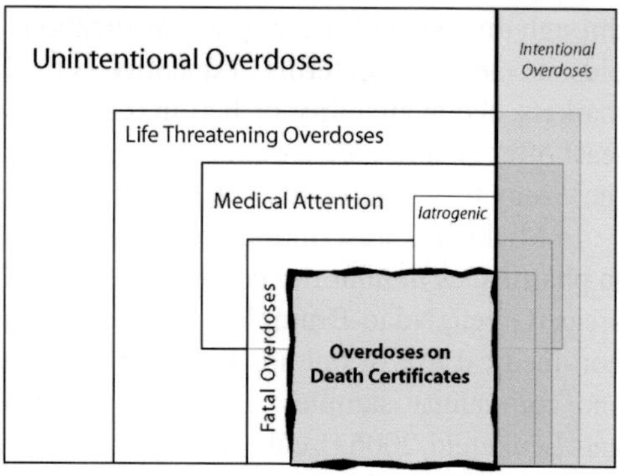

Fig. 11.1 Schematic view of the relationships between intentional and unintentional poisonings and deaths reported to vital statistics reporting systems (death certificate data)

can also introduce bias (Sorenson, Shen, and Kraus 1997). Coding schema used for death certificates are not designed to distinguish between overdoses of illicit drugs and accidental opioid overdoses occurring during medical care ("iatrogenic"), such as complications from anesthesia during surgical procedures (Bowdle 1998) or in-hospital patient-controlled analgesia (Musshoff, Padosch, and Madea 2005). Therefore, the decedents described in this analysis are likely to comprise a heterogeneous group and the results should be interpreted with caution, especially if the primary interest is in understanding the consequences of opioid abuse.

In this chapter, we present the results of an analysis conducted with publicly available vital statistics data. We dichotomized all accidental opioid poisoning deaths by the metropolitan/non-metropolitan county of residence of the decedent and calculated odds ratios (OR) for contributing toxicology from substances known to increase the risk of opioid overdose (cocaine, benzodiazepines, alcohol and anti-depressant medications). While illicit drugs, such as cocaine, can be hypothesized to be more readily available in urban areas, alcohol, benzodiazepines and anti-depressants should be available in both metropolitan and non-metropolitan areas. We do not expect to see variation between decedents of metropolitan and non-metropolitan counties based on alcohol, benzodiazepine, and anti-depressant toxicity.

Data and Methods

Data Sources and Definitions

Death certificate data are routinely collected by state health authorities and reported to the National Center for Health Statistics (NCHS) at CDC, as mandated by federal regulation. Deaths from unnatural causes, such as poisonings, are referred to coroners and medical examiners in accordance with state laws. Underlying and contributing causes of death are attributed after investigation, including the identification of toxic substances that were involved (Stephens, Jentzen, Karch, Wetli, and Mash 2004). Since 1999, NCHS has recoded the cause of death literal entries from state death certificates using the International Classification of Disease, 10th revision (ICD-10) [World Health Organization (WHO) 1992] to create the multiple cause of death mortality file issued by the Division of Vital Statistics, Hyattsville, MD. Additional sociodemographic data and details of the occurrence are also captured. The files include all deaths occurring within the United States, including territories and possessions (but do not include deaths of US citizens and armed forces outside of the United States). Data files used in this project, for the years 1999 to 2003, were obtained from the repository of the National Bureau of Economic Research (National Bureau of Economic Research 2006).

The study period was restricted to years when ICD-10 was used to code death certificate data by NCHS, allowing us to distinguish between underlying causes of death due to different types of opioids. The following toxicology codes are available for opioids: opium (T40.0), heroin (T40.1), other opioids (T40.2), methadone (T40.3), and other synthetic narcotics (T40.4), while some additional opioids are coded in the category other and unspecified narcotics (T40.6). The T40.2 code contains the majority of opioids available for medical use in the United States, except methadone (T40.3) and fentanyl, propoxyphene, meperidine and buprenorphine (T40.4); the methadone code does not distinguish between methadone prescribed for pain control and methadone used in the management of opioid dependence. Fatal toxicity due to cocaine is coded as T40.5 and benzodiazepines as T42.4.

There exists no standardized method for identifying unintentional opioid poisoning deaths using ICD-10. We utilized a previously published case definition appropriate for national surveillance of death certificate data to identify unintentional opioid poisoning deaths (Jauncey, Taylor, and Degenhardt 2005). Jauncey and colleagues used a "core" and an "extended" surveillance definition. Both definitions are designed to exclude deaths due to suicides, iatrogenic exposure, euthanasia, pediatric exposure, and assaults. The definitions differ in that the extended definition also includes deaths where the involvement of narcotics is indicated, but opioids are not specifically described (group 4, below). The extended surveillance definition (as opposed to the core definition) was chosen for this study as it was likely to capture the largest number of accidental opioid poisoning deaths. Jauncey et al. also restrict age at death to between 15 and 54 years.

As displayed in Fig. 11.2, the surveillance definition includes decedents in the following groups: (1) underlying cause of death mental or behavioral disorder due to opioids F11.0–F11.9; (2) underlying cause of death accidental poisoning by narcotics/psychodysleptics X42, with contributing cause of death due to opioid toxicity T40.0–4 and T40.6; (3) underlying cause of death accidental poisoning by other and unspecified drugs X44, with contributing cause of death due to opioid toxicity T40.0–4 and T40.6; (4) underlying cause of death mental and behavioral disorder due to multiple drug use F19.0–F19.9, with either contributing cause of death due to opioid toxicity T40.0–4 and T40.6, or F11.0–F11.9 (Jauncey, Taylor, and Degenhardt 2005).

In order to test the hypothesis that there are geographic differences in the nonmedical use of prescription opioids at a national level, we employed the US Office of Management and Budget's definitions of metropolitan and non-metropolitan

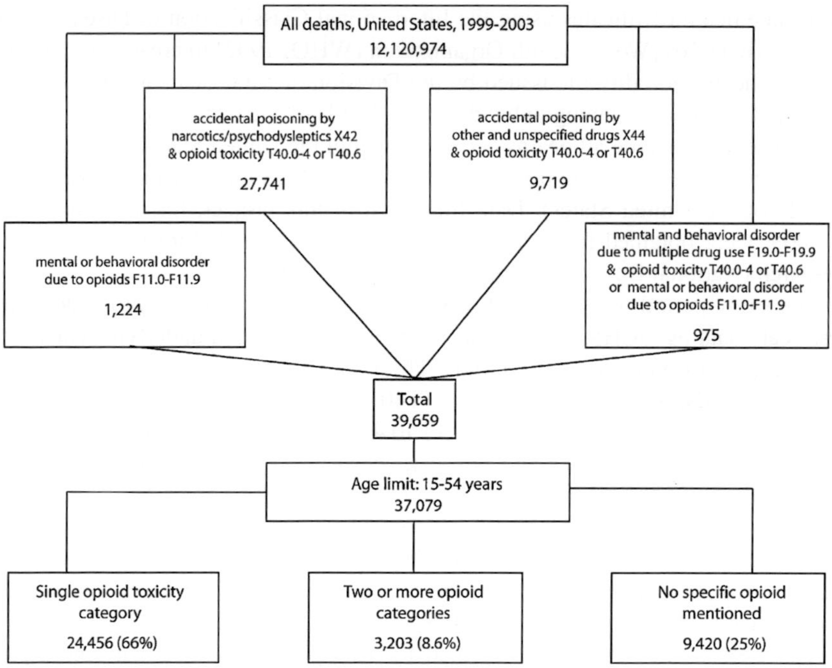

Fig. 11.2 Application of surveillance definition and sample frequencies for unintentional poisoning deaths (death certificate data), United States, 1999–2003

counties (Office of Management and Budget 2004). Briefly, they are, for a metropolitan county: a central county with (1) one or more urbanized areas each having a population of 50,000 or more residents; plus (2) any outlying counties in which at least 25% of the working age population commute to the central county for work or in which 25% of the outlying county's workers commute from the central county. A non-metropolitan county is classified as either a "non-metropolitan, micropolitan" or "non-metropolitan, non-core" county. Non-metropolitan, micropolitan counties have one or more urban clusters (towns) of 10,000–49,999 persons. Both metropolitan and micropolitan areas can have one or more counties. Non-metropolitan, non-core counties contain no town (urban cluster) of at least 10,000 people. The metropolitan/non-metropolitan classification of the county of residence of the decedent was encoded by NCHS in the public use data files. Federal Information Processing Standards (FIPS) codes for counties with less than 100,000 residents were suppressed by NCHS due to confidentiality concerns. While county-specific death certificate data would allow for more nuanced analysis, repeated written requests for these data to NCHS went unanswered.

For the few counties that switched metro/non-metro designation during the study period, the 2003 classification was used (United States Department of Agriculture 2004). It was assumed that annual changes in population in this age group would have a negligible effect on reported rates.

Statistical Analysis

Metropolitan/non-metropolitan county of residence, as indicated on the death certificate, was used to dichotomize decedents; relative effect estimates were calculated with outcomes being toxicology due to alcohol, cocaine, benzodiazepines and anti-depressants. These substances were examined because their presence is a recognized risk factor for opioid overdose (Chan, Stajic, Marker, Hoffman, and Nelson 2006). The county of residence was chosen since it preceded exposure, whereas county of occurrence (of death) would not always. Using county of residence also allowed us to compute rates. Comparisons between metropolitan areas were made using PROC LOGISTIC in SAS version 9.0 (SAS Institute Inc., Cary, NC, USA). Crude OR and 95% confidence intervals (CI) are presented. Adjusted OR were not calculated due to peculiarities of the data; the majority of variables available from death certificate are intermediates on a causal pathway between exposure (place of residence) and outcome (toxicology, causes of death) and thus cannot be used for adjustment in multivariate models. Adjustment for demographic characteristics is also not warranted since it is unclear if age, sex, or race directly influences the choice of residence between metropolitan and non-metropolitan counties (i.e., effects exposure), although sex differences in opioid analgesia have been documented (Fillingim and Gear 2004). Additional analyses were conducted in JMP version 6.0.0 (SAS Institute Inc., Cary, NC, USA) and maps were created in ArcMap 9.1 (ESRI, Inc., Redlands, CA, USA). Descriptive statistics are also presented for three topics related to geography and migration that are often not described in the scientific literature on drug overdose: geographic distribution of opium deaths, poisonings in US territorial possessions and deaths occurring in the United States among residents of foreign countries.

Results

Between 1999 and 2003, a total of 12,120,974 deaths were reported in the United States. During these 5 years, the number of deaths meeting the extended surveillance definition for accidental opioid poisoning was 37,079. Of these, toxicity due to at least one specific opioid (via T40 codes) was mentioned in 27,659 deaths (74%), see Fig. 11.2. The number of unintentional opioid poisoning deaths in the United States increased 57%, from 5,987 in 1999 to 9,377 in 2003, while the US population grew by 6.6% during the same period. The rate of overdose death from unintentional opioid poisoning among 18–54 year olds was 4.26 per 100,000 per year. The ICD-10 codes for accidental poisoning by narcotics (X42) and accidental poisoning by other and unspecified drugs (X44) accounted for over 94% of deaths meeting the surveillance definition, see Fig. 11.2.

Most decedents were male (75%), white non-Hispanic (76%), high school graduates (74%) and a plurality were single (45%), see Table 11.1. Hispanics (of any race) and black non-Hispanics had similar numbers of deaths, despite comprising

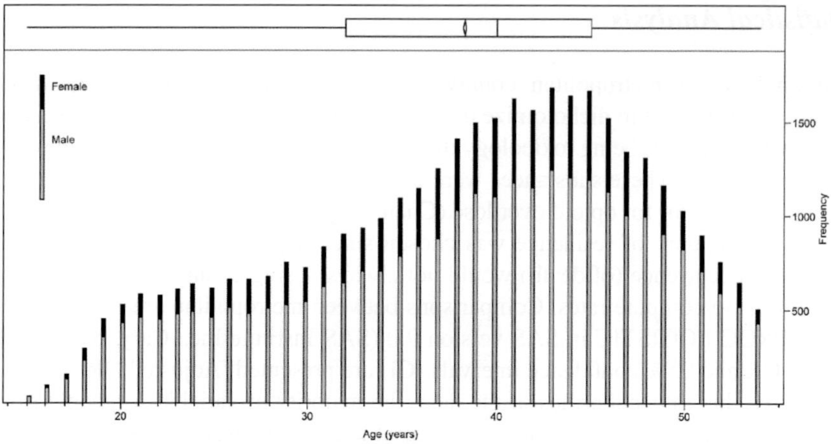

Fig. 11.3 Gender and age distribution of unintentional opioid poisoning deaths (death certificate data), United States, 1999–2003

different proportions of the general population. The mean age at death was 38 years, with a standard deviation of 9.3 years, and a median of 40 years, see Fig. 11.3. The age distribution was bimodal, with peaks in the early twenties and mid-forties, with skewness of -0.45 and kurtosis of -0.67. Males comprised a greater proportion of accidental opioid poisoning deaths at younger ages, with the proportion of females increasing steadily from the mid-thirties onward.

Metropolitan/Non-Metropolitan County of Residence

Most decedents were residents of metropolitan areas 31,784 (86%), versus 5,225 (14%) in non-metropolitan areas, with crude averages of 6,357 and 1,045 deaths per year, respectively, see Table 11.1. By comparison, 80.3% of the United States residents lived in metropolitan areas and 19.7% in non-metropolitan areas, according to Census 2000. This corresponded to 4.51 per 100,000 among residents of metropolitan areas per year and 3.14 per 100,000 among residents of non-metropolitan areas per year. Therefore, by both absolute and relative measures, metropolitan areas had greater burdens of opioid poisoning mortality.

Between 1999 and 2003, the rate of fatal unintentional opioid overdose among 15- to 54-year-old residents of metropolitan counties increased from 3.85 to 5.45 per 100,000 per year, see Table 11.2. Among residents of non-metropolitan areas, the increase was from 1.65 to 5.01 per 100,000 per year. The greatest increase occurred between 2002 and 2003 in non-metropolitan areas, with the rate going from 3.86 to 5.01 per 100,000 per year.

Most decedents died close to home, that is, the county of residence and county of occurrence were the same (73%), and there was no difference in this distribution between metropolitan and non-metropolitan counties. Differences between

Table 11.1 Sociodemographic characteristics of accidental opioid poisoning deaths, by metropolitan classification of county of residence (death certificate data), United States, 1999–2003[a]

	County of Residence					Total	
	Metropolitan[d]		Non-Metropolitan[e]		Missing	*n*	%
	n	%	*n*	%	*n*		
Sex							
Male	23,906	75	3,675	70	65	27,581	75
Female	7,878	25	1,550	30	5	9,428	25
Missing	0		0		0		
Race/ethnicity							
White, non-Hispanic	22,768	72	4,673	90	19	27,441	76
Any race, Hispanic	4,165	13	304	5.8	4	4,469	12
Black, non-Hispanic	4,138	13	112	2.2	40	4,250	12
Other	365	1.2	117	2.2	1	482	1.3
Missing	348		19		6		
Education[b]							
HS graduate	17,882	75	2,456	72	32	20,338	74
HS not completed	6,077	25	968	28	15	7,045	26
Missing	7,825		1,801		23		
Marital status							
Single, never married	14,528	47	1,740	33	31	16,268	45
Married	7,886	25	1,755	34	13	9,641	27
Widowed or divorced	8,552	28	1,704	33	13	10,256	28
Missing	818		26		13		
Place of death							
Residence	13,045	41	2,477	47	28	15,522	42
Hospital or nursing home	11,310	36	1,819	35	2	13,129	36
Other	7,392	23	922	18	40	8,314	22
Missing	37		7		0		
Residence status[c]							
Residents	23,241	73	3,854	74	0	27,095	73
Intrastate residents	6,820	21	1,019	20	0	7,839	21
Interstate residents	1,723	5.4	352	6.7	0	2,075	5.6
Foreign residents					70	70	0.19
Missing	0		0		0		

[a]Deaths meeting extended surveillance definition of Jauncey et al., see text for details. Percentages may not add to 100 due to rounding.
[b]HS, High school; for 2003 only HS graduates include decedents having attained graduate equivalence degree (GED).
[c]Residence status indicates whether the decedent was a resident of the county of residence. For US island territories, "foreign resident" indicates that the decedent was not a resident of the island territory of occurrence, but does not mean that they were residents of a foreign country other than the United States.
[d]Metropolitan county of residence, as defined by Office of Management and Budget, see text.
[e]Non-metropolitan county of residence, as defined by Office of Management and Budget, see text.

metropolitan and non-metropolitan counties of residence were notable for race/ethnicity, marital status, and place of death. The difference in race/ethnicity is likely a reflection of the greater proportion of white non-Hispanic residents in non-metropolitan areas. The reasons for a higher proportion of deaths among

Table 11.2 Counts and rates of opioid overdose deaths per 100,000 population, 15–54 years old, by metropolitan/non-metropolitan county of residence of decedent, United States, 1999–2003[a]

	Metropolitan Counties		Non-Metropolitan Counties	
Year	Deaths	Rate	Deaths	Rate
1999	5,419	3.85	550	1.65
2000	5,537	3.93	714	2.15
2001	5,750	4.08	1, 008	3.03
2002	7,377	5.24	1, 285	3.86
2003	7,679	5.45	1, 666	5.01

[a]Deaths meeting extended surveillance definition of Jauncey et al., see text for details. Percentages may not add to 100 due to rounding.

single, never married individuals in non-metropolitan areas remain unclear. Data on educational attainment were frequently missing, with more uncertainty in non-metropolitan areas. Other variables had similar missing data patterns for metropolitan and non-metropolitan areas.

The frequencies with which particular types of opioids were listed as an underlying cause of death is presented in Table 11.3 and Fig. 11.4, along with the metropolitan/non-metropolitan status of the county of residence of the decedent. The numbers presented in the table and figures are tabulations of any death for which the specific toxicology code was found in death certificate data; decedents with

Table 11.3 Numbers of deaths with underlying cause of death due to opioid toxicity, by metropolitan classification of county of residence (death certificate data), United States, 1999–2003[a]

	Heroin[b]		Other Opioids[e]		Methadone[f]		Synthetic Narcotics[g]		No Opioids Listed[h]		Total	
Year	M[c]	N-M[d]	M	N-M	M	N-M	M	N-M	M	N-M	M	N-M
1999	1,594	116	1,691	228	520	80	274	91	1,901	91	5,422	550
2000	1,579	109	1,645	322	598	151	334	83	1,841	109	5,543	714
2001	1,509	99	1,848	500	858	253	414	139	1,608	111	5,756	1,008
2002	1,770	108	2,461	562	1,416	390	575	168	1,787	176	7,382	1,287
2003	1,755	114	2,599	712	1,720	603	652	257	1,599	163	7,681	1,666
Total	8,207	546	10,244	2,324	5,112	1,477	2,249	738				
Avg./year	1,641	109	2,049	465	1,022	295	450	148				

[a]Deaths may be represented more than once if multiple opioids were specified.
[b]International Classification of Disease, 10th revision (ICD-10) code T40.1 listed as a contributing cause of death.
[c]M, metropolitan county of residence, as defined by Office of Management and Budget, see text.
[d]N-M, non-metropolitan county of residence, as defined by Office of Management and Budget, see text.
[e]Contributing cause of death ICD-10 code T40.2; includes prescription opioids other than methadone and fentanyl.
[f]Contributing cause of death ICD-10 code T40.3; does not distinguish between methadone dispensed for pain management and opioid dependence.
[g]Contributing cause of death ICD-10 code T40.5; includes fentanyl
[h]Contributing cause of death ICD-10 code T40.0.

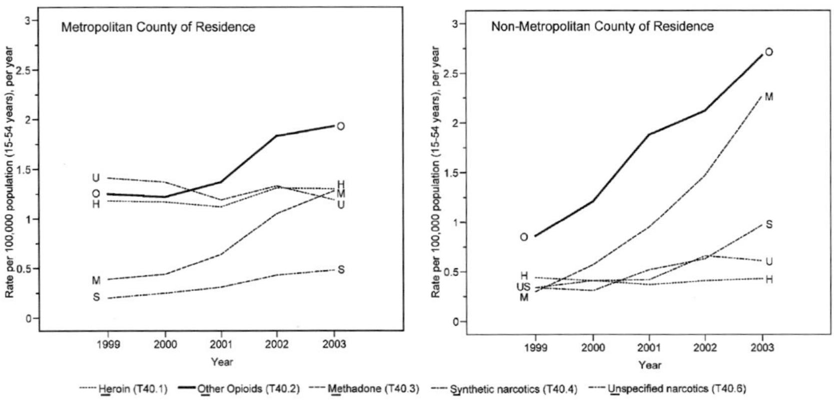

Fig. 11.4 Geographic variation in unintentional opioid poisoning mortality, by type of opioid and metropolitan/non-metropolitan status of county of decedent (death certificate data), United States, 1999–2003

multiple opioid toxicology codes are represented more than once, affecting approximately 10% of decedents (see below). Data on whether an autopsy was performed were only available for 2003. In that year, 72.9% had autopsies performed, 10.2% did not, and the autopsy status was unknown for 16.9%. Despite the lower proportion of deaths with unknown opioid toxicology in non-metropolitan areas ("U" Lines, Fig. 11.4), residents of non-metropolitan areas with opioid poisoning were less likely to have had an autopsy performed, OR = 0.67 (95% CI: 0.57, 0.79).

Use of multiple substances is a risk factor for accidental opioid poisoning. A single opioid was mentioned for 24,456 (66%) deaths, two opioids were mentioned in 3,118 (8.4%) deaths, and three or more opioids in 84 deaths (0.23%). No opioids were specified for 9,420 deaths (25%) meeting the extended surveillance definition. Of single opioid deaths, 30% also mentioned either cocaine or benzodiazepines as an underlying cause of death. Of deaths meeting the extended surveillance definition, cocaine was implicated in 8,387 (23%) deaths during the study period, benzodiazepines in 4,184 (11%) deaths, and anti-depressants (tricyclic and selective serotonin reuptake inhibitors) in 2,725 (7.3%). Acute acetaminophen toxicity is of concern since prescription opioids commonly include non-opioid pain relievers. However, only 1.01% of deaths mentioned acetaminophen toxicity as a contributing cause of death. Alcohol was mentioned in 7,298 (20%) of deaths.

There were many differences observed between decedents of metropolitan and non-metropolitan counties. The odds of female decedents being residents of non-metropolitan areas were greater than them being residents of metropolitan areas, OR = 1.28 (95% CI: 1.20, 1.36). Age differences were not substantial between decedents in metropolitan (mean 38.4 years) and non-metropolitan (mean 37.7 years) counties. Cocaine toxicity was much more likely to be mentioned as a contributing cause of death in metropolitan areas, OR = 2.60 (95% CI: 2.37, 2.84), as was alcohol, OR = 1.34 (95% CI: 1.23, 1.45). Benzodiazepine toxicity was

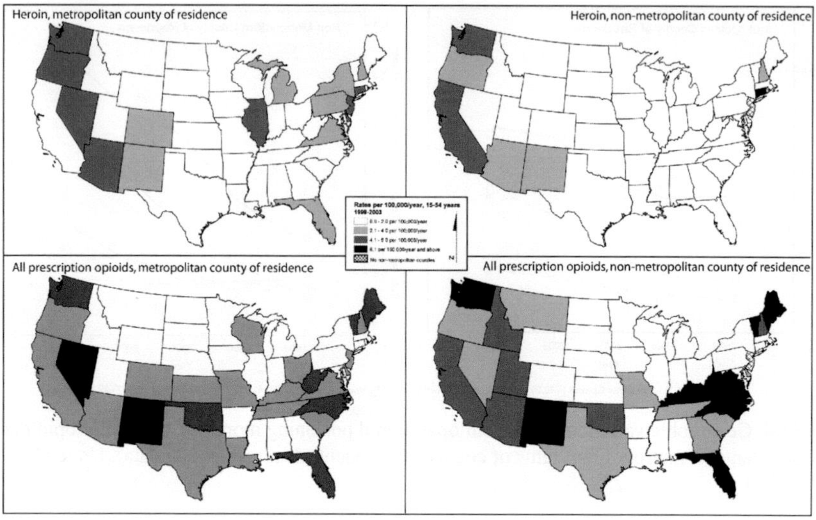

more likely in non-metropolitan areas, OR = 1.56 (95% CI: 1.44, 1.70), as was toxicity due to anti-depressants, OR = 1.42 (95% CI: 1.28, 1.57). Despite the low numbers of deaths involving acute acetaminophen toxicity, residents of rural areas were much more likely to exhibit toxicity to this substance, OR = 2.04 (95% CI: 1.61, 2.58).

Opium was listed as an underlying cause in nearly three deaths per year in the United States, occurring among residents of states following no predictable geographic pattern: Georgia, Hawaii, Indiana, Michigan, Minnesota, Mississippi, New Hampshire, Ohio, Pennsylvania, and Texas. Nearly all opium deaths were among decedents classified as white non-Hispanic (78%). Of the 13 opium deaths, five involved other opioids. Among the eight deaths where the only opioid mentioned was opium, four also involved cocaine. External validation is required to determine whether these deaths were correctly classified as being opium-related.

All deaths occurring in American Samoa, the Northern Mariana Islands, Puerto Rico and the Virgin Islands were to residents of places other than territory in which it occurred. Deaths in these places were due predominantly to heroin or other opioids, with heroin mentioned more commonly. Guam reported no deaths due to opioids during the study period.

Seventy deaths occurred during the study period among residents of foreign countries. Only half of the foreign residents had a specific opioid associated with their deaths, making it impossible to draw conclusions of whether their deaths were caused by heroin or prescription opioids without further data manipulation. However, New York was the state of occurrence for 32 (46%) of these deaths, with 21 occurring in that state in 2003 alone. These 70 individuals were removed from the analysis using county of residence that follows.

Discussion

This study confirms recent findings noting an increase in deaths in the United States from unintentional opioid poisonings (Mueller, Shah, and Landen 2006; Paulozzi, Budnitz, and Xi 2006) and provides more details on the geographic distribution of the deaths. Much of this increase during the study period was due to an increase in mentions of deaths involving prescription opioids and methadone, as well as an increase in opioid-related deaths, in which a specific substance was not indicated. Although the prevalence of medical use of opioids in the United States has increased markedly since the early 1990s (Gilson, Ryan, Joranson, and Dahl 2004), there appears to have been a dramatic increase in mortality due to opioids during the study period. Decedent characteristics observed in this study were similar to other studies describing overdose deaths due to prescription opioids in the United States during the preceding three decades (Finkle, McCloskey, Kiplinger, and Bennett 1976; Henderson 1991) although crude rates of mortality are higher now than that had been reported in the past (Paulozzi, Budnitz, and Xi 2006).

A few differences in unintentional opioid poisoning mortality were observed between metropolitan and non-metropolitan areas. Mortality due to heroin was seen predominantly in metropolitan areas although rates of heroin deaths remained fairly stable during the study period for both types of places. Our analysis confirms that deaths attributable to prescription opioids and methadone have risen in metropolitan areas since 2001 (Paulozzi 2006); however, the rate of increase in unintentional opioid overdose death was greater in the non-metropolitan areas where it was almost exclusively due to the prescription opioids. However, residents of metropolitan counties had a higher proportional burden of opioid overdose deaths than non-metropolitan county residents; this finding is somewhat at odds with assertions by previous researchers that suburban and rural areas have a disproportionately higher burden of non-medical use of opioids (Cicero, Surratt, Inciardi, and Munoz 2007). It is interesting to note that by 2003, the rates of opioid overdose death among residents of metropolitan and non-metropolitan areas had become similar (5.45 and 5.01 per 100,000 15–54 year olds per year, respectively), despite having been distinctly different in 1999 (3.85 and 1.65 per 100,000 15–54 year olds per year, respectively). Perhaps the outcry over non-medical use of opioids in rural areas can be partially explained by the rapid increase in mortality rates. Alternatively, opioid overdose deaths in non-metropolitan areas may attract more attention from the community than in urban areas due to the nature of social structures. However, in this study, metropolitan areas had a greater burden of overdose deaths.

The proportion of deaths, in which specific opioids were not specified, was much greater in metropolitan areas than non-metropolitan areas. The reasons for this difference are not clear. This may have been due to differences in caseload or an effect of the publicity generated about prescription opioid misuse in rural areas.

Our original research hypothesis was that there would be no variation between metropolitan and non-metropolitan areas by alcohol, benzodiazepines, and antidepressants. This hypothesis was not supported by our findings. Notable differences in contributing toxicology between decedent residents of metropolitan and non-metropolitan areas were observed for all non-opioid substances examined. Higher likelihood of cocaine involvement in metropolitan areas is not surprising, given the suspected distribution of the availability of the drug (for an example in American adolescents see: O'Malley, Johnston, and Bachman 1991). However, it is unclear why alcohol would be more commonly found among decedents in metropolitan areas. Similarly, it is unclear why decedents in non-metropolitan areas would have higher odds of toxicity from benzodiazepines and anti-depressants. One study from Alberta, Canada suggested higher rates of benzodiazepine prescriptions among residents of urban areas (Hagen et al. 2005), but there is no compelling reason believed that this holds true for the entire United States. Three possibilities come to mind in trying to explain differences in metropolitan versus non-metropolitan benzodiazepine and anti-depressant use, none of which have been well studied: non-metropolitan counties have a higher burden of depressive disorders, physicians in non-metropolitan areas are more likely to prescribe medicines for outpatient management of depressive disorders, or there is greater non-medical use of these medications in non-metropolitan areas. Results from the National Comorbidity Survey (NCS) suggest that mental health services are increasingly being provided by general medical providers (Wang et al. 2006). While board-certified pain physicians are under-represented in rural areas (Breuer, Pappagallo, Tai, and Portenoy 2007), general practitioners are also taking a larger role in the management of painful chronic conditions (Sinatra 2006). Combined with the fact that medical attention in non-metropolitan areas is mainly delivered through primary care providers, these observations may offer paths for additional exploration. The association between alcohol and opioid overdose may be a function of underlying differences in drinking patterns between urban and rural areas. One study in southern states found that the rural residence was a protective factor for at-risk drinking (Booth and Curran 2006). Another study with similar findings for the entire United States showed that the prevalence of heavy and binge drinking was more prevalent in urban counties (Jackson, Doescher, and Hart 2006), but that annual prevalence rates were increasing more rapidly in rural areas.

The county of residence and county of occurrence of death was the same for the vast majority of decedents. Only 6% of deaths occurred in a state other than the decedent's state of residence, suggesting that interventions should be targeted locally where overdose deaths most frequently occur.

Variations in coding practices for drug poisoning deaths may have influenced the distributions of deaths observed in this study, as have been observed for other fatal conditions (Chen, Walker, and Tong 2002). Although a comparison of the New Mexico medical examiner database with national vital statistics records suggested very high concordance, the generalizability of these findings is unknown without external validation and/or chart review, including the influence of electronic reporting systems (Landen et al. 2003). Decedent residents of non-metropolitan areas were substantially less likely to have autopsies performed, yet "unspecified" opioids were rarer in these counties. This apparent contradiction might suggest basic differences by geography in case of ascertainment.

Globally, heroin is believed to be the opioid most commonly associated with poisoning mortality (Drummer 2005). The year 2001 appears to have been a turning point for opioid overdoses in metropolitan areas. Deaths due to heroin in non-metropolitan areas stayed fairly in level during the entire study period, with a slight decrease in 2001, concurrent with increases in deaths due to other opioids, methadone and synthetic narcotics. After 2001, deaths from prescription opioids continued to rise in both metropolitan and non-metropolitan counties, concurrently with prescriptive use of opioids. Over the study period, deaths from unspecified opioids appeared to decrease slightly in metropolitan areas while in non-metropolitan areas their numbers increased. The overall proportion of deaths without a specified opioid was much lower in non-metropolitan counties. This may be due in part to the difficulty in differentiating heroin and morphine deaths (there being more heroin in urban areas), or additionally due to greater case loads for toxicology results for autopsies in urban areas.

There was remarkable similarity between the shapes of the curves over time for deaths due to prescription opioids, methadone, and synthetic narcotics, in both metropolitan and non-metropolitan areas. Increases in mortality due to prescription opioids and methadone were steeper in metropolitan areas during the study period. Although ICD-10 codes to not distinguish the formulation of methadone involved (e.g., tablet or liquid), the similarity in mortality curves suggests that the previously noted phenomenon of methadone overdose deaths (Ballesteros et al. 2003) is more closely linked to problems with prescription opioids used in the management of pain than methadone prescribed for the management of opioid dependence. The temporal sequence of emergence will be explored in a future analysis.

Approximately, 10% of deaths involved multiple types of opioids. This proportion is lower than that expected by the authors. This may be due in part to the difficulty in distinguishing between types of opioids in toxicology screens. For example, codeine is metabolized to morphine and a toxicology screen searching only for morphine would not be able to distinguish between the two opioids. Similarly, standard serum toxicology cannot distinguish between different formulations containing the same opioid substance; controlled-release oxycodone (OxyContin®) will appear the same as an immediate release oxycodone (Percocet® or Tylox®) although specialized and intensive tests can be used to identify specific formulations in research settings (Cone et al. 2004). The ICD-10 structure strongly also influences these results, since prescription opioids are condensed into three codes (T40.2–T40.4). The role of polysubstance use in accidental poisoning is also confirmed by this analysis, with more than half of all opioid deaths also involving cocaine, benzodiazepines, alcohol or anti-depressants. This may in part explain the high proportion of opioid overdose deaths occurring in the presence of medical attention.

The racial/ethnic distribution of accidental opioid overdose mortality mirrors national demographics of these groups, with an under-representation of Asian Americans. The increased availability and abuse of prescription opioids among Hispanics has been reported in inner cities (Vivian, Saleheen, Singer, Navarro, and Mirhej 2005), and our analysis confirms this and suggests that the phenomenon may be more widespread than previously reported.

The occurrence of accidental opioid poisoning deaths dropped off rapidly beyond 45 years of age, possibly due in part to competing hazards resulting from continued substance use (Poser, Poser, and Eva-Condemarin 1992) and increased burden of comorbidities incipient with advancing age. Interestingly, the use of opioids for the management of chronic pain is likely to increase at ages beyond the peak middle-aged deaths, suggesting that the populations using opioids for medical and non-medical purposes may be somewhat distinct, especially at younger ages before the onset of chronic severe illnesses.

Previous studies have described opioid overdose deaths in individual counties and states; however, interstate differences in opioid mortality have not been previously examined for the entire nation. Many explanations for state-level variation exist and need to be tested formally. For example, the observed variation may be influenced by the availability of purer heroin in the northeast, Medicaid formulary reimbursements for particular chronic pain medications, the presence of prescription monitoring programs, higher supply of opioids due to pharmacy robberies, small-scale international smuggling in border states, law enforcement actions, sanctions by state medical licensure boards, and variations in the medical use of benzodiazepines and anti-depressants. No clear pattern was evident among border states, suggesting that the influence of terrestrial international smuggling may not be reflected in opioid overdose. This raises the question whether the 70 foreign residents who died from accidental opioid poisoning may have been involved in trafficking ("body packing") or unused to the purity of heroin and/or the strength of prescription opioids in the United States. The spate of foreign resident poisonings in New York in 2003 bears further investigation.

Residents of metropolitan and non-metropolitan areas were equally likely to be declared dead in a hospital or nursing home as opioid overdoses are reversible with opioid agonists including naloxone (Dixon 2007). Naloxone hydrochloride is an opioid antagonist that reverses respiratory depression associated with excessive consumption of opioids. It is a prescription medication but is not a controlled substance; it does not have psychoactive properties and has few side effects not directly attributable to withdrawal from the initial opioid agonist exposure. Naloxone is routinely used to revive patients after surgical procedures requiring anesthesia and administered in emergency departments to reverse respiratory depression among poisoning victims. Therefore, it is surprising that more than a third of all accidental opioid overdose deaths occurred in the presence of medical supervision. Many explanations are plausible: there may be inherent difficulties in reviving some patients who have overdosed, comorbid complications may have prevented the effectiveness of naloxone, or overdoses may have been reported to medical personnel too late for resuscitation. While paramedics are allowed to administer naloxone under standing orders from prescribers, other emergency personnel (fire and basic emergency medical technicians) are not routinely allowed to administer the antidote; expansion of authority to administer naloxone may be of interest in rural areas where volunteer emergency services are common. A handful of pilot projects training drug users and their family and peers to administer naloxone have shown efficacy in reversing opioid overdoses and could be considered for expansion (Baca and Grant 2005;

Beletsky et al. 2007; Galea et al. 2006; Maxwell, Bigg, Stanczykiewicz, and Carlberg-Racich 2006; Piper et al. 2007; Sporer and Kral 2007), and over-the-counter sales have been proposed (Strang, Kelleher, Best, Mayet, and Manning 2006). Experience in the management of opioid dependence with buprenorphine in France combined with extensive harm reduction strategies appears to have significantly reduced that country's heroin overdose rate (Carrieri et al. 2006). The nature of the Drug Abuse Treatment Act of 2000 offers the flexibility of extending access to buprenorphine maintenance pharmacotherapy in rural areas (McCarty, Rieckmann, Green, Gallon, and Knudsen 2004).

Heroin overdoses appear to be more common in metropolitan areas, paralleling the other illicit drug cocaine; however, to definitively arrive at this conclusion, one must account for the quarter of opioid-related deaths in which a specific opioid was not noted. Methods for this process could include multiple imputation and will be the subject of a forthcoming publication from the authors.

The definition of unintentional opioid mortality used in this analysis is different from those employed by other researchers examining the occurrence of poisoning deaths using ICD-10 codes among US death certificate data (CDC 2004; Paulozzi, Budnitz, and Xi 2006). The extended surveillance definition used in our study is specifically intended to identify accidental opioid overdose deaths and is suited to the hypotheses under investigation in this chapter (Jauncey, Taylor, and Degenhardt 2005). The extended surveillance definition captures more unintentional opioid poisoning deaths than if solely using the ICD-10 codes X40–49 for unintentional and Y10–Y19 undetermined exposures. However, in the United States, the age distribution suggests that increasing the included age to 60 years may capture non-medical use more completely, see Fig. 11.3.

Future analyses are planned, which will expand on this work and further characterize the relationship between geography and overdose deaths. In particular, we will examine the contributing causes of death among accidental opioid decedents for variations by geography. We will also pursue obtaining county-specific mortality data from NCHS for a more advanced spatial cluster and pattern analysis. Data of this resolution may reveal if distance to medical services has an impact on the fatal overdoses, the impact of major transportation routes, and correlations with other geographical features affecting supply and distribution of prescription and illicit opioids.

Conclusion

This study presents the geographic variations among unintentional opioid poisoning deaths in the United States. Decedents' demographic characteristics reflected the overall composition of the United States and were comparable to observations from previous decades. However, the number of opioid poisoning deaths increased rapidly during the study period in both metropolitan and non-metropolitan areas. Accidental overdose deaths occurred in every state and were due primarily to heroin and prescription opioids, but with substantial contributing toxicity from alcohol, cocaine, benzodiazepines, and anti-depressants. Reasons behind the dramatic increase in deaths from unintentional opioid poisoning ought to be explored further, including the causes of state-level and county size variation.

Acknowledgment In memory of Anthony Givens, the authors thank Hannah Gould, Traci Green and Doug Kramer for helpful comments and review of the manuscript. N. Dasgupta would also like to thank Sue West and the GSK/UNC Center for Excellence in Pharmacoepidemiology and Public Health for supporting this research. Portions of this chapter were presented at the NIDA/AAG Symposium on Geography and Drug Addiction, March 2006, Chicago, Illinois and the International Conference on the Reduction of Drug Related Harm, Warsaw, Poland, May 2007. John S. Brownstein is at Harvard-MIT Division of Health Sciences and Technology, Children's Hospital Boston, Harvard Medical School, Boston, Massachusetts.

Chapter 12
Spatial Patterns of Clandestine Methamphetamine Labs in Colorado Springs, Colorado

Max Lu and Jessica Burnum

Abstract Methamphetamine has become the most dangerous drug in the United States. More than half of the methamphetamine used is thought to be smuggled from Mexico, while the other half is manufactured domestically by clandestine methamphetamine labs that have sprung up all over the country. The ease of making the drug using over-the-counter medicines and household chemicals has encouraged many people to set up methamphetamine labs in their residences. In Colorado Springs, an urban area of about 360,000 people, the number of seized methamphetamine labs rose rapidly after the mid-1990s, from four labs in 1997 to 138 labs in the peak year of 2002. Altogether, 497 labs were seized between 1999 and 2005. Like other crimes, methamphetamine labs are not randomly distributed across space; rather, their distribution pattern is shaped by factors that may explain why an individual would want to start a methamphetamine lab and by those characteristics that make a neighborhood attractive as a place to produce methamphetamine. The spatial analysis of methamphetamine lab distribution in Colorado Springs shows that the methamphetamine labs are clustered roughly in and around the downtown area. They tend to be found in neighborhoods with a young and predominantly white population, small household size, and low educational levels. The distribution of methamphetamine labs also appears to have shifted northward over the 1999–2005 period. Such knowledge may assist law enforcement in their fight against the scourge.

Introduction

Methamphetamine is a highly addictive central nervous system stimulant with multiple street names such as "speed," "chalk," "crank," "crystal," and "ice." The drug may be injected, snorted, smoked, or ingested orally, with its stimulant properties

M. Lu
Department of Geography, Kansas State University, Manhattan, KS 66506, USA. Phone: +1-785-5323413;
e-mail: maxlu@ksu.edu

Y.F. Thomas et al. (eds.), *Geography and Drug Addiction*,

similar to adrenaline. One dose of methamphetamine (about 0.25 grams) lasts for 6 hours or more (KCI, n.d.; San Francisco AIDS Foundation 2007). Users initially experience feelings of increased alertness, well-being, exhilaration, or euphoria, high energy levels, loss of appetite, and a general sense of well-being. As the effects of the drug lessen, these intense and positive feelings will fade to extreme fatigue, depression, agitation, and violence. Feelings of panic, paranoia, hallucination, rage, seizures, and strokes can occur from extended methamphetamine use (NIDA 2007; ONDCP 2005). Since the drug alters the natural brain chemistry, addiction is very strong and hard to escape.

Once limited to specific segments of the population, methamphetamine has become the most dangerous drug in the United States. In 2002, 5.3% of the total US population reported having used methamphetamine in their lifetime (ONDCP 2005). About 1.5 million people regularly use this highly addictive narcotic. In 2000, law enforcement seized 7,436 pounds of methamphetamine nationwide. The drug has ruined many families and lives and has resulted in enormous social and economic costs.

Unlike other illicit drugs like cocaine and heroine, methamphetamine is easy to manufacture in make-shift labs. The ingredients are readily available household chemicals such as acetone, ether, iodine, lithium, methanol, muriatic acid, red phosphorus, lye (sodium hydroxide), drain cleaner (sulfuric acid), and brake cleaner (toluene), anhydrous ammonia (farm fertilizer), and over-the-counter cold and allergy medicines that contain pseudoephedrine. Producers can find recipes for "cooking" methamphetamine on the Internet. The ease of making methamphetamine has encouraged many users to establish labs in the United States in order to meet their own needs and sell for profits. The startup cost of a small lab is generally a few hundred dollars in chemicals and supplies, from which a producer can make thousands of dollars worth methamphetamine (KCI, n.d.; PDFA 2005). In the last decade or so, tens of thousands of methamphetamine labs have been seized by law enforcement in the United States. They come in various shapes and forms, from suitcase size labs in car trunks, to those in garages, basements, and warehouses. Labs have been found in many different places, but most of them were located in residential areas (KCI, n.d.). The clandestine nature of the labs makes it difficult for law enforcement agencies to track them. Often labs are uncovered when law enforcement officials go to homes that emit unusual or strong odors, have blacked out or foiled windows, generate excessive and odd trash, constant activity, and suspicious behavior, or when they investigate incidences of domestic violence, explosion and fire (KCI, n.d.).

The methamphetamine consumed in the United States comes from two sources: Mexico and domestic clandestine labs. By some estimate, about half to 80% of the drug is smuggled from Mexico. In the mid 1990s, Mexican drug traffickers started to dominate the production and distribution of methamphetamine in the United States. They operate "super labs" that are capable of producing at least 10 pounds of methamphetamine in a 24-hour period. Those labs are often located along the Mexican border with the United States and in California. Methamphetamine produced in Mexico enters the United States through ports of entry in California, especially San Ysidro (USDEA, n.d.; ONDCP 2005). Clandestine labs across the

United States (sometimes called "mom-and-pop" labs due to their small size) represent only a minor source of methamphetamine in the United States, but they pose a significant threat to public safety and environment (NDIC 2003). Many chemicals used to produce methamphetamine are highly flammable or toxic. The process of "cooking" methamphetamine also releases toxic and hazardous gases and waste. Usually for every pound of methamphetamine produced, five to seven pounds of toxic waste are produced (KCI, n.d.; Reinertson-Sand 2006). Exposure to these toxic substances may cause respiratory and eye irritations, headaches, dizziness, nausea, and shortness of breath among law enforcement officials and other first response personnel, not to mention people present in the homes with methamphetamine labs. Children in homes with labs are at extreme risk of serious illness, injury, or death due to toxic chemical exposure. Explosion and fire caused by botched operations are not uncommon, which sometimes result in fatalities. Methamphetamine lab operators may dump toxic waste down drains, onto the ground, along rural roads, and sometimes in other neighborhoods where it may be overlooked, which often contaminate soil, rivers and streams, groundwater, and public sewer systems (KCI, n.d.; NDIC 2003).

The purpose of this study is to analyze the spatial patterns of methamphetamine labs and the roles contextual socioeconomic characteristics play in their distribution in Colorado Springs, Colorado – a medium-sized city with just under 370,000 people. We selected Colorado Springs for this study for three reasons. First, after the mid 1990s, Colorado Springs experienced a rapid increase in seized methamphetamine labs, though the number has declined in recent years after reaching a peak in 2002. Understanding the distribution of the labs and the factors that may have affected their patterns may generate important insights into the social and demographic characteristics of individuals who engage in this dangerous and illegal activity and what characteristics make a place attractive to producing methamphetamine. Second, the Colorado Springs Police Department (CSPD) has made available on its website the addresses of seized clandestine methamphetamine labs from 1999 on. The data provide a unique opportunity for analyzing the spatial patterns of the methamphetamine labs in the city. Third, the second author of this paper is familiar with Colorado Springs, having lived there for 11 years. Local knowledge of the study area is very important in order to make sense of the patterns revealed by the data and to interpret the results properly.

The analysis uses the addresses of seized methamphetamine labs in Colorado Springs from 1999 to 2005, available on the CSPD website (www.springsgov.com). These addresses, as well as addresses for 11 superstores (Wal-Marts, Targets, and Sam's Clubs), 2 major universities (University of Colorado – Colorado Springs, and Colorado College), and 4 CSPD stations are geocoded in ArcGIS. Of the 497 lab addresses available, 398 addresses, or 80% of the total, geocoded successfully. Only large retail establishments such as Wal-Mart, Target, and Sam's Club stores are included as superstores because they carry large quantities of necessary supplies at attractive prices that may be used to manufacture methamphetamine. Their large size also provides the buyer a sense of anonymity. The ArcGIS Spatial Analyst distance-to-point (straight line) tool was used to calculate the distance from each census tract centroid to the nearest superstore, university, and police station. Socio-economic data for Colorado Springs census tracts are from the US Census.

In the remainder of the paper, we first provide a conceptual framework for understanding methamphetamine lab distribution. We then describe the spatial pattern of the methamphetamine labs seized in Colorado Springs during 1999–2005 period. Specifically, we will show using nearest neighbor analysis (NNA) that the methamphetamine labs are clustered. The study also analyzes by means of Poisson regression the effects of socioeconomic characteristics and the locations of police stations, superstores, and universities on the distribution of methamphetamine labs at the census tract level. Our hypotheses are that, just like other criminal activities, methamphetamine labs are not randomly distributed across the city, and their distribution pattern is shaped by various socioeconomic and geographic factors. The results show that several variables indeed have statistically significant effects on the distribution of the methamphetamine labs in Colorado Springs.

Conceptualizing the Spatial Distribution of Methamphetamine Labs

Geographic studies of crime have generally focused on identifying crime hotspots and the contextual socioeconomic and geographic variables that make some locations more prone to crime than others (Chainey and Ratcliffe 2005). These studies have used several conceptual frameworks to guide empirical analysis. Since domestic methamphetamine production is a relatively new phenomenon, little research has been done on their spatial distribution. Existing literature on methamphetamine has dealt with addiction (Knowles 1999; Maxwell and Spence 2005; Rawson et al. 2004), policy and policing (Boerl et al. 2006; Hohman et al. 2004), health and environment hazards (Brouwer et al. 2006; Colorado Department of Public Health and Environment 2003; Cunningham and Liu 2003), and the relation between methamphetamine use and other crimes (Bower 2003; Swartz 2005). Though little research has been done on the factors that may shape the geography of methamphetamine production, conceptual frameworks that have been developed to explain other types of crime are instructive.

For methamphetamine labs to appear, there needs to be a convergence of three elements: a demand for methamphetamine, motivated offenders, and an attractive environment in which to manufacture the drug. While there are many theories related to crime in society, such as routine activities theory (Malczewski and Poetz 2005), self-control theory and broken window theory (Doran and Lees 2005), social disorganization theory is the most relevant to this research, as it connects crime to socioeconomic variables. The social disorganization theory, developed in 1942 by Clifford Shaw and Henry McKay, suggests that crime is a result of the failure of community structure to recognize the shared values of its citizens and maintain effective social controls (Andresen 2006). The theory has notably impacted subsequent research (Cahill and Mulligan 2003; Browning 2002; Kelly 2000). Three primary factors may lead to social disorganization: economic status, ethnicity, and residential mobility. The assumption is that communities with lower incomes have less money for formal controls and community organizations; communities with greater heterogeneity and more diverse backgrounds are less likely to come together in groups; and communities with greater mobility are less likely to establish networks and community relations. The resulting breakdown of social structure increases the likelihood of crimes. The more immense the breakdown is the higher the expected crime. Methamphetamine production is hypothesized to be similarly influenced by social and community stability as a result of issues such as economic status, ethnicity, and transient population.

Crime has a geography (Andresen 2006; Chainey and Ratcliffe 2005). We must consider the place where crime occurs. The locations of crime often represent the communities most affected by social disorganization. Many studies have found crime rates to be higher in neighborhoods with low income and higher percentages of minority population (Ackerman 1998; Andresen 2006; Brown 1982; Buonanno and Montolio 2008). It is a common belief that criminal activity, like methamphetamine production, is concentrated in neighborhoods of low socioeconomic status, great ethnic diversity, and large proportion of temporary population (Ackerman 1998; Cahill and Mulligan 2003). Other studies have examined the relationship between crime and proximity variables, such as distances to major transportation routes, to downtown, to alcohol serving establishments (Brown 1982; Groff and La Vigne 2001; Kumar and Waylor 2003; Voltz 2000). Voltz's (2000) analysis of heroin and amphetamine markets shows a link between heroin suppliers, but not amphetamine suppliers, and major roads and railroads, which confirms previous findings that methamphetamine suppliers were less likely to be associated with major arterial roads (Eck 1995). Whether this finding holds for methamphetamine manufacturing is worth investigating.

Methamphetamine in Colorado Springs

Colorado Springs is located in central Colorado just east of the Rockies and at the base of Pikes Peak, encompassing an area of 186.1 mi^2 (481.8 km^2). With a population of 369,815 people in 2005, it is also the second largest city in the state, next only to Denver, which is approximately 60 miles to the north. As an amenity and recreation-rich medium-sized city, Colorado Springs holds many meanings to people. For some, it is a popular vacation destination. For others, it is an ideal location to raise a family. The amazing scenery, exciting activities, and supportive communities notwithstanding, Colorado Springs has its dark side, just like many other places. Sitting in the Garden of the Gods, you may not guess it. Looking down on the city from the top of Pikes Peak you would not suspect it. Driving around town you probably would not see it either. But methamphetamine production and addiction has become a serious problem in Colorado Springs, a growing plague hidden in the veins of the city.

Methamphetamine was once used mainly by specific subgroups of the U.S. population such as members of outlaw motorcycle gangs (OMGs) in the west coast. It entered Colorado in the 1990s, first hitting the streets of the Denver metro area in 1994. After that, methamphetamine rapidly overtook cocaine to become the drug of choice for many people in Colorado due to its more intense high at the same cost (KKTV 2005). The National Drug Intelligence Center (NDIC) reported in 2003 that methamphetamine has become the primary drug threat to Colorado (NDIC 2003).

Most of the methamphetamine available in Colorado is produced and distributed by Mexican drug traffickers in Mexico and in southwest states in the United States, particularly California and Arizona, but Caucasian criminal groups and OMGs also produce and distribute methamphetamine in the state (NDIC 2003). Colorado Springs has a reputation for the production of high-potency methamphetamine (ONDCP 2005) and is one of the regional distribution centers of methamphetamine for the west. Prices for methamphetamine in this region range from \$90 to \$125 per gram, \$700 to \$1,200 per ounce, and \$9,000 to \$15,000 per pound (ONDCP 2005). OMGs such as the Banditos and the Sons of Silence, which is the fifth largest OMG with its national headquarters in Colorado Springs, are active distributors of methamphetamine at the wholesale and retail level in Colorado Springs. Hispanic street gangs like Sureños and West Side Varrios, and African American Street gangs such as the Ruthless Ass Gangsters Crips also distribute methamphetamine at the retail level in Colorado Springs (NDIC 2003).

Although perhaps up to 80% of methamphetamine available in Colorado Springs is believed to be produced in Mexico, California, and Arizona, local methamphetamine labs also produce a significant amount of the drug. Law enforcement agencies report that locally produced methamphetamine generally has a higher purity (as high as 90% pure for crystal methamphetamine) than that brought in from outside (NDIC 2003). Methamphetamine labs were almost unheard of in the early 1990s. After the mid 1990s, the number of seized labs rose rapidly, from four labs in 1997 to 138 in the peak year of 2002 (Fig. 12.1). Since then, the number of methamphetamine lab seizures has been on the decline. In 2005, only 18 labs were seized. Most of the labs seized in Colorado Springs were small, capable of producing small quantities of methamphetamine. They were set up by addicts and local independent dealers to produce the narcotic to satisfy their own needs and to sell for a profit to fund their addiction. Repeat offenders are common because people cannot kick the habit.

The recent sharp decline in seized methamphetamine labs may be the result of several factors: an increased awareness among the general public that has deterred methamphetamine production; laws that restrict the sales of cold and allergy medicines containing pseudoephedrine – an important precursor to methamphetamine; and perhaps also the improved ability of methamphetamine lab operators to conceal their illegal activity. But the decline appears to have not made any dent in methamphetamine supply in the city. Whatever decrease in locally manufactured methamphetamine is being offset by Mexican sources. In 2003, the CSPD's Metro Vice, Narcotics and Intelligence (VNI) Division seized nearly

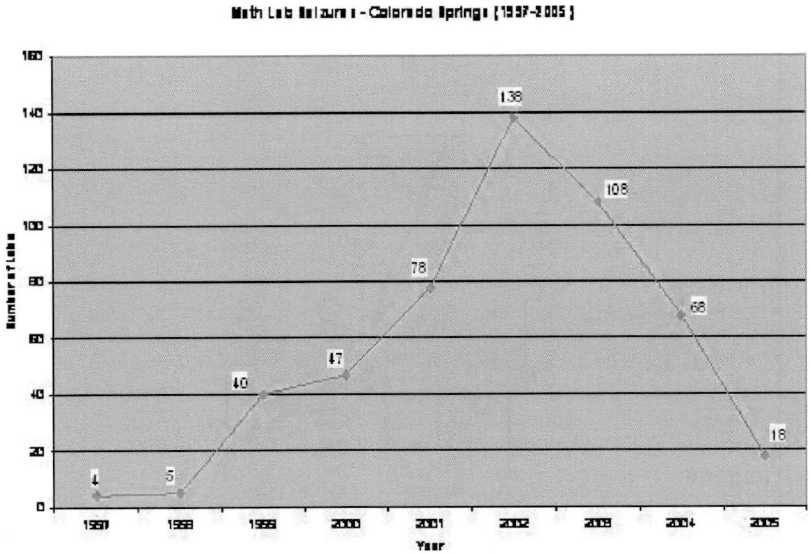

Fig. 12.1 Methamphetamine lab seizures in Colorado Springs, 1997–2005 [Source: Data are from the Colorado Springs Police Department website (www.springsgov.com)]

4,000 grams of methamphetamine. The number increased to 15,000 grams in 2005 and 25,000 grams in 2006, despite drastic declines in lab seizures (KKTV 2007).

An examination of the methamphetamine labs seized in each year from 1999 to 2005 shows that they are mostly located in and near central Colorado Springs, southeast of Colorado College and the central business district, close to the downtown (Fig. 12.2). The downtown area is a mixture of commercial and residential property. Many of the residential neighborhoods around downtown experience more crimes than other areas and are considered to be of low socioeconomic status. Additionally, Acacia Park, located in downtown between Nevada Avenue, Tejon Street, Platte Avenue and Bijou Street, is known for its criminal activity, including drug transactions in addition to being a popular location for concerts, fairs, and outdoor markets.

Methamphetamine is related to other criminal activities, particularly property crimes and identity theft, in Colorado Springs. The focus of methamphetamine users and producers is to obtain money and continue the cycle of addiction and production (KKTV 2005). The relationship between methamphetamine and identity theft is receiving national media attention (Schabner 2005). The majority of identity theft rings are related to methamphetamine and in some places, as many as 95% of identity theft cases are related to methamphetamine addiction. Drug user will steal mail and ID items from cars and purses in order to trade the information for more methamphetamine (Sullivan 2004). To increase public awareness of the problems caused by methamphetamine and help police combat the methamphetamine problem in Colorado Springs, the city produced a documentary – "Methamphetamine – A Social Plague." The documentary was aired

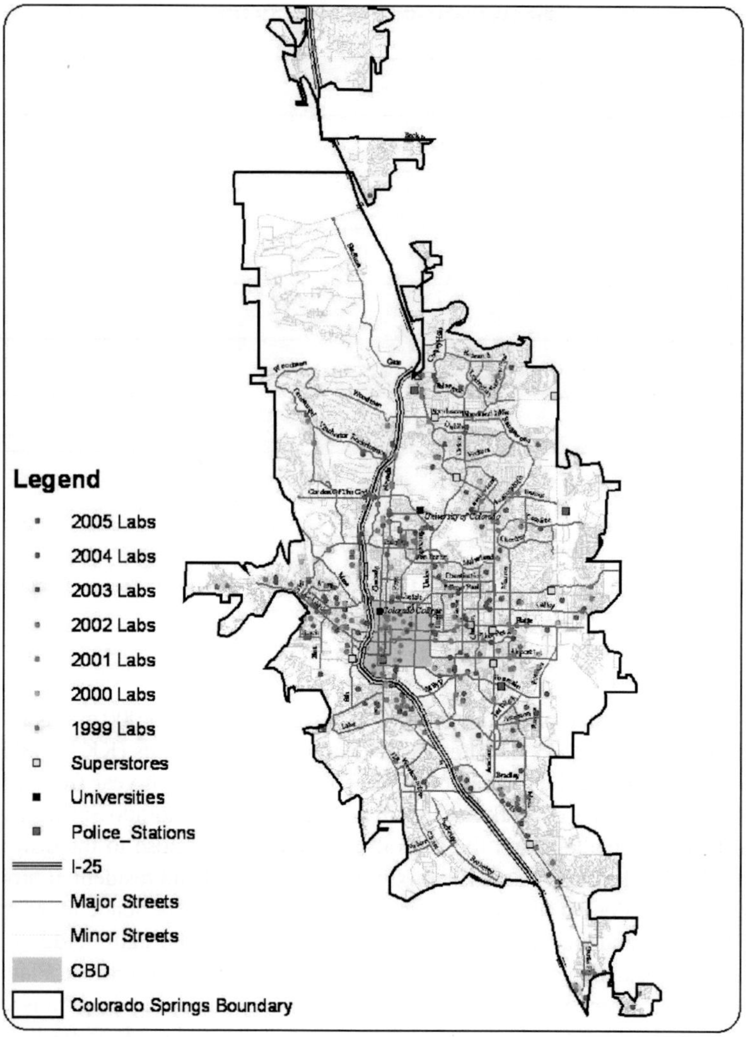

Fig. 12.2 Distribution of the seized methamphetamine labs in Colorado Springs, 1999–2005 (See also Plate 14 in the Colour Plate Section)

simultaneously on June 6, 2005 by all five commercial television stations in Colorado Springs.

Spatial Patterns of Methamphetamine Labs in Colorado Springs

To test if the seized methamphetamine labs are clustered in space or randomly distributed, NNA is carried out. NNA is a widely used technique for spatial pattern analysis, including crime patterns (Chainey and Ratcliffe 2005; Ratcliffe 2005). It calculates the expected mean distance between the locations of methamphetamine labs based on the assumption that the points are randomly distributed and compares that distance with the observed mean distance. If we use (x, y) to denote the coordinates of nearest neighbor labs i and j in a Cartesian space, then the distance between them, d_{ij}, is calculated using the formula:

$$d_{ij} = \sqrt{(x_i - x_j)^2 + (y_i - y_j)^2}$$

The observed mean nearest neighbor distance is simply $d_{\text{obs}} = \frac{\sum d_{ij}}{n}$, where n is the number of the methamphetamine labs in the study area. Given the size of the study area, A, the expected mean nearest neighbor distance under the assumption of random distribution is given by

$$\overline{d}_{\text{exp}} = \frac{1}{2}\sqrt{\frac{A}{n}}.$$

The ratio between the two (observed mean distance/expected mean distance), $\overline{d}_{\text{obs}}/\overline{d}_{\text{exp}}$, is called the nearest neighbor index (NNI). If NNI is <1, then the point pattern shows signs of clustering, and if the NNI is >1, the pattern is dispersed or random. The technique allows us to test if the clustering is statistically significant (Chainey and Ratcliffe 2005; Wong and Lee 2005).

The results of the NNA show that while the expected average distance between the methamphetamine labs is 810.2 m (about 0.5 miles), the actual average distance observed is 260.2 m (0.16 miles). The NNI is 0.32, which means that methamphetamine labs in Colorado Springs are clustered. Furthermore, the Z-score is -26 standard deviations, indicating that the clustering is statistically significant at the 0.01 significance level.

To examine the shift in the general distribution of the methamphetamine labs, we also calculated their mean center of distribution in each year. The mean center, or center of concentration, for a set of methamphetamine labs is their average coordinate values, that is, $\overline{x} = \sum \frac{x}{n}$, $\overline{y} = \sum \frac{y}{n}$ (Wong and Lee 2005). The mean center is useful to show the overall central focal point of the methamphetamine labs, but caution should be exercised in interpreting the results because different sets of locations may generate the same mean center and furthermore its location is very sensitive to outliers (Chainey and Ratcliffe 2005, p. 121).

The mean centers of the methamphetamine labs for the 1999–2005 period showed a northward shift over time with the exception of 2002, when the mean center moved south to nearly the same location as the mean center for 2000 (Fig. 12.3). Northern Colorado Springs consists of middle to upper class areas, not typically characterized by high levels of crime. The northward movement of the mean center may indicate the spread of methamphetamine labs into those areas.

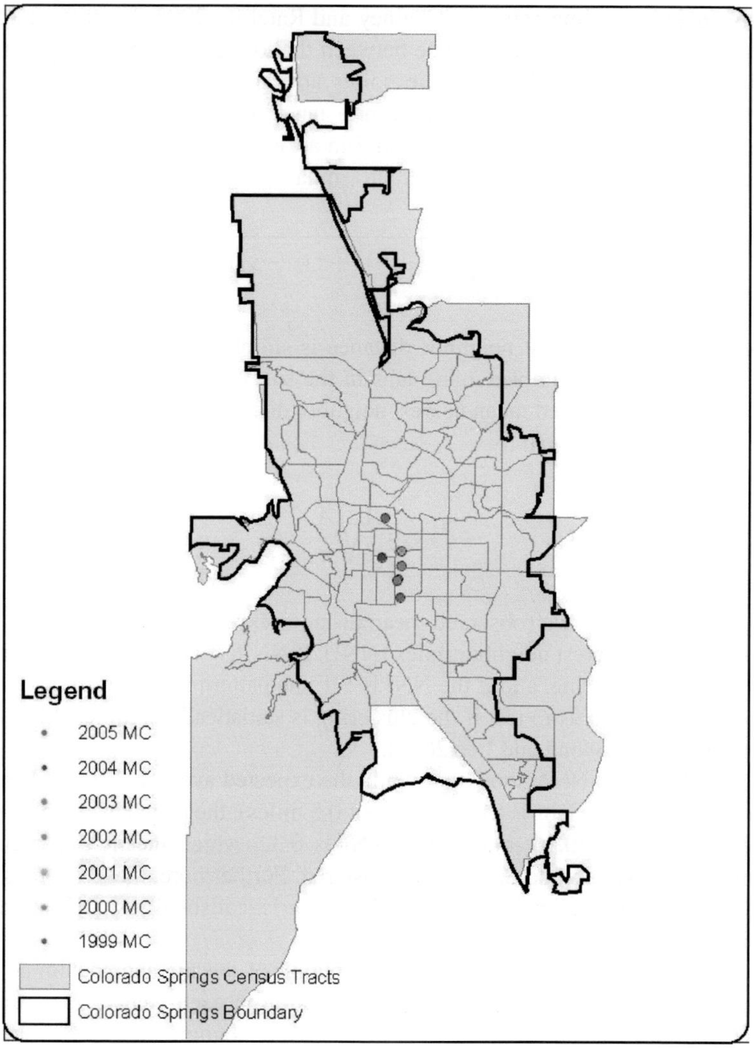

Fig. 12.3 Mean centers of the methamphetamine labs in Colorado Springs

Effects of Contextual Variables on Methamphetamine Lab Distribution

One objective of this study is to examine if there is a geography to clandestine methamphetamine manufacturing, in other words, whether methamphetamine labs are more likely to be established in certain neighborhoods due to their specific socioeconomic and geographic characteristics as implied in the social disorganization theory. Little research has been done on this topic. Anecdotal evidence seems to indicate that methamphetamine in Colorado Springs is a middle-class problem that knows no racial or gender-based boundaries. For example, Sgt. Terry Curry of the Colorado Springs Police Department's Metro VNI Division states that "We [have] busted labs in the Broadmoor, and the Briargate, and the B street area... and the inner city of Colorado Springs... some of the not more affluent areas. It doesn't matter. It could be anywhere..." (KKTV 2005). The Boradmoor area is one of the wealthiest neighborhoods in the city with large grand homes surrounding a five star golf resort. Briargate, located in northern Colorado Springs, is considered an upper middle class area with expensive homes, one of the best school districts, and many recreational amenities. Rundown housing, low-income neighborhoods, and high crime rates characterize the B street area and the inner city. What we wanted to do in this study is to examine if this is borne out in the data.

The spatial patterns of methamphetamine labs may be examined at the census tract level. Colorado Springs has 109 census tracts, most of which are located completely within the city boundaries but some census tracts on the fringes extend beyond. During the 1999–2005 period 79 census tracts had methamphetamine lab seizures though in most cases only a few labs were discovered (Figs. 12.4 and 12.5). The census tracts along I-25 and the downtown tend to have more methamphetamine labs than do those located elsewhere. Two census tracts, one located north of the downtown and the other south, had the greatest number of methamphetamine lab seizures, both at 15, during the study period. A cursory examination of their socioeconomic characteristics indicates that relatively low median household income levels seem to be their only common characteristic. The northern census tract has a population of 3,768 with 20% minority, a median age of 37.4 years, and an

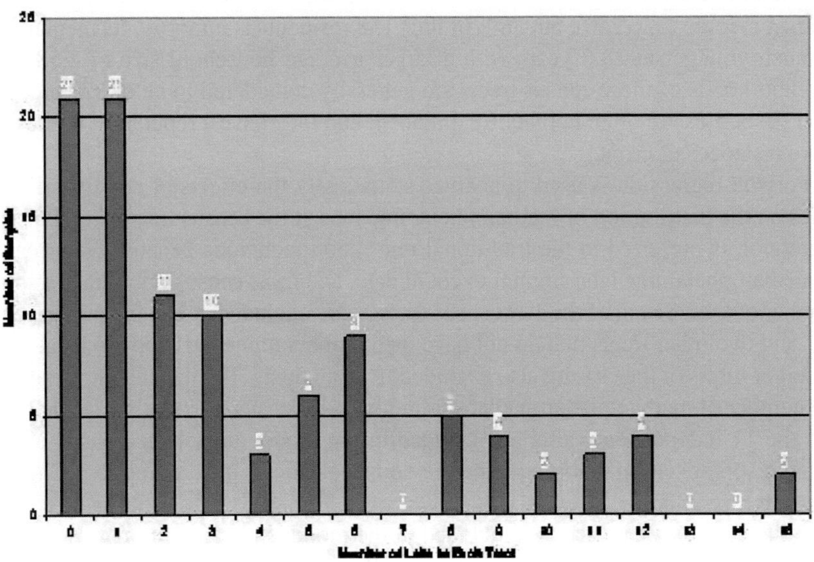

Fig. 12.4 Number of methamphetamine labs by census tract

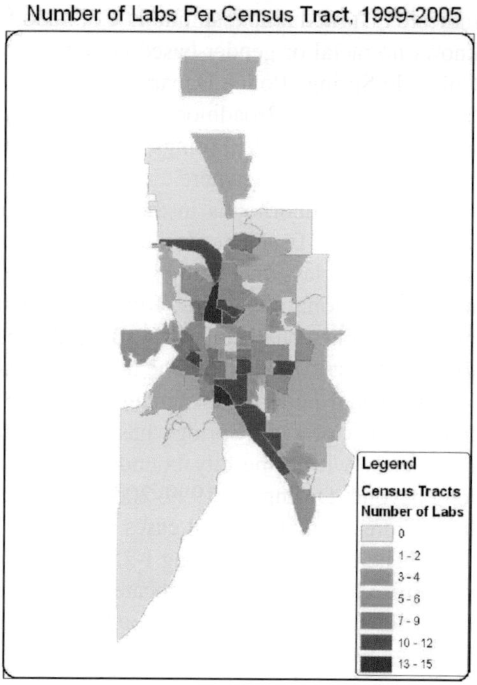

Fig. 12.5 Numbers of methamphetamine labs per census tract, 1999–2005

average household size of 1.99 persons. Its median household income is $26,848 (the number for the whole city is $50,667). The southern census tract has a similar median income ($26,250), but 38% of its 7,158 residents is minority. Its median age is much younger at 28.6 years with a larger average household size of 2.58. The neighborhoods of these census tracts are generally considered to be of low income, their properties are often not well-maintained, and they have a reputation for having more instances of crime.

Poisson regression is used to analyze statistically the effects of contextual variables on the distribution of methamphetamine labs at the census tract level. Poisson regression is preferred to the traditional regression technique because occurrences of methamphetamine labs amount to count data, and most census tracts have a small number of labs. Seven of the 109 census tracts were excluded due to incomplete data sets and two other tracts that do not have methamphetamine labs and are located on the outer limits of the city are also excluded from analysis. The dependent variable is the number of methamphetamine labs found in each census tract during 1999–2005, and the 11 independent variables included in the model are selected based on the social disorganization theory and the previous work on crime distribution (Chainey and Ratcliffe 2005).

Table 12.1 lists the variables included in the model and their summary statistics. The 2000 total population is included as an offset variable to control for the varying population sizes of the census tracts. The Poisson regression analysis was carried

Table 12.1 Summary Statistics of the Regression Variables

Variable	Mean	Minimum	Maximum
Number of methamphetamine labs (dependent variable)	3.8	0	15
Total population in 2000	4,498	1,411	8,743
Percent minority population	19.2	4.0	50.1
Median age	34.4	20.9	51.6
Average household size	2.6	1.7	3.6
Percent rental	37.7	4.0	98.0
Median household income	47,474	14,700	99,432
Percent with HS education or less	32.8	8.0	65.0
Median age of the structures	32.7	8.0	66.0
Median rent	754.3	423.0	2001.0
Distance to store (m)	3305.4	429.2	18020.2
Distance to university (m)	6431.6	214.6	24628.7
Distance to police (m)	4525.1	479.9	18673.6

out using a 2004 free trail version of the statistical and power analysis software, NCSS, downloaded from the NCSS website (www.nccs.com/poisreg.html).

The resulting model has a Pseudo R^2 value of 0.49, indicating that the model was effective in predicting 49% of the sample variations. Five of the eleven independent variables are statistically significant at either the 0.05 or 0.01 significance level (Table 12.2). Not surprisingly, the effect of the offset variable – 2000 total population – is positive because a census tract with a larger population will have more methamphetamine labs than one with a smaller population, *ceteris paribus*. Contrary to findings in other crime studies (e.g., Ackerman 1998; Brown 1982; Groff and La Vigne 2001), the proportion of minority population in a census tract affects the number of seized methamphetamine labs negatively, that is, census tracts with larger percentages of minority population are associated with fewer seized methamphetamine labs. Median age of population also has a negative relationship with the number of labs in a census tract. These findings are consistent with the observation that methamphetamine users and producers are generally young and white (ONDPC

Table 12.2 Regression Coefficient Estimates

Independent Variable	Coefficient	Standard Error	Probability
Intercept	4.78	1.45	0.00
1. 2000 Total population	0.00	0.00	0.00
2. Percent minority	−0.03	0.01	0.02
3. Median age	−0.08	0.02	0.00
4. Average HH size	−1.64	0.41	0.00
5. Percent rental	−0.01	0.01	0.16
6. Distance store	−0.00	0.00	0.85
7. Distance university	−0.00	0.00	0.16
8. Distance police	0.00	0.00	0.55
9. Median HH income	0.00	0.00	0.09
10. Percent no college	0.06	0.01	0.00
11. Structure age	0.01	0.01	0.11
Dispersion Phi	–	2.41	–

2003). The average household size and the number of methamphetamine labs are also negatively related. As average household size increases the number of labs tends to decrease. The percent of population with high school education or less has a positive effect. This may be because individuals with low educational attainment often have fewer employment opportunities and are more likely to engage in risky activities such as drug use and methamphetamine production.

Interestingly, the analysis shows that the relative location of census tracts to superstores (Wal-Marts, Targets, and Sam's Clubs), universities, and police stations, percent rental properties, median household income, and median structure age do not have statistically significant effects on methamphetamine lab activity at the census tract level. We can only speculate on the reasons for these results. Being a medium-sized city, different parts of Colorado Springs are all within a reasonable distance from a superstore; hence the store locations do not make much difference to where one decides to operate a lab. The distances to universities did not matter probably because most residential areas in Colorado Springs are not in proximity to the two college campuses included in the study. While students may be potential customers of methamphetamine producers, they are less likely to make the drug by themselves due to their usually shared or group living arrangements. Structure age does not have a significant effect probably because it may not imply a particular type of housing. For example, some of the older neighborhoods are rundown and less costly, other older neighborhoods, such as the Broadmoor area, are upper-class with large, expensive homes.

Conclusions

This study analyzed the spatial patterning of seized methamphetamine labs in Colorado Springs, Colorado by means of NNA, mean center, and Poisson regression. The results show that methamphetamine labs in Colorado Springs are clustered, roughly in and around the downtown area. Over the 1999–2005 period, the mean center of the labs experienced a northward shift, which may indicate that methamphetamine labs have gradually moved into the middle-class, more "respectable" neighborhoods. The distribution of the methamphetamine labs at the census tract level is affected by several socioeconomic variables such as proportion of minority population, median age of the population, household size, and educational attainment.

The findings of this study provide insights into the kind of people that are likely to engage in the dangerous and illegal activity and the characteristics that make a place more attractive for manufacturing methamphetamine. Generally speaking, methamphetamine labs tend to be found in neighborhoods with a young and predominantly white population, small household size, and low educational levels. Such knowledge may in turn assist law enforcement in their battle against methamphetamine and related crimes. Also, the distribution of methamphetamine labs changes over time. This may be a result of diffusion, or production displacement. In a study on methamphetamine use in Nebraska, for example, Herz (2000) points out that more intensive law enforcement in one area may displace methamphetamine to other areas. In Colorado Springs, the distribution has shifted northward over time. It would be interesting to see if the lab distribution shifts further as law enforcement forces become more active in relatively high risk areas.

One difficulty of analyzing the patterning of methamphetamine labs is that the individuals running methamphetamine labs try hard to hide them. The available data include only those labs that have been discovered and seized by law enforcement. It is not certain if the distribution of such seized methamphetamine labs represents that of all methamphetamine labs out there. A Colorado Springs Police Officer speculated that for every lab uncovered, ten more labs remain unknown (KKTV 2005). Also, the pattern of seized methamphetamine labs is likely affected by spatially differential levels of law enforcement. Further research may examine additional information regarding methamphetamine labs, such as the type of labs (large or small? in car trunks, warehouses, apartments, or single family homes?), the kind of individuals involved, primary reasons for production, how labs came to be seized, and so on, to better understand the variables affecting lab locations.

Chapter 13
A Therapeutic Landscape? Contextualizing Methamphetamine in North Dakota

Kevin Romig and Alex Feidler

Abstract Crystal methamphetamine production and consumption has rapidly expanded over the last 10 years throughout much of the Great Plains. One of the states experiencing tremendous growth is North Dakota. Distinct spatial patterns exist when examining methamphetamine crimes of production and consumption, creating a geographic duality in the landscape of abuse. In terms of treatment, a cursory inspection might suggest North Dakota to be a therapeutic landscape due to its rural nature and slow pace. A more detailed investigation reveals that few treatment facilities exist internally, and North Dakota regularly exports offenders who seek treatment. This chapter highlights the problems faced by the state in limiting crystal methamphetamine production and consumption and suggests applying a more holistic approach to confronting this issue.

Introduction

North Dakota is a rather quiet, sparsely populated, rural, Midwestern state with historically low levels of crime. According to the 2000 Uniform Crime Report, North Dakota was ranked the safest state in the nation. North Dakota was ranked 50th (the safest ranking) in overall crime, violent crime, robbery, and aggravated assaults (Table 13.1). However, North Dakota has the third highest rate of clandestine methamphetamine lab seizures in the country. This is no insignificant statistic; clandestine lab seizures are seen as a major problem facing many parts of the country. The US Department of Justice reports that, "methamphetamine is the most significant drug threat to North Dakota and is the drug-related investigative priority for federal, state, and local law enforcement officials" (2002). According to North Dakota Attorney General Wayne Stenehjem, methamphetamine is "the single most serious law enforcement issue North Dakota is facing – and has ever faced." In a locale

K. Romig
Department of Geography, Texas State University-San Marcos, TX, USA

Y.F. Thomas et al. (eds.), *Geography and Drug Addiction*,

Table 13.1 Comparison of crime rates between North Dakota and adjacent states

State	Crime Index	Rank	Violent Crime Index	Rank
North Dakota	2288.1	50	81.4	50
South Dakota	2319.8	49	166.8	47
Minnesota	3488.4	32	280.8	38
Montana	3533.4	31	240.6	44

Source: Uniform Crime Reports 2000.

that is quite inexperienced in dealing with significant crime issues, the onslaught of methamphetamine is staggering.

The production, sale, and use of illegal drugs stresses a community in a number of different ways, be it crime, rehabilitation measures, law enforcement expansion, or a decline in the public social environment. Methamphetamine use is a particularly serious problem in some rural areas, many of which lack the infrastructure necessary to engage a major drug problem. For example, many rural locales do not have decent access to nearby treatment providers or the expertise to sufficiently respond to methamphetamine abusers. Similarly, law enforcement officials in rural areas lack the training and financial resources to house addicted inmates, recognize meth addiction or production facilities, and pay for laboratory cleanup costs associated with methamphetamine manufacturing in their communities. In rural settings, clandestine drug labs tend to be smaller scale labs. However, these smaller scale labs are sometimes more of a problem to the community and law enforcement largely because they are inefficiently run. Due to these inefficiencies, rural clandestine drug labs can cause three main types of harm: (1) physical injury from explosions, fires, chemical burns, and toxic fumes; (2) environmental hazards due to the numerous amounts of wastes produced; and (3) child endangerment since many of the labs run out of the home or garage and kids are exposed to highly toxic substances (Scott 2000).

The phenomenon of small, rural clandestine labs is one that can be found in many parts of North Dakota. Of the 46 laboratories seized in 2000 by the North Dakota Bureau of Criminal Investigation, 22 were located in rural areas, 17 were in urban areas, and 7 were located in areas classified as small towns (US Department of Justice 2002). Since 2000, the number of seized laboratories has increased (to a high of 297 in 2003) and has shown a growing tendency to move out of the urban areas and into rural ones. One of the most taxing aspects of tackling the methamphetamine epidemic is the clandestine lab cleanup that follows a seizure. According to the US Department of Justice (2002), the national average cost of cleaning one site is \$5000; however, costs can exceed \$100,000 for larger sites. To put that into perspective; in 2003, there were 297 methamphetamine lab seizures in North Dakota, which would mean their cleanup would cost taxpayers around \$1,485,000. However, if damage done by a lab was particularly extensive, the cost can go up substantially, being that the upper limit for cleanup costs can well exceed the \$5000 average. Given North Dakota's small population and number of taxpayers, this is a significant financial burden.

While crimes involving the production of methamphetamine have caught the attention of law enforcement, local consumption of the drug is also a source of concern. Over the four-year period, arrests have gone from 369 in 2000 to 932 in 2003. As the maps show, the spatiality of arrests as defined by clandestine lab seizures and possession arrests are distinctly different, meaning that there is high likelihood that many of the drugs being consumed by offenders are produced and trafficked in from outside the state. The possession arrests have consequences on the health landscape of the state. Methamphetamine is a highly addictive substance, and ridding addiction and addictive behaviors is extremely important in suppressing local demand for the drug.

This chapter has two goals. First, we intend to illustrate the pattern of methamphetamine offenses in North Dakota as visualized by criminal arrests. In this vein, we will address why there is a significant spatial disparity when mapping production versus consumption arrests according to known police records, and why this is significant. Secondly, we will describe the treatment provision linking with the literature on therapeutic landscapes and explain how North Dakota will not be able to properly combat drug abuse without consistent and better treatment facilities and providers. To better theorize this situation, we believe the theory of neoliberalism is critical in better understanding the issue of treatment in North Dakota and discuss how this is emblematic of similar healthcare issues in rural America in terms of access and funding.

Literature Review

To contextualize this research, this chapter briefly reviews three bodies of literature: crime mapping, cultural geography focused on cultural landscape, and therapeutic landscapes from health geography and healthcare literature. Early manifestations of crime mapping were performed by Pauly, McEwen, and Finch (1967) who were interested in computer methods of visualizing crime patterns in Saint Louis. With the advent of geographic information technology, the applications within crime mapping have exploded (Harries 1999). Crime mapping has been one of the most successful applied geography fields in illuminating the positives of spatial thinking. However, the level of mapping applications and detail is constrained by the granularity of the data. Like any other system, better data inputs will result in more useful visualization (Star and Estes 1990). The data received from the North Dakota Bureau of Investigation had only a ‘county’ geographic identifier, so the mapping was limited to mainly choropleth techniques since point data was not available (Slocum et al. 2005). This also limited the types of spatial analyses we could perform beyond broad comparisons of regions within the state. While a regression model including demographic information would have been useful as a predictive tool (Butler 2000), the data did not lend itself to that level of analyses because of uneven demographic patterns within counties.

The idea of a cultural landscape dates back to the seminal writings of Carl Sauer. The landscape refers to generalizations derived from the observation of different geographical scenes (Sauer 1925). The idea of culture was then applied to the landscape, meaning that certain cultural areas would produce and reflect differing landscapes and landscape ideals. This can vary from the specific way a Mormon landscape is laid out with certain fencing styles and street layout (Francaviglia 1978) or how people of the Upper Great Plains use weather and the harsh climate to define themselves as tough survivors (Norris 1993). Each landscape has cultural identifiers that can be interpreted or read. Numerous cultural landscapes then overlap each other and become a collection of lands that are interrelated and part of a broad system (Jackson 1984). The cultural landscape or setting is imperative in understanding the poignancy of this issue. North Dakota is an American anomaly. Most of the state is losing population; hospitals are closing due to not enough demand for services; and local law enforcement has not received much training in identifying drug abuse because there has been little history of narcotic abuse in the state. The rise of methamphetamine has devastated a cultural landscape that is short on medical and economic capital to address this issue.

The impetus of this chapter is in linking the literature on therapeutic landscapes to identify endogenous reasons as to why this drug has had a crippling effect on North Dakota. Social geography has increasingly interested the subjective meanings of places and examining links between the material and spiritual significance of place as identified by residents (Williams 2002). The idea of a therapeutic landscape is a place or situation in which a social environment is optimal for treatment and healing. This is often enhanced by areas of low stress and calmness derived from a location sheltered from the normal 'high octane' American scene. They are also believed to have an enduring reputation for achieving physical, mental, and spiritual healing (Williams 1999). Therapeutic landscapes can be more nature-focused sites of healing (Palka 1999), locales that have clusters of rehabilitation facilities to promote a landscape of healing (Wilton and DeVerteuil, 2006), or landscapes of commodification where healing places are marketed through imagery and lore (Gesler and Kearns, 2002). The rural aspect of North Dakota's cultural landscape is inescapable. Theoretically, North Dakota has many intrinsic characteristics of a therapeutic landscape in its slow pace and rural nature. Our interest is in defining whether North Dakota or parts thereof are a therapeutic landscape for recovery from such an addictive substance like methamphetamine.

Methods

In order to better understand the spatial aspects of methamphetamine activity in North Dakota, ArcGIS, a Geographic Information System software program, was implemented to aid in visualizing the data received from the North Dakota Bureau of Criminal Investigation. Creating and implementing data layers such as methamphetamine-related arrests, lab seizures, as well as certain demographic aspects of North Dakota provide a more holistic picture of the spatiality of this phenomenon. Methamphetamine abuse is not evenly distributed through the state of North Dakota. Some counties have unusually high abuse rates while others do not have a single documented case of a methamphetamine related arrest or a clandestine lab seizure. Certain counties are more prone to production, while others seem to be more prone to distribution and usage. To better understand the methamphetamine phenomenon in North Dakota, the next section attempts to answer two questions: (1) What is the spatial pattern of methamphetamine abuse according to known police records in North Dakota? (2) Why is it such a problem in North Dakota in a broad crime context? These questions will help understand the spatiality of methamphetamine across the state's landscape.

To investigate the overall landscape of methamphetamine in North Dakota, a combination of archival research and informal interviews with law enforcement and institutional officials involved in methamphetamine prosecution and treatment helped elucidate a clearer picture of the landscape of methamphetamine in North Dakota. We accessed data from the North Dakota Bureau of Investigation on methamphetamine-based arrests (possession) for the years 2000–2003 as well as data on clandestine laboratory seizures for 2000–2005. GIS was used to better spatially organize the data to determine trends and patterns across different areas and comparing said patterns with population trends. Before this exercise, methamphetamine-related data had not been effectively mapped at any scale in North Dakota. While the county is not the ideal level of granularity, at this point, it was the best geographic identifier we could apply since many of the arrests took place in rural locales that had no formal address, and law enforcement in the state is not equipped with GPS. Additionally, local newspaper accounts of the surge in methamphetamine crime ground the research to place. To enhance the secondary data, the authors interviewed public officials from local police to members of the Attorney General's office in Bismarck, and those interviews were conducted between June 2005 and May 2006.

Spatiality of Arrests

When examining the spatial pattern of methamphetamine-related arrests, our research discovered that they are more likely to take place in counties that are urban, closer to or containing a larger city, highly populated, higher per capita income, have a larger proportion of vacant houses, a higher percentage of the population in their late twenties, and a growing population (Figure 13.1). Methamphetamine-related arrests and arrest rates tend to be highest in counties with southeast and central parts of North Dakota, where some of the larger population centers are situated, such as Fargo, Bismarck, and Jamestown. This leads to a conclusion that much of the substance consumed in the state is not manufactured here due to the accessibility of southeast corner of North Dakota to drug trafficking routes.

The general trend for clandestine laboratory seizures in North Dakota is that they are more likely to take place in counties that are more rural, further away from larger cities, and have smaller populations. These counties, in general, tend to be located in the northwestern corner of the state. Williams county is the most interesting case

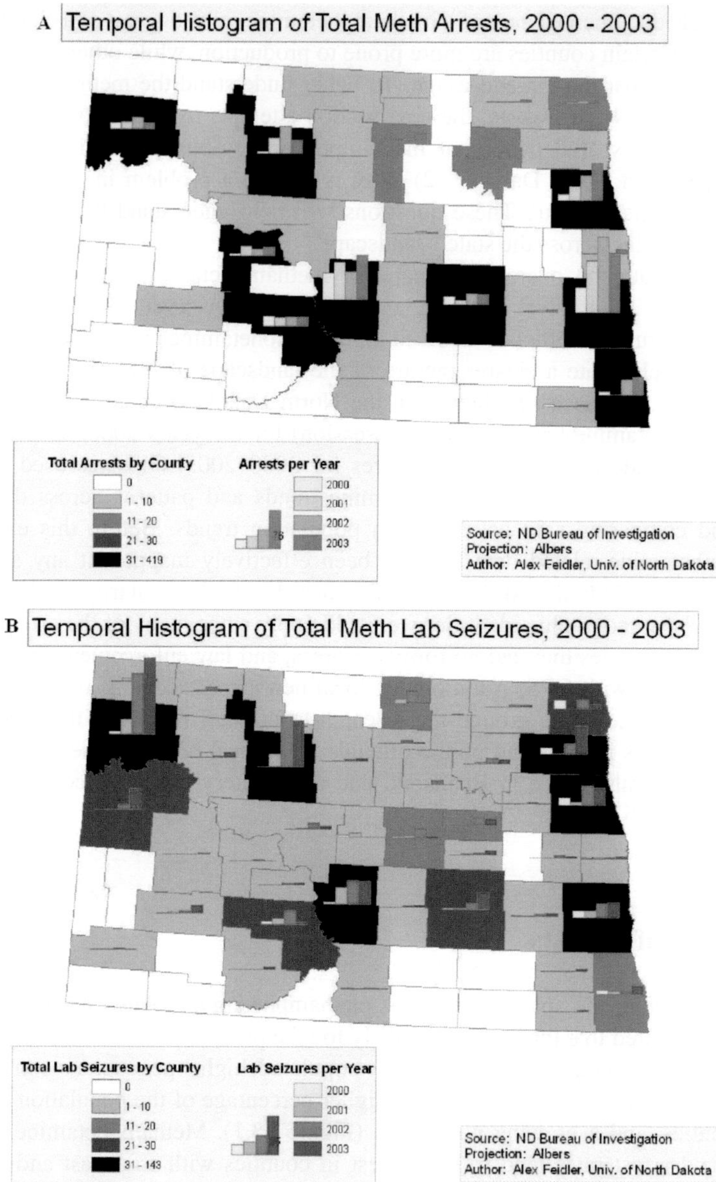

Fig. 13.1 (A) Map showing the growth of methamphetamine arrests in North Dakota over a four-year period; (B) map displaying the growth of clandestine laboratory seizures in North Dakota (See also Plate 15 in the Colour Plate Section)

because it has such a high seizure rate as well as one of the higher arrest rates. Other rural counties had no methamphetamine arrests over this four-year period as defined by known police records. With no documented methamphetamine-related criminal activity, these counties are statistically saying that they are the perfect place to 'not find methamphetamine'. There are two possible explanations for this anomaly. The first is that production is taking place in these counties, but they have gone undetected thus far. The second explanation is simply not all rural communities are prone to methamphetamine production at this time. Most of the counties that are both 100% rural and have no documented methamphetamine activity tend to be clustered together, particularly in the southwestern corner and southern border of the state.

The State Responds

North Dakota is proactively responding to increased methamphetamine crime through a number of programs and policies. In the northeast counties of Pembina, Walsh, and Cavalier counties, locks have been placed on anhydrous tanks so that anhydrous ammonia cannot be stolen from farms to manufacture methamphetamine. These locks are free to farmers who participate in this program (Figure 13.2). Another ingredient cracked down upon was pseudoephedrine, commonly found in the over-the-counter drug 'Sudafed'. On June 1st of 2005, North Dakota HB 1346 took effect, which limits the amount of methamphetamine precursor drugs that can be sold at one time and requires written documentation of the purchase in a retail log book (Indiana Criminal Justice Institute 2005). Although more organized groups of manufacturers can find ways around this, either by obtaining the ingredients illegally or using a number of people to buy the precursors at a number of different locations, it has shown to be successful.

Fig. 13.2 Anhydrous ammonia tanks in a rural part of the state. Photo: Kevin Romig

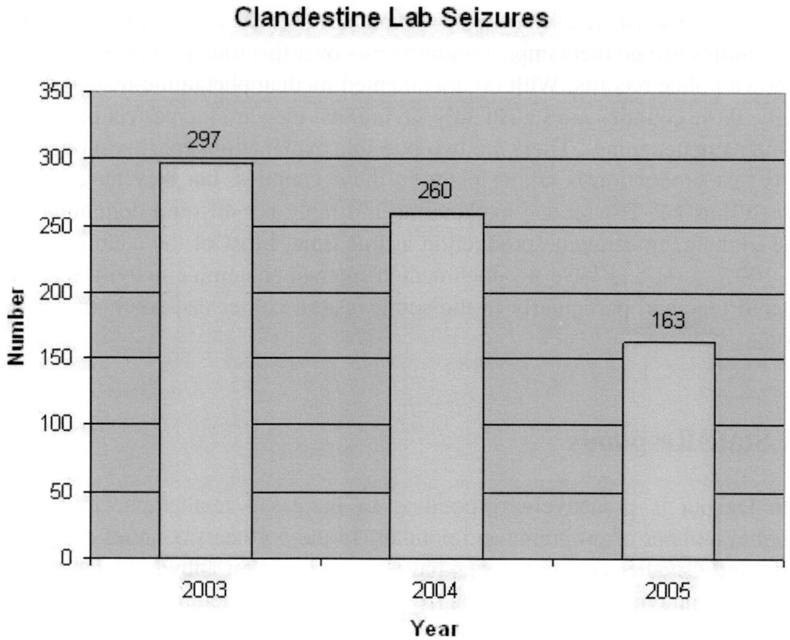

Fig. 13.3 The decline in clandestine laboratory seizures in North Dakota after new laws and restrictions go into effect
Source: North Dakota Bureau of Investigation.

In addition to the restriction placed on many of the chemicals needed to properly manufacture methamphetamine, the North Dakota's Organized Drug Enforcement Task Force (OCDETF), along with local drug task forces, battles trafficking within the state. One of the more significant ways this can be accomplished is by imploring local law enforcement to call a regional Task Force member to the scene if an offender is thought to be under the influence of methamphetamine. North Dakota state law now allows officers the ability to arrest persons under the influence of methamphetamine for possession of a banned substance since finding proof of substances in the body is now considered possession. These combined crack-downs on illicit methamphetamine have actually reduced the number of clandestine lab seizures from 297 in 2003 to 260 in 2004, and most recently to 163 in 2005 (Figure 13.3). It has also provided a stern warning to users that law enforcement is proactively engaged in deterring the production and consumption of methamphetamine across the state.

A Therapeutic Landscape?

While the numbers of clandestine laboratories has subsided due to law enforcement and government efforts, the consumption of methamphetamine as defined by possession offenses has not seen such a significant decrease. The state is seeing an increase in the numbers of persons seeking treatment for methamphetamine addiction. In 2003, the arrest rate was 72/100,000 while the rate of persons seeking treatment was 44/100,000. This gap is surprising. This likely has to do with the lack of available rehabilitation facilities in the state. Also, according to some addicts, the treatment programs in North Dakota are not as successful as other more stringent programs elsewhere in the Northern Plains (Bakken 2006). This incongruity in rates between possession arrests and treatment may also be a function of the number of repeat offenders who have incredible difficulty breaking their addiction.

In the widely read book *Dakota: A Spiritual Geography*, Kathleen Norris paints a bucolic yet sparse picture of life in the Great Plains. It is an autobiographical look at the process of how and why she moved to the Upper Plains to learn more about herself. As a clergy person in more urban settings, she tended to lose focus on what was important or the larger picture of life and how she understands her place in the universe, but in the Dakotas, she was able to notice more, feel more, and get to know her surroundings in a more visceral and meaningful way. The rural landscapes are calm, friendly, and spiritual. She writes, "Dakotans know why they like living here, where life is still lived on a human scale" (Norris 1993, p. 35). As compared with many of the faster-paced, urbanizing landscapes in the United States, much of North Dakota remains an anachronism in its focus on agriculture and rural character. Drug abuse is intrinsically thought of as an urban situation. The pressures of a stressful urban life become overwhelming and persons fall into abusive lifestyles in order to deal with those stressors. This is why North Dakota was so unprepared to deal with the methamphetamine onslaught. While binge drinking has been and continues to plague many communities across the state, drugs have always been a problem somewhere else. In its halcyon tranquility, bucolic North Dakota *should be* a landscape of recovery or a therapeutic landscape. In the Northern Prairie, people should not fall through the cracks because life is lived on the human scale, meaning that relationships, hugs, and handshakes take precedence over case law, mistrust, and political correctness. If anything, people in North Dakota have always been proud of their lack of crime, but the rapid rise in methamphetamine has proven to be a significant radical departure from normality.

The gap in the methamphetamine abuse cycle *should* be able to be closed endogenously as law enforcement procedures crack down on supply, and a landscape of healing should help recovery and curb demand. However, this is not the case. For 6 years, a Drayton, North Dakota resident abused methamphetamine. He entered three different treatment programs in North Dakota with the longest successes lasting 1 week (Bakken 2006). After nearly dying of an overdose in the emergency room in 2003, he served 4 months in prison. He attempted to recover in North Dakota, but found no success with rehabilitation in the seemingly therapeutic landscape. His prison release was predicated on his family sending him to a privately run rehabilitation facility in Brookings, South Dakota. This was a 12-month, faith-based treatment program with strict discipline and structure guiding individuals toward a better life (Teen Challenge of the Dakotas 2006). As of May 2006, this former abuser had been sober for 21 months and felt no significant cravings or desires to use methamphetamine again (Bakken 2006). While this scenario depicts a success story in treating methamphetamine addiction, the questions remain of, why did he respond so much better to a private facility in another state? And what does this say about the healthcare or therapeutic landscape in North Dakota?

A critical issue is access to effective healthcare and related facilities. This is a significant social and economic problem in the state of North Dakota, limiting the ability of this calming landscape to produce therapeutic successes. When asked why there are not more successful treatment programs in North Dakota, a representative of Teen Challenge said, "North Dakota does not have the critical mass for a program to be sustainable." While much of this restricted access to modern treatment programs and facilities relates to an agricultural state with low population growth and a sparse population, it could be argued that equal amounts of state resources should be directed at rehabilitation of methamphetamine as law enforcement programs dealing with supply issues. This is not the case however, and North Dakota is not an anomaly. This is part of the broad neoliberal movement where the government and public agencies attempt to rid themselves of the provision of social goods and services (Greene 2002). If regional faith-based providers can do a decent job in rehabilitating offenders, why should the state pour slim public resources into social rehabilitation programs? Secondly, should the state be spending funds in rehabilitating delinquents who fall into a life of abuse? It is much more politically palatable in a conservative state to spend health capital on the large elderly population aging in place across many of the small towns throughout the state than abusers who are seen as burdens to the society. Faith-based programs are not the only proven successful programs in facilitating recovery of methamphetamine addiction, yet these are the most widely reported successful programs in the Dakotas (Bakken 2006). The state should be more proactive in offering better alternative programs for individuals who may not respond as well to a faith-based program. If the state is truly interested in combating the methamphetamine epidemic, they need to attack the problem more holistically and offer more funding to treatment programs to optimistically reduce the number of repeat offenders in the state.

This also speaks to another critical social issue in the Great Plains, which has access and accessibility to modern healthcare facilities. As market forces continue to make urban healthcare more lucrative and efficient, this is pulling needed resources away from many rural areas in the United States. According to the Center for Rural Health, more than 75% of the state is deemed medically underserved, and the center is involved in programming to deal with medical staff shortages across the state. Most of the top regional medical students are lured to the Twin Cities, Denver, or elsewhere to practice medicine. While this situation is important for the overall benefit of the society, it is critical when considering the rise of the methamphetamine epidemic over the past 10 years. Many of the primary treatment facilities for methamphetamine addiction are local and regional hospitals or sheriff department holding facilities which barely have the capability to offer helpful detoxification programs (Ness 2005). If neoliberalism is defined as the introduction of market discipline in the operation of social goods and services, there is likely not enough demand for private vendors to provide better access for many aspects of healthcare outside some of the more urban areas in the state. As the aging population of North Dakota continues to make the interstate commute to Rochester, Minnesota, for contemporary medical services, it will also be likely that substance abusers will also have to seek alternatives away from their native calming and quiet landscape.

This illuminates an important conundrum in understanding North Dakota's cultural landscape. While the social environment, with its slower pace and life, "Lived on the human scale," is seemingly conducive for therapeutic healing from drug abuse, the economic and population characteristics of North Dakota make it disadvantaged when it comes to treating abusers. Without an economic 'bottom line', this would be a positive place to try to avoid pressures and situations that caused or incited abusive behavior. While the majority of facilities are non-profit, they are not anti-profit and have to think about locating in more effective areas for treating broader markets. Due to neoliberal practices in healthcare provision, state and local governments do not attempt to bridge this gap. This being the case, North Dakota is not an effective therapeutic landscape for recovering for serious health issues because of the difficulty in accessing modern services.

Conclusion

This research attempts to better understand the methamphetamine phenomenon in North Dakota by answering two questions: (1) What is the spatial pattern of methamphetamine abuse according to known police records in North Dakota? and (2) Is North Dakota an effective therapeutic landscape for recovery from methamphetamine addiction. Based on the results of this study, methamphetamine arrests in North Dakota are more likely to take place in counties that are urban, closer to or containing a larger city, and highly populated. These counties, in general, tend to be located in the eastern and southeastern parts of the state. Counties with high methamphetamine-related arrests are where the substance is being sold and consumed, though not necessarily produced. The trend for clandestine lab seizures in North Dakota is that they are more likely to take place in counties that are more rural, further away from larger cities, and have smaller population. These counties, in general, tend to be located in the more northern and northwestern areas of the state.

Why is methamphetamine such a problem in North Dakota in a broad crime context? This question is more difficult to answer and is addressed less directly than the first question by the analysis for this study. However, based on the data that has been analyzed and the review of the literature, the methamphetamine phenomenon in North Dakota can be explained by three different factors. The first is the relatively easy access to the chemicals needed to physically manufacture the substance, particularly anhydrous ammonia. The vast majority of communities in America do not have readily available access to an abundant source of anhydrous ammonia with low rates of law enforcement. However, North Dakota communities do. The second reason is the unusual remoteness of the North Dakota landscape relative to other American landscapes. This 'remoteness' is important for two reasons. First, remote or isolated environments are difficult for law enforcement and communities to efficiently monitor for clandestine criminal behavior. Second, the remoteness of the North Dakota landscape, in addition to the long harsh winters (which most other states never have to face), lends itself to a landscape where certain people are susceptible to idleness, boredom, and despair. These symptoms can be a gateway for drug experimentation and abuse. The final reason why methamphetamine is such a problem in North Dakota is due to the social and cultural difficulty in adjusting to confront and combat the problem. First of all, North Dakota does not have a history of drug problems. Because of this, the state as a whole was largely unprepared for this outbreak. Drug problems tend to be associated with urban areas, not rural ones, and by most standards, North Dakota is an overwhelmingly rural state. Another point is that the North Dakota landscape is predominantly one where people are passive and non-confrontational in nature (Norris 1993).

As mentioned earlier, North Dakota is ranked extremely low in the Uniform Crime Reports indexes for criminal violations. In fact, North Dakota is considered to be the safest state in the nation (Uniform Crime Reports 2000). Within the broad neoliberal era, the issue of treatment or rehabilitation also challenges state and local resources. Specialized methamphetamine addiction treatment is not readily available in many areas, and the burden is placed on private care facilities often in other states. Since meth is a highly addictive substance, the problems will be difficult to reduce without proper treatment. While many of the cultural characteristics of North Dakota might suggest it is a therapeutic landscape with the low stress level and calm lifestyle associated with an agrarian experience, the landscape of healing and recovery for methamphetamine is often very far from home for North Dakota's methamphetamine addicts. While this state meets many of the qualifications for being a landscape of healing, there is not a critical mass for an infusion of private treatment facilities to the area due to issues with a small population, including number of addicts and supply of trained employees in rural areas. Until North Dakota deals more directly with the issue of rehabilitation and treatment, the overall methamphetamine problem will not subside. While the substance may not be as likely to have been produced locally, if addicts need the substance, cartels and dealers will find these people who desire to consume methamphetamine..

Although the rural Midwest is traditionally not associated with drug activity, it is hoped that this study will aid in better understanding processes occurring in such a landscape. The factors mentioned above that make North Dakota susceptible to methamphetamine abuse can also be found in other plains. Even in 2006, actions are being taken by surrounding states to combat methamphetamine in inventive ways. For instance, The Montana Meth Project runs graphic ads on television to aggressively dissuade teens from getting involved with the drug (MontanaMethProject.com 2006). Similar measures are likely to be taken in North Dakota. Conventional drug trafficking efforts that work in the more urban metropolitan areas will not always carry over and be effective in the rural environment. The nature of the problem is different. The culture, the population densities, and the relationship with the physical environment are often reversed when compared to big cities. Understanding these aspects of the North Dakota landscape will help to better understand and fight the methamphetamine phenomenon throughout the Northern Great Plains.

Chapter 14
Are Spatial Variables Important? The Case of Markets for Multiple Drugs in British Bengal

Siddharth Chandra and Aaron Swoboda

Abstract Results from the preliminary analysis of a dataset consisting of population-level statistics for opium and three forms of marijuana for districts in the province of Bengal in British India for the period 1908–1928 are presented. The findings shed light on the importance of geographic phenomena for the economic analysis of drug consumption, and on the economic characteristics of opium and marijuana consumption. Exploratory spatial data analyses reveal spatial clustering for a number of variables that are important for understanding drug consumption. In addition, there is evidence of substitutability between marijuana and opium, and within the three different forms of marijuana. Finally, the consumption of all four products is responsive to changes in their prices.

Introduction

This chapter explores the role of spatial aspects of drug consumption in the broader analysis of the economics of drug consumption and addiction. The aim of the overarching project of which this chapter is part is to advance our knowledge of the behavior of multiple drug-consuming populations. We do this with a series of analyses using a unique recently-discovered and extraordinarily rich and reliable dataset on the consumption of multiple addictive substances collected at a time when the consumption of these substances was legal. When it is collected and entered in its entirety, the dataset will contain annual statistics for a large number of districts from British India over an approximately three-decade period in the early 20th century. These data will include population-level information on the consumption and prices of alcohol (in multiple forms and at multiple proof strengths), opium, and cannabis in three separate forms, namely *charas* (hashish), *ganja*, and *bhang*. *Bhang* is the leaf of the marijuana plant, *ganja* consists of dried parts of the plant (including the bud with the THC-rich resin, THC being the acronym for tetrahydrocannabinol,

S. Chandra
Graduate School of Public and International Affairs, University of Pittsburgh, 3601 Wesley W. Posvar Hall, Pittsburgh, PA 15260, USA
e-mail: schandra@pitt.edu.

Y.F. Thomas et al. (eds.), *Geography and Drug Addiction*,

the main psychoactive component of hemp drugs), and *charas* consists primarily or wholly of the THC-rich resin.[1] In addition, corresponding information on wages, the cost of living, taxation, and a variety of other pertinent economic and non-economic phenomena are being collected and entered. The above data are combined with historical spatial data for the districts for which they are available.

In this pilot study, the results of a preliminary analysis of a subset of the larger dataset are presented. These data consist of population-level statistics for opium and the three forms of marijuana for districts in the province of British Bengal for the period 1908–1928 (Fig. 14.1). This pilot dataset enables the pursuit of a number of lines of enquiry, of which the analysis of substitution and complementarity of opium and marijuana (in its three different forms) with each other and the

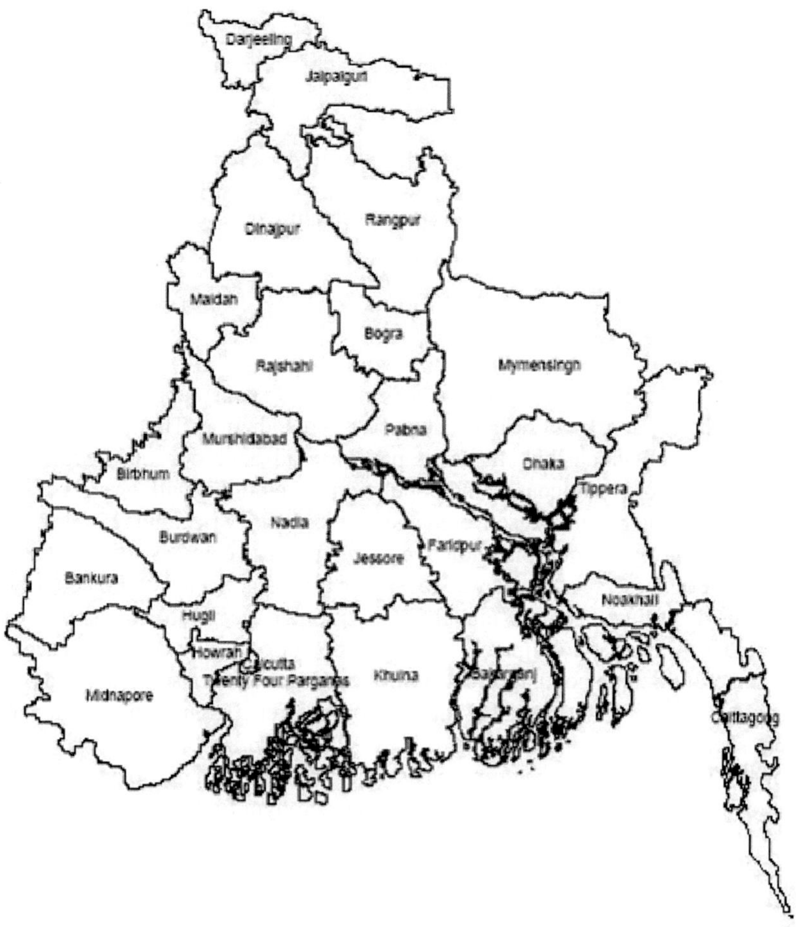

Fig. 14.1 Districts of Bengal in 1919

price sensitivity at the population level of marijuana and opium consumption in the presence of the other drugs is the focus of this study.

Spatial variables can play an important role in understanding the etiology of drug addiction for a variety of reasons. First, space is inherently important in understanding the consumption of most agricultural commodities. To cite just one case-specific example, because different parts of Bengal show differing degrees of suitability for the growth of wild and cultivated cannabis, the former of which was used as a substitute for the cannabis being sold by the government (for which the sales data are recorded), different districts of Bengal are likely to show differing degrees of price responsiveness of consumption of legal marijuana.[2] To the extent that marijuana was a substitute for opium, estimates of the price-responsiveness of opium may also be affected by the availability of wild marijuana, which depends on geography. Further, variations in other phenomena, measured or unmeasured, that may affect patterns of consumption also coincide with geography. One example is population; in keeping with the related literature on the economics of consumption of psychoactive substances, consumption is operationalized in per capita terms. Another example is inflation; the prices of the drugs are normalized for inflation in this study. To the extent that changes in population (because of epidemics, for example) and measures of inflation (because of the often localized nature of food shortages, for example) may vary systematically with geography, these considerations should be included in the analysis as well. A related reason for the inclusion of the spatial dimension in any model of drug consumption is the possibility of measurement error. In the case of Bengal, for example, Calcutta, one of the most urbanized and densely populated districts of Bengal, is bounded by a number of relatively sparsely populated districts. 24-Parganas, the district to the southeast of Calcutta, for example, consists primarily of the Ganges delta (also known as the Sunderbans), most of which is very sparsely populated. The majority of the population of 24-Parganas is concentrated in a relatively urbanized belt that borders Calcutta, giving a picture of the district that is not entirely accurate.[3] To some extent, including spatial variables can help to mitigate this inaccuracy. The aim of this chapter is to shed light on the importance of these and other geographic phenomena for the economic analysis of drug consumption.

In light of the continuing interest in policies relating to the use of psychoactive substances in general and marijuana in particular, one cannot help but note the relative lack of systematic analysis of the economics of marijuana use. While a number of studies exist, some even based on reliable statistical data, that characterize the economic properties of other substances such as opium, tobacco, and alcohol, little is known about how populations of marijuana users respond to changes in the price of marijuana or in general economic conditions.[4] There is a good reason for this difference. Opium, for example, was sold in a number of Asian colonies (including the Netherlands Indies and Japanese Taiwan) in the late 19th and early 20th centuries under tightly controlled government-administered systems. Because these governments kept close tabs on quantities sold, sale prices, numbers of users, and other relevant data, they created statistically viable data sets that, with the development of the requisite methodologies, were ready to be analyzed, even though opium itself had been prohibited by the time these methods were developed.[5] Unlike opium, however, the body of data on marijuana consumption is small. When it was legal in the early 20th century, governments paid scant attention to it, and neglected to follow its consumption in as great detail as they did that of opium. As a consequence, after it became illegal in the mid-20th century, precious little information about its economic properties was available to scholars for statistical analysis.

Late-colonial British India is an exception to this lack of attention to marijuana consumption. In 1893–94, the *Indian Hemp Drugs Commission Report* (Hemp Report) was published following over a year of detailed research on a variety of aspects of marijuana consumption in India.[6] The Hemp Report was written in response to questions regarding the legal status of marijuana and whether the then-legal drug should be made illegal. Following this report, in the early 20th century, detailed statistics on consumption, prices, and a variety of other significant variables were collected, which enable a careful examination of the economic properties of marijuana.

A secondary goal of this chapter is to provide some background into the consumption of marijuana and opium in late-19th and early-20th century British India in general and in undivided Bengal in particular. Using statistics from reports produced on the state of opium and marijuana consumption in Bengal in the early 20th century, we illuminate some of the questions that were being raised at the time, using econometric methods that were developed many decades later. A specific question that will be addressed is the significant issue of whether behavior in relation to the different forms in which hemp drugs were available at the time was the same and therefore, by implication, whether the categorical approach taken by the British government to keep the consumption of marijuana in *all* its forms legal was justified. Because the analysis is preliminary in the sense that we do not fully incorporate the spatial dimension, alcohol data, or wage data into the econometric analysis, we will limit our interpretations of the data, reserving more conclusive assertions for a series of future and more comprehensive studies.

In the early 20th century, the consumption of marijuana and opium was legal and widespread – from Asia to North America, these substances were being consumed under a variety of regimes, from the strictly government controlled to the laissez-faire. In a variety of Asian colonies, including the Dutch East Indies, British Malaya, Japanese Taiwan, and French Indochina, to name but a few, large quantities of opium and/or marijuana were being sold, usually through government-controlled markets for the benefit of government coffers. British India was no exception. A century later, at the beginning of the 21st century, the situation is changed. Marijuana, opium, and their derivatives are considered by many to be harmful, and their use is widely banned.

The remarkable turnaround in attitudes toward and laws pertaining to psychoactive substances over the past century has not diminished interest in their study, however. For one, there is irrefutable evidence that the consumption of these substances continues, perhaps even unabated. Some of these substances, including opium, are considered to be more addictive than others, such as marijuana, adding a layer of complexity to the notion of "substance abuse," a term that has come to be uniformly applied to the excessive use of any of these substances. Additionally, debates continue about the advisability of maintaining the illegal status of some of these substances, and marijuana in particular. In fact, in some countries, as also in some states of the USA, the possession of small quantities of marijuana is no longer a criminal offense, and the partial or complete legalization of the substance is the subject of many a heated debate in a number of state legislatures.[7]

At the scientific level, opium and marijuana are interesting for a variety of reasons. Morphine, which is extracted from opium, is used worldwide as a painkiller in a variety of medical traditions, including the allopathic tradition. In the Ayurvedic system of medicine, marijuana and opium are ingredients in a long list of remedies for ailments ranging from the digestive to the sexual.[8] Because little is known about the interactions of these two drugs with each other, or about how addictive they are, however, it is difficult if not impossible in the current state of knowledge to weigh the pros and cons of using these substances or their derivatives. In the case of marijuana, for example, there is some debate about whether marijuana is addictive, and if so, to what degree it causes psychological and/or physical dependence. The econometric analysis of the behavior of populations that simultaneously consume both types of substances can shed light on some of these questions. In the following sections, we will lay out some of the background for political and social conditions under which opium and marijuana were being consumed in British India. Following this, we will present results of a preliminary and pilot analysis of the data for Bengal using exploratory geographic and econometric methods.

Historical Context: The Reports of the Late 19th Century

Background to the Marijuana "Problem": The Indian Hemp Drugs Commission Report

In 1893, the British Government commissioned a report on Indian "hemp drugs."[9] The goal of this exercise was to inform drug policy in India. Specifically, the question to be answered was whether the government should keep Indian hemp drugs legal for general (medicinal, recreational, and religious) use, as was the case at the time, or whether it should restrict consumption to only necessary uses such as medical therapies or, at the extreme, even ban them outright. The report originated as the result of pressure from British parliamentarians on the government in London to review its policy governing the sale of marijuana to its Indian subjects. Partly as a result of the influence of the religious (missionary) establishment in India, segments of which strongly opposed the free use of these substances, the issue was pressed until the report was commissioned.[10]

The resulting Hemp Report, published in 1894, was a painstakingly detailed analysis of the consumption of marijuana in India, including content on methods of cultivation of the hemp plant, the production of consumer products from the plant, the geography of its cultivation, the position of various forms of marijuana products in the cultural and social milieu of the time, and so on. Because it had the advantage of being written at a time when the drug was legal, and for which plenty of data were therefore readily available, it is still considered to be a policy document of tremendous value.[11] While the Hemp Report does not contain data that are statistically viable, it is an excellent contextual foundation on which to build and interpret any statistical picture of marijuana consumption in late-colonial British India.

The Hemp Report concluded with the view that the sale of "hemp drugs" in British India was neither harmful nor culturally unacceptable. In fact, because the drugs had medicinal value, and had been in use for centuries in religious and cultural ceremonies, banning their use would have created disturbances in society, something that the authorities were keen to avoid. A scholar on the subject of the report, James Mills, has recently challenged the legitimacy of the recommendations of the Hemp Report on the grounds that it was written by a group of people who were intensely loyal to the British government.[12] Given the high stakes in the form of revenue losses that the government stood to incur should hemp drugs be banned, these individuals naturally advocated their continued legal sale. In sum, Mills argues, profit triumphed over principle, and marijuana continued to remain legal in the ensuing decades.

Background to the Opium "Problem": The Report of the Royal Commission on Opium, 1893–1895

A parallel development in the area of opium policy and the different interpretation that it has received makes for a very interesting comparison with the Hemp Report. This difference of interpretations also sets the stage for the analysis presented in the second half of this chapter.

In 1893, in response to pressures not dissimilar from those in the context of marijuana, the British parliament commissioned a report on the legal sale and taxation of opium in British India. The resulting *Report of the Royal Commission on Opium, 1893–95* (henceforth Opium Report) published in 1895, presented a thorough analysis of the consumption of opium in India, including content on methods of cultivation of the opium poppy, the production of consumer products from the plant, the geography of its cultivation, the position of various forms of opium products in the cultural and social milieu of the time, and so on.[13] Like the Hemp Report, because it had the advantage of being written at a time when opium was legal, and for which plenty of data were therefore readily available, it is still considered to be a policy document of great value. While the Opium Report does not contain data that are statistically viable, like the Hemp Report, it is an excellent contextual foundation on which to build and interpret any statistical picture of opium consumption in British India. Together with the Hemp Report, the Opium Report resulted in a variety of policy resolutions, the contents of which often found their way into local and regional laws.

The Opium Report concluded with the view that the sale of opium was neither harmful nor culturally unacceptable in British India. In fact, it asserted that opium had religious and medicinal value. A scholar on the subject of the report, John

Richards, has recently supported the findings of the report on the grounds that it was a justifiable attempt to keep the "cultural imperialism" of some of the more religiously-motivated and contextually ignorant British colonials at bay. In sum, he argues, preservation of Indian culture triumphed over imperialism, and opium continued to remain legal in the ensuing decades.[14]

A Comparison of the Hemp and Opium Reports

A striking feature of the marijuana and opium reports is the degree to which they resembled each other. Both reports were very thorough in the information they gathered – each report comes with half a dozen or so voluminous appendices including reams of testimony given by people involved, directly, or indirectly, with the consumption or trade of the substances.[15] The reports were published within a year of each other. They were commissioned as a parliamentary response to the same prohibitionist forces. Both commissions were dominated by, and the ensuing reports written by, British loyalists, who were likely cognizant of the significant potential losses from the prohibition of the substances. While Indians were represented on both commissions, they had little control over and say in the outcome of the reports. And both reports came to the same conclusion, i.e., the continued regulated (and heavily taxed and therefore profitable) sale of the substances to Indian subjects.

By contrast, the differences between the reports appear to be so small as to be almost superficial. One was about "hemp drugs," while the other was about opium, and one (i.e., the Opium Report) considered in more detail the foreign trade in the substance, while the focus of the other (the Hemp Report) was almost wholly domestic.[16] On a few other details too, the reports differed, but in form and spirit they were the same.

Far more striking than the similarities in the two reports are the differences in the interpretations that the two aforementioned scholars have provided of them. Mills rejects the conclusions of the Hemp Report on the grounds that it was motivated by financial considerations. Richards welcomes the conclusions of the opium report on the grounds that it rejected cultural imperialism in favor of a more culturally harmonious status quo. How is one to reconcile these two radically differing judgments of two reports that were so similar in tenor? The answer lies in a closer reading of the reports. In addition to providing a general if preliminary characterization of the economics of simultaneous marijuana and opium and consumption, in this chapter, we will focus on the Hemp Report and demonstrate, using econometric methods, the validity of this closer reading of the report.

The Indian Hemp Drugs Commission (hence Hemp Commission) consisted of a president (British), six members, three of whom were British and three Indian, and a British secretary. While the report concluded in favor of the maintenance of the legal regime, it contained dissenting opinions by two of the three Indian members (forming a majority of the Indians on the Hemp Commission).[17] These opinions are remarkable in that they showed a nuanced (and, as will be seen, relevant even in modern India) understanding of the position of marijuana in Indian society, and they actually challenged the financial interests of the British authorities. If any pressure had been brought to bear on the Indian members of the Commission to fall in line with the recommendations of their British superiors, then it worked on only one of the three Indians, making the two notes of dissent all the more remarkable.

Both dissenting opinions drew distinctions among the different forms in which marijuana was consumed, namely *bhang*, *ganja*, and *charas*. While the Hemp Report had a tendency to pigeonhole the three major forms of marijuana under the broad heading of "hemp drugs," it made clear the fact that, while these substances were all derived from the same plant, they had differing effects on consumers. Both dissenters recommended the continued legal use of *bhang*, because it was the weakest form of the drug, it did not seem to have deleterious effects on its consumers, and it was widely used for medicinal and religious purposes. Both recommended the gradual and eventual prohibition of *ganja* and *charas* based on the view that these forms of "hemp drugs" were more harmful in their effects (no doubt due to their higher potency), were not widely used for religious purposes (unlike *bhang*), and their consumption was generally frowned upon in Indian society. They proposed gradual prohibition because of the negative impact that rapid prohibition would have on the part of the economy that benefited from the production of these drugs.[18]

The positions of the dissenters, whose opinions were arguably the most culturally nuanced and were, in addition, informed by the same facts as those that informed their British comrades on the Commission demonstrates the partial validity of Mills' position on hemp drug policy in British India. Mills is right to suspect the profit motive of the British – if profits were not part of the equation, then the stronger forms of the drug should have been banned, assuming that they were different from *bhang* in their effects on consumers.[19] Unfortunately, these (and *ganja* in particular) were precisely the forms of the drug that yielded the greatest profits to British coffers. In large swaths of India, *bhang* grew wild and, while it was considered, for good reason, to be of significantly inferior quality (in the sense of lower THC content) to cultivated *bhang*, it posed a potential threat to any hemp drug enterprise that depended solely on *bhang*. Because of the importance of the resin for the potency of the *ganja* and *charas* forms of the drug and the necessity to cultivate marijuana in order to derive high-potency *ganja* and *charas*, however, these two forms of the drug lent themselves much more to control, and hence to profit in a legal and controlled regime.[20] Little wonder then that the voices of two of the three Indian members of the Commission went unheeded.

The Richards argument applied to the hemp drugs context is also partially correct. The prohibitionist forces were being culturally imperialistic, at least to the extent that they advocated the prohibition of *all three* forms of the drug. If the members of the Commission, British and Indian alike, agreed on one issue, it was the maintenance of the legal status of *bhang*. Categorical prohibition would have altered this, imposing foreign norms on the indigenous culture, which is what Richards means when he uses the term "cultural imperialism" in the context of opium.[21]

Of course, the above analysis takes as given the assertions of the dissenting Indian majority on the Commission, i.e., that *ganja* and *charas* were different from *bhang*. This is not at all obvious, given that the intoxicating substance, namely THC, was present in all three forms of the drug. In the second part of this chapter, the question of whether *bhang*, *ganja*, and *charas* can be treated as different drugs from a behavioral perspective is taken up. The responses of consumption of these drugs to changes in their price and in the prices of the other drugs are estimated and summarized. If the drugs appear to be substantially different in their economic properties, then we have evidence (over and above the qualitative evidence with which they supported their dissenting views) in support of the distinction made by the dissenters on the Hemp Commission.

Preliminary Analysis of the Drug Data

Data Description

Prior to the statistical analysis of the data, an overview of the population and the consumption and price statistics pertaining to opium and marijuana in Bengal are presented. The current dataset includes price and consumption data for opium, *bhang*, *charas*, and *ganja* as well as other related variables for 27 districts over 21 years (1908–1928).[22] Because we are working with pre-partition Bengal the area covered currently straddles Bangladesh and the state of West Bengal in India.[23] The 27 districts are: Bakarganj, Bankura, Birbhum, Bogra, Burdwan, Calcutta, Chittagong, Dacca, Darjeeling, Dinajpur, Faridpur, Hooghly, Howrah, Jalpaiguri, Jessore, Khulna, Malda, Midnapore, Murshidabad, Mymensingh, Nadia, Noakhali, Pabna, Parganas, Rajshahi, Rangpur, and Tippera (see Figure 14.1 above). We currently have 525 total district year observations instead of $27 \times 21 = 567$ because we are missing all data for the years 1908–1910 for 14 districts. It should also be noted that we do not have any marijuana price data for these years (1908–1910). In terms of the price data, in most years, the retail price per unit weight (usually a *seer*), is provided for all the regions.[24] All prices are denominated in *rupees* and *annas*.[25]

Primary Variables

The first challenge is missing data for our primary variables, prices and consumption of the four drugs. Of the eight variables of interest (prices and consumption for opium, *bhang*, *charas*, and *ganja*), we have approximately 74% of the total possible data. We are missing consumption data for 50 *bhang* and 278 *charas* observations, and price data for 1 opium, 114 *bhang*, 318 *charas*, and 74 *ganja* observations. The missing consumption data for *bhang* and *charas* are assumed to be 0 (none officially consumed). Table 14.1 displays the summary statistics for drug consumption. *Ganja* consumption is highest by weight (and expenditure), followed by opium, *bhang* and *charas*. Table 14.2 displays the summary statistics for the

Table 14.1 Drug consumption descriptive statistics

	Minimum	First Quartile	Median	Third Quartile	Maximum	Missing Observations	Standard Deviation	Mean
Opium	77	513	842	1,617	32,010	0	3,352	1,916
Bhang	4	75	200	906	20,710	50	2,692	1,065
Charas	2	14	27	82	1,815	278	295	125
Ganja	274	1,481	2,010	3,344	22,560	0	3,291	3,149

Table 14.2 Drug price descriptive statistics

	Minimum	First Quartile	Median	Third Quartile	Maximum	Missing Observations	Standard Deviation	Mean
Opium	7,680	12,290	19,970	24,960	32,640	1	6,560	18,580
Bhang	576	1,440	2,880	3,840	3,840	114	1,249	2,621
Charas	7,104	14,590	17,280	23,040	30,720	318	6,058	18,860
Ganja	3,840	9,600	12,960	16,800	20,160	74	4,425	13,320

drug prices, adjusted for inflation. Opium and *charas* are the most expensive drugs, followed by *ganja* and *bhang*.

We balance the panel by estimating the missing values using seemingly unrelated regressions. Each dependent variable is estimated using a time trend with district specific slopes and intercepts. This process is equal to performing ordinary least squares in each district, regressing the dependent variable on a time trend. We estimate missing observations for the price of opium (1 obs), price of *bhang* (71 obs), price of *charas* (128 obs), price of *ganja* (51 obs), and the price of rice (76 obs), which is used as an index of inflation. Not all missing values are estimated. There are still missing values for *bhang* and *charas* prices in those districts that did not have any sales of these drugs. Specifically, Darjeeling has no *bhang* and *charas* price data and Jalpaiguri has no *bhang* price data for any year.

Other Variables

We estimate the district population in each year using the census population data in years 1891, 1901, 1911, 1921, 1931, and 1941.[26] This exercise is complicated by the global influenza epidemic that struck British India in 1918–1919. It is well-established that British India experienced the highest death toll of any country or territory as a result of this event. While estimates of the death toll in India vary widely, they are in the range of 12 million to 20 million, or approximately 50% of the worldwide figure.[27] Because the disease entered India through Bombay on the west coast of India, Bengal, in the eastern part of India, was not hit as hard as some of the western provinces. That said, the census data for 1911 and 1921, which provide the closest census population figures before and after the year of the epidemic, suggest that parts of Bengal were in fact severely affected by the disease. Specifically, Fig. 14.2a shows that 11 districts in the dataset experienced declines in

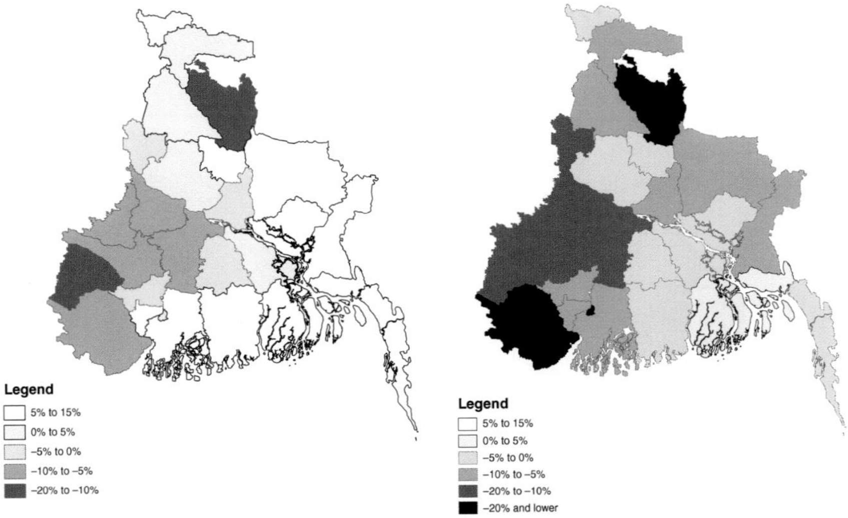

Fig. 14.2 (**a**) Percentage change in population in Bengal from 1911 to 1921 (census years straddling the influenza pandemic); (**b**) percentage change in population in Bengal from 1918 to 1919 (estimates for the influenza pandemic) (See also Plate 16 in the Colour Plate Section)

population over the entire decade. A number of additional districts show very slow rates of growth.

Therefore, our estimation procedure proceeds as follows. First, we assume an exponential rate of population growth, $P_{t+1} = P_t \times (1 + r)$, where r is the growth rate and t is the time period. Then, rather than assuming a constant growth rate for each district over the entire time frame, we allow for a structural break in the year 1919 to account for the influenza pandemic. We estimate a growth rate for each district before 1919 using the 1891–1911 census data and a growth rate after the pandemic using the 1921–1941 census data.[28] The growth rates are displayed in Table 14.3. These growth rates are used in conjunction with the census data to estimate the district population over all years. This exercise also enables us to estimate the drop in population between 1918 and 1919 as a result of the influenza epidemic. The results, displayed in Fig. 14.2b, appear to be consistent with the census data displayed in Fig. 14.2a. Finally, the rates of population growth (or decline) in Figs. 14.2a (1911–1921) and 14.2b (1918–1919) show strong evidence of spatial clustering, as demonstrated by the univariate Moran's I statistic, which is significant at the 1% level for both variables.

The other variable of interest is the price of rice. Because Bengal was (and still is) a heavy rice consuming region, the price of rice is used as an indicator of the cost of living.[29] The rice price data are available annually at the district level. The variation in rice prices over time is clearly non-linear, with at least four inflection points. Therefore, we estimate the 142 missing instances of rice prices in the data set using a 5th degree polynomial. However, this model results in seemingly extraordinary estimates for prices in the later years (>1925). This aspect of the study will be

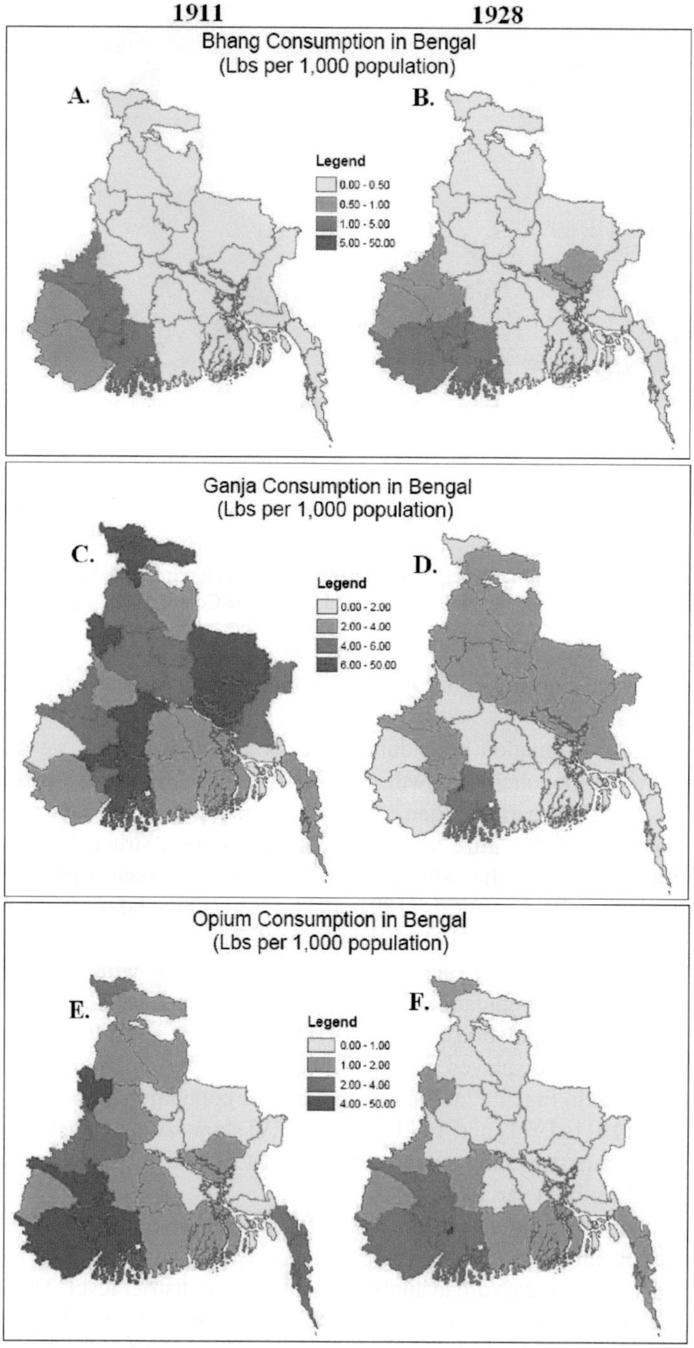

Fig. 14.3 (**a**) Bhang consumption in Bengal – 1911; (**b**) bhang consumption in Bengal – 1928; (**c**) ganja Consumption in Bengal – 1911; (**d**) ganja Consumption in Bengal – 1928; (**e**) opium consumption in Bengal – 1928; (**f**) opium consumption in Bengal – 1911 (Lbs per 1,000 population) (See also Plate 17 in the Colour Plate Section)

Table 14.3 Population growth rates before and after the 1919 influenza pandemic

	1918 and Before (%)	1919 and After (%)
Bakarganj	0.5	1.2
Bankura	0.3	1.2
Birbhum	0.8	1.0
Bogra	1.1	1.0
Burdwan	0.5	1.4
Calcutta	1.7	3.4
Chittagong	0.8	1.1
Dacca	1.1	1.5
Darjeeling	0.8	1.5
Dinajpur	0.7	0.6
Faridpur	0.6	1.3
Hooghly	0.3	1.2
Howrah	1.1	2.0
Jalpaiguri	1.4	0.5
Jessore	−0.4	0.7
Khulna	0.3	1.4
Malda	1.1	1.0
Midnapore	0.4	5.8
Murshidabad	0.4	1.3
Mymensingh	1.3	1.1
Nadia	−0.1	1.3
Noakhali	0.8	1.5
Pabna	0.2	1.0
Parganas	1.1	1.8
Rajshahi	0.3	0.2
Rangpur	1.6	0.7
Tippera	1.7	1.3

refined in future iterations. All nominal (i.e., Rupee-denominated) variables were adjusted for inflation using this variable.[30]

We now proceed to a spatial and econometric characterization of the data, with a focus on the figures for population and consumption of opium, *bhang*, *charas*, and *ganja*.

Models of Drug Consumption

As discussed earlier, the dissenting members of the Hemp Commission emphasized the distinct nature of the different forms of marijuana, namely *bhang*, *ganja*, and *charas*. As a consequence of this emphasis, their concluding opinions (to prohibit *ganja* and *charas* and to keep *bhang* legal) were at odds with those of the British-dominated majority on the Commission, which took a categorical (i.e., all or nothing) approach to the issue. Were the dissenters justified in treating *ganja* and *charas* differently from *bhang*? While there is substantial evidence that they were justified on religious and social grounds, the question of differences in the substances at a more basic behavioral level still stands. In particular, because all three forms of marijuana contain THC as the main psychoactive ingredient, it is possible that all three forms produced similar behavior among users. The goal of this section is to analyze the data on the substances for evidence of differences and similarities. To this end, standard models of consumption of *bhang*, *ganja*, *charas*, and opium are estimated.

In keeping with the literature on how changes in prices affect the consumption of the drugs under study, we use per capita consumption data rather than aggregate consumption. Therefore, we normalize the consumption data by the estimated population in each district. Figure 14.3 shows per capita consumption of *bhang*, *ganja*, and opium for the start- and end-years (i.e., 1911 and 1928) for the pilot data used in this chapter. Table 14.4 demonstrates that there is evidence of spatial clustering of per capita consumption of the drugs under study, suggesting that geography may be important. Specifically, for half of the variables, Moran's I statistic suggests clustering and is statistically significant. In addition, other important variables such as population density and the percentage change in population during the influenza epidemic (1918–1919) show evidence of spatial clustering. These findings indicate in favor of modification of the econometric models for spatial phenomena in future iterations of this research.

For each good, the model was written as a variant of

$$C_{it} = \alpha + \beta_1 \text{BHANGPRICE}_{it} + \beta_2 \text{GANJAPRICE}_{it} + \beta_3 \text{CHARASPRICE}_{it} + \beta_4 \text{OPIUMPRICE}_{it} + e_{it}$$

where C_{it} is natural logarithm of per capita consumption of *bhang*, *ganja*, *charas*, or opium in district i at time t, BHANGPRICE_{it}, GANJAPRICE_{it}, CHARASPRICE_{it}, and OPIUMPRICE_{it}, are the natural logarithm of the real price of *bhang*, *ganja*, *charas*, and opium and e_{it} is the random error term. α is the regression constant, and $\beta_1 - \beta_4$ were the coefficient estimates as written above. Because the markets were monopolized by the government, the issue of endogeneity of consumption does not arise.[31]

Table 14.4 Clustering patterns of drug consumption

Variable	Year	Moran's I	z-Score	Significance Level (0.01, 0.05, or 0.10)	Clustered or Dispersed
Opium consumption	1911	0.026	3.423	0.01	Clustered
	1928	0.074	5.439	0.01	Clustered
Bhang consumption	1911	−0.009	1.750	0.10	Dispersed
	1928	0.062	5.153	0.01	Clustered
Ganja consumption	1911	−0.036	0.135	Not significant	Neither
	1928	0.017	2.568	0.05	Clustered
Charas consumption	1911	−0.070	−1.912	0.10	Dispersed
	1928	−0.047	−0.412	Not significant	Neither

Consumption is measured in pounds per 1,000 population.

Preliminary Regression Results for the Pilot Data

As a preliminary step, two categories of models were estimated for each substance. These were fixed effects models and random effects models with district-specific effects. In each category, two models were estimated, one each with and without a time trend. Finally, these four models were estimated using a restricted dataset for which only those observations for which *bhang* and *charas* data were available were used, and a larger dataset for which all observations were used by dropping the prices of *bhang* and *charas* from the model. The results demonstrated that the fixed effects specification with a time trend is usually superior to other specifications.[32] The results of the regressions for opium, *ganja, charas*, and *bhang* are presented in Tables 14.5–14.8.

Table 14.5 Opium regression results (Fixed effects with time trend model)

Variable	Estimate	*Std.Error*	t-Value	Pr(>|t|)
Intercept	67.79	6.29	10.78	0.00
Opium price	−0.70	0.10	−7.20	0.00
Ganja price	0.38	0.12	3.20	0.00
Charas price	0.13	0.11	1.14	0.25
Bhang price	0.03	0.04	0.73	0.47
Bankura	0.12	0.08	1.65	0.10
Birbhum	0.48	0.08	6.19	0.00
Burdwan	0.95	0.08	12.48	0.00
Calcutta	3.18	0.07	42.78	0.00
Dacca	−0.34	0.08	−4.51	0.00
Dinajpur	−0.25	0.08	−3.16	0.00
Hoogly	1.49	0.07	20.40	0.00
Howrah	1.29	0.08	16.46	0.00
Khulna	0.06	0.08	0.73	0.46
Malda	0.67	0.08	8.45	0.00
Midnapore	1.18	0.07	16.14	0.00
Murshidabad	0.23	0.08	2.93	0.00
Mymensingh	−1.29	0.08	−16.00	0.00
Nadia	0.09	0.08	1.19	0.24
24-Parganas	1.48	0.07	20.06	0.00
Rangpur	−0.51	0.08	−6.33	0.00
Year	−0.04	0.00	−11.95	0.00
Summary statistics				
Residuals	SE residual	0.22	DF residual	300
Regression	F-statistic (21,300)	311.60	p-value	0.00
N	N	322	N of FE	17
Specification	LR test	112.23	p-value	0.00
Fit	AIC	173.07	R^2	0.96
	BIC	199.49	Adj. R^2	0.95
	log Lik	−79.53		

Table 14.6 Ganja regression results (Fixed effects with time trend model)

Variable	Estimate	*Std.Error*	*t*-Value	Pr(¿—t—)
Intercept	59.40	5.89	10.08	0.00
Opium price	−0.09	0.09	−0.96	0.34
Ganja price	−0.22	0.11	−1.95	0.05
Charas price	0.04	0.11	0.41	0.68
Bhang price	0.09	0.04	2.42	0.02
Bankura	−0.18	0.07	−2.49	0.01
Birbhum	0.49	0.07	6.82	0.00
Burdwan	0.77	0.07	10.81	0.00
Calcutta	2.47	0.07	35.46	0.00
Dacca	0.72	0.07	10.14	0.00
Dinajpur	0.63	0.07	8.53	0.00
Hoogly	0.87	0.07	12.75	0.00
Howrah	0.83	0.07	11.28	0.00
Khulna	0.07	0.07	1.01	0.31
Malda	0.80	0.07	10.66	0.00
Midnapore	−0.12	0.07	−1.79	0.07
Murshidabad	0.37	0.07	4.89	0.00
Mymensingh	0.82	0.08	10.89	0.00
Nadia	0.31	0.07	4.28	0.00
24-Parganas	1.32	0.07	19.04	0.00
Rangpur	0.31	0.08	4.16	0.00
Year	−0.03	0.00	−11.25	0.00
Summary statistics				
Residuals	SE residual	0.21	DF residual	300
Regression	F-statistic (21,300)	146.70	p-value	0.00
N	N	322	N of FE	17
Specification	LR test	97.54	p-value	0.00
Fit	AIC	104.50	R^2	0.91
	BIC	130.92	Adj. R^2	0.91
	log Lik	–45.25		

Table 14.5, the opium model, shows an own-price elasticity of -0.7. This is consistent with earlier findings on opium price elasticities. There is also evidence that *ganja* and opium are substitutes. Further refinement of the models to include more observations and spatial and other variables such as wages will shed more light on this and other possible drug interactions.

Table 14.6 contains the results of the *ganja* model. The own price elasticity of *ganja* is low, at -0.22. Aside from this finding, little can be said about cross-price elasticities at this time aside from a small substitution effect from *bhang*. The results for *bhang* are shown in Table 14.7. The own-price elasticity of *bhang* is low, at -0.33. Table 14.8 summarizes the *charas* model. Because of the small quantities of *charas* being consumed in British Bengal and the correspondingly large fluctuations in consumption in percentage terms compared to opium, *ganja*, or *bhang*, these results should be interpreted with caution. *Charas* has a high own-price elasticity of -0.79, is a substitute for *ganja*, and complements *bhang*. We are currently collecting data for provinces in which *charas* consumption was relatively high, and expect that

Table 14.7 Bhang regression results (Fixed effects with time trend model)

Variable	Estimate	*Std.Error*	*t*-vValue	Pr(¿—t—)
Intercept	−179.10	19.09	−9.38	0.00
Opium price	−0.17	0.29	−0.59	0.56
Ganja price	0.18	0.39	0.47	0.64
Charas price	0.25	0.37	0.67	0.50
Bhang price	−0.33	0.13	−2.48	0.01
Bankura	1.00	0.20	5.07	0.00
Birbhum	2.40	0.19	12.46	0.00
Burdwan	2.48	0.20	12.70	0.00
Calcutta	4.79	0.18	26.46	0.00
Dacca	−0.58	0.19	−3.08	0.00
Dinajpur	−0.06	0.20	−0.29	0.77
Hoogly	1.22	0.18	6.63	0.00
Howrah	0.51	0.35	1.44	0.15
Khulna	0.90	0.20	4.58	0.00
Malda	1.23	0.31	3.97	0.00
Midnapore	−0.33	0.19	−1.71	0.09
Murshidabad	0.94	0.21	4.60	0.00
Mymensingh	0.07	0.35	0.20	0.84
Nadia	1.25	0.20	6.32	0.00
24-Parganas	1.64	0.18	9.03	0.00
Rangpur	−0.57	0.42	−1.37	0.17
Year	0.09	0.01	8.82	0.00
Summary statistics					
Residuals	SE residual	0.53	DF residual	216
Regression	*F*-statistic (21,216)	85.07	*p*-value:	0.00
N	*N*	238	*N* of FE	17
Specification	LR test	86.86	*p*-value	0.00
Fit	AIC	521.57	R^2-Squared:	0.89
	BIC	545.88	Adj. R^2-Squared	0.88
	log Lik	−253.79		

analyses of those data will yield more robust and perhaps different estimates of the price elasticity of *charas*.

In addition to the findings discussed earlier, Table 14.5–14.8 demonstrate further possible evidence of the importance of space in understanding the economics of consumption of *bhang*, *ganja*, *charas*, and opium. Specifically, the majority of the district fixed effects are statistically significant, suggesting the possibility that geographic variables may be an important addition to the models. For this additional reason, the results should be interpreted with caution, pending further broader and deeper analyses of these and additional data, which we expect will be forthcoming soon.

The preliminary analysis suggests that *bhang*, *ganja*, and *charas* (and opium, for that matter) elicited very different behavioral patterns. These patterns are summarized in Table 14.9. First, different forms of marijuana show differing degrees of own-price responsiveness. Second, there is no symmetry of substitution effects, which one would expect if the different forms of marijuana were being consumed in the same manner and for the same reasons. For example, while the price of *ganja* is significant and positive in the *charas* consumption model, the price of *charas* is

Table 14.8 Charas regression results (Fixed effects with time trend model)

Variable	Estimate	*Std.Error*	*t*-Value	Pr(¿—t—)
Intercept	32.50	10.03	3.24	0.00
Opium price	−0.17	0.16	−1.07	0.29
Ganja price	0.82	0.19	4.33	0.00
Charas price	−0.79	0.18	−4.41	0.00
Bhang price	−0.23	0.06	−3.60	0.00
Bankura	1.57	0.12	13.29	0.00
Birbhum	1.70	0.12	13.97	0.00
Burdwan	2.39	0.12	19.94	0.00
Calcutta	4.82	0.12	41.21	0.00
Dacca	1.33	0.12	11.18	0.00
Dinajpur	−1.38	0.12	−11.17	0.00
Hoogly	2.53	0.11	22.11	0.00
Howrah	2.71	0.12	22.02	0.00
Khulna	0.64	0.12	5.20	0.00
Malda	−1.53	0.13	−12.16	0.00
Midnapore	2.08	0.12	18.06	0.00
Murshidabad	0.32	0.13	2.51	0.01
Mymensingh	−0.82	0.13	−6.49	0.00
Nadia	1.03	0.12	8.54	0.00
24-Parganas	2.86	0.12	24.65	0.00
Rangpur	−1.52	0.14	−11.08	0.00
Year	−0.02	0.01	−4.13	0.00
Summary statistics				
Residuals	SE residual	0.35	DF residual	294
Regression	*F*-statistic (21,294)	360.30	*p*-value	0.00
N	*N*	316	*N* of FE	17
Specification	LR test	120.77	*p*-value	0.00
Fit	AIC	359.90	R^2	0.96
	BIC	386.19	Adj. R^2	0.96
	log Lik	−172.95		

not significant in the *ganja* consumption model. The complementarity effects are, likewise, far from uniform. In sum, there is a clear evidence in support of the implicit assumption in the dissenting notes in the Hemp Report that *bhang*, *ganja*, and *charas*, although originating from the same plant (i.e., hemp) and containing THC as the main psychoactive substance, were qualitatively different in their usage and impact on users.

Future Research Directions

In a variety of ways, this is a pilot study. The dataset that will ultimately be used to analyze relationships between the consumption of the various drugs sold in British India will be larger in numbers of observations and in geographic scope than this dataset. Second, additional variables including wage or income data will be used

Table 14.9 Summary of the properties *of Bhang, Ganja, Charas*, and Opium

Property	Substance			
	Cannabinoids (increasing strength, from left to right)			Opioid
	Bhang	*Ganja*	*Charas*	Opium
Own-price responsive	Yes**	Yes*	Yes***	Yes***
Substitute A†	None	*Bhang***	*Ganja****	*Ganja****
Substituted B†	*Ganja***	*Charas****, *Opium****	None	None
Complement A‡	None	None	*Bhang****	None
Complemented B‡	*Charas****	None	None	None

"None" may indicate an inconclusive result that will be sharpened by analysis of the expanded dataset.
***Parameter estimates significant at the 1% level.
**Parameter estimates significant at the 5% level.
*Parameter estimates significant at the 10% level.
† The price of the 'row' drug is significant in the model of consumption of the 'column' drug.
‡ The price of the 'column' drug is significant in the model of consumption of the 'row' drug.

in future analyses. Further, based in part on the findings of this pilot study, which show a strong spatial character for some of the key variables in the analysis, we will include spatial considerations in final econometric models to be estimated using the expanded dataset. Many additional concepts, which are not discussed here for lack of space, can be analyzed using these data, including (1) the addictiveness (using economic models of addiction) of each of these substances in a multi-substance setting using legal data, (2) the price sensitivity at the population level of marijuana, alcohol, and opium consumption in the presence of different combinations (these vary depending on the province being studied) forms and strengths of the same and other substances, (3) the analysis of differential behavior depending on the strength of some of these substances (such as alcohol based on alcohol content and cannabinoids based on form, each of which has a different potency (i.e., *charas* vs. *ganja* vs. *bhang*)), and (4) whether the consumption of one or more substances or forms of substances tends to systematically precede in time the consumption of one or more of the other substances or forms of substances.

Conclusion

In this chapter, using a small pilot dataset and simple econometric models, a simple and perhaps obvious but nevertheless significant point has been demonstrated in the context of marijuana. The preliminary evidence indicates that drugs containing the same basic psychoactive substance can, depending on their form and content, produce very different effects on the consumer. This is, furthermore, manifested in econometrically identifiable behavior. For this reason, the tendency of the

British authorities to pigeonhole all forms of marijuana into one legal category was arguably counterproductive from a consumer welfare perspective. In advocating the continued legal sale of *bhang* but the prohibition of *ganja* and *charas*, the Hemp Commission dissenters were making nuanced judgments about behavior and their culture.[33] Their judgments are amply validated by the fact that, after India won her independence over 50 years after they penned their dissenting opinions, policy toward marijuana gradually evolved to the point that it is now legal only in the *bhang* form and illegal in the *ganja* and *charas* forms.

In historical context, the *Indian Hemp Drugs Commission Report* is a remarkable document. It contains a wealth of information on many aspects of marijuana production and use in British India. This information suggests that a nuanced approach to the management of the consumption of hemp drugs in British India would have been preferable to the categorical (and financially motivated) approach taken by the British-dominated majority on the Commission. Ultimately, the outcome of the *"Hemp Report"* holds a cautionary message that is paralleled in similar historical works on other substances.[34] Governments can, for financial or other reasons, pursue policies relating to addictive substances and activities that may suboptimal in the sense that, given their social and cultural milieus, they are overly liberal or restrictive.

Acknowledgment The authors gratefully acknowledge support from the National Institute on Drug Abuse (Grant 1R21DA020160-01A1) and the assistance of Andrea Arkin, Benjamin Colaiaco, Katy Collins, Michael Siciliano, and David Totten in the collection and preparation of the data used in the analysis. A variant of this chapter was also presented as "The Demand for Marijuana and Opium in Early Twentieth Century India" at American University in Washington, DC, as part of the Washington Area Economic History Seminar on Friday, October 13, 2006. The geographic data are based in part on modifications of a geographic dataset supplied by ML Infomap (New Delhi, India).

Notes

1. For detailed definitions and descriptions of the three main different forms of marijuana studied in this chapter, see the *Indian Hemp Drugs Commission Report*, henceforth "Hemp Report" [Kaplan (1969) reprint], p. 59. For research into the potency of marijuana, see Mikuriya and Aldrich (1988).
2. In this chapter, the terms "cannabis" and "marijuana" are used interchangeably to refer to all three forms of the drug. Where a specific form (i.e., *bhang*, *ganja*, or *charas*) is discussed, the specific term for that form is used.
3. Figure 14.1 shows the districts of Bengal. Because Bengal in the early 20th century was part of British India, it had not been partitioned by the British into Bangladesh and the state of West Bengal in India. Hence, the data being analyzed in this study straddle the modern countries of Bangladesh and India.
4. For opium, see Van Ours (1995), Liu et al (1999), and Chandra (2000). For tobacco and alcohol, there is a much larger literature, typified by such studies as Becker, Grossman, and Murphy (1994), Chaloupka (1991), Goel and Morey (1995), and Grossman et al. (1998). Most of the marijuana studies to date rely on individual level self-reported data that are,

in most cases, highly unreliable themselves. For data quality issues and problems relating to the economics of marijuana use, see the data descriptions in Cameron and Williams (2001), Chaloupka, Grossman, and Tauras (1999), Desimone and Farrelly (2003), DiNardo and Lemieux (1992), Nisbet and Vakil (1972), Pacula (1998), Thies and Register (1993), and Williams (2004).

5. Cigarettes and alcohol are, by contrast, legal in a number of countries. Hence, economists have been able to analyze the impact of price (and therefore tax) changes on these substances using recent data. It should be noted in the context of opium that no spatial models of opium consumption have been estimated to date. This is a gap in the literature that will hopefully be filled as part of this project.
6. See Indian Hemp Drugs Commission (1894).
7. A series of recent events illustrates the currency and importance of this debate, and demonstrates the need for an understanding of the behavior of drug-consuming populations in a regime of legalization. In June 2005, for example, the House of the Rhode Island General Assembly voted 57-10 in favor of legalizing marijuana for medical uses. Following a veto by Governor Carcieri, the Senate overrode the veto with a 28-6 vote. At the time the first draft of this chapter was being prepared, and pending a similar expected overriding vote by the House, Rhode Island's status as the 11th state to legalize marijuana for medicinal use was imminent. At the same time, the U.S. Supreme Court ruled that "state laws sanctioning medical marijuana use provide no defense against enforcement of federal anti-drug laws by federal agencies" (Lader 2005). And, as these legal events were unfolding, a report entitled "The Budgetary Implications of Marijuana Prohibition" (Miron 2005) demonstrating the significant budgetary benefits of marijuana legalization, and arguing for the inclusion of this economic consideration in any comprehensive debate on the pros and cons of legalization, was published.
8. See, for example, Pandey (2005).
9. "Hemp drugs" were defined as the drugs produced from the *cannabis sativa* or *cannabis indica* plant [*Hemp Report*, Kaplan (1969) reprint, pp. 16–17].
10. For a summary of the origins of the Indian Hemp Drugs Commission, see Mills (2005).
11. The number of hits on any leading search engine on the worldwide web in response to a search for the title of the report is testimony to this. The report has, at one point or another, also been cited in most national-level debates on the (il)legal status of marijuana, and continues to be so.
12. Mills (2005), concluding paragraph. See also Mills (2003).
13. See Great Britain Royal Commission on Opium (1894–1895).
14. Richards (2002), pp. 418–420.
15. The Indian Hemp Drugs Commission, for example, took testimony from 1,193 witnesses [*"Hemp Report"*, Kaplan (1969) reprint, p. 12].
16. As a source of export revenues, opium was far more important than "hemp drugs." Among the beneficiaries of this trade were China, the Dutch East Indies, French Indochina, and Siam.
17. For these dissenting opinions, written by Raja Soshi Sikhareshwar Roy and Lala Nihal Chand, see the notes appended to the end of the report. [*"Hemp Report"*, Kaplan (1969) reprint, pp. 363–477].
18. *"Hemp Report"* [Kaplan (1969) reprint], p. 379 (Roy) and *"Hemp Report"* [Kaplan (1969) reprint], p. 436 (Nihal Chand).
19. In a later section, this assertion will be tested.
20. The most THC-rich resin is found on female marijuana plants that have not had a chance to produce seeds. In order to ensure this, an elaborate process of eliminating male plants from any plantation in its early stages to prevent pollination and, therefore, seed production, had to be undertaken. This could only happen in controlled and cultivated circumstances. For details in the historical context, see *"Hemp Report"*, pp. 59–84.
21. Richards (2002), p. 420.
22. Data sources will be provided on request.

23. In 1947, Bengal was partitioned into East Pakistan (later Bangladesh) and Indian (West) Bengal.
24. The units of weight used in British Bengal were as follows: 1 *maund* = 40 *seers* and 1 *seer* = 2.057 pounds.
25. Prior to the introduction of the metric system in independent India, one *rupee* consisted of 16 *annas*.
26. See India Census Commissioner (1891, 1901, 1911, 1921, 1931, 1941).
27. Patterson and Pyle (1991), pp. 14–15.
28. We do not use the census data for the censuses prior to 1891 because of questions about their reliability.
29. See Bengal, Department of Agriculture (1920). This is standard practice in the economic history of rice-consuming Asia. See, for example, Mansvelt and Creutzberg (1978).
30. While daily nominal wages obtained from quinquennial wage surveys for the years 1911 and 1916 could be interpolated for the other years in the decade and used as indicators of income, we plan to obtain additional wage data before including this economic variable in the analysis. Missing wage data have not been estimated due to the extreme care that must be taken given the high numbers of missing observations.
31. A similar (implicit) assumption is made by scholars who have analyzed opium data in colonial Asia [Van Ours (1995), for example].
32. For example, with only one exception (the bhang model with fixed effects and no charas price), the time trend was statistically significant. Further, the bhang and charas prices are usually significant in the models in which they were included. Therefore, in this chapter, we will focus on the fixed effects models with the time trend using the restricted dataset so that bhang and charas prices are included in the analysis.
33. This nuanced approach is not unlike the approach adopted in some countries in South America, in which the coca leaf, which contains psychoactive substances, is legal, while its much more powerful and dangerous derivative, cocaine, is prohibited.
34. Rush (1990), Trocki (1990), and Chandra (2000), for example, demonstrate the willingness of governments to pursue profit over principle in the matter of drug policy in historical context.

Colour Plates

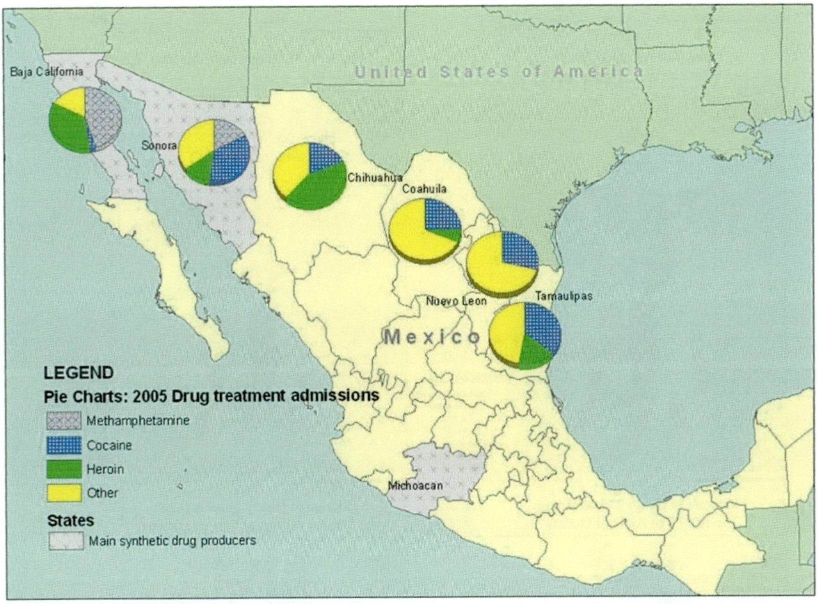

Plate 1 See also Fig. 3.2 on page 37

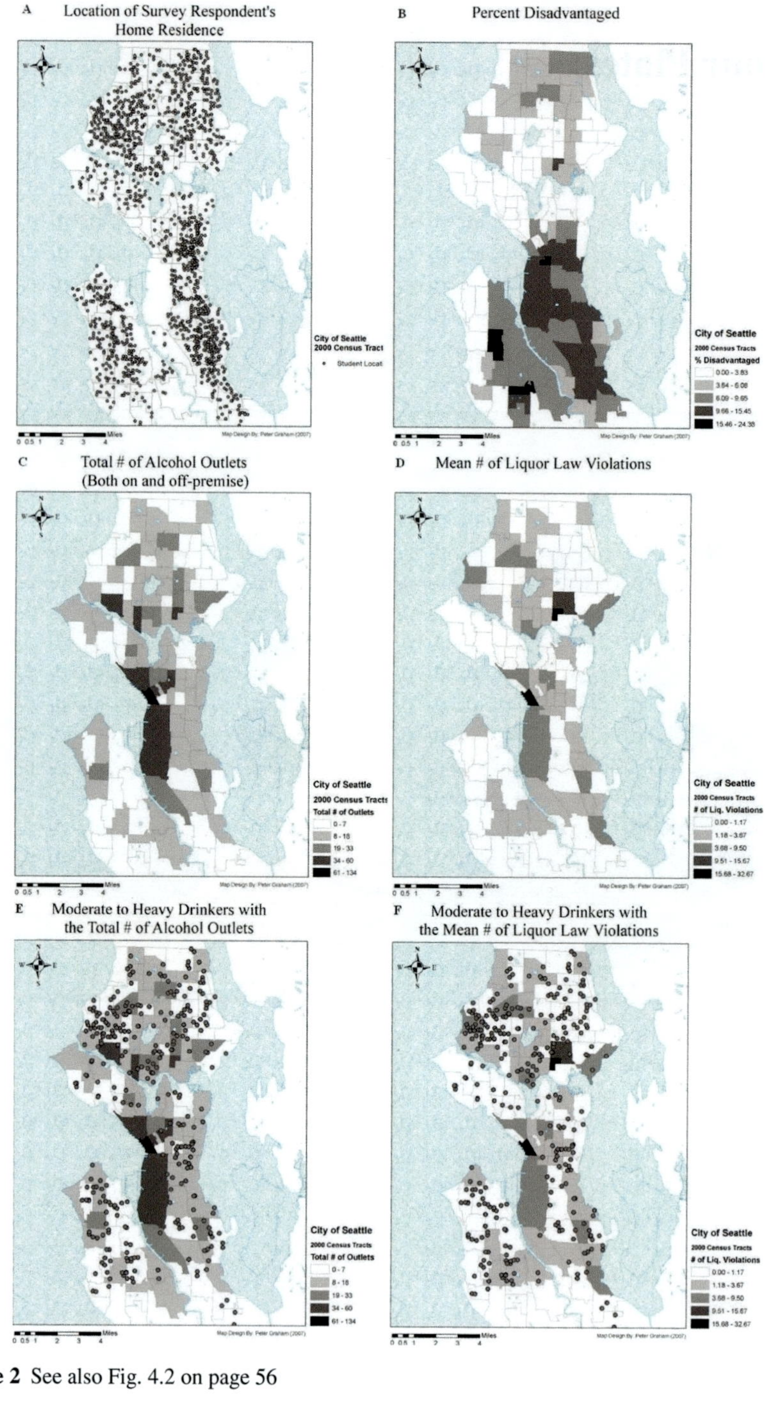

Plate 2 See also Fig. 4.2 on page 56

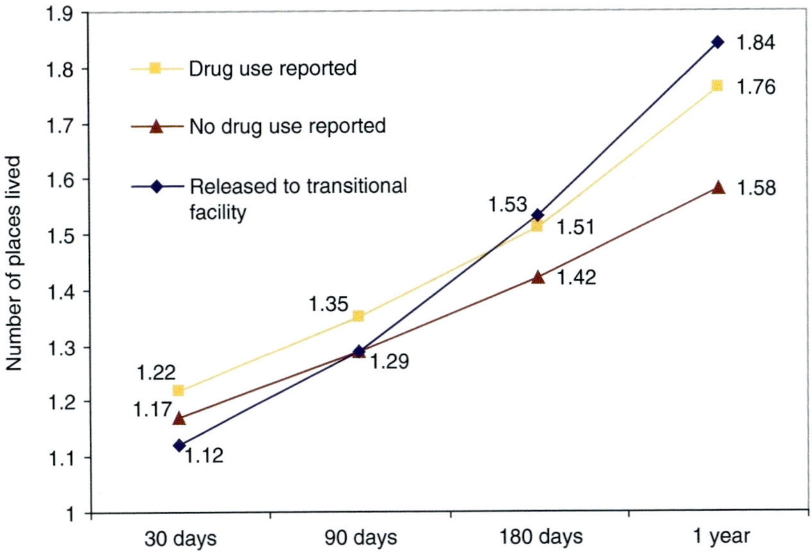

Plate 3 See also Fig. 6.1 on page 99

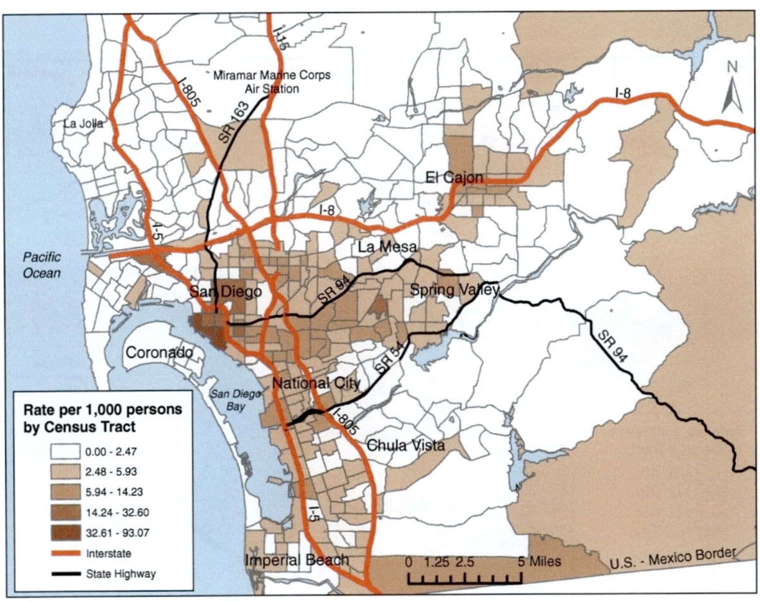

Plate 4 See also Fig. 6.2 on page 101

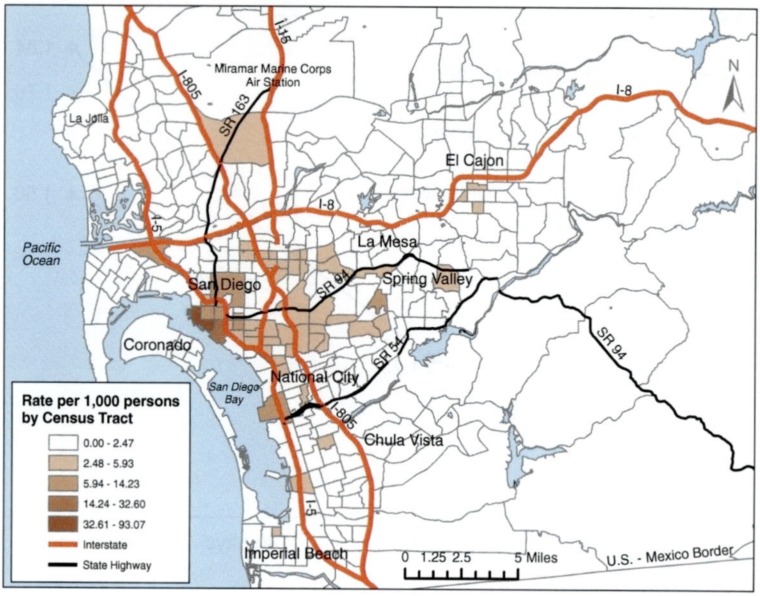

Plate 5 See also Fig. 6.3 on page 102

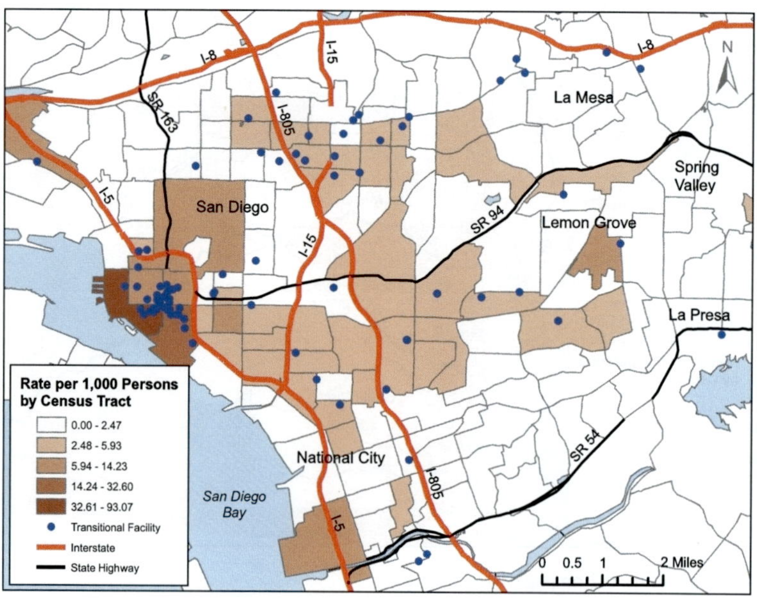

Plate 6 See also Fig. 6.4 on page 103

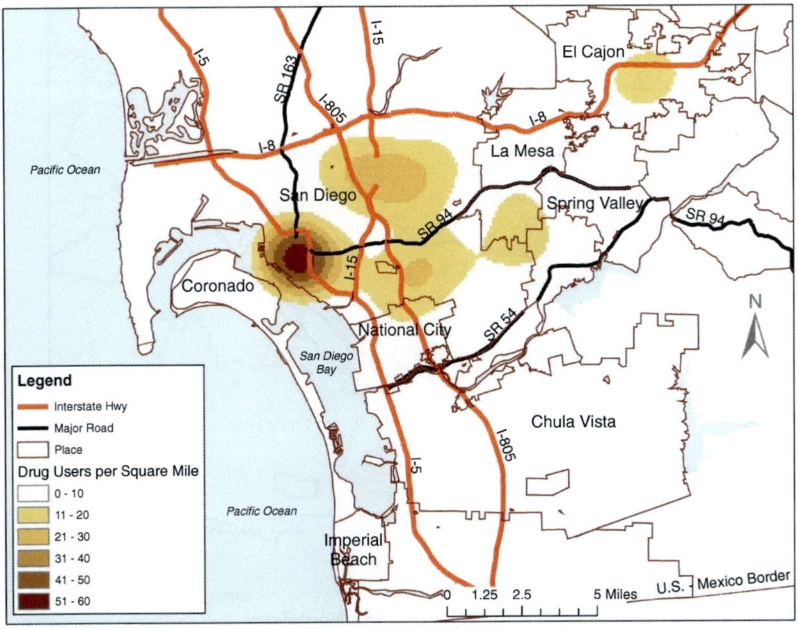

Plate 7 See also Fig. 6.5 on page 104

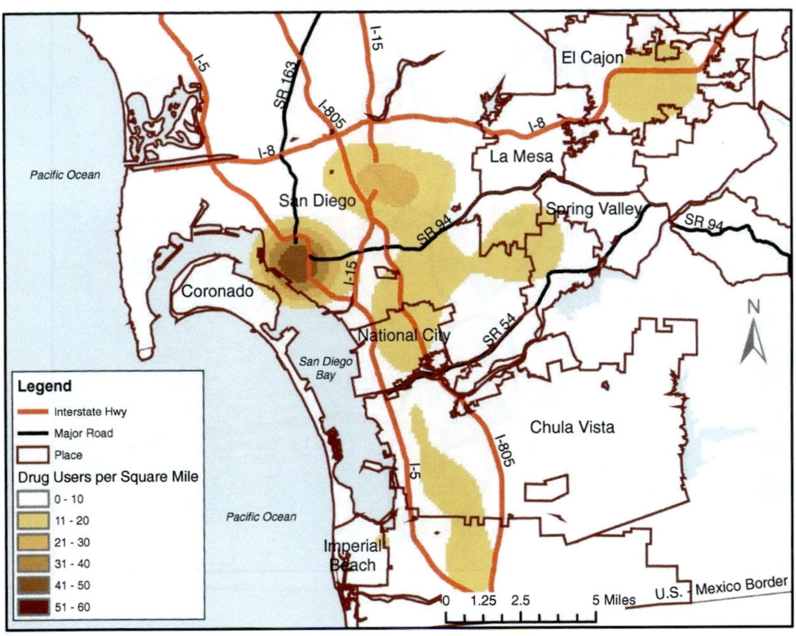

Plate 8 See also Fig. 6.6 on page 105

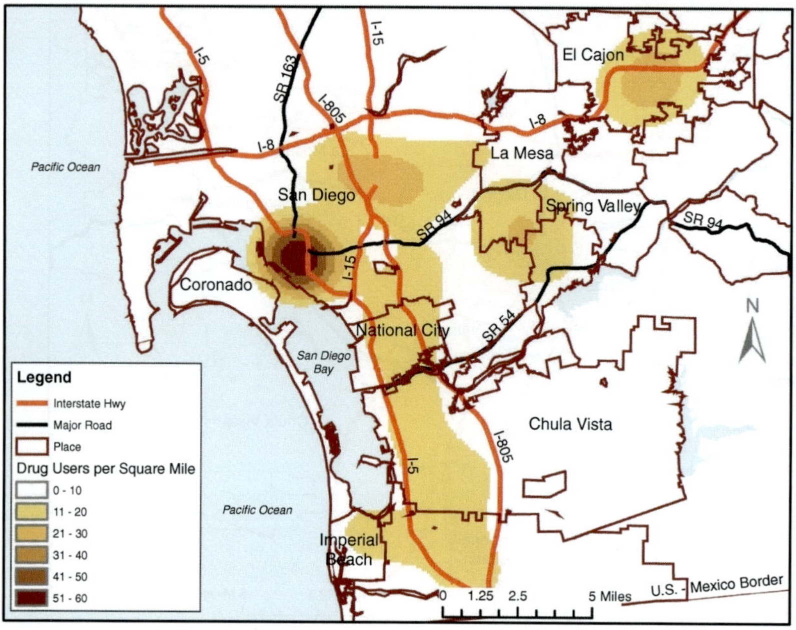

Plate 9 See also Fig. 6.7 on page 105

Plate 10 See also Fig. 6.8 on page 107

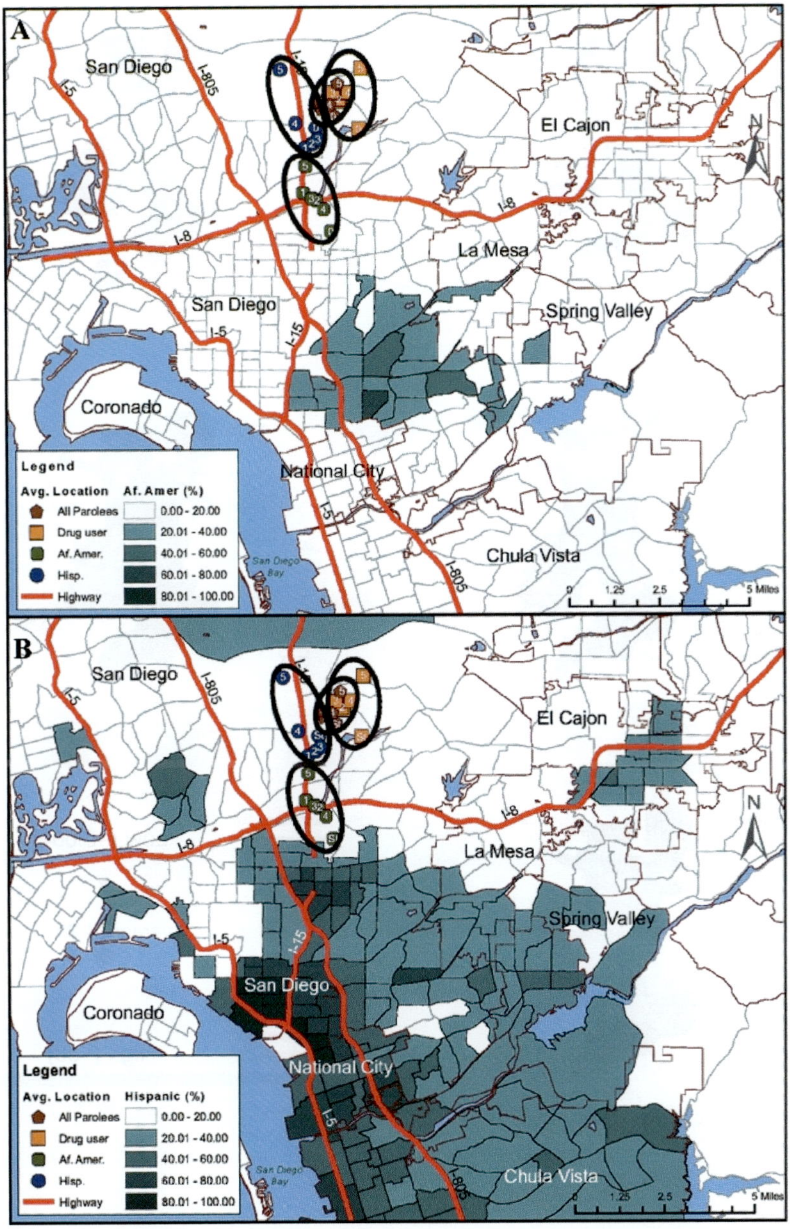

Plate 11 See also Fig. 6.9 on page 108

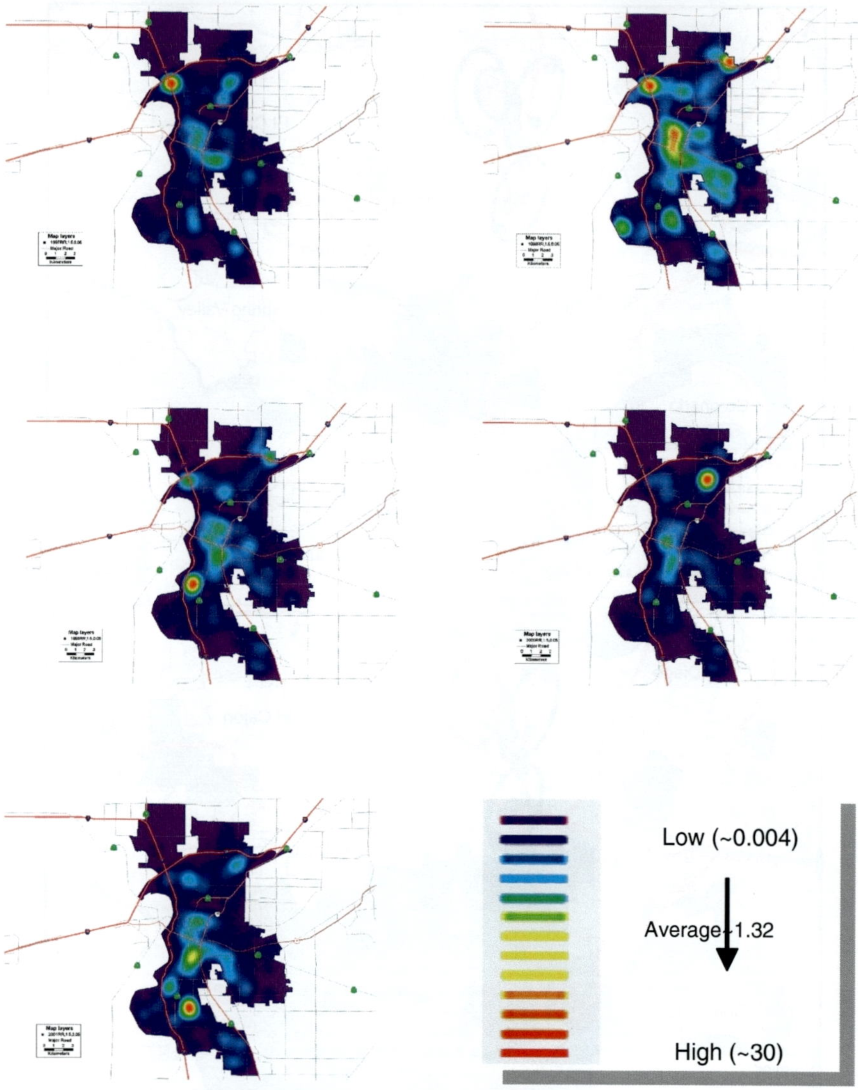

Plate 12 See also Fig. 7.1 on page 128

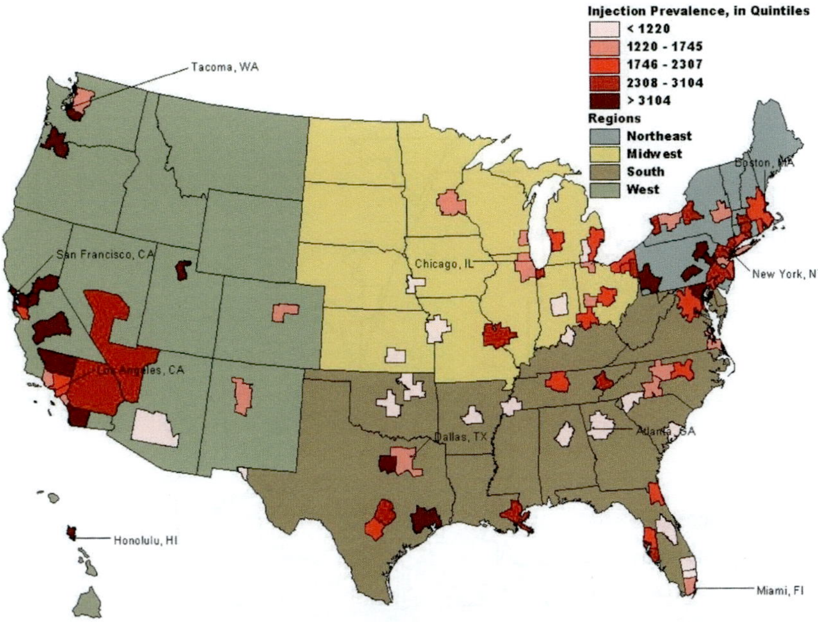

Plate 13 See also Fig. 9.1 on page 154

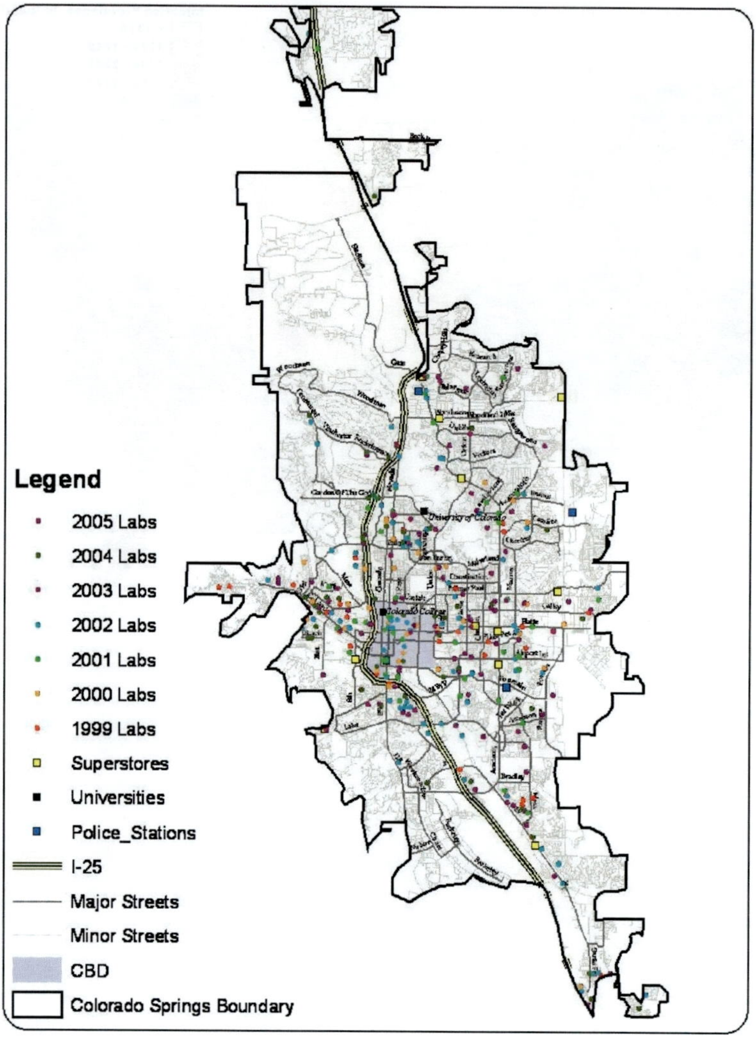

Plate 14 See also Fig. 12.2 on page 200

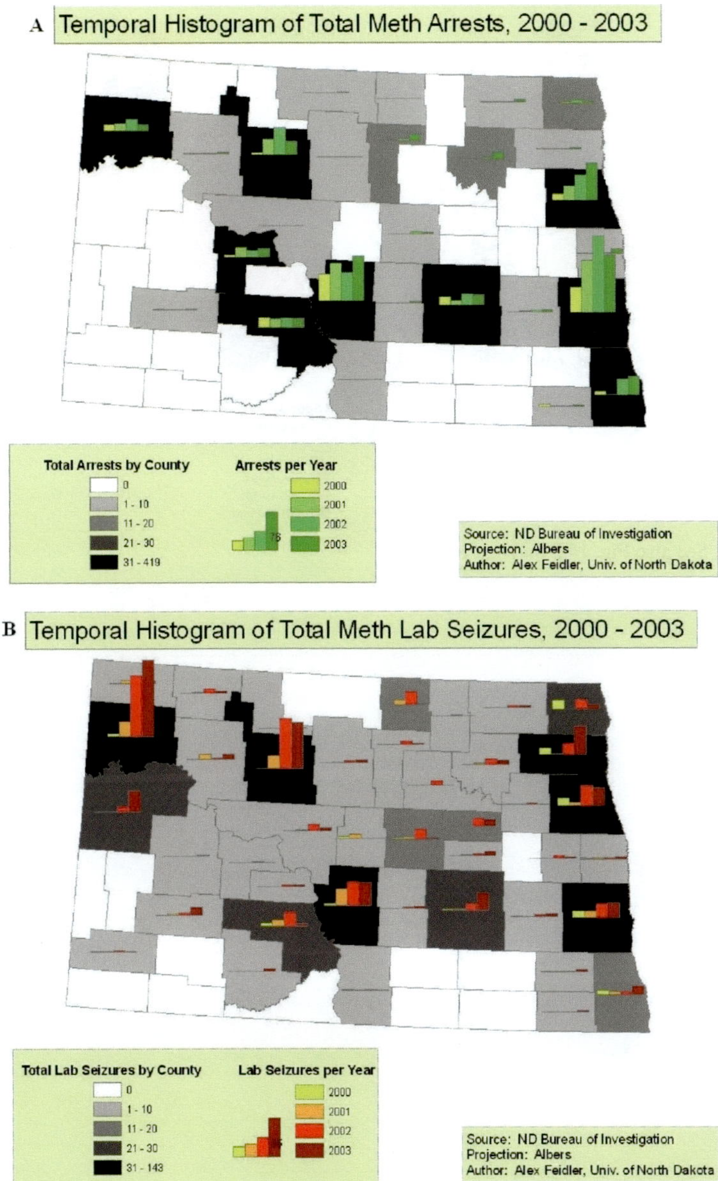

Plate 15 See also Fig. 13.1 on page 214

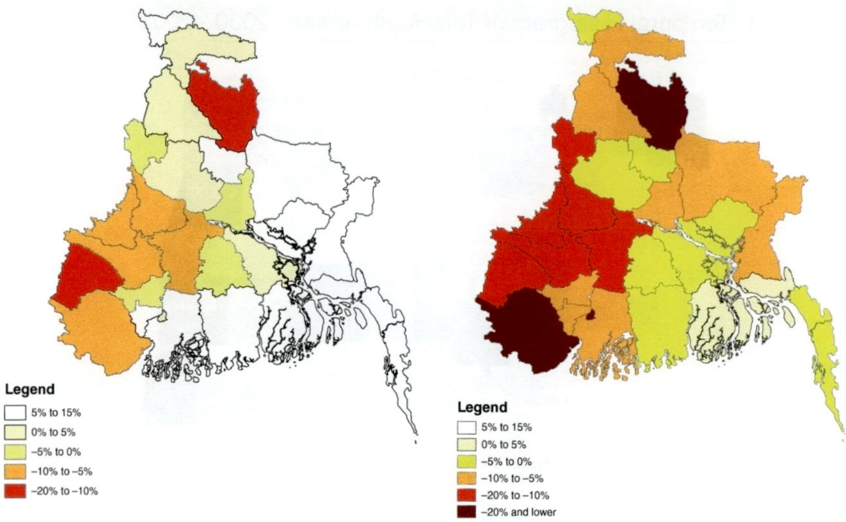

Plate 16 See also Fig. 14.2 on page 231

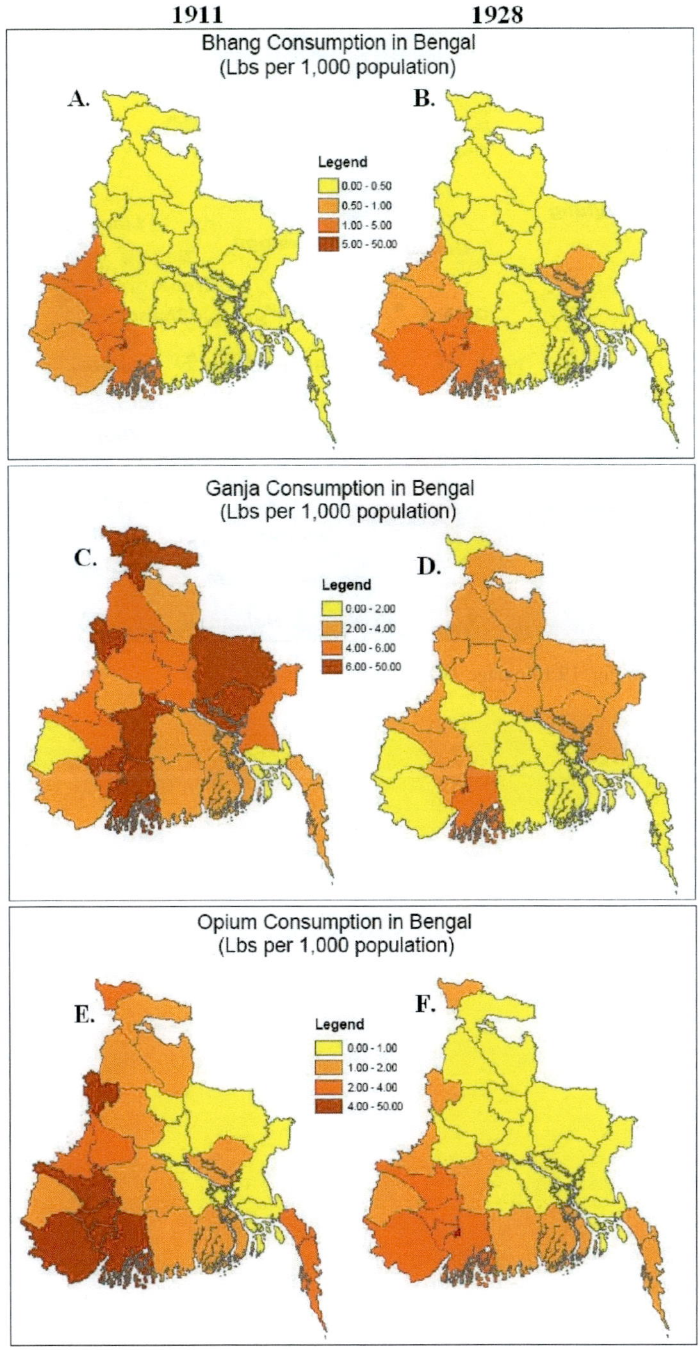

Plate 17 See also Fig. 14.3 on page 232

Plate 18 See also Fig. 19.1 on page 310

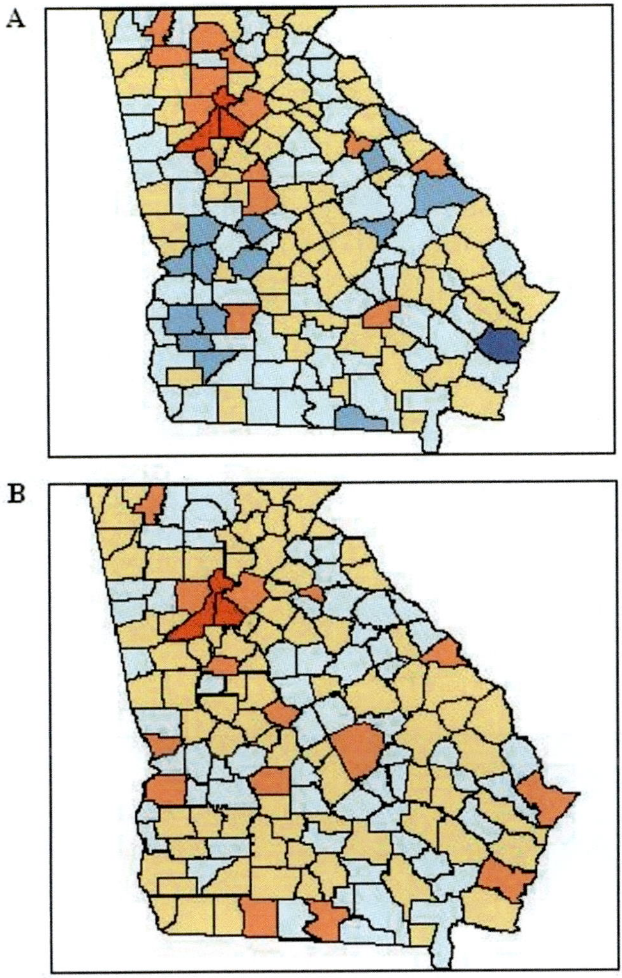

Plate 19 See also Fig. 22.1 on page 363

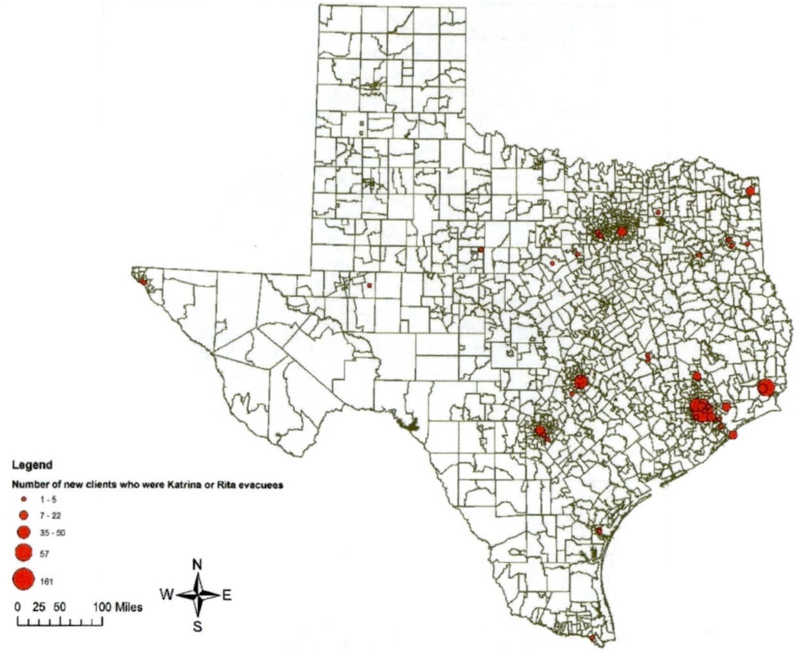

Plate 20 See also Fig. 23.1 on page 376

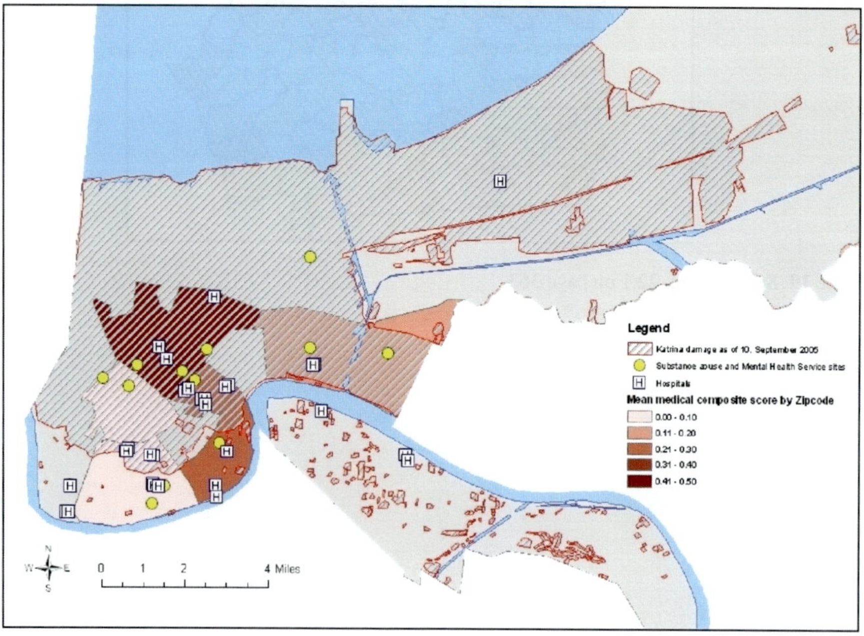

Plate 21 See also Fig. 23.2 on page 381

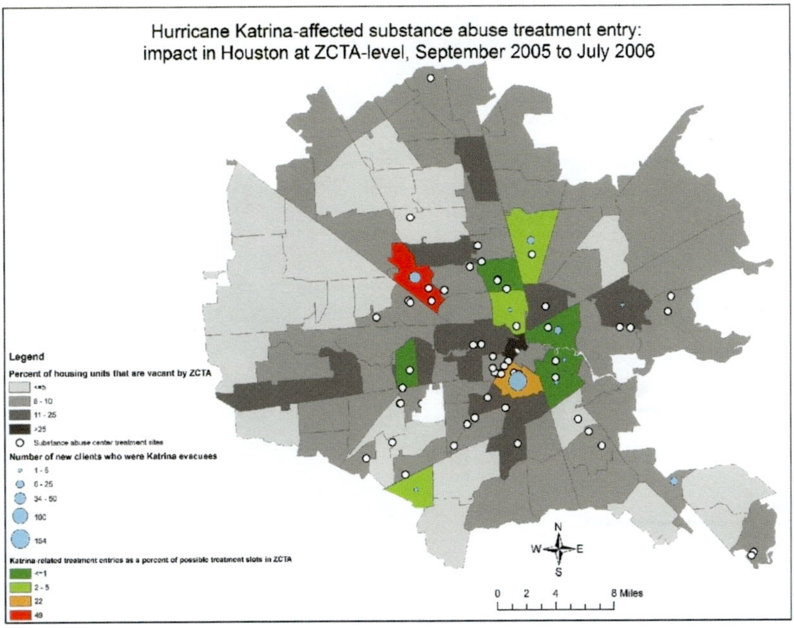

Plate 22 See also Fig. 23.3 on page 382

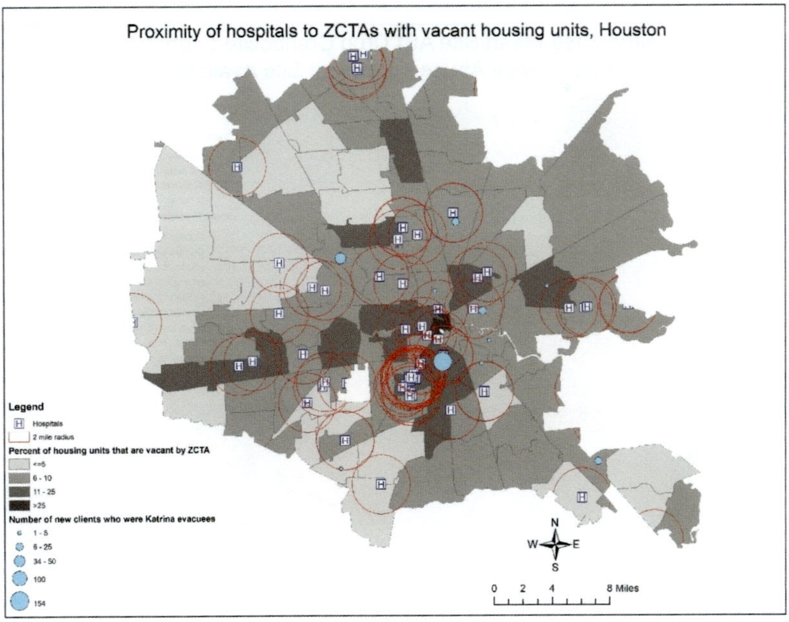

Plate 23 See also Fig. 23.4 on page 383

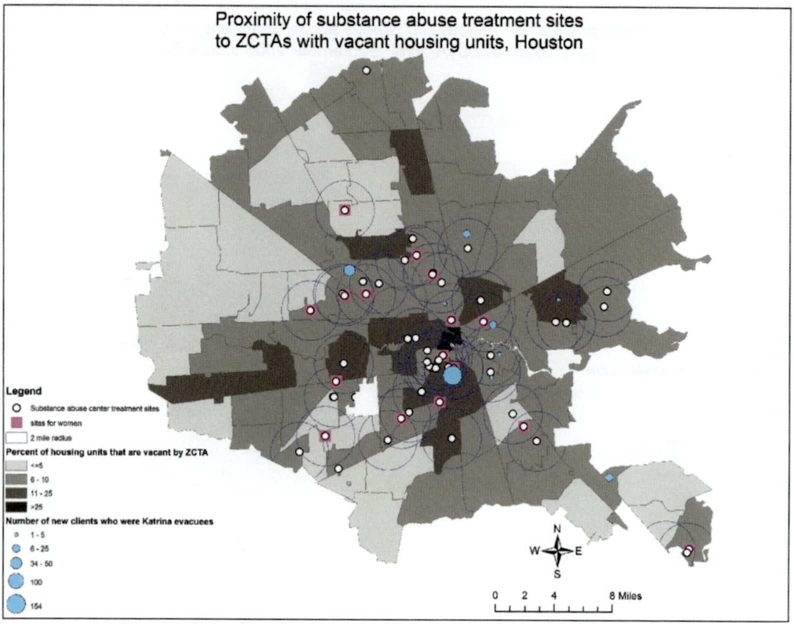

Plate 24 See also Fig. 23.5 on page 383

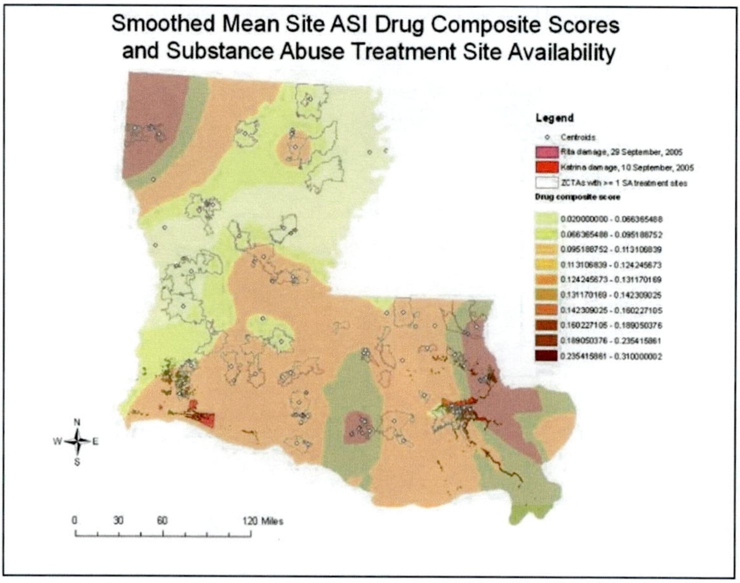

Plate 25 See also Fig. 23.6 on page 385

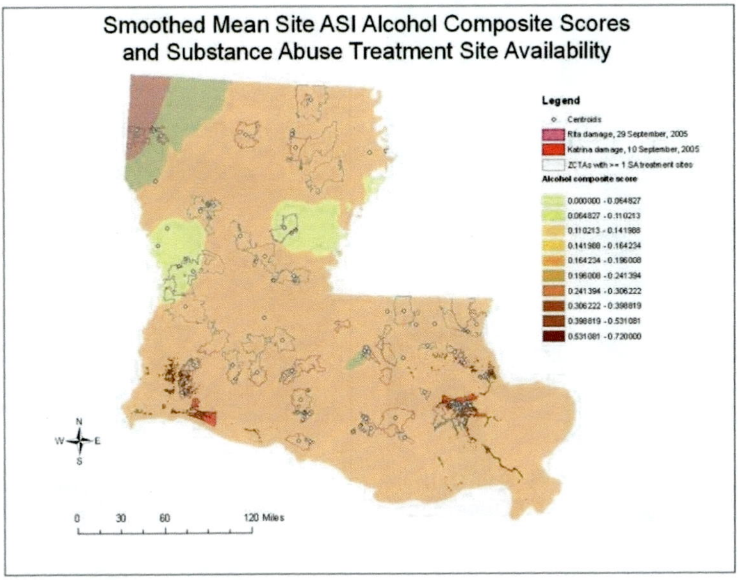

Plate 26 See also Fig. 23.7 on page 385

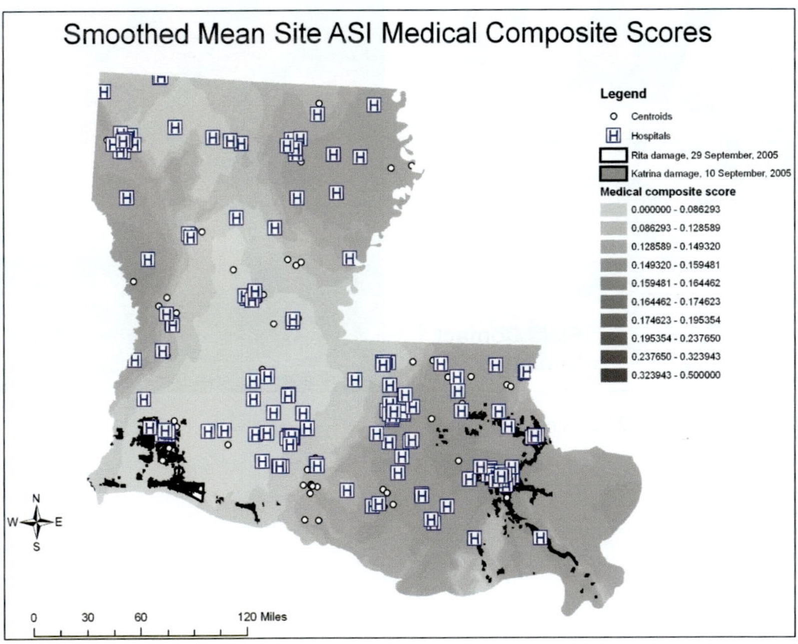

Plate 27 See also Fig. 23.8 on page 386

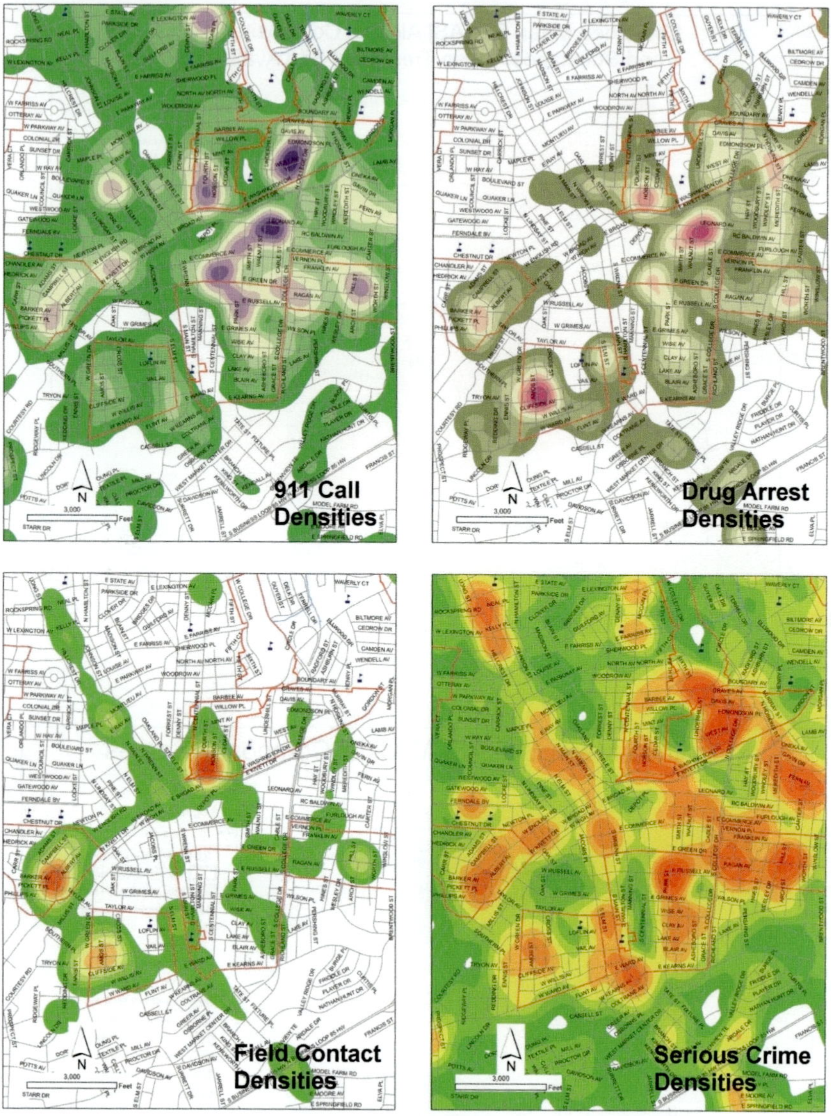

Plate 28 See also Fig. 24.2 on page 399

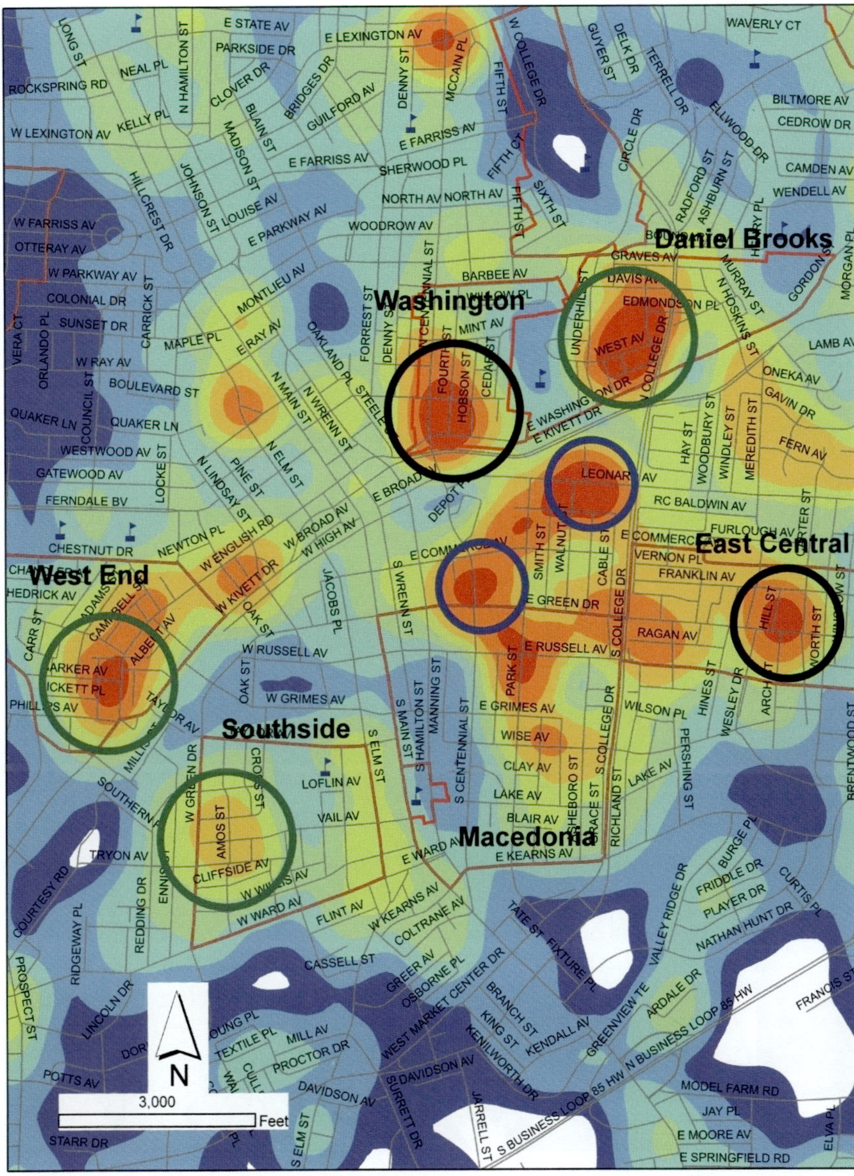

Plate 29 See also Fig. 24.3 on page 400

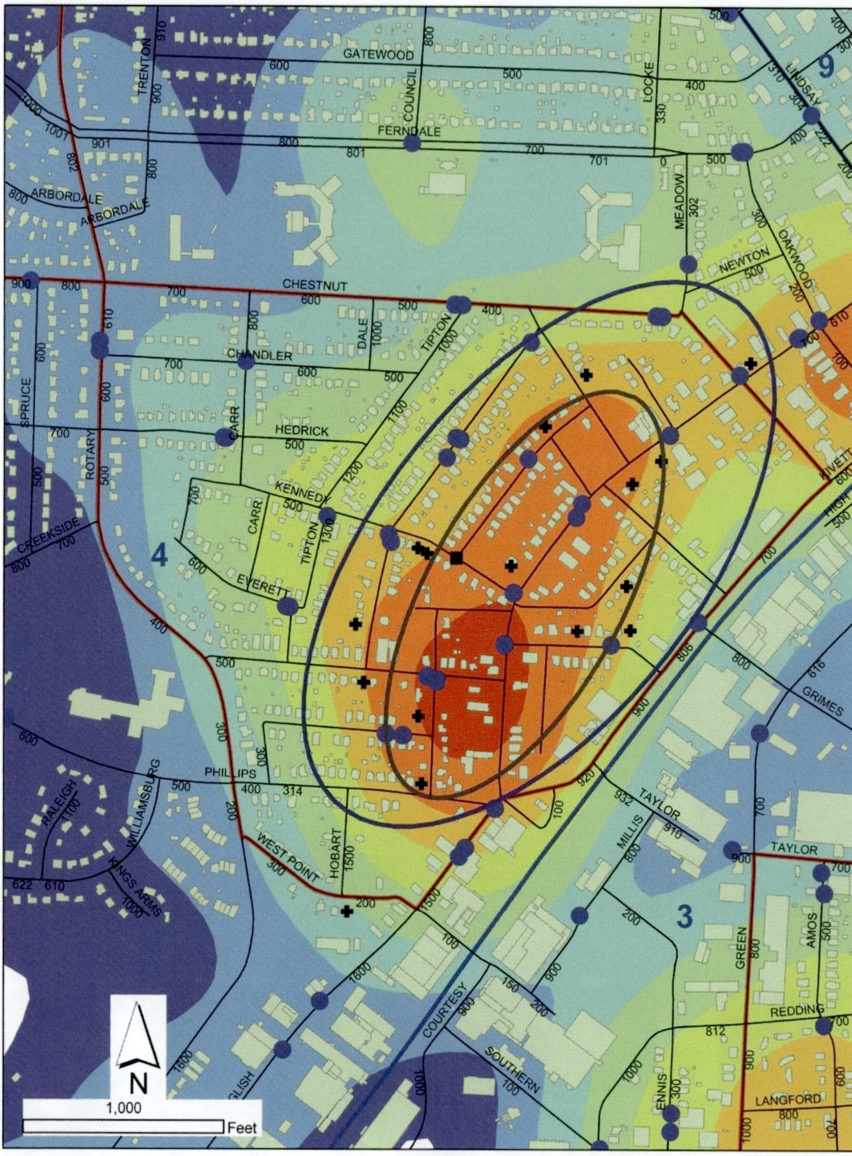

Plate 30 See also Fig. 24.4 on page 403

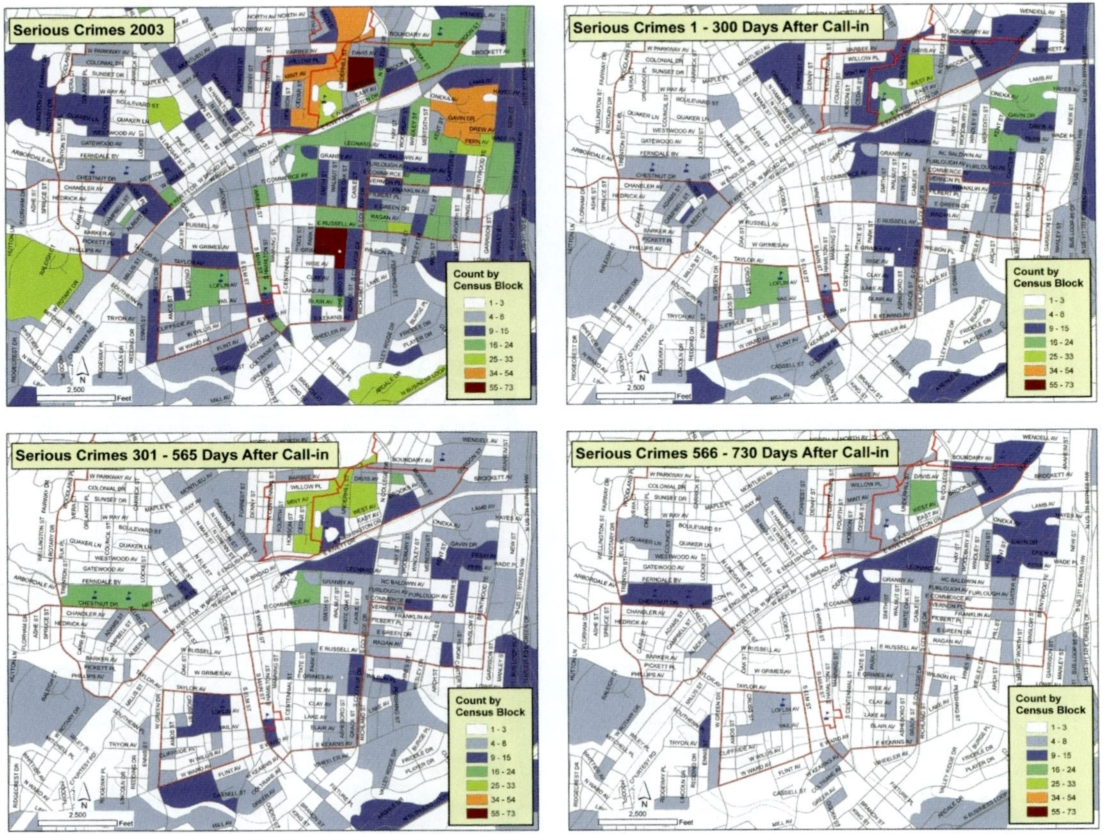

Plate 31 See also Fig. 24.5 on page 407

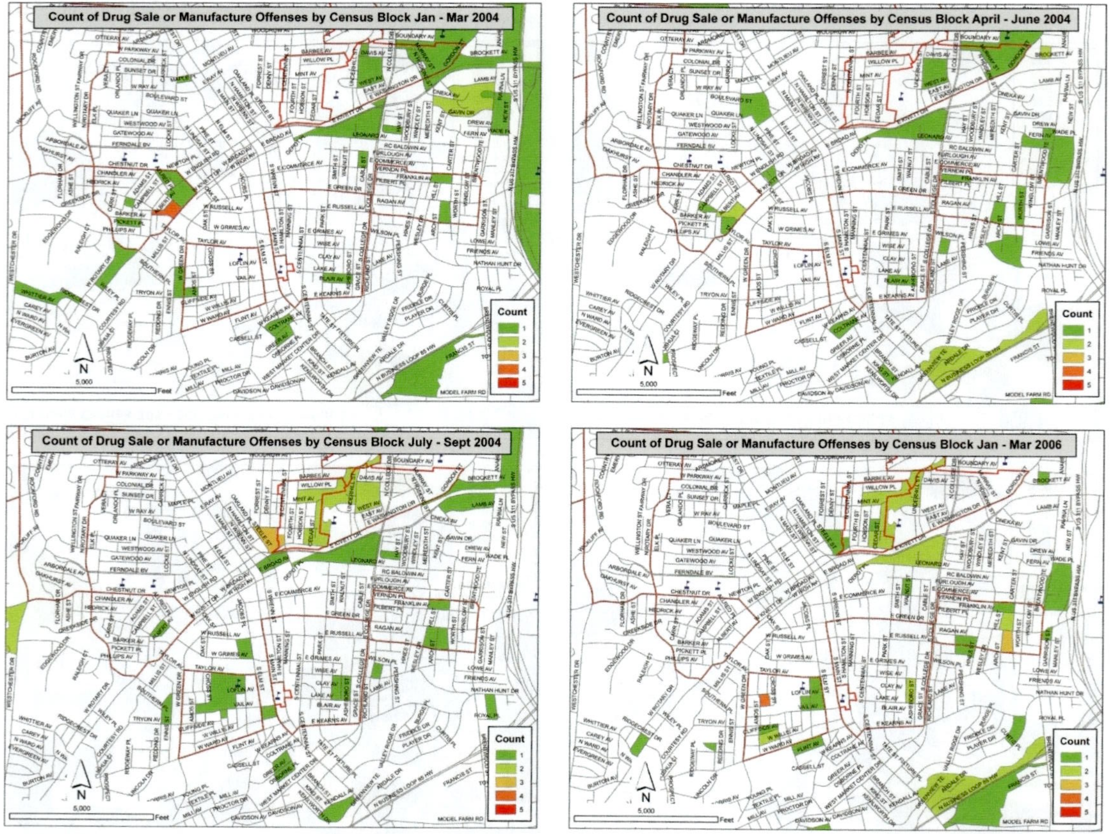

Plate 32 See also Fig. 24.6 on page 408

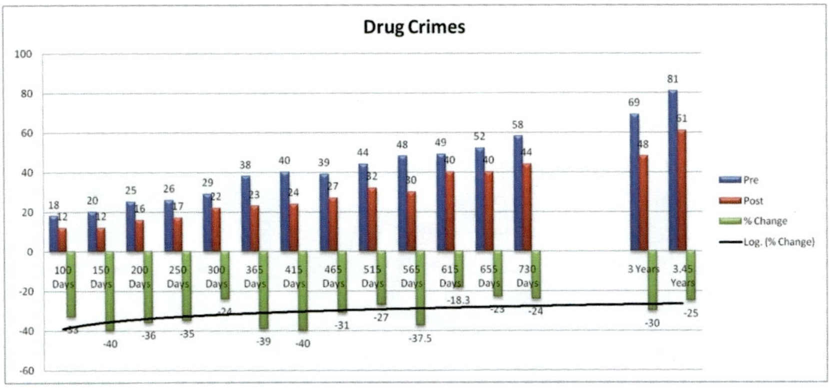

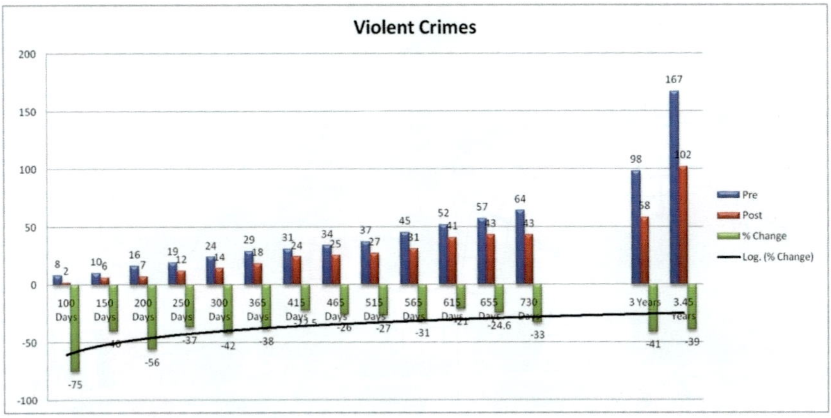

Plate 33 See also Fig. 24.7 on page 409

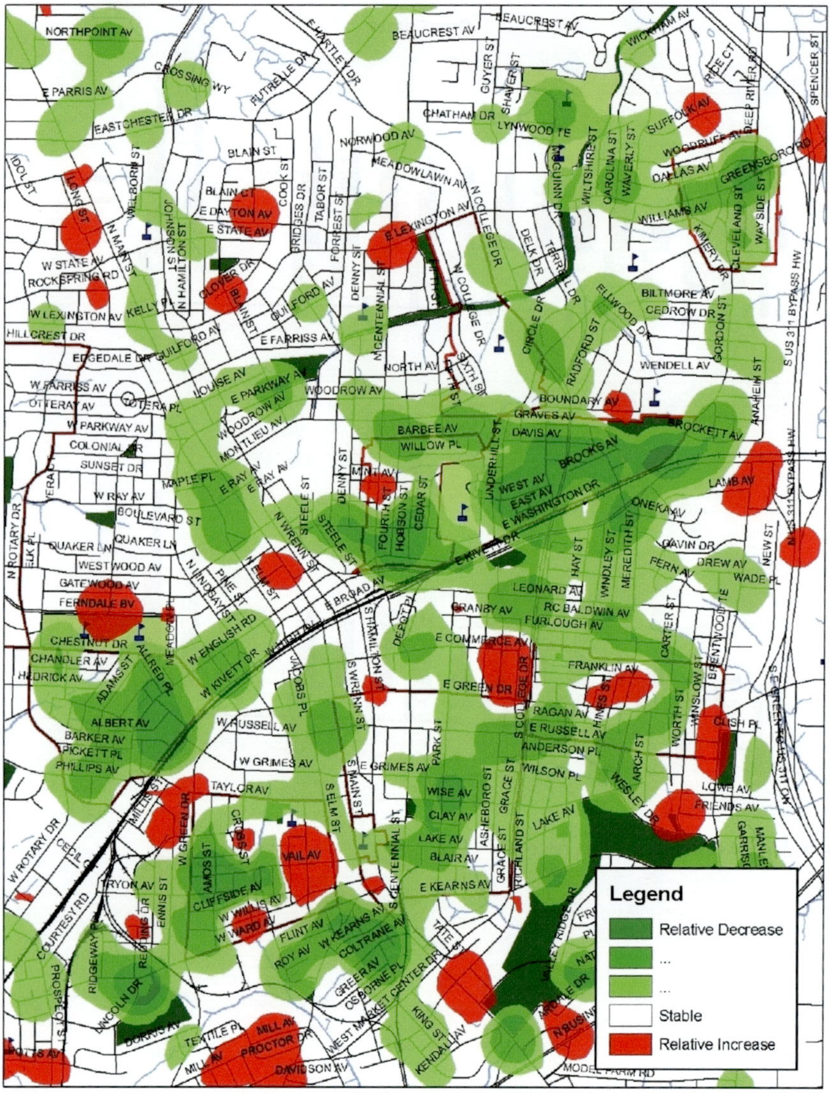

Plate 34 See also Fig. 24.8 on page 410

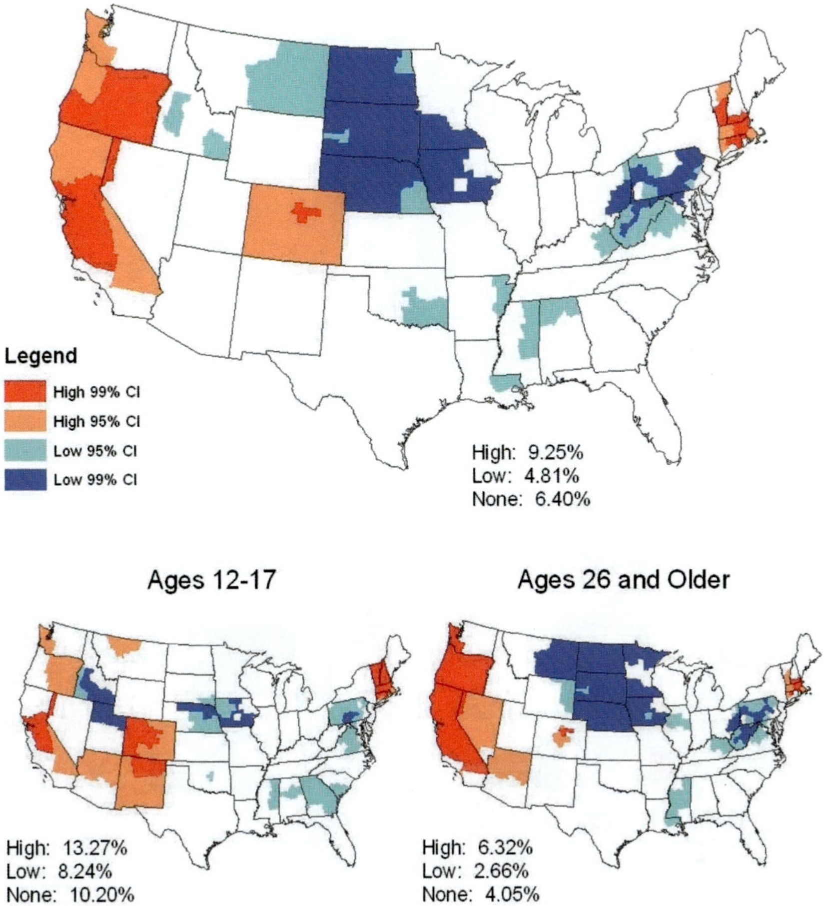

Plate 35 See also Fig. 25.1 on page 422

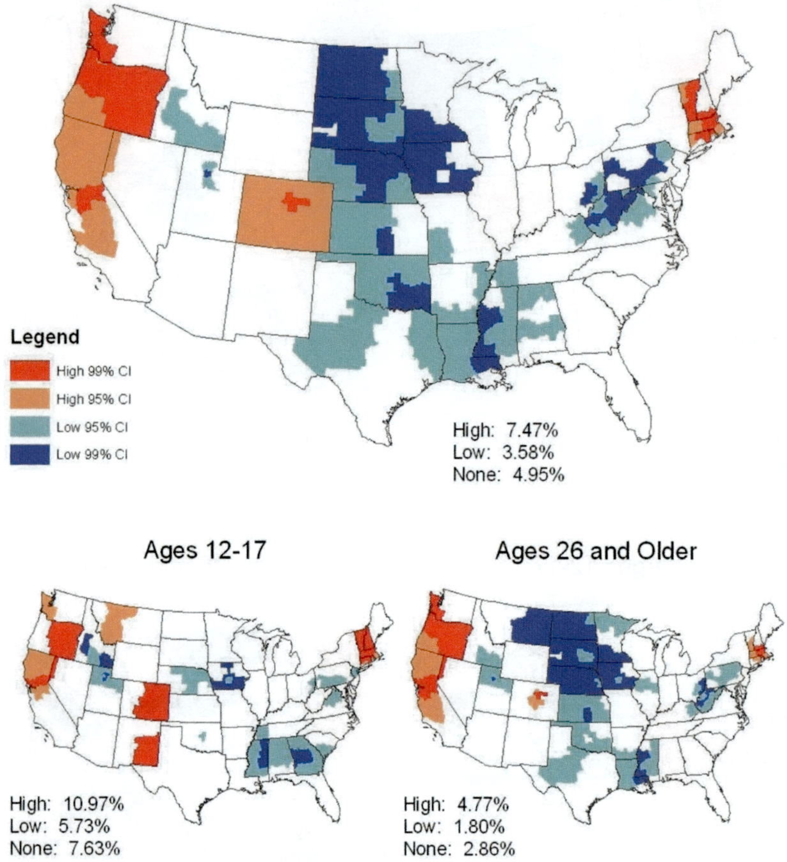

Plate 36 See also Fig. 25.3 on page 423

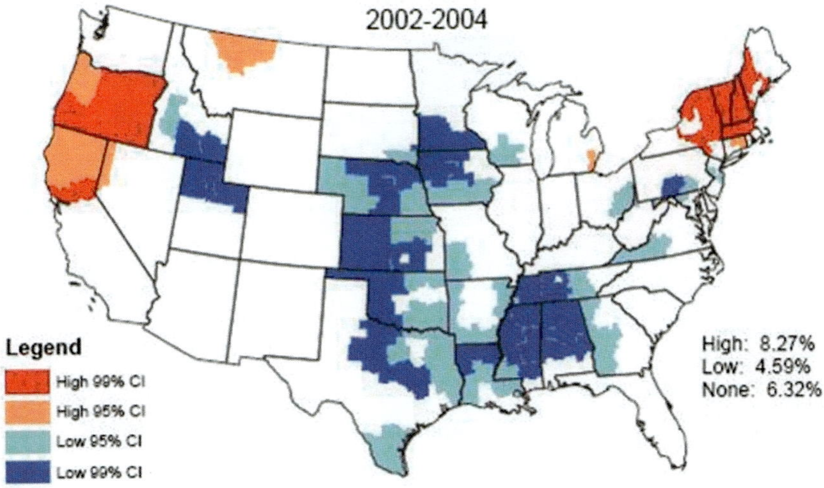

Plate 37 See also Fig. 25.4 on page 424

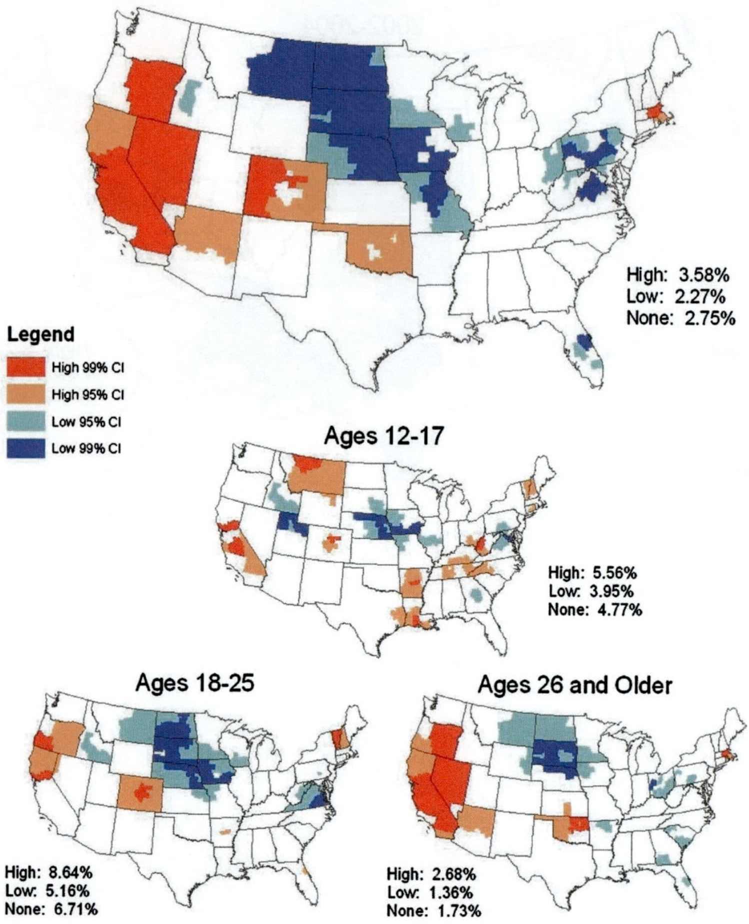

Plate 38 See also Fig. 25.5 on page 425

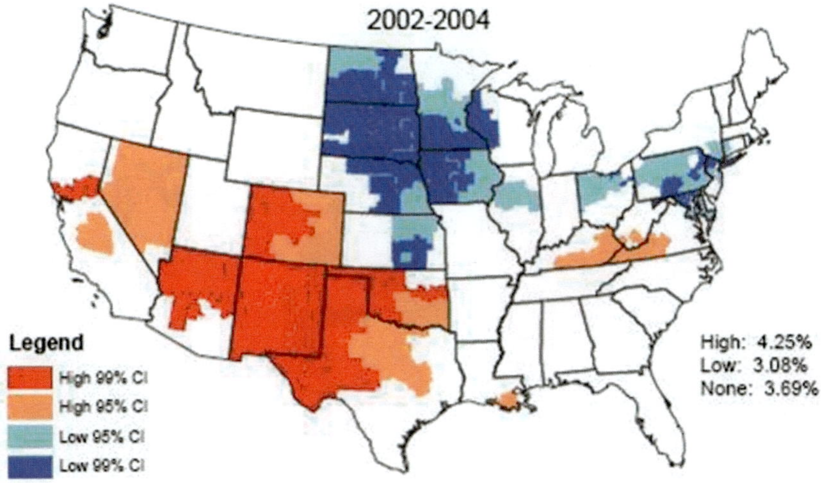

Plate 39 See also Fig. 25.6 on page 426

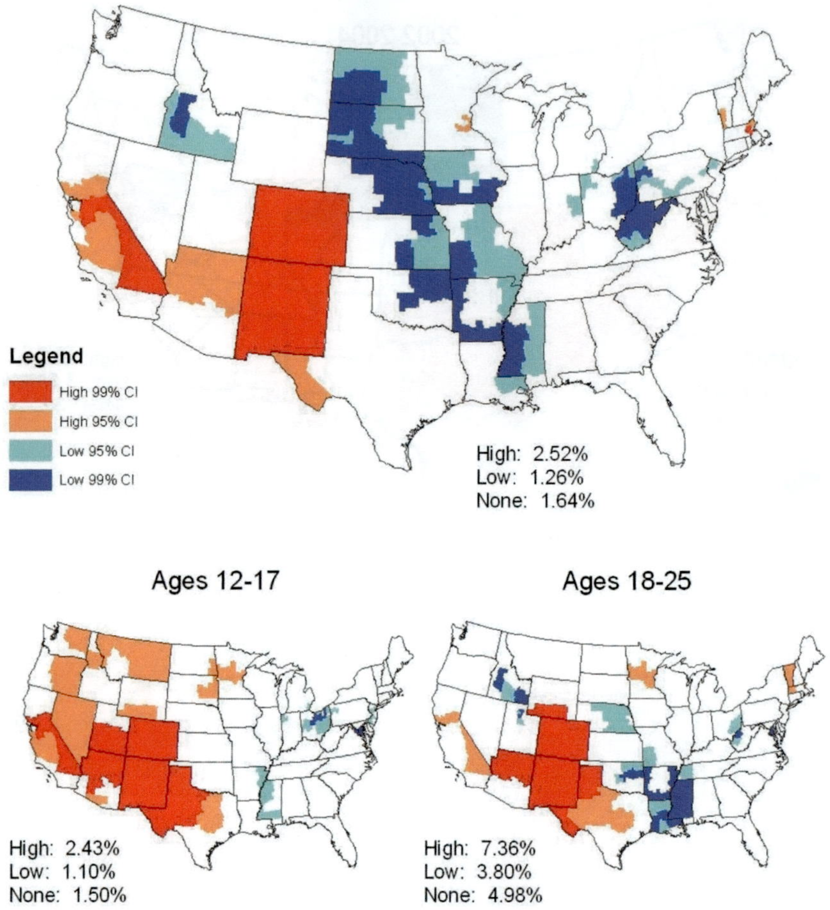

Plate 40 See also Fig. 25.7 on page 427

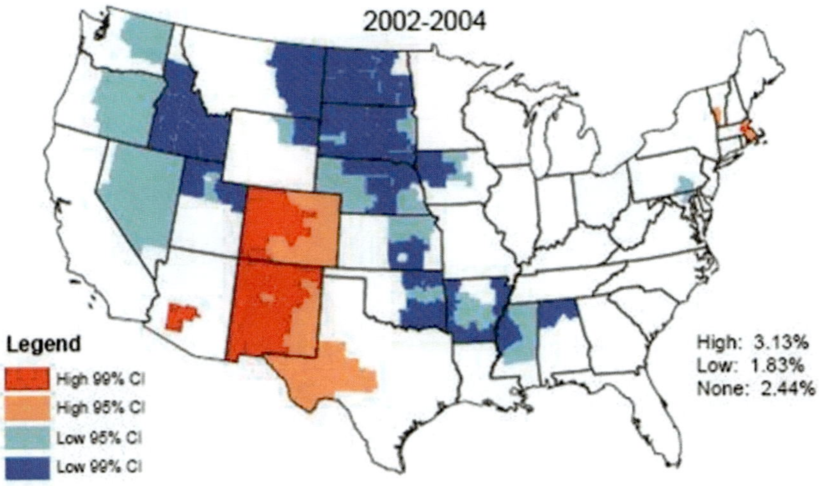

Plate 41 See also Fig. 25.8 on page 428

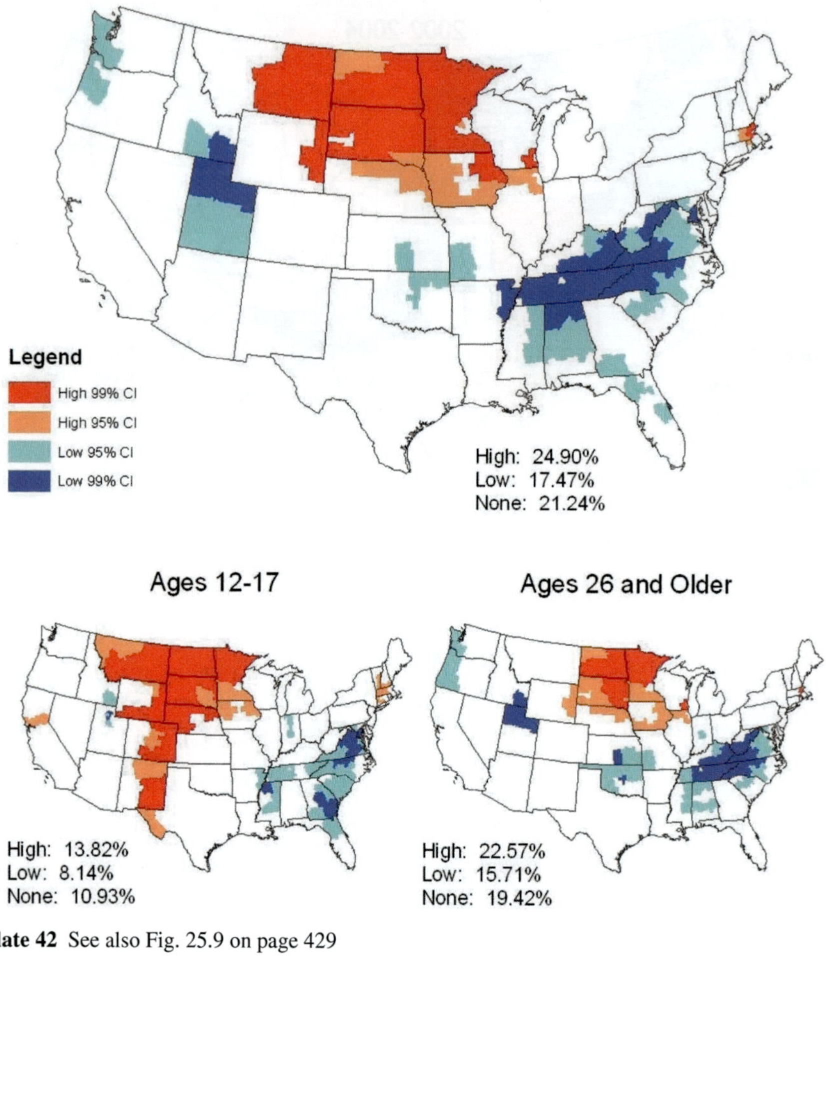

Plate 42 See also Fig. 25.9 on page 429

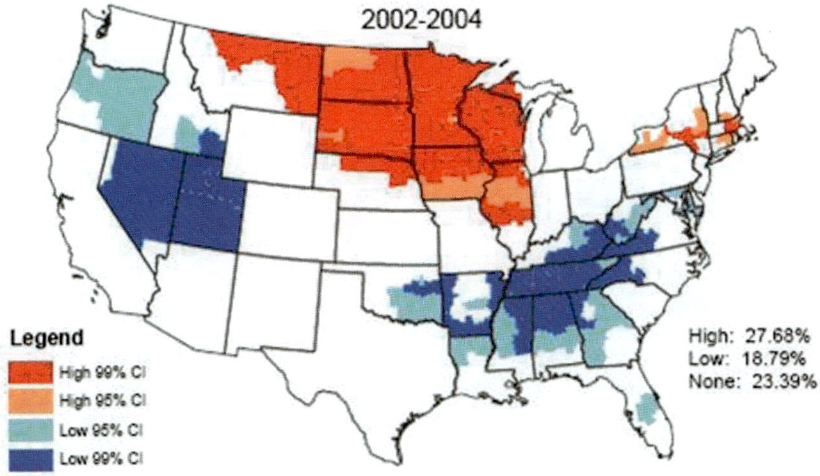

Plate 43 See also Fig. 25.10 on page 430

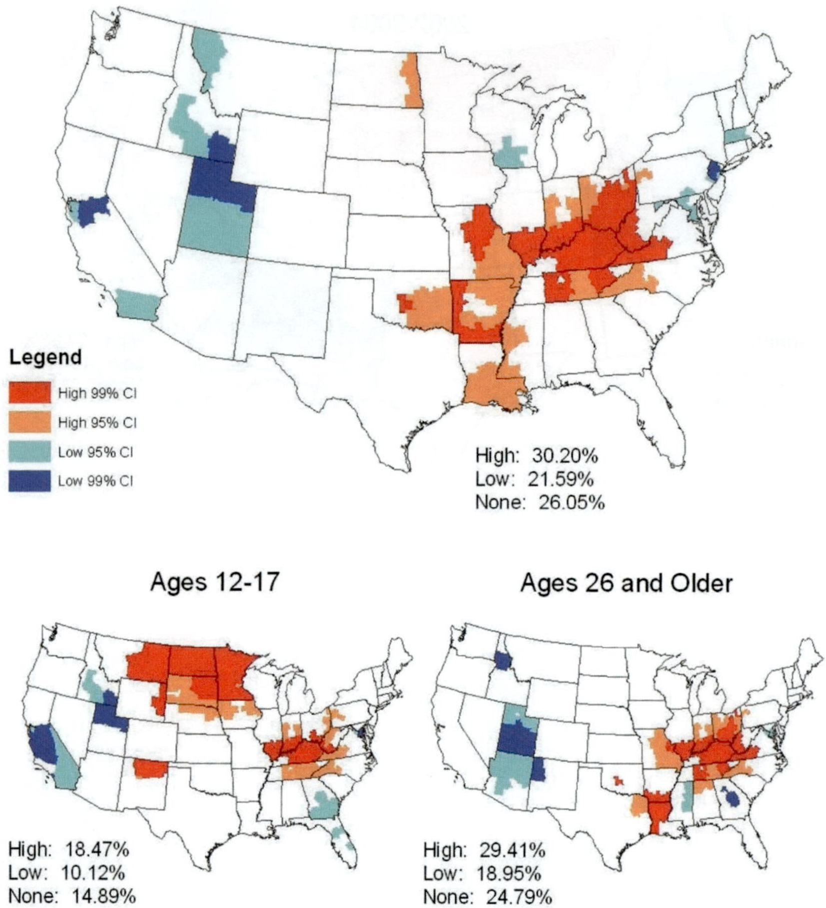

Plate 44 See also Fig. 25.11 on page 431

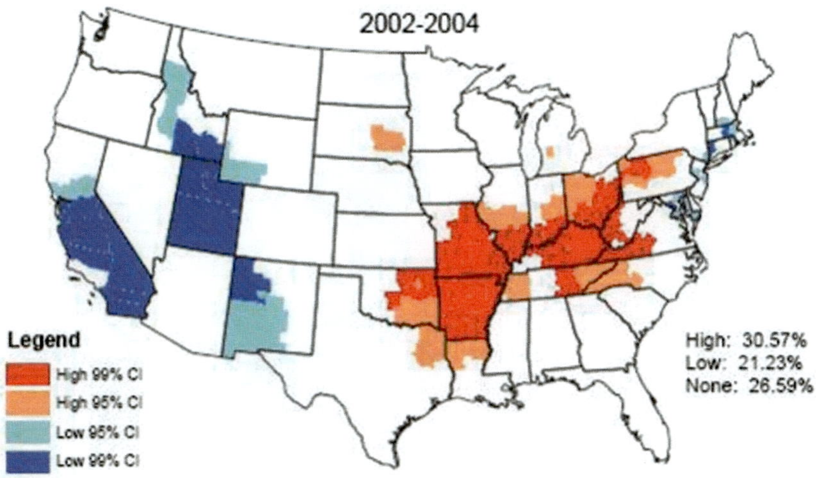

Plate 45 See also Fig. 25.12 on page 432

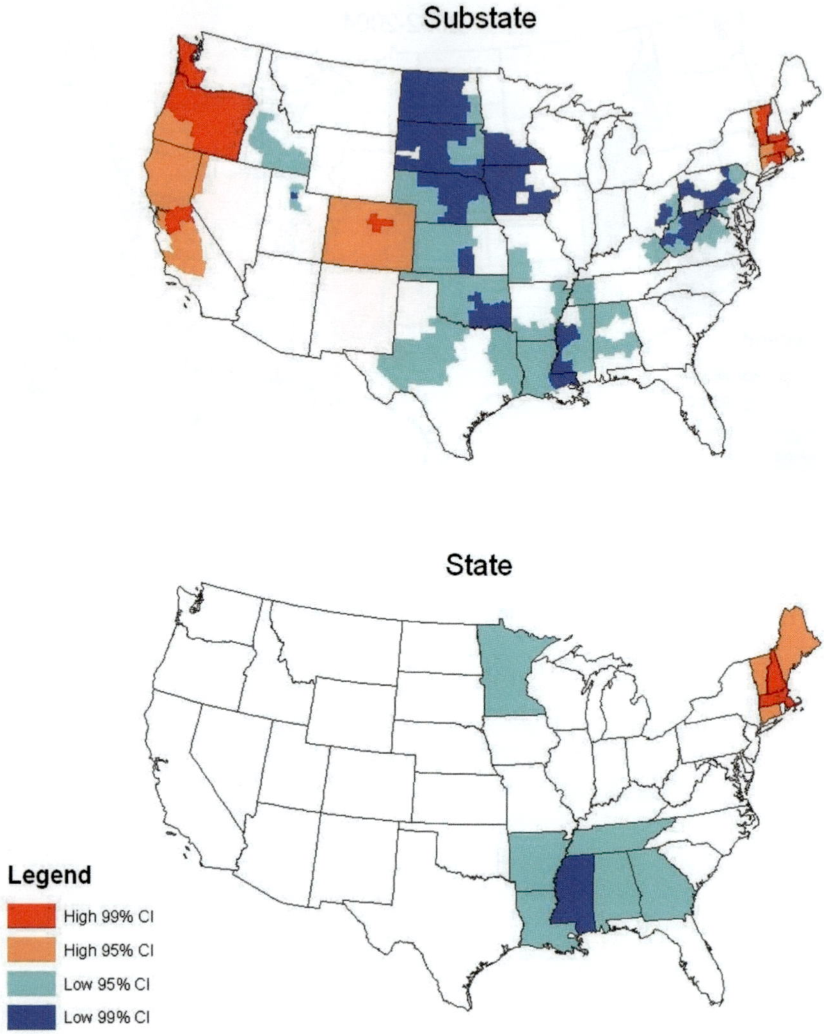

Plate 46 See also Fig. 25.13 on page 433

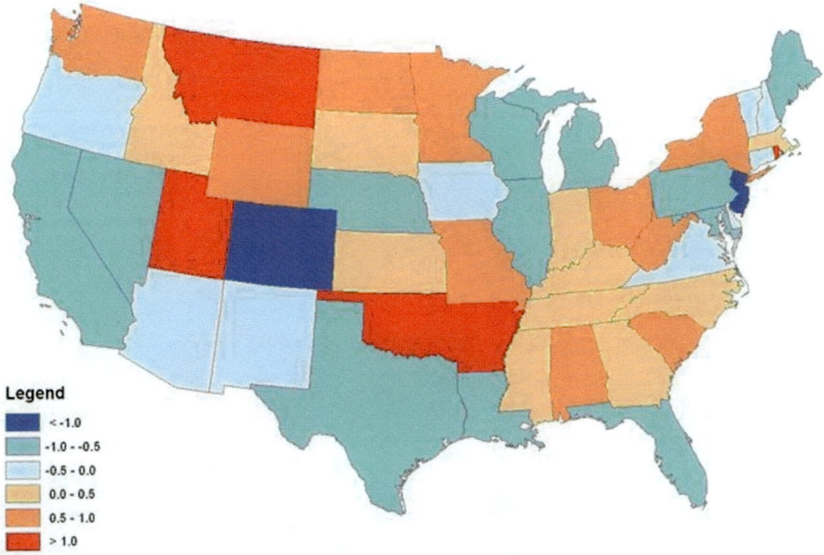

Plate 47 See also Fig. 25.14 on page 434

A

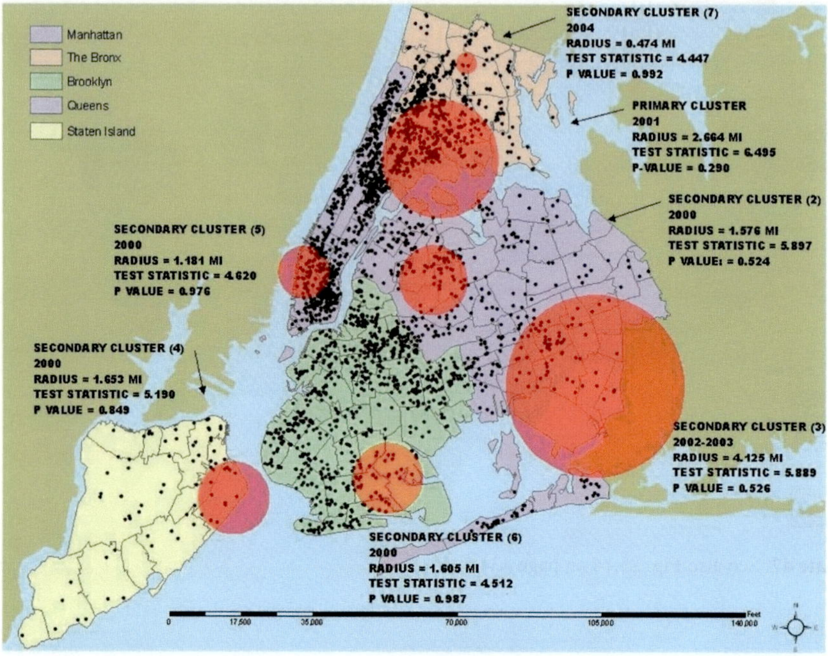

B

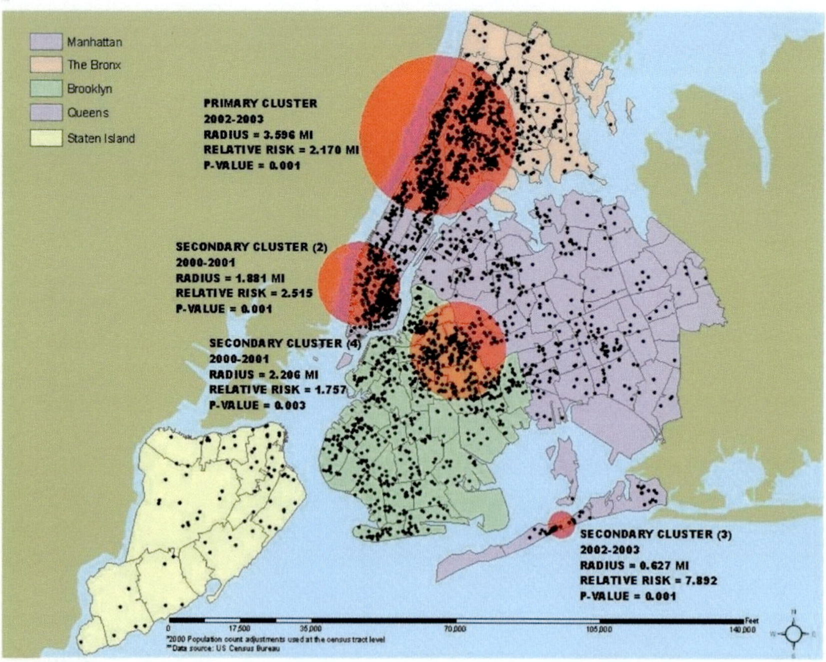

Plate 48 See also Fig. 27.2 on page 455

Chapter 15
Human Immunodeficiency Virus (HIV) Infection Rates and Heroin Trafficking: Fearful Symmetries

Chris Beyrer

Abstract There is mounting evidence that the spread of the human immunodeficiency virus (HIV) is associated with heroin trafficking routes. The relationship between the two is best illustrated by the routes leading from the two primary regions for the illicit opium poppy cultivation and heroin manufacture: the Golden Triangle of South-East Asia and the Golden Crescent of Central Asia. The producers in the Golden Triangle are the Lao People's Democratic Republic and Myanmar, and those in the Golden Crescent are Afghanistan and Pakistan. Together, those States accounted for perhaps 80–90 percent of the world heroin supplies in 2007. HIV outbreaks resulting from unsafe injection practices among injecting drug users (IDU) in trafficking zones have been documented in Myanmar itself, in Belarus, China, India, Indonesia, the Islamic Republic of Iran, Malaysia, Pakistan, the Russian Federation, Tajikistan, Thailand, Ukraine, Uzbekistan, Vietnam, and in several States in Eastern Europe. Heroin trafficking in those States has led to serial epidemics: first of heroin use, then of injection, then of blood-borne pathogens, including hepatitis C and HIV. Ethnic and trade relationships in heroin trafficking zones appear to facilitate such epidemics, as does drug testing by petty traders in market nodes. Policy responses, or the lack thereof, have increased the vulnerability of users and their communities, as have the limited drug treatment options available in those zones. While "supplyside" approaches, including interdiction and policing, are likely to continue, those HIV epidemics will require improved drug treatment, access to HIV prevention services, including harm reduction, and new approaches to the prevention of HIV in areas where heroin trafficking occurs.

Introduction

There is growing evidence that the epidemic spread of human immunodeficiency virus (HIV) infection occurs in tight and complex relationship to heroin-trafficking routes (Yu et al., 1999; Beyrer et al., 2000; Quan et al., 2000). The relationship is

C. Beyrer
Director, Johns Hopkins University Fogarty AIDS International Training and Research Program, Baltimore, MD, USA

Y.F. Thomas et al. (eds.), *Geography and Drug Addiction*,

most clearly illustrated by the routes leading from the two primary regions for illicit opium poppy cultivation and heroin manufacture: the Golden Triangle of South-East Asia and the Golden Crescent of Central Asia. The mechanisms that lead to those HIV outbreaks are just beginning to be understood, as are the special vulnerabilities of communities in trafficking zones. The principal heroin producers in the Golden Triangle are the Lao People's Democratic Republic and Myanmar and those in the Golden Crescent are Afghanistan and Pakistan. Together, those States accounted for perhaps 80–90 percent of the world heroin supplies in 2002 (U.S. Department of State Bureau of International Narcotics and Law Enforcement Affairs, 2002). While those four countries generate most of the world's heroin, the HIV epidemics resulting from unsafe heroin injection practices have largely been seen in the neighboring countries or in the countries of destination further afield. The spread of HIV is well documented in countries in and around the Golden Triangle, such as in Myanmar itself, in China, India, Malaysia, Thailand, Vietnam and, more recently, in Indonesia (Kato et al., 1999; Piyasirisilp et al., 2000; Monitoring the AIDS Pandemic, 2001; Chelala and Beyrer, 1999; Shao et al., 1998; Crofts, Reid, and Deany, 1998). For countries in and around the Golden Crescent, the data are only now emerging and the HIV epidemics are much newer: the epidemic spread of HIV, hepatitis C, or both appears to be under way in Belarus, India, the Islamic Republic of Iran, Pakistan, the Russian Federation, Tajikistan, Ukraine, Uzbekistan, and several States in Eastern Europe (Monitoring the AIDS Pandemic, 2001; Gillis and Mubbashar, 1995; Dehne et al., 1999; Bobkov et al., 1998; Rhodes et al., 1999; Khanina et al., 2000). In virtually all the studies that have investigated HIV and hepatitis C virus infection among injecting drug users (IDU) in those regions, hepatitis C virus has been shown to be far commoner. The prevalence of hepatitis C virus among IDU generally reaches 80 percent or higher, which is a function of the very high transmissibility of this agent through parenteral exposure. However, it is the subtyping of HIV that has proven more useful, thus far, as an investigative tool to understand trafficking.

The third important area for illicit poppy cultivation is the New World, in Colombia, Mexico, and Peru. While significant for the markets of North and South America, it will not be discussed in the present chapter as it is less well understood in terms of interactions between trafficking and HIV spread in the production zones.

Licit poppy cultivation for pharmaceutical opiate derivatives such as morphine, codeine, and Demerol is centered in Tasmania, Australia, which accounts for about 50 percent of licit world production, and in India and Turkey and has not been associated with heroin manufacture or trafficking, or with the spread of blood-borne infections. Opiate derivatives remain an important class of analgesics and are widely and generally safely used, with minimal public health effects. It is the illicit nature of criminal production and distribution and the rapid uptake of heroin use, injection and the unsafe use of equipment by young people in vulnerable communities along the trafficking routes and in destination markets further afield that has led to the fearful symmetry of heroin trafficking and the spread of HIV.

The present chapter seeks to clarify the degree to which the existing data demonstrate the relationship between heroin trafficking and HIV and to establish how much is known about the mechanisms that spread HIV. It also seeks to describe the tools for understanding and responding to that relationship that are available to researchers, policy makers, clinicians, and others. Finally, the chapter attempts to explain why societies from Ukraine to Vietnam have been so vulnerable to the interactions between heroin trafficking and the spread of HIV and to examine what can be done to reduce the harm resulting from them.

Injection Drug Use and the Spread of HIV

In 2000, an international research team published the findings of its investigations of the causes of HIV infections along the four heroin trafficking routes leading from the Lao People's Democratic Republic and Myanmar to China, India, Thailand, and Vietnam (Beyrer et al., 2000a, b). The findings had been obtained through the use of molecular epidemiology, Land Remote Sensing Satellite (Landsat) technology, qualitative research methods, and epidemiologic review. By using deoxyribonucleic acid (DNA) fingerprinting technology, the team was able to show that HIV viruses from the blood of infected IDU could help trace heroin routes. Heroin users and petty traders helped the research team to understand how heroin use spreads in communities and pinpointed the key roads, villages, towns, and cities through which heroin from the Golden Triangle was moving, leaving the legacies of addiction, and acquired immunodeficiency syndrome (AIDS). Since the publication of those findings, several other groups have independently investigated those zones, confirming the general hypotheses made by the team and documenting the further spread of infections and their impact on communities in China, Myanmar, and Vietnam (Quan et al., 2000; Piyasirisilp et al., 2000; Motomura et al., 2000). The fearful symmetry of heroin trafficking and the spread of HIV can perhaps best be illustrated by the HIV epidemic in Yunnan Province of China, which is east of Myanmar and the first destination for overland exports of heroin from Myanmar to the rest of China. The farthest province from China's booming coastal cities, Yunnan has the highest HIV infection rate in China (Shao et al., 1998). The outbreak began in ethnic minority communities in three mountain districts along the China–Myanmar border in the early 1990s, notably among the Kachin and Wa ethnic groups (Wu et al., 1996). As heroin spread among the young people of Yunnan and a rapid transition to injecting took place, there was a predictable rise in HIV infection. Equally predictable was the subsequent spread of infection to the non-IDU sexual partners, wives, and children of the largely young adult male IDU population. Yunnan now has the most mature HIV epidemic in China. Intense security and interdiction activities (supported partly by the United States of America) have failed to control the heroin trade. But it could be argued that at the current stage of the spread of the HIV epidemic in Yunnan, the virus has moved well beyond the IDU population. A similar situation obtains in Manipur State, north-east India, and on the western border of Myanmar: the early outbreak of HIV infection among IDU there has led to a disseminated HIV epidemic and the highest HIV prevalence by province in India (Panda et al., 2001). The present chapter will attempt to establish whether or not such situations are typical.

Observation of the rate of HIV and AIDS infections in Belarus, China, Kazakhstan, the Russian Federation, Tajikistan and Ukraine or further east in Indonesia, the Islamic Republic of Iran, Malaysia, and Vietnam reveals an epidemiologic picture of HIV in 2002 that is striking. What these diverse States have in common is that in each, the majority of reported HIV infections and AIDS cases in 2001 were attributed not to sexual transmission, which is the predominant mode in Africa, but to parenteral infection, exposure through needle-sharing behaviors among IDU (Crofts, Reid, and Deany, 1998). While the numbers of IDU in any one of those States may not be large on a population basis, those States represent enormous young populations, many of which have rapidly rising substance abuse rates. One example is Vietnam, a country with a population of over 78 million, where IDU accounted for 88 percent of all reported HIV infections in 2000 and where heroin trafficking from the growth zones of the Lao People's Democratic Republic has led to a dramatic increase in use among young people (Quan et al., 2000).

In other regions and countries, the majority of HIV infections are not among IDU, but IDU spread has played an important role in HIV epidemics. HIV infections among IDU were reported to the World Health Organization by 52 countries in 1992; the number of reporting countries grew to 114 by the year 2000, underscoring the global nature of IDU risk and spread (Joint United Nations Programme on HIV/AIDS, 2000). IDU-related outbreaks have also played a key role in the dynamics of the spread of HIV, notably in introducing HIV into populations, as shown by the early epidemiology of HIV in Thailand and Vietnam, where IDU were the first group in which HIV outbreaks occurred (Quan et al., 2000; Weniger et al., 1991). Such outbreaks have also played an important role in the dissemination and dispersion of novel HIV-1 subtypes. One example is the recent explosive spread of the HIV-1 subtype A virus in the Russian Federation and Ukraine and the outbreak of a B/C recombinant HIV strain that has been documented among IDU in southern and western China (Yu et al., 1999; Lukashov et al., 2000).

HIV Infection Rates in South and South-East Asia

IDU played a crucial role in the spread of HIV in South and South-East Asia in every country with a significant HIV epidemic except, perhaps, Cambodia (Beyrer et al., 2000). States experiencing epidemics that were initially IDU-related or are predominantly IDU-related include China, India, Malaysia, Myanmar, Thailand, and Vietnam (Beyrer et al., 2000). IDU were the first group in which the epidemic spread of HIV was detected in China, Indonesia, Malaysia, and Myanmar (Crofts, Reid, and Deany, 1998) and in all those outbreaks trafficking in heroin from the Lao People's Democratic Republic or Myanmar, or both, was involved (Beyrer et al., 2000).

The HIV epidemic in Thailand is among the best documented in Asia. The epidemic spread of HIV in that country was first detected among IDU in Bangkok in 1988 (Weniger et al., 1991). It was an explosive outbreak with clear links to incarcerated IDU and occurred initially among low-income, ethnic Thai, male urban residents. However, HIV spread rapidly among Thai IDU nationwide and within a year HIV rates of 20–40 percent were the norm. The IDU-related HIV epidemic was followed by a heterosexual outbreak of HIV that was larger and involved many more people. However, while the rate of infection of heterosexuals and other groups at risk declined after the period 1995–1996, that of IDU did not decline (Thailand, Ministry of Health, 2001).

Similarly, in Malaysia and Vietnam, IDU were the first group in which the epidemic spread of HIV was detected. However, unlike in Thailand, IDU have remained the predominant risk group affected by HIV in Malaysia and Vietnam, accounting for roughly 60–70 percent of cumulative infections in those two countries by 2000 (Joint United Nations Programme on HIV/AIDS, 2000). The heroin in Malaysia appears to be transported by sea along that country's long coast on the Andaman Sea (U.S. Department of State Bureau of International Narcotics and Law Enforcement Affairs, 2002). The Lao People's Democratic Republic was identified as the likely source of this heroin and molecular-typing data confirm that HIV infections have spread north from this zone in Vietnam into the neighboring Guangxi Province of China (Kato et al., 1999). China's emerging epidemic remains overwhelmingly due to needle-sharing among IDU and the three most HIV-affected provinces of China (in order of prevalence, Yunnan, Xinjiang, and Guangxi) have all experienced IDU-related outbreaks along the major heroin routes (Yu et al., 1999; Shao et al., 1998).

Myanmar is somewhat a different case. It is the region's major producer and a major consumer of heroin (Chelala and Beyrer, 1999). Trafficking within Myanmar from the heroin manufacturing areas in the Shan and Wa hills in the far north-east of Myanmar has led to a national outbreak of HIV. The United Nations International Drug Control Programme (UNDCP) and the Ministry of Health of Myanmar identified drug use rates among township adults of 2–25 percent in 1995, one of the highest rates worldwide (Thaung, Gyee, and Kywe, 1996). IDU were the first group in which HIV was identified, in 1989, and remain at extraordinary risk, with HIV prevalence rates of 60–95 percent nationwide (Myanmar, Ministry of Health, 2000). It is unclear how heroin is moved within Myanmar, but it is known that a major city in the trade is Mandalay, the largest city in northern Myanmar. Petty traders from Manipur State in India travel inland across Myanmar from India to buy high-grade No. 4 heroin in Mandalay and trucks carrying heroin to India leave from Mandalay as well (Beyrer et al., 2000). HIV rates among IDU in Mandalay have been consistently in the 60–80 percent range since at least 1995. Mandalay is in a region that is too hot and at too low an altitude for poppy cultivation.

The outbreaks of HIV infection among IDU in South-East Asia have several features in common. First, they have been explosive: HIV prevalence among Bangkok IDU went from 2 to 40 percent in 6 months in 1989. Second, they have been transnational: both China and India have had their highest prevalence zones along their borders with Myanmar (Yunnan Province and Manipur State, respectively). Third, they have led to further spread among non-injecting populations, initially the sexual partners of IDU, as has been documented in China, India, and Thailand. Fourth, these epidemics have proven difficult to control owing to government policies on injection drug use, the status of drug treatment in the affected States and the limited HIV prevention measures targeting IDU.

The Golden Crescent: HIV Infection Rates and Central Asian Heroin

Much less is known about heroin and HIV epidemics in the Golden Crescent than is known about those in South and South-East Asia. For most States affected by heroin from Afghanistan and Pakistan, HIV spread is a more recent event and many have little data or research capacity. However, the little that is known suggests another region of fearful symmetry.

Several tools can be used to measure poppy production, but arguably the most accurate is Landsat satellite technology, which measures crop densities (U.S. Department of State Bureau of International Narcotics and Law Enforcement Affairs, 2002). United States intelligence agencies have used Landsat to assess poppy cultivation, estimate opium base harvests and calculate heroin yields (10 kg of opium base gives roughly 1 kg of refined heroin). In 1996, after the establishment of Taliban rule, the estimated yield in Afghanistan was 200 tons (United States of America, Department of State, Bureau of International Narcotics and Law Enforcement Affairs, 2002). By 1999, Afghanistan was manufacturing 450 tons of heroin per year and had become the world's largest single manufacturer in a multibillion-dollar industry. Poppy growing appeared to cease in 2000 after the leader at that time, Mullah Muhammed Omar, delivered an edict on the subject, but stockpiled heroin reserves held by producers and traffickers apparently ensured that the supply was maintained despite the ban on poppy cultivation. In 2002, the new Administration and its allies acknowledged that a reduction in poppy cultivation could only be achieved through a long-term process of agricultural reform and development and the extension of government control across the vast rural areas of the country. For the short-to-medium term at least, Afghanistan will remain a significant producer. The second largest grower in the region, Pakistan, produces about 20 tons of heroin a year, roughly equal to the production of the Lao People's Democratic Republic and mostly in the remote tribal zones along the Afghan border in the North-West Frontier Province (United States of America, Department of State, Bureau of International Narcotics and Law Enforcement Affairs, 2002). Those areas are only marginally under federal control, very underdeveloped and likely to remain dependent on poppy cultivation for some years.

The HIV-related repercussions of heroin exports from Afghanistan and Pakistan are only now beginning to be understood, as nascent HIV epidemics take hold in a region for which data have been sparse, but that was thought to have been relatively spared from HIV. The Islamic Republic of Iran and Pakistan appear to be two of the major overland routes for the trafficking in heroin from Afghanistan (Monitoring the AIDS Pandemic, 2001). While HIV prevalence is low in both States, Pakistan had estimated a 3 million heroin addicts in 2000 and has suffered great social harm as a consequence (Gillis and Mubbashar, 1995). The Islamic Republic of Iran led the world in 1999 in narcotics seizures by volume (U.S. Department of State Bureau of International Narcotics and Law Enforcement Affairs, 2002). The country also has an enormous epidemic of heroin use among its young people (Monitoring the AIDS Pandemic, 2001). The Iranian Government is deeply concerned about this and it was a primary source of tension and border conflict while the Taliban regime was in control in Afghanistan. A nascent epidemic of HIV among Iranian IDU appears to have begun in the period 2000–2001, with recent reports of very high rates of HIV infection among incarcerated IDU in Tehran (up to 67 percent in one facility) (Monitoring the AIDS Pandemic, 2001).

The countries most affected by HIV are the Russian Federation and the two former Soviet republics of Belarus and Ukraine (Dehne et al., 1999; Rhodes et al., 1999). In its *Report on the Global HIV/AIDS Pandemic, 2000*, the Joint United Nations Programme on HIV/AIDS (UNAIDS) identified those three States as having the fastest-growing HIV epidemics in the world (Joint United Nations Programme on HIV/AIDS, 2000). More than 75 percent of all infections in the Russian Federation and its neighbors in 2000 were owing to injecting drug use (Dehne et al., 1999; Bobkov et al., 1998; Burrows, 2001). The far east of the Russian Federation has been particularly affected. The Irkutsk region of Siberia, around Lake Baikal, has the highest rate of HIV infection in the Russian Federation after Moscow and, again, more than 80 percent of the HIV infections reported in Irkutsk have been among IDU (Russian Federation, Ministry of Health, 2002). Kazakhstan, too, has seen a recent outbreak of drug use and HIV infection, although it is unclear whether the source of the trafficking route is China to the east or Afghanistan to the south.

Poppy Cultivation and Politics

While the HIV epidemic is a new challenge to the Russian Federation, the trafficking connections of the Golden Crescent are not new, having existed during the long Afghanistan war with the former Union of Soviet Socialist Republics (USSR), when poppy cultivation by the Mujahidin was tolerated by the West because the anti-Soviet forces had no other exports comparable to heroin in terms of value and ease of transport (Elliot, 1999). The consequent high rates of heroin use and addiction among Soviet forces engaged in the Afghan conflict were a predictable outcome and helped undermine support for the war among those troops, their families, and Soviet citizens.

The poppy farmers of Afghanistan are largely subsistence farmers who sell opium as a cash crop to supplement minimal incomes. As in the Golden Triangle region, the real profits of heroin come not from farming, but from trafficking and it is among the trafficking networks that real revenues accrue (Elliot, 1999). But part of the legacy of war is the local expertise in poppy cultivation and the production and sale of narcotics. The Afghan war, which the Soviets lost, appears to have brought heroin first to the dispirited troops and then to Moscow. Trafficking links may therefore be a legacy of the long struggle of the people of Afghanistan, although that remains speculation.

Another heroin-related epidemic is currently being experienced by China. The Xinjiang Uighur Autonomous Region is the only Muslim majority region of China. Xinjiang shares borders with Afghanistan, Kazakhstan, and the Russian Federation

(Siberia) and is linked to the rest of China by the Silk Road. It also has China's second highest rate of HIV infection by province, after Yunnan in the far south (Joint United Nations Programme on HIV/AIDS, 2000). More than 78 percent of the HIV infections in Xinjiang are owing to injection drug use involving heroin. Tragically, more than 90 percent of IDU in the two largest cities in Xinjiang are ethnic Uighur, which means that the HIV infections in this large province are largely among young Muslims (Shao et al., 1998).

Belarus, China, the Islamic Republic of Iran, Kazakhstan, Pakistan, the Russian Federation, and Ukraine are all experiencing outbreaks of heroin use among their young people and all now appear to have HIV epidemics related to that use. Heroin exports from the Golden Crescent are at the root of these complex new problems. These are regional challenges, but they point to a global problem that ties the Golden Crescent to the Golden Triangle: illicit heroin revenues. In 2000, Afghanistan was the world's poorest State, on paper, and Myanmar was designated a "least developed country" by the United Nations. Afghanistan was almost entirely dependent on donor aid in 2002 and had essentially no foreign reserves, a bankrupt treasury and limited licit exports. The economics of the drug trafficking networks based on the Golden Crescent are not known, but it is known that taxes on poppy farmers and protection money from drug traffickers were among the main sources of revenue for both the Taliban and the Northern Alliance before the interim Government led by Hamid Karzai came to power. In both Afghanistan and Myanmar, heroin has enabled the purchase of weapons on the black market and funded militias, insurgencies, and crime (U.S. Department of State Bureau of International Narcotics and Law Enforcement Affairs, 2002). Afghanistan has the potential to grow other crops, including grain and orchard fruits, but those would require irrigation systems and access to markets, which remains a huge challenge for much of the country. In Myanmar, the poppy-growing regions have been at war with the central Government virtually since the departure of the British after the Second World War (Beyrer, 1998). In order to achieve a reduction in the opium supply from those regions, viable alternative economies for the rural poor will need to be established, which will take time, and sustained donor investment, and the nurturing of stable, functioning civil societies will be necessary. If Afghanistan were once more to descend into civil strife and warlordism, it is likely that heroin production would revive. Indeed, as in Myanmar, it is in the interests of the narcotics cartels and the corrupt leaders that they have supported that civil society should fail; a chilling reality, given the wealth and power that heroin revenues have already generated.

Mechanisms on the Ground

Before discussing policy responses to the interactions of heroin and HIV, it would be useful to consider some mechanisms of those interactions which bear on the spread of HIV and how best to curtail it. One obvious feature shared by all of the primary trafficking zones out of the Triangle and the Crescent is geographic: overland heroin is moved almost exclusively across remote border regions, generally mountain and forest zones adjacent to the hills where poppy will grow. The illegal and clandestine nature of this industry demands such remote areas. Indeed, as a former director of UNDCP has pointed out, there are very few regions remote and lawless enough to support a major heroin industry. It is surely no coincidence that it is Myanmar, secretive, closed, and ruled by junta; the isolated Lao People's Democratic Republic under its Communist Party; Afghanistan with its decades of strife; and the tribal zones of Pakistan that became the world's leaders in heroin production: these are almost the only places in the world closed enough to sustain the heroin industry.

A second shared feature is ethnicity. Those areas generally have in common populations who are ethnic minorities or tribal groups, or both (Beyrer, 1998). In South-East Asia they are virtually all hill-dwellers, whereas the majority populations of the Lao People's Democratic Republic, Myanmar, Thailand, and Vietnam are lowlanders and rice cultivators (Beyrer, 1998). One especially important factor is that those groups tend to straddle national borders. Therefore, in Yunnan Province of China, both sides of the China–Myanmar border lands are farmed by ethnic Kachin and Wa, and family, language, and trade links long pre-date heroin trafficking. Another example is the ethnic Manipuris of Manipur State in India, who are Tibeto-Burmans like the majority population of Myanmar, not ethnic Indians, and can move easily into Myanmar to access the heroin markets in Mandalay (Khomdon Singh Lisam, 1998).

For HIV to spread along trafficking routes, local people have to use the drugs. Qualitative work in China, India, Myanmar, and Vietnam has suggested a direct mechanism for the "exchange" of HIV-1 subtypes (known as clades or strains). This mechanism relies on the fact that many petty traders in the region are also users, who support their own habits by purchasing and selling small amounts of heroin. In at least four States, it is known that those petty traders typically self-test heroin purity by injecting themselves. Since travelling across those zones with injecting equipment is an obvious sign of intent to use drugs, they rarely have their own equipment. On the border between China and Vietnam, for example, traders typically cross the mountains from China, stay the night with their contacts in Vietnam, and share drugs and equipment before making purchases. The very low genetic diversity of strains in this region suggests rapid spread of only one viral subtype, a molecular feature favored by this kind of direct spread. Major traffickers moving heroin in consignments weighing hundreds of kilograms or more have very different ways to move their product, including trucking, sea, and air routes. But it is likely that the spread of HIV in overland regions occurs on a more local person-to-person basis. A recent report from Yunnan Province found that 75.9 percent of a large series of IDU in south-eastern Yunnan were of Han Chinese ethnicity which led the authors to conclude that the epidemic in Yunnan was no longer confined to non-Han ethnic minorities (Zhang et al., 2002).

A fourth mechanism is also likely to be important, though somewhat variable. Along at least some of the major trafficking routes, services for truckers have developed, which, in addition to fuel, food and lodging, often include sex services. In South-East Asia, those sex services generally consist of roadside brothels, karaoke parlors, bars, and so forth. In Central Asia, they may be less apparent, but still available, or may have young male sex workers, as in the trucking industry in Pakistan. Those border zone sex service venues can overlap with drug trafficking and provide another mechanism by which HIV could spread where heroin and other contraband are moved. On the borders between China and Myanmar and between Myanmar and Thailand, women and girls are trafficked on the same routes, and indeed by some of the same trafficking networks, as heroin (Beyrer, 2001).

The interaction of heroin trafficking and sex industry-related HIV risks can also be found in the special economic zone of Pingxiang on the highway and in the trains crossing from Vietnam to China (Yu et al., 1999). Pingxiang was one of the first Chinese cities to experience a rapid HIV epidemic among IDU and molecular work has confirmed the cross-border nature of this epidemic (Kato et al., 1999). But Pingxiang also has a booming sex trade on the Chinese side of the zone. As many as 19 separate brothels in a four-street radius in the trucking zone in Pingxiang were counted in 2000, each with 10–30 women and girls working in it. HIV rates had remained low among those women until 2001, though there now appears to be increasing prevalence (Dr. Chen Jie, unpublished data). In settings like Pingxiang, sex workers and their clients in border and trafficking zones may be key "bridge" populations from IDU to wider networks of people at sexual risk.

Policy Responses

IDU outbreaks associated with heroin trafficking have proven difficult to prevent or control. In the major production zones and in the wider affected regions, treatment and prevention programs for drug use were limited before the spread of HIV (Poshyachinda, 1993). That remains largely the case: across the whole of Asia the only place where evidence-based heroin treatment and methadone maintenance therapy are available on demand to drug users is the Hong Kong Special Administrative Region of China. This is tragic, given that there is a large and growing international evidence base for success in prevention of HIV infection and other blood-borne diseases among IDU (Gray, 1995). While the majority of published reports have been from the developed world, principally Australia, North America, and Western Europe, there have been several reports of pilot projects and successful programs in Asia, including reports from India, Nepal, Thailand, and Vietnam. Much of that work has focused on harm reduction and needle and syringe exchange programs, the basic tools of most reported interventions. In 1998, the *Journal of Substance Use and Misuse* published its "Bibliography on Syringe-Exchange References", which included several hundred published reports on those interventions and the debates that they generated (Des Jarlais and Friedman, 1998).

Successful needle exchange programs have been conducted in Australia, the Netherlands, the UK and Northern Ireland. In the largest analysis published, the incidence of HIV increased by about 6 percent per year in 52 cities without needle exchange programs and decreased by 5.8 percent per year in 29 cities with needle exchange programs (Substance Use and Misuse, 1998). The New York City needle exchange programs have been studied in prospective cohorts: lower rates of incident HIV infection were documented among IDU using needle exchange programs (1.4–1.6 percent per year) than among those who did not attend needle exchange programs (5.3 percent per year, 95 percent confidence interval: 2.4–11.5) (Hurley, Jolley, and Kaldor, 1997). Long-term methadone maintenance therapy has been shown to reduce HIV risk behaviors, in particular needle use, and there is strong evidence that methadone maintenance therapy prevents HIV infection among IDU.

Where harm reduction and methadone maintenance therapy are available, as they were to many IDU in the United States HIV Network for Prevention vaccine preparedness studies, sero-incidence can be low (Des Jarlais et al., 1996). In that study, HIV incidence among homosexual and bisexual men between 1995 and 1997 was measured at 1.55 per 100 person years, while among male IDU the rate was 0.38 per 100 person years.

The Thai Working Group on HIV/AIDS Projections has recently published projected scenarios for the Thai epidemic (Seage et al., 2001). They found that a decline in needle-sharing from 20 to 10 percent among Thai IDU would avert 21,774 new infections by 2006 and 81,761 infections by 2020. That would constitute the single largest number of infections averted for any one intervention strategy. By 2006, roughly 3800 of the expected 22,000 infections nationwide would be averted by that intervention alone. Vietnam has reported on the feasibility of needle exchange programs and on pilot needle exchange programs in the cities of Hanoi and Ho Chi Minh (Thai Working Group on HIV/AIDS Projections, 2001). While they did not measure impact, they were able to conclude that needle exchange programs were feasible, but that they required acceptance by the community and by the police to be sustained. Needle exchange programs have also been implemented in India, notably in New Delhi and Manipur State, where high rates of IDU behavior are common.

Taken together, those studies all support the contention that harm reduction and needle exchange programs are effective prevention tools, and that they might have an effect on heroin-related epidemics in trafficking zones. Yet those approaches have been little used in the fight against HIV/AIDS.

It is difficult to imagine a public health tool with reasonable evidence of efficacy that has generated as much debate as have prevention programs for IDU. A review of the literature suggests three principal problems with the implementation of harm reduction approaches and needle exchange programs (Quan, et al., 1998; Lurie and Drucker, 1997). First, they have repeatedly been seen as condoning or facilitating injecting drug use, making them politically unpopular outside the prevention community. Second, they have faced legal, security, and policy challenges, since they require "safe" domains of interaction with active IDU. A third challenge, where needle exchange programs have been implemented, is the coverage rates of needle exchange programs for IDU populations. One exception to the lack of programs in affected areas has been the rapid implementation of needle exchange programs in the Russian Federation and the Commonwealth of Independent States, established in partnership with the Open Society Institute of the George Soros Foundation, Médecins Sans Frontières, the ministries of health of many affected nations and the

United States Agency for International Development. The Open Society Institute has supported the establishment of more than 140 needle exchange programs in the Russian Federation alone and that effort has the potential to reach levels of coverage which might control HIV among IDU.

Conclusions

Individuals, communities, and countries that have the misfortune to be on major heroin-trafficking routes faced multiple epidemics in 2002. Those epidemics began with heroin use, heroin injection, and then HIV infections. While the clear long-term goal for all the States involved is to be free of drug trafficking, the realities of the current political and development situations of the major producers, most notably Afghanistan and Myanmar, suggest that narcotics-based economies will be with the world for some time to come. In the short to medium term, a public-health-based approach would be to minimize the health impacts of heroin trafficking by working with affected communities. Such approaches could include reducing heroin addiction through improved treatment and support for IDU and reducing the spread of HIV among those who continue to inject by expanding harm reduction and needle exchange programs. The prevention of spread beyond IDU suggests that that may be critically important to the prevention of wider epidemics of HIV/AIDS. A clear priority for further research and programs are the front-line Central Asian States and around the Golden Crescent: the Islamic Republic of Iran, Kazakhstan, Kyrgyzstan, Pakistan, Tajikistan, Turkmenistan, and Uzbekistan. Those must be considered States that are at very high risk of an explosive spread of HIV in the coming years and that could benefit from the programmatic and research experience that have elucidated the heroin and HIV interactions of the Golden Triangle.

Chapter 16
Metropolitan Area Characteristics, Injection Drug Use and HIV Among Injectors

Samuel R. Friedman, Barbara Tempalski, Hannah Cooper, Spencer Lieb, Joanne Brady, Peter L. Flom, Risa Friedman, Karla Gostnell and Don C Des Jarlais

Abstract Epidemics, behaviors, and programs to change behaviors and ameliorate epidemics are shaped by the characteristics of geographically and socially defined communities. This chapter presents the rationale, methods, and selected findings from a study of injection drug users, HIV, and services for drug injectors in the 96 largest metropolitan areas in the United States. It presents data that show that metropolitan areas vary widely in the prevalence of injectors in their populations; in HIV prevalence among injectors; and in the percentage of injectors who are in drug abuse treatment. Furthermore, theoretically specified locality characteristics, such as inequality, legal repression of drug users and others, the degree of popular organization and mobilization for helping drug users, fiscal constraints and others, help predict the values of these variables in metropolitan areas. These findings help us to identify metropolitan area characteristics, including some that can be changed by public authorities or as a result of popular demand and social movements, which can be targeted for intervention to address drug-related health issues. Future research on the social and geographic causation of injection drug use, of its sequelae, and of programs like drug treatment and syringe exchange is clearly warranted, and should include research interventions that change metropolitan area characteristics in ways that reduce drug-related problems.

Background

Many behavioral and epidemiologic processes are deeply shaped by the places in which they occur. Geographical locations in human society and the behaviors and epidemics that take place in them are imbued with historically and socially shaped

---

S.R. Friedman
National Development and Research Institutes, Inc., New York, NY, USA

Y.F. Thomas et al. (eds.), *Geography and Drug Addiction*,

structures, meanings, and resources (Tempalski 2007). As this chapter will show, the amount of injection drug use, services for injection drug users (IDUs), and HIV among drug injectors in a metropolitan area is related to the area's social, economic, and political conditions.

This chapter is an overview of the rationale, methods, and selected findings of the Community Vulnerability and Response to IDU-Related HIV project. This project is a study of how metropolitan area characteristics are related both to a variety of measures of community vulnerability to HIV among drug injectors and, also, to policy and program variables that may affect HIV in the community. Thus, it is a study of how different characteristics of one type of geographic unit, the metropolitan area, are related to each other.

The project was initially funded in 2000 as a cross-sectional study of the 96 largest metropolitan statistical areas (MSAs) in the United States (as of 1993). It has since been re-funded as a longitudinal study. As of this writing (2007), we are in final stages of creating estimates of the numbers (and population percentage) of IDUs, of racial/ethnic population densities of IDUs, and of HIV prevalence among IDUs in each MSA for each year, 1992–2002. We will study the predictors of change in these and other variables, including change in the presence or extent of syringe exchanges, drug abuse treatment coverage, and hard drug arrests per capita. Here, we present some findings about estimates we made for the year 1998 during the first phase of this project.

Characteristics of a Locality Predict Many Health Outcomes

Research that compares localities has a long tradition in public health. Most of this research has been cross-sectional, although in some cases independent variables have preceded dependent variables in time. Statistical associations have been found between state, MSA or neighborhood characteristics (like income inequality, per capita income, or poverty) and all-cause mortality, age-specific mortality, sedentary activity, medical care expenditures, low birth weight, malignant neoplasms, coronary heart disease, homicide, violent crime, social pathology, smoking, arrests related to illegal drugs, drug use, heroin, opioid, or narcotics addiction, initial cocaine usage (and perhaps the staging of drug "epidemics"), TB, and AIDS (Bell et al. 1998; Brugal et al. 1993; Chein et al. 1964; Faris et al. 1939; Hsing 1996; Kaplan et al. 1996; Kennedy et al. 1996; Lynch et al. 1998; Nurco 1972; Nurco et al. 1984; Petronis and Anthony 2003; Redlinger and Michel 1970; Wallace and Wallace 1998). Multilevel analyses have shown that individuals' behavior and health may be influenced by their social context, with neighborhood characteristics being related to low birth weight, preterm birth, and individual tobacco smoking (Ahern et al. 2003; Diez-Roux 2000; Duncan et al. 1999; Kawachi and Berkman 2003; O'Campo et al. 1997).

Thus, there is considerable evidence that the characteristics of different places may create spatially bounded socioepidemiologic processes that affect human health. Since IDUs are at high risk for HIV, hepatitis B and C, and many socially related problems, it is important to study what causal factors are related to how many IDUs are there in a geographic area and how widespread HIV is among them.

Here, we briefly present selected analyses of predictors of (1) HIV prevalence among IDUs; and (2) the population density of IDUs (IDUs per 10,000 population) in large metropolitan areas of the US in 1998. We also summarize published findings about a related issue (Friedman et al., 2007 IJDP): the extent and predictors of the proportion of IDUs in a metropolitan area who were in drug abuse treatment in 1998.

Methods

"Sample" and its Statistical Implications

We operationalize "place" as MSA. The sample is the 96 largest MSAs in the United States in 1993. MSAs are contiguous counties that contain a central city of 50,000 people or more and that form a socioeconomic unity as defined by commuting patterns and social and economic integration within the constituent counties (Office of Management and Budget 2000; US Bureau of the Census 1998). This chapter thus studies a "population" rather than a sample, so there is no sampling error (though there is measurement error). Whether statistical inference is relevant is debatable. Some researchers studying similar populations use "p-values" or "confidence intervals" as heuristic devices to avoid over-interpreting model parameters (Friedman 1977a,b; Kaplan et al. 1996; Lynch et al. 1998; McCarthy et al. 1988; Turk 1977). Other analysts might view the population as a random sample of "possible universes;" in this interpretation, "pseudo-confidence intervals" have a probabilistic interpretation.

In some analyses, missing values on one or more variables reduce the N below 96.

Variables

Since the derivations of both dependent variables (injectors per capita and HIV prevalence among injectors) discussed in this chapter have been described elsewhere, we do so only briefly in this study.

Dependent Variables

Drug injectors per capita in the MSA population in 1998 was estimated in a three-step process (Friedman et al. 2004). The number of persons who had injected drugs in the USA in 1998 was first estimated by adjusting and averaging others' prior estimates (SAMHSA 1992; Holmberg 1996). This number was allocated to each MSA using four multipliers (using data on drug injectors among drug abuse treatment populations, HIV counseling and testing clients, and AIDS cases, and estimates of numbers of injectors and HIV prevalence among them in 1993 (Holmberg 1996)). These four estimated numbers of injectors in each MSA were then averaged; and the mean divided by the MSA population.

HIV prevalence (defined as the proportion of IDUs who are HIV-positive) among IDUs in 95 MSAs in 1998 was estimated by taking the mean of two estimates (Friedman et al. 2005). (1) The first estimate was calculated by modifying CDC Voluntary HIV Counseling and Testing data to correct for their inherent underestimation of prevalence. Research-based data on HIV prevalence for 25 MSAs were used to calculate regression equations to perform these adjustments. (2) The second estimate was based on methods developed by Lieb et al. (2004). Briefly, the estimated total number of HIV-positive IDUs (including those who are also men who have had sex with men) living in an MSA was designated as $\boldsymbol{k}$ (and estimated by adjusting data on AIDS cases). The estimated numbers of IDUs (Friedman et al. 2004) (**a**) and the estimated HIV prevalence among IDUs (**b**) were variables related by the function, $\boldsymbol{k} = \boldsymbol{ab}$; thus, $\boldsymbol{b} = \boldsymbol{k/a}$.

Treatment coverage is the ratio of the number of IDUs in drug abuse treatment in each of 94 metropolitan areas to the number of drug injectors in the area as estimated above. Data on the number of IDUs in treatment in each MSA came from the Uniform Facility Data Set [US Department of Health and Human Services. Office of Applied Studies, Substance Abuse and Mental Health Services Administration (SAMHSA) 1999], which provides data based on a survey of each publicly or privately funded facility in the country that provides substance abuse treatment.

Independent Variables

Almost all of the independent variables precede the dependent variables in time so that the temporal sequence is correct. (This does not, of course, take account of the high degree of autocorrelation over time in many of these variables.)

1. Unemployment rate in 1990. A number of studies have found that economic conditions are associated with rates of substance use and/or HIV prevalence (Friedman et al. 2000; Geronimus 2998; Selik et al. 1988, 1989).
2. Three measures of legal repressiveness: (a) arrests for possession or sale of heroin or cocaine (1994–1997), taken from US Federal Bureau of Investigation (FBI) data; (b) police *employees* per capita (1994–1997), taken from FBI data (Police Employees Data; County-Level Detailed Arrest and offense Data); and (c) "corrections" expenditures per capita (1997), taken from United States Census Bureau data on government finances (US Census Bureau 1992). Arrest of drug users may be an indicator of pressures on police; and the fear of arrest may encourage drug users to become or remain drug injectors and also may lead injectors to inject less safely (Aitken et al. 2002; Bluthenthal et al. 1999a,b; Cooper, Moore, Gruskin, and Krieger 2005; Maher and Dixon 1999). Police *employees* per capita may reflect a public willingness to spend money and person-power on policing. It may also indicate more direct effects on HIV risk; for example, Corey et al. (2005)

found that police presence, as distinct from arrests, was associated with less use of syringe exchanges in Philadelphia. Correction expenditures are an indicator of public willingness to spend resources on local incarceration and probation systems as well as an indicator of the number of people arrested and the average time they spend in jail before and after trial, which would tend to increase fear of arrest and thus increase the risk of using drugs by injection in unsafe ways.

3. Proportion of the MSA population who are black (US Census Bureau 2003). Many studies have found that black injectors are more likely than other injectors to be HIV infected and/or to have AIDS (Friedman et al. 1987; Novick et al. 1988; CDC 2001, 2002; Selik et al. 1988, 1989); and earlier research from this project shows that higher percentages of black populations than of whites in these metropolitan areas are injection drug users (Cooper, Friedman, Tempalski, Friedman and Keem 2005).
4. For analyzing HIV prevalence, IDUs per 10,000 population in 1993 (Holmberg 1996) was also used as a control variable. It was a predictor of HIV prevalence among injectors in 1993 (Friedman et al. 2000).
5. Two measures of structural racism in 1990 as embodied in residential segregation (Massey and Denton 1992).
   - Black/white residential dissimilarity index.
   - Hispanic/white residential dissimilarity index.
6. Income inequality: ratio of total income of all households in the upper 10% to the total income of the bottom 10% (1989).
7. Region. US regions differ politically and culturally, and on the mean values of both dependent variables. On the other hand, their relevance for this project is unclear. We hope to determine predictors of IDUs per capita and of the proportion of IDUs who are infected with HIV in order to develop ways to reduce these health problems. The physical location of an MSA, however, cannot be changed. Thus, for us at least, finding that region is a significant predictor of one of our dependent variables just opens up more questions about what it is about a given region that leads to these effects. In order to make our categories for regions more homogeneous politically, culturally, and economically, prior to these analyses US Census categories for region were adjusted by moving Maryland, Delaware, and Washington, DC, to the Northeast Region; Texas to the West; and Oklahoma to the Midwest. Midwest was treated as the reference category because it had the lowest mean value on drug injectors per capita. (It had the second lowest mean value, 4.85%, on HIV prevalence, which was not statistically different $\{p[t] = 0.59\}$ from the mean 4.56% HIV prevalence in the West.)

Statistical Analysis

Since the unit of analysis in this study is the metropolitan area, dependent variables are rates for a given metropolitan area. Correlation and linear regression are used to estimate associations among variables. Standardized coefficients (betas) are reported to facilitate comparisons of magnitudes of association. Statistical analyses were done in SAS version 9 (SAS Institute 2004).

Results

Description of Statistical and Geographic Distributions of Dependent Variables

Table 16.1 presents data on the distributions of IDUs per 10,000 population, HIV seroprevalence among IDUs, and drug abuse treatment coverage for IDUs. MSAs vary considerably on all of these measures, with the MSA with the lowest value having approximately one-tenth the value of the highest MSA for IDUs per 10,000 and for HIV prevalence. The range is even greater for treatment coverage.

Figures 16.1, 16.2, and 16.3 show how these variables are distributed across metropolitan areas in the USA. As Fig. 16.1 shows, the metropolitan areas with higher prevalence of IDUs per capita seem to fall mainly in the Northeast coast down through Virginia and in an arc from New Orleans through Texas, New Mexico, Arizona and on up the West Coast. HIV prevalence is considerably higher in the metropolitan areas near New York City and up and down the coast from it; with a secondary concentration in Florida (see Fig. 16.2). Treatment coverage for IDUs is low. Only 9 of 94 metropolitan areas provide treatment to one IDU in five or more; these are primarily old industrial areas (Buffalo/Niagara Falls, Detroit, Gary, New Haven/Bridgeport/Danbury, New York, Providence/Warwick, and Scranton/Wilkes Barre/Hazelton) although Nassau/Suffolk and Salt Lake City/Ogden also provide this level of coverage (see Fig. 16.3).

Predictors of HIV Prevalence of IDUs

To begin with, we present preliminary results that we have conducted in efforts to understand how prior results on the relationship of "legal repressiveness" to HIV prevalence in MSAs might be illuminated by incorporating measures of inequality as prior predictors in a causal change. The previous analyses (Friedman et al. 2006, *AIDS*) showed that three different facets of "legal repressiveness" were independently and positively associated with subsequent (1998) HIV levels among IDUs in these MSAs: mean arrests for hard drugs per capita, 1994–1997; (2) Police

Table 16.1 Distributions of dependent variables

Variable	Mean (Std Dev)	Median (IQR)	Min, Max
IDUs per 10,000 population, 1998	66 (33)	60 (42–87)	19, 173
HIV prevalence among IDUs, 1998	7.7% (5.5%)	5.5% (3.9–9.7%)	2.4%, 27%
Percent of IDUs in drug abuse treatment, 1998	10.2% (6.8%)	8.6% (5.7–13.8%)	1.1%, 39.3%

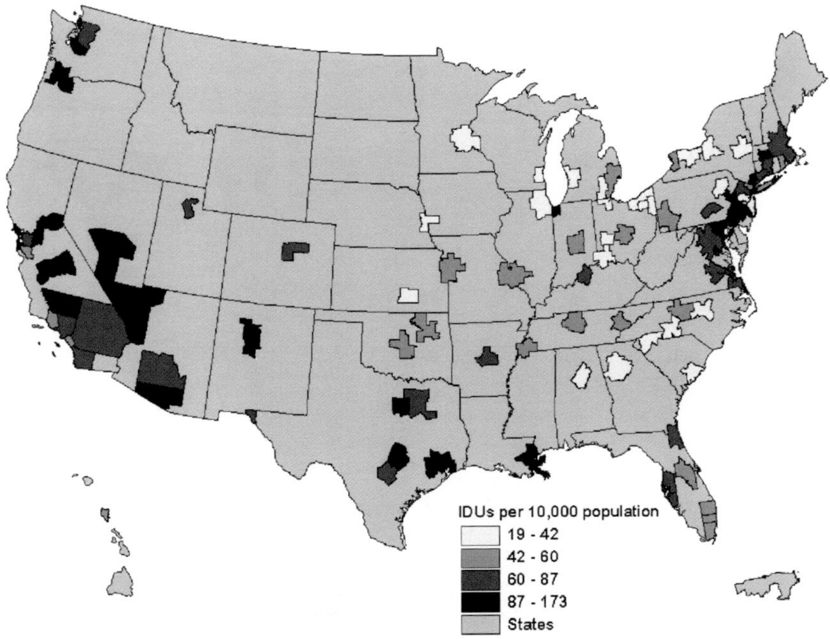

Fig. 16.1 96 large US metropolitan areas, 1998

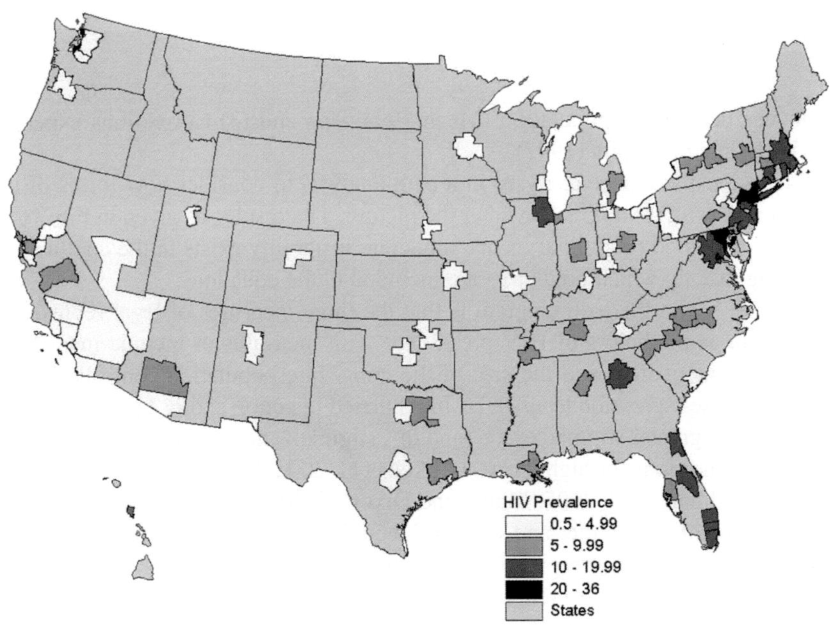

Fig. 16.2 Map of HIV prevalence (%) among IDUs in 95 large US metropolitan areas, 1998

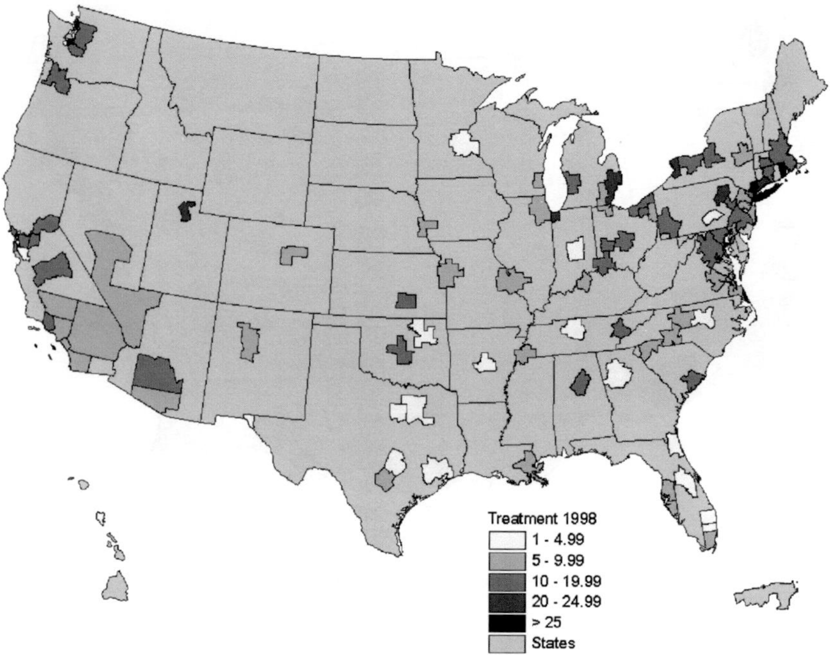

Fig. 16.3 Percent of injection drug users in drug abuse treatment in 94 large metropolitan areas, 1998

employees per 10,000 population (Mean 1994–97); and (3) Corrections expenditures in dollars per capita (1997).

Here, we extend these analyses in a path analysis to consider how forms of inequality and/or structural racism enter the picture. The results, as given in Fig. 16.4, must be considered preliminary since moderate instability exists in the coefficients depending exactly which predictors are included in the equations.

What we see in the path diagram is that the three measures of legal repressiveness remain as predictors of HIV prevalence with measures of income inequality, racial/ethnic inequality, and the size of the non-white population controlled. This may be because repression leads IDUs into hurried injection and/or because it is accompanied by greater stigmatization (and this stigmatization creates social and psychological conditions for higher risk) (Aitken et al. 2002; Bluthenthal et al. 1999a,b; Cooper, Friedman, et al. 2005; Maher and Dixon 1999). Residential racial segregation, the percent non-white, and income inequality are all associated with higher subsequent levels of legal repressiveness, although the magnitude of these associations seems to vary and different independent variables are related to different measures of legal repressiveness. One measure of racial segregation, the Black/white residential dissimilarity index, was also related to HIV prevalence directly as well as indirectly (through police employees per capita).

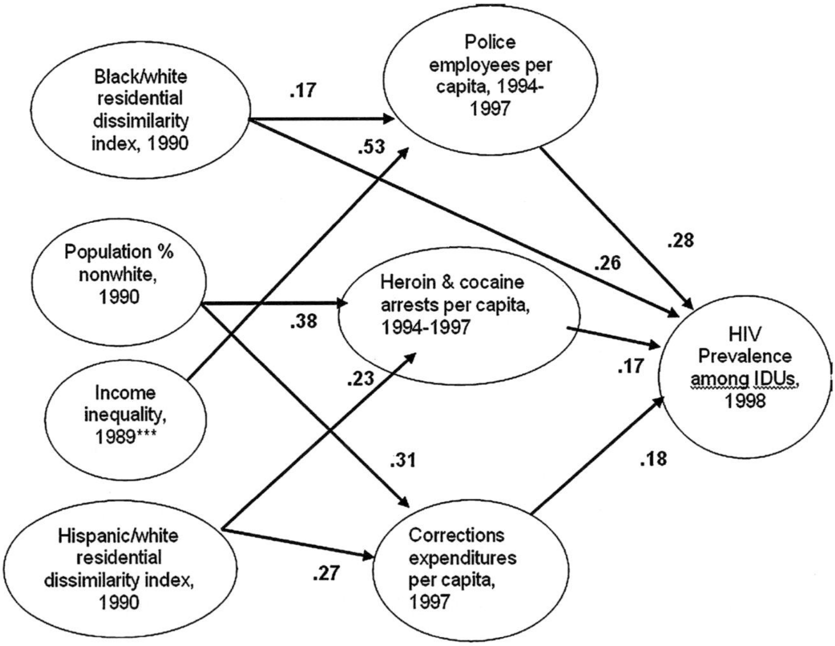

*These paths are considered preliminary because there is moderate instability in paths with modifications of the variables included in the model.
**Paths with p >= 0.10 not shown.
***Ratio of income of households in the upper 10% to the income of the bottom 10% (1989)

Fig. 16.4 Preliminary path analysis

Taken together, these data suggest that there may be a causal pathway of structural human rights violations (institutional racism and, arguably, income inequality) to legal repressiveness, and thence to higher HIV prevalence among IDUs.

Predictors of Population Prevalence of IDUs

Turning next to the characteristics of metropolitan areas that predict the population density of IDUs in 1998, Table 16.2 presents the results of theoretically-guided stepwise and backwards linear regression. The extent of unemployment is positively related to IDUs per capita. MSAs in the West have relatively more IDUs per capita, and those in the Midwest have fewer, than do those in the Northeast. There is a question about whether "Region" really tells us anything useful. What we really want to know (but have not been able to explain yet) is what it is about "region" that is associated with IDU population density. We will explore this further later in the project.

Table 16.2 Predictors of IDUs per 10,000 Population

	Adjusted Beta
Unemployment (%), 1990	0.29*
Midwest	–0.41*
South	–0.17
West	0.26*
R^2	0.46

*$p < 0.05$.

Predictors of Treatment Coverage of IDUs

Here, we simply report on findings from another paper (Friedman et al., 2007). (Since we do not report any new analyses of treatment coverage, we did not include descriptions of additional independent variables to our variable descriptions in Methods section; see the cited paper for this). Drug treatment coverage for IDUs was quite limited, with the median metropolitan area providing treatment to approximately 1 in 12 IDUs. In these conditions of shortage, an indicator of epidemiologic need (the per capita extent of AIDS among IDUs) did not predict treatment coverage. There is some indication that, given the shortage, competition for access by non-injecting drug users may limit the extent to which IDUs receive treatment. Stringent metropolitan finances (higher long-term governmental debt per capita) were associated with less treatment coverage. Political variables (racial structures, the presence of organizations that support drug treatment, and budget priorities) also appear to be important determinants of treatment coverage for injectors.

Limitations

These findings are subject to a number of limitations. First, causal mechanisms are hard to study at a single level of analysis since both higher-level and lower-level variables may affect observed relationships.

Although almost all independent variables precede the dependent variables in time, all variables are subject to considerable temporal autocorrelation and, in some cases, likely two-directional causation. Thus, causal inference would have been stronger if longitudinal data had been used. Such analyses are planned for the relatively near future, including further study of the possibly two-directional relationships between legal repressiveness and injectors per capita.

Some of the prediction equations may be mis-specified by leaving out important predictors or due to weaknesses in the variables we have. For example, in Fig. 16.4, where our explorations found some instability of results, residential racial segregation is only one aspect of structural racism, although a very important one, so it is possible that another dimension of institutional racism might underlie the results observed; and it is worth noting that police employees per capita is not the same as police on duty in drug-using areas or in drug squads which may or may not have attenuated the effects of this variable.

Discussion

It is clear from these analyses that a number of metropolitan area characteristics are related to the subsequent prevalence of drug injection in the population, to HIV prevalence among IDUs and to treatment coverage for IDUs. The extent to which this reflects place as a concrete realization of various socioeconomic and political characteristics, geographic location as a cultural location, and geographic diffusion of behaviors and of HIV among localities will be important to study in our longitudinal analyses in the second phase of the Community Vulnerability project.

Our preliminary findings about structural inequality, legal repressiveness, and HIV prevalence among IDUs, if confirmed by additional analyses, have serious implications. First, they suggest that legal repressiveness is associated with higher HIV prevalence among IDUs in US MSAs; and is not associated with lower rates of IDUs per capita. These results also suggest that institutionalized racism and income inequality may produce the impetus for governmental repressiveness at the metropolitan level. They further suggest that programs and social movements to reduce or eliminate structural racism, inequality, and legal repressiveness have, in addition to whatever value they have in their own right, a role to play in the fight to contain HIV/AIDS and perhaps other infectious diseases. The presence of organizations or movements that support drug treatment also suggests the importance of political dynamics in shaping drug-related outcomes and programs. Further research might use historical and time-series research, as well as sociopolitical experiments, to investigate these hypotheses.

Thus, the findings in this chapter suggest that MSA characteristics, including some that can be changed by public authorities or as a result of popular demand or social movements, may be part of the causal chain that shapes the extent and patterns of drug use and related diseases, as well as of related services.

In conclusion, then, further research on social and geographic causation of drug-related problems is clearly warranted. This should include developing new models of intervention that change MSAs or their environments in ways that reduce drug-related problems.

Acknowledgment We would like to acknowledge US National Institute on Drug Abuse grants R01 DA13336 (Community Vulnerability and Response to IDU-Related HIV project) and 5T32 DA07233 (Behavioral Sciences Training in Drug Abuse Research, MHRA). US Center for Disease Control and Prevention for providing data, including Abu Abdul-Quader, Bob Byers, Patricia Fleming, Stacie Greby, Robert Janssen, Steven Jones, Meade Morgan, Scott Holmberg, and Xen Santas. Michael Ratcliffe, US Census Bureau. Cathie Alderks and Coleen Sanderson, US Office of Applied Studies, SAMSHA. Elizabeth Lambert, National Institutes on Drug Abuse.

Chapter 17
Factors Influencing Drug Use and HIV Risk in Two Nicaraguan Cities

Michele G. Shedlin, Rita Arauz, Pascual Ortells, Mariana Aburto and Danilo Norori

Abstract This chapter presents the results of two related exploratory, qualitative studies on drug use and HIV risk conducted in the cities of Managua and Chinandega, Nicaragua between 2002 and 2005. The objectives of this research were to: identify methods of reaching drug using populations in the country; provide an initial description of the patterns of drug use, emphasizing regional differences; explore the relationship between drug use and HIV transmission; and provide preliminary recommendations for the development of drug use and HIV prevention efforts and for future research directions. The study designs included ethnographic observation and interviews to assist in gaining entry into drug-using communities, in-depth interviews with drug users and traffickers (121) and focus groups (13) with sectors of the population likely to provide different perspectives on the research domains: health professionals working with high-risk behavior groups, female sex workers, gay men, university students, taxi drivers, injection drug users, and family members of drug users. Drug use, the availability of drugs and distribution were reported, and included: ubiquitous drug supplies; the involvement of all social strata; the impact of crack on drug-use patterns; concerns about use by children and youth; well-established local distribution mechanisms; group drug purchase and sharing, and (limited) needle use and equipment sharing. Sexual risks included unprotected sex with partners and sex for drugs and/or drug money. The lack of drug prevention education in the community and schools, and limited treatment resources were also reported. Conclusions highlight the need for public and policy acknowledgement and response regarding drug use, and the link between HIV/AIDS and drugs in the country.

M.G. Shedlin
Leavell Professor of Nursing & Health Sciences, Co-Director, Hispanic Health Disparities Research Center, College of Health Sciences, University of Texas at El Paso, 1101 N. Campbell St., Rm. 716, El Paso, Texas 79902
e-mail: mshedlin@utep.edu

Y.F. Thomas et al. (eds.), *Geography and Drug Addiction*,

Factors Influencing Drug Use and HIV Risk in Two Nicaraguan Cities

Data gleaned from health surveys and HIV studies, as well as media reports, increasingly identify drug use as a growing problem in Nicaragua. Poverty, international drug traffic, and the lack of information, education, and drug treatment add to the potential for drug use being an increasingly important factor in HIV transmission. However, no large-scale studies on drug use have been carried out in the country, nor has any significant government or international funding been directed at HIV/AIDS behavioral research. These qualitative behavioral studies funded by the National Institute on Drug Abuse (NIDA) (Grant No. R03 DA16163, M. Shedlin, PI and an international supplement to P-30 DA011041-06S2) were a direction new to health research in the country. The objectives of the studies were twofold: (1) to identify, document, and describe drug use in the cities of Managua, the capital, and Chinandega, closer to the Honduran and Salvadoran borders, and (2) to support increased awareness of the relationship of drug use and HIV transmission in Nicaragua.

It was clear during the design of this research, however, that any investigation of drug use and HIV/AIDS in the country would need to place these two emerging epidemics in the context of two decades of war, a prolonged economic crisis and a large population increase. These factors all contributed to a marked deterioration of the social indicators of a population of 5.1 million people, 53% of which is under 18 years of age. To explain further, Nicaragua is the third poorest country in the Americas, with a per capita gross national product of $453. Poverty affects 2.3 million persons, 831,000 of whom live in extreme poverty, mainly in the Central and Atlantic regions. The breakdown of income distribution shows that 45% of all income goes to the richest 10% of the population, while only 14% goes to the poorest (United Nations Development Program [UNDP] 2000). Although the unemployment rate is officially estimated at 12%, underemployment may be as high as 14%, and higher in the Atlantic region (El Centro Nicaragüense de Derechos Humanos [CENIDH] 2004).

Nicaragua's main challenge is to overcome inequity and poverty, which affect children and women most severely. One of every three children has some degree of chronic malnutrition and 9% suffer from severe malnutrition. The maternal mortality rate (MMR) of 150 per 100,000 live births is unacceptably high. In the Atlantic region and areas difficult to access, MMR may be twice as high as the national average. Adolescent pregnancies account for one of every four births nationally. Safe water and sanitation coverage continues to be low, particularly in rural areas and those with dispersed populations (United Nations Children's Fund [UNICEF] 2004).

Geopolitical, Social, and Cultural Factors Influencing the Nicaraguan HIV Epidemic

The onset of the HIV epidemic was more recent in Nicaragua than in neighboring countries due to an 8-year-long civil war, an economic embargo imposed by the United States, and a policy of semi-isolation. With a change of government in 1990, Nicaragua opened itself to international contact, including people arriving or returning from countries where HIV is prevalent such as Honduras and the United States (Low et al. 1992). However, while Nicaragua's epidemic began later, it is believed that all the preconditions exist to bring about a serious HIV/AIDS epidemic (MCP 2002). Arauz, Ortells, Morales, Guevara, and Shedlin (1997) examined demographic and social indicators and concluded that conditions existed for an epidemic of "major dimensions" in Nicaragua.

The USAID National HIV Assessment by Sanchez, Shedlin, and Araica (2000) came to the same conclusion and listed the following preconditions:

- Displacement due to civil war and natural disasters.
- Increased prostitution in areas with a high concentration of transient workers.
- Indicators that drug trafficking is influencing local drug use, especially among youth.
- A large gay population that is encouraged to stay hidden.
- Men-who-have-sex-with-men (MSMs) who form the bridge between MSMs and heterosexual communities.
- Migrants and other highly mobile populations.
- Shared borders with countries that have major HIV epidemic, e.g., Honduras.
- A large youth population often lacking the information necessary to avoid sexually transmitted infections (STIs) and HIV.
- Catholic Church and Christian evangelical groups deeply antagonistic to HIV/AIDS prevention education.
- Lack of a condom culture that supports condom acceptance and use.

Several older studies have indicated that the general population lacked some basic knowledge about HIV/AIDS. For example, Low et al. (1992) discussed a 1989 Knowledge, Attitude, and Practical (KAP) study of adults in Managua that revealed that more than 90% knew the major routes of HIV transmission, yet approximately half believed kissing and mosquitoes could spread HIV. Most believed that isolation of HIV+ individuals and control of immigration were ways to control the epidemic (Low et al. 1992). A 1995 study with university students illustrated their lack of knowledge of the difference between HIV and AIDS and ambivalence about individuals with HIV (Cisneros Santamaria 1995). While general knowledge about HIV/AIDS and condoms was good, the actual use of condoms remained low. The Demographic and Health Survey (USAID 1998) found that six of every 10 women knew they could prevent getting HIV by using a condom and one-fourth thought they could prevent infection by having only one partner. These numbers were similar to answers given by the men, 65% and 22%, respectively. Yet, only 16% of sexually active men reported using condoms occasionally (Ministerio de Salud [MINSA] 2000). In 2000, one of every 10 women had heard of AIDS and knew of ways to prevent transmission; however, only 2% of women said they were currently using protection (MINSA 2002).

The Ministry of Health and AIDS Control and Prevention conducted HIV prevalence surveys in 1996 among female sex workers (FSWs) and MSMs. They found an HIV prevalence rate of 1–2% among FSWs and MSMs (MINSA 2000) suggesting a prevalence rate in the general population of $<1\%$, and placing Nicaragua at the level of a nascent epidemic using World Bank criteria. However, there were many concerns about the accuracy of these statistics. Arauz, Ortells, Morales, Guevara, and Shedlin (1997), for example, discussed the reasons why seroprevalence data did not "correspond to the reality" of HIV prevalence in the country, including lack of testing services and a focus on high-risk behavior groups that omitted the general heterosexual population (p. 123). A more recent USAID HIV/AIDS Assessment also identified serious constraints on the utility of existing prevalence data (Sanchez et al. 2000).

The first case of AIDS in Nicaragua was diagnosed in 1987. In 2002, the Central American Multicenter HIV/STI Study found an HIV prevalence rate of 0.3% in sex workers in Managua, Corinto and Bluefields ($n = 463$) and an HIV prevalence of 9.3% in MSM in Managua ($n = 199$). It was also reported that among MSMs who had anal sex with occasional partners in the last 30 days, only 61% of the 62 MSMs responded that they had used protection (MINSA 2003).

As of March 2006, Nicaragua reported 2,116 registered HIV infections (MINSA 2006). Ninety-four percent of the HIV infections in Nicaragua are transmitted through heterosexual contact. Individuals who consider themselves homosexual and MSMs, contributed 26% of the total 94% of HIV infections transmitted through heterosexual contact (MINSA 2006). HIV transmission through sharing syringes for injecting drugs contributed to 3% of the HIV infections, perinatal transmission of infection accounted for 2.8%, and blood transfusions contributed to only 0.2% of the infections. The epidemic remains primarily in the main urban and economic centers (MINSA 2004). The United Nations Joint Programme on HIV/AIDS (UNAIDS) estimates that there are 6,400 persons living with HIV in Nicaragua, 0.2% of the adult population (UNAIDS 2004). A more realistic estimate is believed to be between 24,160 and 36,240 based on registered deaths and a 60% under-registration estimated by the Ministry of Health (MINSA 2004). And as is true for other countries in the region, HIV infection among housewives is rapidly increasing. The US Agency for International Development (USAID) reports that the risk of HIV infection among housewives in Chinandega is twice as that among commercial sex workers in the area (USAID 2004).

The National Program for the Prevention and Control of HIV/AIDS and Sexually Transmitted Infections in Nicaragua was implemented in 1998. Currently, antiretroviral treatment in the country is financed exclusively by the Global Fund. Nicaragua has been granted approximately US$ 10.1 million in funding to be spent over 5 years through the Global Fund to fight AIDS (Global Fund 2006). The only hospital administering antiretroviral medications is in the capital, Managua; however, MINSA is planning to decentralize distribution. Nicaragua is one of the few Latin American countries in which antiretroviral treatment is not available to the majority of the population which needs it. This lack of access to antiretroviral treatment is compounded by stigma and marginalization affecting HIV positive individuals in Nicaragua (USAID 2004). In 2005, the World Health Organization estimated that 1,000 individuals in Nicaragua needed antiretroviral treatment (World Health Organization 2005).

The international community is also assisting Nicaragua to develop effective prevention strategies. The Center for Communication Programs at the Johns Hopkins Bloomberg School of Public Health and Fundación Nimehuatzin, for example, are collaborating on a multisectoral community mobilization project. This model was validated in Chinandega and is now being implemented in nine additional locations with funding from Luxemburg. Health communication initiatives are supported in school, the local health systems, and local government (Center for Communication Programs 2005).

Geopolitical Factors Influencing Drug Use in Nicaragua

The problem of drug use in Nicaragua began primarily as one of geography because Nicaragua's Atlantic Coast lies on a transshipment route between drug producing and consuming countries. When boats laden with drugs are about to be apprehended, they jettison their illicit cargo. Fishermen find the packages floating on the water, and children retrieve them from the beaches. Of course, drugs enter across land borders as well. The returns from selling drugs are far greater than can be earned from fishing or agriculture. "People get hooked, first economically, then physically" (Hamilton 2004).

The Nicaraguan legislature passed a comprehensive Drug Law (#177: The Narcotic Drugs, Psychotropic Substances and Other Controlled Substances Act) in May, 1994, but drug trafficking and drug consumption continued to increase. "Every day it is more evident that the inclusion of the country in the international narcotraffic route is accompanied by an increase in the local use of drugs" (Van Wichen, Largaespada, Ormel, and Montedeoca 1995). A police report cited in *La Nación* in 1999 already noted that the presence and use of drugs in Nicaragua was rapidly growing (La Nación 1999). The National Police estimated more than 12% of crimes in Nicaragua were drug-related and were on the increase, a 202% increase since 1990. As the article quoted:

> "The tentacles of drug trafficking and its related crimes are drawing into their decadent trap thousands of young people, with repercussions of violence and thefts, from which not even the families of the addicts can escape, awakening awareness of the seriousness of this phenomenon in more and more layers of society" (p. 15).

Anecdotal accounts in the media continued to appear, especially on trafficking and the consequences of drug use by youth. However, few studies and little official information are available on drug *use*. A UNICEF study in 1996 found the use of drugs was even then endemic, the majority of the respondents reporting at least 2 years of drug use. Glue sniffing was reported for street children and crack was another street alternative. The study noted, in addition, that the level of drug consumption in the marginal neighborhoods had grown to "worrisome levels" with children initiating drug use at home (Santamaria 1997). A study of Managuan youth, carried out also in 1996 by the Centro de Investigaciones de la Comunicación reported that 92% believed more drugs were being consumed than ever before; that marijuana and cocaine were the drugs most frequently consumed; and almost 19% reported using drugs even then, almost a decade ago (Chamarro1996). In a study carried out by Fundación Nimehuatzin, sex workers revealed personal drug use with clients, and reported that "most" clients used drugs. They noted that drug use appeared to increase dramatically after 1990, and consisted mainly of marijuana, cocaine, and crack cocaine. The use of injected heroin and cocaine was said to be beginning; however, these women were able to describe injection drug use with great detail. Sex workers on the Atlantic Coast emphasized the impact of crack on women. Gay men in the study were reluctant to admit use, but it was clear that "normal" use included forms of both cocaine and marijuana. MSMs also reported injection drug use (Arauz, Ortells, Morales, Guevara, and Shedlin 1997). The National Survey of Adolescents and Youth identified that 88% of adolescents and youth considered drugs a serious issue (Agudelo 1999).

Description of the Study

The research was carried out between 2002 and 2005 in Managua, the capital city, and Chinandega by an interdisciplinary team organized by Fundación Nimehuatzin, a Nicaraguan NGO with long experience in reaching high-risk behavior populations in the country. Data obtained include: the sociodemographic characteristics of drug users; the context of drug use and HIV risk; norms regarding drug use, sexuality and sexual behavior (e.g., partner relationships, transactional sex, MSMs, high-risk practices); differences in drug and sex-related protective and HIV-risk behaviors; perceived changes in drug use patterns over time and factors influencing these changes; and knowledge and perception of risk of HIV.

Among the objectives of the research were: (1) identify methods of reaching drug-using populations in Nicaragua; (2) provide an initial description of regional differences in patterns of drug use in the country; (3) explore how drug use influences HIV transmission; and (4) develop institutional capacity to conduct research, especially in the NGO sector, in drug and HIV behavioral research.

These objectives reflected the need for data on the relationship between drug use and HIV which are almost entirely missing in Nicaraguan HIV research, prevention planning, and policy. They also reflected a response to the need for capacity building in substance abuse research, not only at the government levels, but also at the community level by agencies most able to reach drug users to obtain data and to implement prevention interventions.

Study Locations

Data collection was carried out in Managua, the country capital and the largest city, and Chinandega, a city on the northwest border, contiguous with Honduras. After the destruction of Managua in 1972 by a major earthquake, rebuilding has been disorganized and only recently has the center of the city begun reconstruction and re-population. The city currently has more than a million and a half inhabitants who are predominantly Mestizo. Because of migration from the rural areas and the economic crisis, unemployment is high (UNDP 2000). Male and female prostitution, as well as the number of homeless women and children on the street is increasing, as is the problem of youth gangs involved in crime and drug dealing. There are 10 public hospitals in Managua and a similar number of private clinics. There are many identified barriers to HIV testing, among which is the limited availability of counseling and testing services (Fundación Nimehuatzin 2004).

Chinandega has a population of 122,000 in the urban center (Population Figures 2006). It is located 137 km from Managua and has fluid communication with the Capital and with towns and cities in Honduras. The major health problems include malaria vivax, STIs, and dengue. There are two hospitals, one health center, 10 health posts, and six private clinics in the area. Recent reports provided by the Health District (SILAIS), cite STIs and HIV as a grave problem in the area (for HIV/AIDS through August 2002, 33.8/100,000, the highest in the country). According to these data, housewives and women workers are the most affected. In December 2000, Chinandega represented 14% of the HIV infections diagnosed throughout Nicaragua, and as of June 2004 the number of diagnosed HIV infections in Chinandega made up 17% of the recorded national total. Chinandega also has high unemployment rates and is also higher when compared with the unemployment rate of Managua.

Methods

Gaining Entry into Drug-Using Communities

In consultation with NGOs and government agencies, the research team identified key informants and gatekeepers to assist in locating areas where drug users could be contacted and the most appropriate, sensitive, and confidential methods for gaining their participation. Because the implementing agency, Fundación Nimehuatzin, had been involved in HIV prevention in these areas, they had already identified individuals and agencies with whom they could work and from whom they obtained guidance for this new research focus. Recruitment sites were street locations, bars and community agencies and treatment programs. The local worker's familiarity with the community also assisted in gaining access to the population as well as providing additional assurances of confidentiality, since Nimehuatzin was well known and trusted in high-risk behavior communities in Managua and throughout the country.

Multi-Method Ethnography

The study involved an ethnographic approach and methods. A key element, especially in studies of high-risk behavior groups and "hidden populations," is sufficient access to understand behavior by observation and interviews (Marshall, Singer, and Clatts 1999). The ethnographic component had several aspects, including identifying locations for recruitment, collecting data, and using multiple data sets (incorporating triangulation of sources and types of data), to examine validity and reliability of findings. Along with interviewing, community observation was undertaken to identify and describe drug use patterns and sexual risk behaviors. In addition to documenting the existence of drug-using communities and describing them, the study was able to identify a range of experiences, attitudes, and beliefs relating to drug use and HIV risk. Qualitative data obtained include: detailed descriptions of situations, events, people, interactions and observed behaviors, and direct quotes from individuals about their experiences, attitudes, and beliefs.

Interviews

Three types of interviews were conducted. Individual semi-structured interviews were carried out with key knowledgeable sources (30) and with drug users/traffickers (121). In addition, focus group sessions (13) were carried out with sectors of the population likely to provide different perspectives on the research domains: health professionals working with high-risk behavior groups, FSWs, gay men, university students, taxi drivers, injection drug users, and family members of drug users.

Analysis of Interviews, Focus Groups, and Observational Data

Data in the form of audiotapes, responses to the semi-structured questionnaires, interview notes, and transcripts of the focus group sessions were obtained. The assessment of the quality and completeness of the data in relation to the key research questions was an on-going process. Because similar issues are researched using different methods and because multiple data sources were involved, multiple data sets were available on the research domains and key issues. Analysis involved the use of Atlas.ti software and SPSS. The classification of evidence from all data sets was organized to identify salient patterns and relationships.

Results

The results discussed represent the reported experience, beliefs, and perceptions of the drug users in individual interviews, the key informants, the focus groups with the seven groups listed above, and observations by the field staff.

Characteristics and Experience of the Sample

Of the 121 drug users and four others who professed being only "traffickers" interviewed individually, 98 (81%) drug users were male, the majority (75%) between

18 and 35 years of age. The 23 women interviewed (19%) were distributed among all age groups within the 18–55 sample. Most (73%) were born in Managua with the remainder born in other cities in Nicaragua, and two in Mexico City. Thirteen percent of the participants were either illiterate or had incomplete primary education; many (42%) had attended, but not completed, high school. Nineteen percent of the sample had attended a university, but not obtained a diploma. Five participants had completed a university degree and four had completed a technical degree. Eighty-three percent of the participants spoke only Spanish and 12% spoke English. The occupations listed were varied, ranging from housewife to economist. Student status was given by 15% and was the most frequently cited occupation. Eight participants stated that they did not have any occupation. More than one-third (34%) stated they did not have any religion; 44% reported being Catholic and 17% "Evangélico."

Partner Status and Children

More than half (66%) were single, 26% were married or in a free union, and 7% were divorced. Only one participant was a widower. More than half (52%) were currently in a sexual union. Forty-three percent did not have any children and 22.3% had only one child. Two older men (49 years and 52 years) and one older woman (54 years) reported 13, 14, and18 children, respectively. (The 23 women reported a total of 120 pregnancies between them.) Of the men and women who had children, only 37% reported having a good or regular relationship with them.

Residence and Travel

All but three participants lived in an urban area, and 79% had lived in the city of the study (Managua or Chinandega) for more than 10 years. Most lived as part of a family unit (26%), with friends/relatives (54%), or with a sexual partner (7%), Four men lived alone (3%), 3% of the participants were homeless and eight persons were incarcerated at the time of interview. Participants reported travel outside the country, primarily Central America and to the United States. Reasons for travel included work, visiting family, tourism, and drug treatment. About one quarter (34%) had lived outside Nicaragua at some time in their lives, and among those 9% had lived in the United States or Canada. Almost 77% had some relative or friends living abroad, and 61% had friends or relatives in the United States or Canada. Thirty-eight percent maintained communication with them regularly.

Incarceration Histories

Of the 121 individuals who responded to the question concerning incarceration, 77 (64%) had been in jail or prison at least once, many of these individuals having multiple incarcerations in Nicaragua and outside the country (the United States, Switzerland, Panama). They listed drug transport (being a "mule"), selling, robbery, gang activity, sex for drugs, prostitution, and arson (the home of a member of another gang) as reasons for their arrests.

Initiation of Drug Use

Most users (88%) initiated their drug using careers before they were 21 years of age, and 56% before the age of 15 years. Family and personal problems and curiosity were the main reasons given for initiating drug use, although a few users mentioned peer pressure/acceptance. Most mentioned friends (84%) in the neighborhood, at school and in the street as those with whom they first used a drug. Only seven individuals (6%) mentioned that they had started using a drug alone. Because of the age spread of the sample and differences in age at first use, the data capture initiation of drug use over an almost four decade span, between 1967 and 2001. Almost two-thirds (61%) reported marijuana as the first drug used. Alcohol was mentioned by only 15% of the sample, cocaine was mentioned by 12% as the first drugs used. Four and 2% respectively of the sample mentioned crack and glue as the first drug used. Collective purchase appeared common, 58% of users responding that they had done so ("echarle la vaca").

Current Drug Use

One hundred and twelve of the 121 study participants mentioned crack as their main drug (93%), some saying that they used it alone, others stating that they combined crack with marijuana, snorted or injected cocaine and alcohol. The majority (86%) also used marijuana regularly, either alone or in combination with crack, cocaine, alcohol, and/or "pills" (pharmaceuticals). Cocaine use was cited by 81%, either alone or in combination with other drugs. Only one person mentioned using cocaine alone. Eleven individuals (16%) had ever-injected an illicit drug; only one man reported current injection (however, a focus group was held with six additional injectors). Private locations for drug use listed by users included homes and hotels. Public places mentioned were: street; parks; discos; bars; "expendios" (places where drugs are sold).

Perceptions of Drug-Related Health Problems

Users described a wide range of health problems that they attributed to their drug use. They cited kidney problems, hepatitis, anorexia, weight loss, nervous system problems, sinusitis, high blood pressure, hallucinations, stomach pain, respiratory problems, cardiac alterations, anxiety attacks, emphysema, hemorrhoids, pulmonary problems, and psychiatric hospitalizations (Managua). The question about drug-related health issues also elicited personal information about drug overdoses. Injuries attributed to drug-related violence were also listed.

Sexual Behavior

Of the one hundred respondents who answered the question about perceived frequency of oral sex 67% stated that they do have oral sex, and 53% of the 95 respondents who answered the question regarding the perceived frequency of anal sex, respectively, indicated that they do have anal sex. Twenty-eight percent of the men interviewed said that they had had sex with other men. Fifty-six percent said they had not used condoms during their last sexual encounter. Only 17 individuals (14%) stated that they always used condoms. Of those responding to the question of engaging in sex for drugs or money (94), 35% said they had done so. Fifty-seven percent of the participants reported that they had sex with individuals in their drug networks. Seventy percent believed they could contract an STI, few stating that their risk was related to drug use. Sixty-seven percent believed that they were vulnerable to HIV infection. Fifty percent of the participants had a friend who died of AIDS. Most of the sample (88%) believed they knew about HIV/AIDS having obtained information from health personnel, NGOs, TV, and friends. Seventy-five percent of the participants agreed that there was a relationship between drug use and HIV infection.

Salient Themes and Issues

Popular Attitudes Relating to Drug Use

Focus group data and individual interviews elicited a historical perspective on drug use in the country, especially the perceived control of trafficking and use (primarily marijuana) during the years of the Sandinista government. After the 1990s, traffic, supply, and use of drugs began to increase and expand to new populations. There was also an increase of new drug forms and patterns of use. Key informants and users all agreed that drugs are now sold and used everywhere in the country, along with a greater salience of the pervasiveness of the problem by the community.

The breakdown of societal values was seen by many respondents as both cause and effect of this incremental leap in drug use during the last decade. As one of the health professionals observed sardonically, "Unfortunately, Nicaragua does not have to envy the other Central American countries (any longer)!" The increase in drug trafficking, corruption of the police, the lack of attention to family and children, and the lack of drug prevention education in the schools were cited as critical factors in this increase in supply, demand, and drug-related violence. As drug use spreads, they said, "there is more evidence of personal and family harm, more fear in the adults that children are being destroyed." University students also saw drug use in the country as a "grave problem."

Family members of drug users discussed stigma and the shame that keeps individuals and families from seeking help. As one focus group participant observed, "the most serious problem is shame, the pain, the suffering, the sadness, and that is what makes us isolate and not share the problem. . .or seek professional help. . .". They also shared their experience with the extreme attitudes of family and friends toward their using drug children and spouses. As one mother said, "the discrimination I've felt toward my daughter by my brothers, my family, my sisters". . .they say "let her die in the street. Why do you go around like an idiot looking for her? Leave her, let her die if that's the life she wants."

This sense of stigma was reflected in the comments of the drug users who said that they are generally seen as "garbage," people "lacking character" who cannot face their problems. People reject us," they said, "those of us who use drugs. . . . we feel badly, ashamed." Some, however, explained that although drug use is still seen as a "moral deficiency," various organizations are working to educate the public that addiction is an illness. Still others stated that it is becoming more common to view users as "normal" because "users go to church, study, play sports and it is common to use drugs. . .this is a change." "There is more tolerance of users," observed another, "it's normal that someone takes out a tube to smoke crack or cocaine. . .that's a frequent thing."

Treatment professionals, on the other hand, were not as optimistic about the attitudes of the health sector toward users. One of the focus group participants stated that the health sector personnel feel that. . . "the best way for those people to be cured is for them to be killed or kill themselves." Even the police, they said, saw their patients (addicts in recovery) as "never getting better, better that they die." "The concept of addiction as sickness, aside from not existing," added one of the providers, "is not accepted (even) when it is explained."

Groups Perceived as Most Affected

Most users interviewed believed that all levels of society and age groups were affected by drugs because "drugs don't discriminate." The professionals in their focus group discussion agreed that similar patterns existed among youth in both public and private schools in the country. All groups, in fact, viewed children and youth as those most vulnerable and most affected by drugs. Taxi drivers described high school and university students as "the worst" (affected), observing the truancy and drug use first hand as they drove around the city. Street children and glue-sniffing youth were described by all sectors interviewed, and are one of the most obvious sights in the main markets of Managua and elsewhere.

Many respondents perceived that "lower classes" were more likely to use crack "because of more violence in their lives," but others saw crack as the drug of choice for all classes. Crack-using sex workers stated that they smoke with each other and that the drug has "no end" without professional help. Use of crack in their community was said to be extensive, e.g. 20–30 women work one of the main tourist hotels in Managua and all use crack. "When I try it," said one of the women, "I want more and more. I am capable, if you forgive me, to stop a client because I don't want to go to the room with him other than to rob him (and continue smoking)". They agreed that many women now do sex work to buy drugs, unlike in the past. "The most beautiful thing is to live without drugs" lamented one of the sex workers.

While gay men were not identified by the users as a drug-involved community, the men themselves said that some use of marijuana and cocaine existed among them, especially on weekends and holidays/parties. Alcohol, however, was said to be the main drug used in the gay community. Alcohol is a more "social" drug, they said. Vanity and the "deterioration of the body" from drug use, they agreed, is a major deterrent to use for them. The upper class was said to be able to afford such drugs as heroin, to buy larger amounts (of cocaine) at a time, and to be more discreet in their use.

Reasons for Initiating Drug Use

The professionals' focus group saw the motivation and initiation of drug use in "emotional reasons," especially "not to feel" hunger, pain, suffering, and to have even a temporary experience of well-being ("bienestar"). Users' answers were in agreement, citing emotional pain caused by such factors as parental abandonment and death, family conflict and disintegration, sexual abuse and hopelessness. However, they also listed: economic pressures (poverty, lack of employment, and opportunities); peer pressure and rebellion ("bad company"); "addictive tendencies" and curiosity/recreation. University students emphasized the lack of opportunities and a vision of the future as major factors in use among youth, as well as the "euphoria of the moment." Family members of users mentioned many of these factors, and as one parent explained, "I think also that drug use is a product of fear, the fear of life, for example fear of what is ahead, so it is a way to avoid what scares you in the future."

For the users, family drug use, including parents, siblings, stepfathers, uncles and cousins was also discussed as influencing their use. "All types of addiction exist in my family," explained one user, "alcoholic uncles, a father addicted to work, my grandfather addicted to gambling and women, and my mother co-dependent."

Perception of Legal and Illegal Drugs

Most users stated that they believed legal and illicit drugs to be equally destructive, including alcohol and tobacco, and that legal drugs led to other drug use. Some discussed the lack of awareness that pharmaceuticals were addictive and denial by those addicted to them, while others stated that legal drugs were "less addictive."

Others stated that "legal drugs are recommended by doctors and cure you, the illegal ones destroy you." The distinction between legal and illegal carried over to beliefs about recovery as well, some respondents saying that "you can leave legal drugs by yourself, the illegal ones, no (because) they are stronger and you need a lot of help."

Local Availability

Focus groups confirmed user perceptions that drug supplies and sales (as well as use) were increasingly less hidden in the community. Taxi drivers (identified by sex workers and university students as people who connect drug sources and buyers as well as being "delivery" resources to homes and motels), described well-known locations where drugs could be obtained all over the city 24 hours a day. They emphasized food marts attached to gas stations as ubiquitous sources for drugs. One taxi driver quoted a recent press report stating that 572 *expendios* (small community locations for drug sales) in Managua had been identified by the police, and noted that he knew of "two or three *expendios* in every neighborhood." Most bars, nightclubs, and casinos in the city were also said to be sources of drugs and places where drugs were used. The gay men stated that although connections could be made in their discos, drugs were not generally available there. They did know of other discos where drugs were sold, however.

Observation of Changes in Drugs Used and Patterns of Use

All of the sectors interviewed had observed continual changes in drug-use patterns. Many noted the greatest change with the introduction of crack in the 1990s. Ecstasy and Ketamine were said to be more recent introductions, the former limited in use by its cost. They noted other changes as well, including more diversity in the places where drugs are used, more women using drugs and at a younger age, the mixing of drugs especially pharmaceuticals and alcohol, and the combination of crack and alcohol because "crack destroys your nervous system and liquor balances you." Cocaine and marijuana together, called "maduro con queso" (a national dish of plantains and cheese) was said to be a popular mixture.

The treatment professionals further described their observations of the closing gender-gap in drug use among students, with more female students now using. They also described an increase in poly-drug use, and stated that crack is now the number one substance in their centers, followed by alcohol, marijuana, and, lastly, cocaine. Although most respondents had not used or seen heroin, including the drug-knowledgeable taxi drivers, heroin was said to be coming into the country and being injected, mostly by the upper class. Some users interviewed, however, had observed heroin use in middle- and lower-class users along with a few reporting their own experience. The sex workers also reported observing clients injecting heroin, but agreed with the professionals that heroin users were still primarily foreigners.

Foreign Influence on Drug Use

The users and key informants saw foreign influence on national drug use from a number of perspectives. Culturally, many agreed that Nicaraguans were influenced by other countries, imitating such things as fashions, holidays, gangs, and drug use. They also noted the influence of foreign drug traffickers and money-launderers (said to be Colombians and Costa Ricans) as well as drug-using tourists from Spain, South America, and the United States. Others, however, stated that while the sequelae of the international drug economy and its players affect Nicaraguan use in various ways, the country has graduated to its own domestic drug epidemic. As one user stated, "Yes, there has been a lot of influence (of foreigners), but now you would be surprised to see how many people use drugs." Another explained, "foreigners do the trafficking, but the use is national." Nicaraguans returning home with new drug habits, including students, and those deported for drug-related crimes were said to influence drug use as well.

Patterns of Drug Use

Users listed alcohol, marijuana, cocaine, crack, pharmaceuticals, and combinations of these drugs as common and currently used by them. Use was reported to be mainly with friends, and less frequently with partners or family members. Adolescents were said to use marijuana, crack, and glue, with marijuana and alcohol together as common. Men were identified as more likely to include alcohol in the regular mix of drugs used. Heroin was said to be increasing in importance with more heroin coming into the country, but still too expensive for most users. Most snorted or smoked their cocaine. Injection was still largely associated with heroin use although some injecting of cocaine was mentioned, especially when discussing drug overdose. University students said that use in their community was mainly marijuana (widespread and frequent use) and alcohol.

The majority of the user/trafficker sample knew or believed drugs were being injected in Managua but did not know about other cities in the country. Of this sample, more than half actually knew someone who injected drugs. They reported knowing of injection use of cocaine, heroin, Ketamine, and pharmaceuticals (especially Demerol). Needles were said to be obtained mainly in pharmacies, and many in the sample had seen injection equipment being shared. Four of the men stated that they had themselves shared injection equipment. Ecstasy was also mentioned, assessed as still too expensive for common use and largely limited to specific night clubs.

Beliefs and Attitudes About Drug Injection

The professionals explained that the negative attitude toward injection and the limited needle drug use in the country had a cultural basis. Injection, they said, was associated with fear and threats instilled in children. In addition, they noted that injection was associated with hospitals and doctors in white coats. On the other hand, they saw illicit drugs historically as coming in not as injected, but as smoked or swallowed. "We are still a bit far from injecting," said one provider, "and if injection has arrived, it has been by external influence."

Most of the comments about needle drug use in the individual user interviews were extremely negative. Most expressed fear of any needles, of injecting anything. They also cited physical risks such as harm to the "blood and brain (brain damage)," inflammation of the injection site, as well as the risk of AIDS. Beliefs about the effects and meaning of injected drugs also appeared to deter injection, such as: injected drugs are stronger; are more addictive; the dose can't be calibrated; a risk of hepatitis; causes perverse behavior; creates more "dependency"/"addiction" means you have "fallen too low"; "takes you to death." Sex workers describing heroin and cocaine-injecting clients said clients shoot up in the hotel room but bring them crack because the clients know they would not inject. The few positive comments included: "Injected drugs calm my problems; it is a rapid way to solve my problems," "...other drugs don't satisfy me anymore;" "...it's the best (thing)... divine."

Knowledge/Experience of Drug Overdose and/or Drug-Related Death

A disturbing finding of this study was that more than half of the user/trafficker sample (55%) knew of someone who had died due to drug-related causes. Most described seeing cardiac arrests after cocaine injecting, one man stating that he knew of eight such deaths, and another citing six friends dead due to a combination of heroin and cocaine and one who died while using crack. Others also said they knew of crack overdoses, one man said to have died from crack because he was old and "that drug is not for old people." Their interviews revealed, in addition, witnessing friends (or multiple people) overdosing but surviving, including their own personal experiences. These were attributed to cocaine injection, heroin injection, and crack. Some individuals who said they had not seen or experienced overdoses knew of friends and neighbors who had died of drug-related "deterioration," accidental deaths due to drug effects or drug-related violence.

Treatment Resources in the Country

A number of religious, 12-step and private NGO programs were cited by the users and the professionals. All agreed that resources were few, many of the programs rudimentary and without trained staff, and that treatment slots were limited. As one parent of a drug-using son explained, "what is most worrisome is that there are no specialized centers in the country, that is to say, specialized therapists that can really provide follow-up or specialized attention to this. That is really worrisome because there are thousands of young people that have this problem."

Less than half the users interviewed had ever been in any type of rehab program anywhere. Providers explained that stigma and societal attitudes toward women addicts and patterns of use were very different from those of men, necessitating different treatment resources and strategies not available in the country. This issue was especially salient in the focus group session of the sex workers who did not perceive availability of any treatment resources for themselves.

Drug Use as Risk for HIV/AIDS

All sectors interviewed were well informed about the risk of HIV from needles and unprotected sex. Many of the users reported knowing someone who had died of AIDS. Although few of these deaths were thought to be related to drug use, the majority of users still perceived themselves at risk for HIV and STIs because of unprotected sex, high-risk behavior partners, and transactional sex. Nevertheless, more than half had not used protection with their last sexual encounter, and a number of them were clear that drug use influenced their ability to use protection. Sex workers also stated that although they always used condoms with clients when not using drugs, they estimated that drug use reduced protection to about 60 percent of the time. University students also saw the relationship between drugs and HIV as sexual risk. One student gave the example of young girls on the street prostituting for drug money as the "most direct form here in Nicaragua because it is rare that anyone injects..."

Summary and recommendations

Many of the preconditions for an HIV epidemic identified earlier in this discussion combine with additional contextual factors to serve as preconditions for a less well-recognized drug epidemic: poverty, displacement due to civil war and natural disasters, increased prostitution, indicators that drug trafficking is influencing local drug use, especially among youth, a large stigmatized gay population, migrants, tourists, and other highly mobile populations, shared borders with countries also involved in the regional drug economy, and a large young population lacking in education, opportunities and future orientation. Included in this long list are also an abundance of drugs easily available, well-established local distribution mechanisms, (perceived) involvement of the authorities in local supply and protection, the lack of drug prevention education in the community and schools, and limited treatment resources.

The contribution of this exploratory research is in the identification of a range of cultural and contextual factors which are influencing patterns of drug use differentially in the country and thus contributing to and shaping Nicaragua's emerging and related HIV and drug use epidemics. The personal characteristics of the users, their beliefs and behaviors, provide historical perspective as well as current information on drug use among low income men and women. Their reporting of ubiquitous drug supplies along with behaviors such as unprotected sex with partners, sex for drugs and/or drug money, cultural acceptability of anal sex, group drug purchase and sharing, and (limited) needle use and equipment sharing, all illustrate factors affecting HIV transmission. The fact that so many had seen drug overdoses and drug and AIDS deaths provides a glimpse into a community not yet documented or publicly recognized.

Focus group discussions with other sectors of the community provided different perspectives on drug use in the country and the capital city, these perceptions and experiences concordant with the user interviews. Health and treatment professionals highlighted geopolitical, social, and cultural issues fostering supply and use. Gay men and sex workers shared their own realities and concerns, including increasing drug involvement; taxi drivers, as self-identified "lay psychologists"/witnesses/local drug supply experts, shared their insights; and families of users vented their feelings of guilt, frustration and anger at the societal norms and conditions (and lack of resources) that have placed and keep their lives in disequilibrium.

The study also identified important differences in drug use and HIV risk between two geographic areas of the country, where history, geography, economics, and culture combine to create different risk environments. In the capital city, Managua, respondents reported greater supply, use, and more varied ways of using crack than in Chinandega. Not surprisingly, there were fewer locations where drugs were sold, and they were less stable and more clandestine in Chinandega, thus influencing supply. The study also identified greater stigma toward drug users, better information about HIV/AIDS and more reported willingness to use condoms in Chinandega. What emerged as most salient in both cities, however, was the unanimous concern for children and youth, and the perception that all strata of Nicaraguan society were losing them to drugs and addiction.

Given these conditions, and the eloquent concerns expressed about the growing use of drugs by different sectors of the community and by the users themselves, it is interesting that acknowledgement of drug *use* and associated health risks appear missing from media attention and public acknowledgment, both nationally and internationally. Concern about drug use has also been missing from the development and implementation of policies and programs at the national and international levels based, apparently on the reputation of the country as merely a bridge for drugs and the erroneous belief that the population remains abstinent.

While this study may not be generalizable to all of Managua and Chinandega, nor to the country as a whole because of the small, qualitative sample, it does, nevertheless, begin to address the urgent need for information regarding drug use and HIV risk. Furthermore, the study has identified regional differences in factors affecting these risks. Based upon these findings, we suggest the following recommendations for national and international policy development and program planning:

- Identify and acknowledge the problem of drug use and addiction.
- Separate trafficking and consumption as policy issues.
- Clearly establish the link between HIV/AIDS and drug use.
- Educate health professionals and policy makers about drug use and addiction.

- Re-enforce traditional values which work to support prevention and treatment.
- Dedicate more resources to the provision of treatment, especially gender appropriate services.
- Respond quickly to the growing drug crisis in youth.
- Develop and implement education and prevention activities in all sectors of the community, not only among the most vulnerable and involved.
- Continue research which will identify the dynamic patterns of drug use and HIV risk in rural and urban areas and among different populations.

Chapter 18
Drug Use and HIV/AIDS: Risk Environments in Post-Soviet Russia

Dominique Moran

Abstract This chapter explores the links between drug use and HIV/AIDS in post-socialist Russia, investigating the connection between drug use, especially Injecting Drug Use (IDU), and HIV/AIDS (Human Immunodeficiency Virus and Acquired Immunodeficiency Syndrome). It considers drug use both in terms of IDU as a direct means of transmission for the HIV virus and non-IDU (including alcohol and marijuana), which has been observed to contribute to risk behaviors, which can lead to transmission of the HIV virus. Russia and other post-socialist states are unusual in the context of the global HIV/AIDS pandemic in the relative importance of IDU as a means of transmission of the virus within their territories, and also in their stance toward harm reduction programs such as needle exchange. The chapter draws upon a wide range of literatures, including sociological literatures dealing with substance abuse in general and in Russia in particular; geographical and area study literatures with a focus on the Russian Federation; research publications on HIV/AIDS in Russia from the epidemiological and public health literatures, as well as policy-oriented and practitioner-focused materials produced by non-government organizations, international donors (e.g., UNAIDS) and policy research institutions.

Introduction

This chapter explores the links between drug use and HIV/AIDS in post-socialist Russia, investigating the connection between drug use, especially Injecting Drug Use (IDU), and HIV/AIDS (Human Immunodeficiency Virus and Acquired Immunodeficiency Syndrome). The chapter considers drug use as a direct means of transmission for the HIV virus, and drug use which has been observed to contribute to risk behaviors, which can lead to transmission. It also considers progression of the virus from "high-risk" populations (including injecting drug users and commercial sex workers) to the mainstream population, and the role of IDU in facilitating this

D. Moran
University of Birmingham, Birmingham, UK

Y.F. Thomas et al. (eds.), *Geography and Drug Addiction*,

"crossover." Drug use is considered in the context of post-Soviet transition; observers have noted societal change in Russia post-1991, with a perceived "relaxing" of social norms, increase in illegal activity, and increase in use of drugs (including alcohol, marijuana and also narcotics) as response to the stresses experienced in the last 15 years.

In considering drug use and the geography of HIV/AIDS in Russia together, this chapter is informed by the work of Rhodes, Singer, Bourgois, Friedman, and Strathdee (2005), who posit a "risk environment" for HIV infection connected with IDU, which is defined as "the space – whether social or physical – in which a variety of factors exogenous to the individual interact to increase the chances of HIV transmission" (p. 1027). This risk environment may be considered to operate at a number of different scales; ranging from the micro-level interpersonal interactions between drug users, through the meso-level of local environmental influences such as the nature of local policing and the location of needle exchange facilities, to the macro level of structural factors such as laws, policies, economic and social conditions, and wider cultural beliefs. It is characterized by the influence of factors such as population movement, neighborhood disadvantage, social norms and networks, law enforcement and policing, and crucially, the interplay both between these *types* of risk, and the *level* of the risk environment (micro/meso/macro).

Drug Use in Russia

For the purposes of this chapter, it is useful to clarify exactly what is meant by the term "drugs". Clearly at its broadest, this term means any biological substance that is not taken for dietary needs. More narrowly, the focus here is on "recreational" (as opposed to medicinal) drugs, which again at a broad level could include alcohol, nicotine, cannabis, and a range of "harder" substances such as methamphetamine, cocaine and heroin, as well as, in the Russian context, a range of other locally produced opiate and amphetamine substances such as *"chornaya", "khimiya", "vint"* and so on. However, given the focus of this chapter, injectable drugs are of primary concern, with other substances such as alcohol and cannabis having secondary importance. For the sake of clarity, therefore, "drugs" will be taken to mean injectable drugs, unless specified otherwise. For the most part, the injectable drug in question is heroin. Heroin came on to the Russian market relatively recently, becoming readily available in cities only in the second half of the 1990s, when it effectively replaced less powerful homemade opiate solutions, anesthetics, and medical preparations which had previously been injected. Heroin has rapidly become the "drug of choice" among drug users (Abdala, Grund, Tolstov, Kozlov and Heimer 2006; Pilkington 2006).

Drug use in Russia has increased considerably since the collapse of the Soviet Union. During the late Soviet period, attention was focussed heavily upon alcohol as the drug whose abuse caused health and social problems. The Soviet government admitted only in the 1980s that opiates and drugs other than alcohol existed within its territory; under Gorbachev's policy of *glasnost* (openness), it finally became acceptable for the Soviet medical literature to acknowledge the increasing problems caused by abuse of narcotics (Conroy 1990). As Platt et al. (2004) note, it is widely believed that since the collapse of the Soviet Union in 1991, and especially over the past 10 years, there has been a sharp increase in IDU in Russia as evidenced by indicators such as a ninefold increase in the number of people attending state-provided drug treatment clinics in the 1990s. Authors such as Shelley (2006) point to Russia's status as a transit country for drugs, where many substances entering the country are also consumed domestically. However, there is very little information pertaining to the prevalence of IDU in the general population. The Russian Ministry of the interior estimates between three and four million injecting drug users, which in a population of 150 million, would yield a figure of 1–2%. Estimates derived from rapid assessment studies of urban areas suggest higher urban prevalence rates between 3% and 5% (see, for example, Dehne and Kobyshcha 2000, Koshkina, Koryakin, and Tsarev 2002).

Considering the geography of drug use in Russia, Shelley (2006) argues that the highest concentrations of usage are in the major cities along the Trans-Siberian railway; Vladivostock, Irkutsk, Ekaterinburg and Moscow, and in the city of St Petersburg, reflecting the fact that internal trade in Russia relies heavily on rail networks. (See Fig. 18.1 for these and other locations noted in this chapter.)

However, there is still a great deal of uncertainty. Despite the unprecedented rise in the use of drugs in post-socialist Russia, prevalence data have been patchy, deriving mainly from national household surveys, and regional level "snap shots". Research has traditionally focussed on Moscow, and it is unclear whether results can be extrapolated to smaller cities and towns. However, focussing on the use of

Fig. 18.1 Orientation map

drugs by young people (aged 14–19) in Russia, a recent study of drug use in three regions has shown considerable geographical variation both in the use of drugs, and in the circumstances under which individuals become drug users. The three regions selected for the study were Krasnodar, Samara, and Komi, which reflect a geographical spread from north to south, and encompass a diversity of drug markets. This study found considerable regional differentiation in drug use, with a very high rate of reported drug use in the Komi region (29%) as opposed to 16% in Krasnodar and Samara regions. This finding was particularly significant as it contradicts "accepted wisdom" in Russia that drug use is strongly associated with supply, and in particular with drug trafficking routes (Pilkington 2005). The Komi region, situated in the Far North of the Russian Federation, is at some considerable distance from the trafficking routes known to operate in the south of the country.

Statistical data pertaining to the regional distribution of drug use in Russia are extremely scarce, and even if such data were available, their accuracy would be suspect. The number of registered drug users is certainly only a fraction of the actual number, but the size of that fraction is not known. Data detailing the number of registered IDUs in Russian regions are not available (Koshkina 2001), and furthermore, there are major disincentives for IDUs to register at their local narcology unit. In Togliatti, for example, registration has a negative effect on the ability to gain employment and find housing, and it increases an individual's chance of being detained by police (Platt et al. 2004). Togliatti's local narcology service obtains most of its registrations through police referral rather than through voluntary registration, and this pattern may well be repeated across Russia. A proxy for these data is the number of recorded crimes associated with illegal narcotics (Figs. 18.2 and 18.3). Such drug crime data have their weaknesses; regional funding for enforcement of

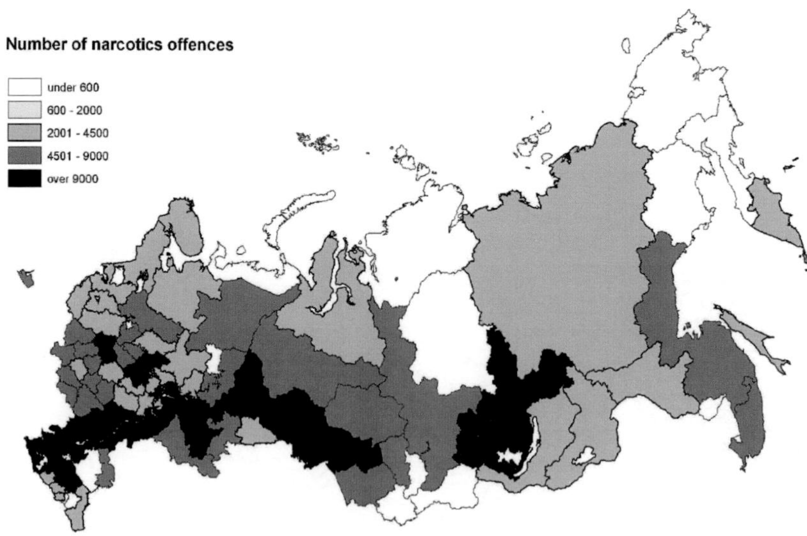

Fig. 18.2 Distribution of narcotics offences by Administrative Unit, 2003

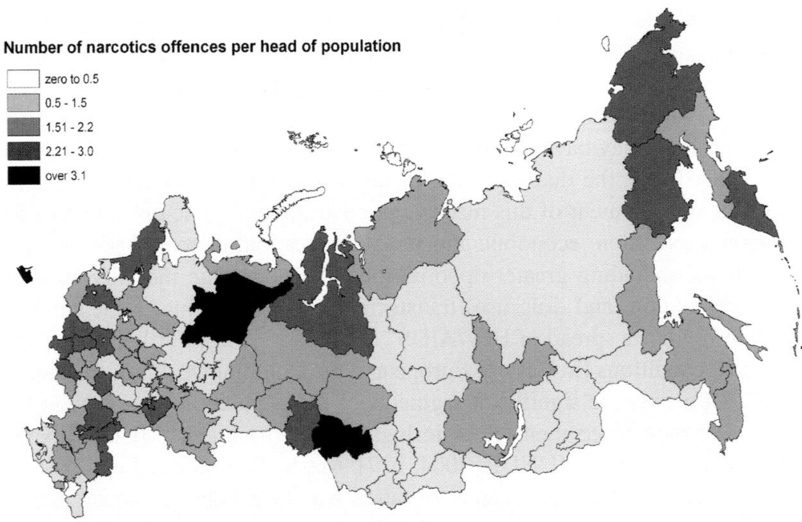

Fig. 18.3 Narcotics offences per capita by Administrative Unit, 2003

legislation relating to narcotics will vary, as will the level of priority afforded to enforcement by regional administrations, and by local police. Regional policy toward harm reduction (including needle exchange) also varies, and all these factors will affect the statistics collected under this heading. The level of drug crime does not, therefore directly corresponds to numbers of IDUs. Figures 18.2 and 18.3 depict the regional distribution of drug crimes and the number of crimes per capita in 2003, respectively. The first map shows that the absolute number of narcotics offences is the highest in the large conurbations and in the south of the country – probably a function of the trade in narcotics entering Russia cross-border and directly to Moscow (i.e., offences pertaining to commercial trade in rather than personal use of narcotics), and of heightened attention to such crimes in these areas. The second demonstrates that the level of narcotic offences per head of the population places the highest figures again in the south and the major urban regions, but also in the Russian Far East and the European North, perhaps suggesting that the distribution of drug users may actually be more even across space than the absolute numbers suggest. Drug crime is positively correlated with the urban distribution of population in ($r = 0.432$, significant at the 0.01 level), suggesting that at the local level, drug users are probably most heavily concentrated in urban areas.

The Geography of the Russian HIV/AIDS Epidemic – The Macro-Risk Environment

Rhodes et al. (2005) argue that the macro-risk environment for HIV transmission among IDUs can be viewed as "comprising large-scale social, physical, economic, organizational and policy systems" (p. 1028). In Russia, HIV is widely perceived to be connected with the experience of post-Soviet transition, in that the circumstances of transition, (sudden and fundamental economic, social and political change), comprise the macro-risk environment. Although the first HIV infections occurred before the end of the Soviet era (Table 18.1), it is argued that "social changes arising from political transition may have contributed to the spread of HIV" (Rhodes and Simic 2005, p. 220). Thus, the disintegration of the Soviet state appears to have set the stage for the development of this macro-risk environment. "The upheavals of Russia's ongoing transition: economic and social dislocation, increased poverty, [and] new freedoms (including greater opportunities for geographic mobility, extramarital sex, prostitution, and drug use) transformed the country into a far more conducive setting for the spread of HIV/AIDS" (Eberstadt 2002). Furthermore, "social and political conditions [in Eastern Europe and the Commonwealth of Independent States (CIS)] encourage significant segments of the population to engage in high-risk behaviors and place themselves in high-risk environments" [United Nations Development Programme (UNDP) 2004, p. 7]. Mashkilleyson and Leinikki (1999, pp. 40–41) contend that "the present situation has deep political, social and economical roots, a pronounced economic stratification of the once uniform society, . . . with . . .consequent unemployment particularly among young people." The authors noted that the resulting rise in rates of crime and prostitution, and increases in alcohol and drug consumption, "influenced the prevalence of risk behavior, particularly of teenagers and young adults." UNDP (2004, p. 12) observes that the CIS region "today, this is predominantly an epidemic among urban, young, male injecting drug users and their sexual partners," with IDU and unprotected sex as the main means of transmission of the HIV virus [European Centre for the Epidemiological Monitoring of AIDS (ECEMA) 2002].

The characteristics of the Russian epidemic fit the pattern of a "developed" Northern Hemisphere country, with a predominantly urban distribution of HIV, as opposed to sub-Saharan Africa's relatively equal impact on rural and urban regions.

Table 18.1 Number of reported HIV infections, 1987–30/06/2006

Year	Newly Diagnosed HIV Infections	Cumulative Total of Infections
1987–1994	887	887
1995	203	1,090
1996	1,513	2,603
1997	4,315	6,918
1998	3,971	10,889
1999	19,758	30,647
2000	59,261	89,908
2001	87,671	177,579
2002	49,923	227,502
2003	36,396	263,898
2004	32,147	296,045
2005	37,287	333,332
30/06/2006	13,492	347,222

Source: *www.afew.org*

In Russia, there is a positive and significant correlation between the proportion of a region's population which is urban, and HIV prevalence, with the relationship particularly strong in the Central and North-western areas of the country, and in the Urals (Moran 2005).

In order to explore the geography of HIV/AIDS in post-socialist Russia (Fig. 18.4), macro-level data can be used to discover whether the perceived links between the post-socialist condition and increased HIV infections have resonance beyond the specific local contexts studied (Table 18.2), and to draw some conclusions about the ways in which this macro-risk environment might "combine with micro-level factors to "structure" the risk environments in which HIV risk and harm is produced and reproduced" (Rhodes et al. 2005, p. 1028). Data pertaining to HIV prevalence in the Russian regions in 2005 has been correlated with socio-economic indicators for years previous to 2005 (given that infections detected by 2005 must have taken place some months or years previously) drawn from the State Statistical Agency. The accuracy of both sets of data is acknowledged to be questionable [Bradshaw and Vartapetov 2003; Center for Strategic and International Studies (CSIS) 2005], and results of analysis must accordingly be treated with care.

Correlations between HIV prevalence and socioeconomic indicators show that there are strong and statistically significant relationships between HIV prevalence and the patterns of regional economic development and domestic population movement characteristic of the post-socialist period, and also with social "dislocation", particularly amongst the age group of the population experiencing most infections. Considering the economic aspect of the post-socialist condition, observers of Russia's HIV epidemic have noted a connection between economic dislocation and stratification, and risk behaviors that facilitate infection. However, risk behaviors do

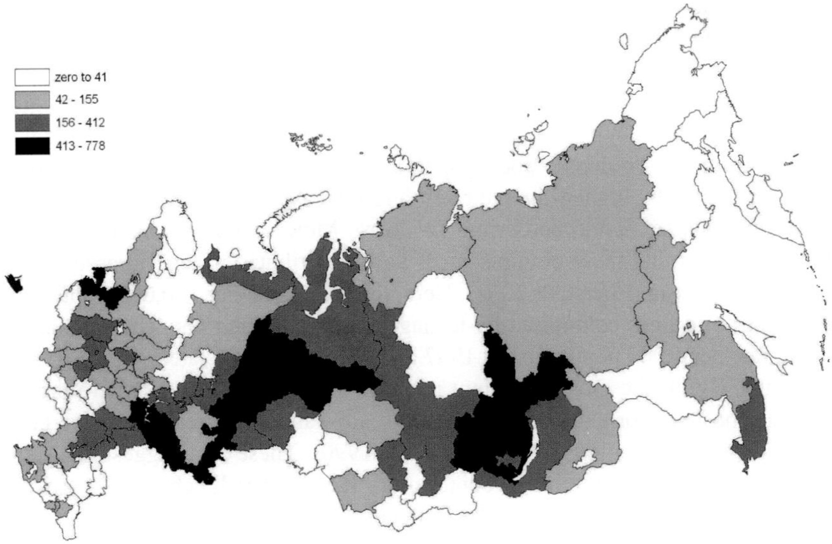

Fig. 18.4 HIV prevalence per 100,000, June 2006

Table 18.2 Correlations between HIV prevalence 2005 and independent variables

	Urban Pop'n 2005	Change in FDI 1995–2003	Unemployment 2003	Teenager Crime 2003	Cars per 100,000, 2003	Kilometers of paved road, 2003
Russian Federation	0.432**	–	–	0.480**	0.380**	0.351**
Urban regions (>60% urban)	0.430**	–	–0.310*	0.410**	–	0.279*
Rural regions (<60% urban)	–	0.626**	–	0.440*	–	0.500**
Low prevalence regions (below 300/100,000)	0.438**	–	–	0.540**	–	0.409**

*Significant at the 5% level
**Significant at the 1% level.
Source: Moran (2005)

not seem to be related to economic factors in a straightforward manner; there are strong correlations between HIV prevalence and indicators of regional economic growth rather than decline, with unemployment and HIV negatively correlated, indicating that high levels of unemployment in Russia coincide with low HIV prevalence. Also, the relationship between HIV and foreign direct investment (FDI) is striking, with long-term increase in FDI positively correlated with HIV prevalence.

Analysis of domestic population movement – one of Eberstadt's "new freedoms" (2002) – and HIV prevalence shows that there are strong and significant correlations with statistics for kilometers of paved road, and number of cars per 100,000 population, indicating that population mobility may be a factor in explaining regional distribution of HIV infections.

Social as well as economic dislocation has been suggested as a contributory factor, and analysis here included crime and divorce rates as proxies for this intangible situation. It has been hypothesized that there has been social dislocation (fragmentation of the more "ordered" society perceived to have existed under the Soviet system) indicated by higher divorce rates, and increased incidences of extramarital sex, commercial sex work, and crime (Walberg, McKee, Shkolnikov, Chenet, and Leon 1998), and that these circumstances have contributed to the increase in HIV infections in Russia (Eberstadt 2002). Here, the crime rate per 100,000 population, the number of crimes committed by teenagers and with their cooperation, and the divorce rate per 1,000 population in 1997 and 2003 were correlated with 2005 HIV prevalence. By far the most important result is for teenager crime, where there is a constant positive correlation at all spatial scales, and an increase in the strength and significance of relationships since the late 1990s. These data suggest that social dislocation could be an important contributory factor for risk behaviors, particular within an age group at increased risk of infection (Moran 2005).

Extension of this analysis shows that the salient features of the macro-risk environment for HIV transmission in Russia include the process of urbanization, particularly in already highly-urbanized regions; domestic population mobility and social dislocation, with the association between HIV and mobility being the most statistically significant. While this macro-scale research confirms that there is a relationship between HIV/AIDS and the "condition" of post-socialism (Moran 2007), a nuanced understanding of local and individual engagements with post-socialism – the micro-level risk environment – is needed in order to uncover how these processes are worked out in everyday lives – the contexts and circumstances which enable and encourage individuals and especially injecting drug users to engage in risky behaviors (Moran and Jordaan 2007).

Drug Use and HIV/AIDS in Russia

A major characteristic claimed for the Russian HIV/AIDS epidemic is the link between HIV and IDU (Kalichman et al. 2000), through the use of contaminated needles and drug use practices that encourage infection. In the public health literature, papers by Rhodes et al. (1999, 2002), and Krupitsky et al. (2004) have described the link between HIV and drug use in Russia and elsewhere as unequivocal. Where the source of Russian infections is known, statistics have variously linked up to three quarters of HIV infections to IDU (ECEMA 2002). Federal AIDS Center reported in 2002 that at least 90% of HIV infections have been linked to IDU (cited in Rhodes et al. 2003). The youth of the HIV epidemic in Russia, in terms of the high proportion of infections amongst individuals aged under 35, is distinctive in comparison with Western Europe and the USA, and potentially linked to drug use, which also tends to be the most common amongst teenagers and young adults. The nature of drug abuse in Russia has changed since the mid 1990s, with drug use becoming more common at a younger age (Kramer 2003); 6% of 15 and 16 year olds in Moscow report having used heroin at least once, whereas in Western Europe comparable figures do not exceed 2% (Paoli 2002).

While relationships between HIV and IDU have been widely observed in individual city studies, including studies of Togliatti (Platt et al. 2004) and of Krasnodar, Nizhniy Novgorod, Rostov, Saratov, Tula, Tyumen', Tver', Irkutsk, and Moscow [Alcabes, Beniowski and Grand 1999; Dehne, Khodakevich, Hamers and Schwartlander 1999; World Health Organization (WHO) 2000] and IDU is a major risk factor, it is difficult to establish a statistical relationship between IDU and HIV prevalence at the national or even the regional level due to the availability and the nature of current data.

While data detailing the number of IDUs in Russian regions is not available, the results of correlations between the drug crime proxy data mapped in Figs. 18.1 and 18.2 and HIV prevalence in 2005 are presented in Table 18.3. Drug crime data themselves have their weaknesses as discussed earlier; in particular, the level of drug crime does not directly equate to numbers of IDUs, and IDU itself does not necessarily result in HIV infection, given the potential for safer injecting practices. However, as in the case of the HIV prevalence data, it is impossible to control for such variation, and correlations should be viewed in this light.

Table 18.3 HIV prevalence/100,000 2005 and number of crimes connected with illegal narcotics 2003

	n	Correlation Coefficient r
Russian Federation	87	*0.418***
Urban regions (65% or over urban population)	57	0.370**
Rural regions (under 65% urban population)	30	0.361
Eleven highest prevalence regions	11	−0.180
Federal Okrugs		
Central FO	10	0.630**
Far Eastern FO	10	0.809**
North Western FO	11	0.655*
Privolga FO	15	0.519*
Siberian FO	15	0.277
Southern FO*	12	0.510
Urals FO*	6	0.456
Seventy-six lowest prevalence regions	76	0.418
Urban regions (65% or over urban population)	47	0.395**
Rural regions (under 65% urban population)	29	0.262
Central FO	17	0.388
North Western FO	8	0.221
Privolga FO	12	0.710**
Siberian FO	13	0.030

None of the 11 top prevalence regions is located in the Southern FO, and there are too few regions in the Urals FO for a correlation removing its three top 11 prevalence regions to be carried out
*Significant at the 0.05 level
**Significant at the 0.01 level
Source: Goskomstat Rossii 2004:328–9, Moran (2005)

At the Russian Federation level and in urban regions, there is a highly significant (1% level) if not particularly strong, correlation between 2005 HIV prevalence and 2003 drug crime, with stronger regional associations in the Central and the Far East regions ($r = 0.630$ and 0.809, respectively, both at the 1% level). However, the most notable result is that for the 11 top-prevalence regions, there is no significant correlation between the two variables (the very weak association is actually negative, $r = -0.180$). Nevertheless, there are three obvious outliers within this group, regions at the extremes of the range of values for each variable. Removing them from the analysis leaves a group of eight high-prevalence regions for which there is a strong and significant (1%) r-value of 0.877. HIV prevalence and drug crime as a proxy for drug use are clearly positively correlated at various levels in the Russian Federation. However, the lack of a consistently strong, positive, and significant correlation between HIV and drug crime could be argued to support the case of observers who contend that whereas almost all infections in the earliest phase of the epidemic have occurred among IDUs, sexual contact may be increasing in importance as a means of transmission (Mashkilleyson and Leinikki 1999; Grassly et al. 2003; Lowndes, Alary, and Platt 2003).

The Risk Environment: Drug Use and Risk Behaviors

The consideration of the macro-scale risk environment presented above concludes that although there are certain characteristics of the condition of post-socialist transition in Russia which seem to have an association with the transmission of HIV (whether sexually or through IDU), a nuanced understanding of the ways in which these circumstances are played out at the local level, via meso and micro-level risk environments, is critical to an understanding of drug use and HIV/AIDS in Russia. Transmission of HIV through needle and syringe sharing is relatively efficient (Royce et al. 1997), and in the Russian context, there has been a preference for injection (over other methods of drug use, although this tendency may be on the wane). Unsafe injecting practices facilitate the transmission of HIV, and individual studies have tracked local increases in IDU followed by (presumably linked) increases in HIV prevalence in Russia (e.g., Krupistky et al. 2006). There is a growing body of knowledge about precisely *how* the injection of drugs is associated with infection, or in other words, about the micro-risk environment in which unsafe injection practices take place. This section focuses on the geographical aspects of drug use, especially but not exclusively IDU, in relation to HIV prevalence; the micro-geographies of drug use, paying particular attention to the ways in which local social and spatial configurations facilitate drug use and risk behavior.

The critical factor making injection practice unsafe is the sharing of injection paraphernalia (i.e., drug preparations, containers, syringes, and needles), with the concomitant risk of contamination with infected body fluids. Such sharing takes place in specific circumstances where the configuration and use of space is highly significant in encouraging sharing. The following examples are drawn from the small but significant number of studies in Russia, which have focussed on this issue.

The first issue is the nature of spaces encountered and used by IDUs. Although the availability of drugs is described as widespread (Rhodes et al. 2003), with users able to access dealers in public spaces such as urban streets and in the courtyards enclosed by apartment blocks, most drug injection seems to take place either in the private spaces of the user's home, sometimes in the company of friends who also use, or at the location of the drug vendor. Clean syringes, while also relatively widely available, are, by contrast, obtained in public spaces such as pharmacies and needle exchanges, where these exist. The process of negotiating the spaces between these three locations is central to understanding the choice to share equipment, especially needles and syringes. In the Russian Federation, the possession of injection paraphernalia where the intention to use drugs is evident, or of quantities of narcotics, is an offence, and those suspected of possession are liable to arrest. Police have been reported to target pharmacies, needle exchanges, and the courtyards enclosed by apartment blocks as likely locations to encounter drug dealers and users. There is also a perception that the apparently "neutral" spaces of needle exchanges are in some way associated with the police. For example, Rhodes et al. (2003) report that the *Tsentralnii* (central) clinic in the city of Togliatti is located within the City Narcology Unit, where there is invariably a police presence as individuals suspected to be drug users are routinely brought in for verification and/or registration.

The location and the apparent appropriation of this clinic by the official authorities serves as a deterrent to drug users needing to obtain clean needles; "These exchanges should be independent – not controlled by the police" (Rhodes et al. 2003, p. 49).

The nature of the location of the needle exchange is also an important factor. The fact that a needle exchange is fixed in space limits its appeal for some drug users; for example, in the city of Barnaul in central southern Russia, the needle exchange operates at a fixed site in the local AIDS Center, and local users find that the cost and inconvenience of traveling to the center discourages their use of it, whereas in other cities, Volgograd and Moscow for example, there are no fixed-site exchanges, services being provided instead by mobile exchanges and outreach workers [Department for International Development (DFID) 2006].

In order to avoid arrest, users prefer neither to travel to the vendor in possession of a syringe, nor to travel home in possession of drugs. There is also a belief that buying a syringe in advance of purchasing drugs places a jinx on the likelihood of a successful subsequent drug purchase, and anecdotal evidence suggests that although new syringes are cheap and readily available, many users follow this superstition (Rhodes et al. 2003). Depending upon individual circumstances, there may also be an urgency to inject as soon as drugs are purchased. The alternative to crossing the neutral space between home and dealer carrying incriminating evidence is either to deposit a syringe at the dealer's premises for use *in situ*, or to re-use needles previously deposited or discarded by previous clients. Two quotations surmise this situation (Rhodes et al. 2003, p. 49):

> I know that I might get stopped by the police on the way. That is why you sometimes have to use other people's syringes. It is mostly because of the fear of the police.
>
> People are afraid to carry syringes on them. They try to hide any evidence that they are drug users. The police can easily arrest you for a few days; no drug user would want that to happen.

The fact that needle *exchanges* are precisely that necessitates the return of used needles to be exchanged for new ones. Some drug users in Barnaul find it difficult to store used needles for this purpose (for example in order to avoid discovery of their drug use by friends and family), and are also reluctant to travel to the needle exchange carrying incriminating evidence of their habit (DFID 2006).

In the user's or the vendor's private space, the social networks that operate around the injection of drugs encourage risky behaviors, and as Rhodes et al. (2003, 2005) argue, situational factors contribute to sharing. There is a predilection toward practices that increase the risk of infectious contacts with contaminated blood, including distribution of drugs in pre-loaded syringes, use of group-prepared drug preparations (especially home-made injectable opiates, where potentially contaminated blood is added to solutions to stabilize them), as well as the social context of drug injecting in particular environments, which increase the potential for multiple transmission events in IDU populations (Rhodes et al. 1999).

Apart from the direct sharing of needles themselves, some "indirect" sharing practices have been found to be common among injecting drug users in the city of Togliatti. Rhodes et al. (2003) found that 73% of IDUs interviewed reported that in the previous week they had drawn their drug solution from a container into which someone else had already put a used needle. This sharing of common containers seemed ubiquitous within a social group (Rhodes et al. 2003, p. 47):

> You fill up from one syringe. This is the person who makes, it, draws it up, and he pours it into a wine glass. Let us suppose I wanted to inject first. I fill up myself, go off and inject. Then the next person draws up. So everyone fills up in turn.
>
> We buy drugs together, prepare them together, everything together.
>
> There were only two syringes for the group... There were around five people... This is a usual situation.

The unsafe injection of drugs is clearly a major vector threat for the transmission of HIV, but the influence of other drugs, including alcohol and marijuana, on risk behavior should not be underestimated. Krupitsky et al. (2004) argue that there is evidence for an indirect role of alcohol use in HIV transmission through the modulation of sexual or IDU-associated risk behavior. Research has shown an association between alcohol use and unsafe sex (McEwan, McCallum, Bhopal, and Madhok 1992), and although this effect has not been widely studied in Russia, the fact that Russia's alcohol consumption is amongst the highest in the world, suggests that this may be an important factor. For instance, looking specifically at alcohol- and drug-dependent (mainly heroin) attendees of a substance abuse facility in the Leningrad region of north western Russia, Krupitsky et al. (2004) found an increase in HIV infection amongst alcohol-dependent individuals. The means of transmission of the virus was not known, but unsafe injecting was unlikely given that none of the individuals affected had a prior record of IDU. Those infected were also from a significantly older age cohort than their drug-dependent counterparts. Whether infection took place through sexual contact or through injection, it might reasonably be assumed that alcohol mediated the decision (or lack thereof) to share equipment or to engage in unprotected sex.

The transmission of the HIV virus from high-risk groups such as injecting drug users into the mainstream population, amongst whom are counted the alcohol- but not drug-dependent, is generally perceived to take place via a "bridge population" – the result of particular social norms and configurations. As Kramer (2005) has argued, drug users themselves fuel this transmission by engaging in sexual relations with non-users, especially when the drug users are also commercial sex workers (CSWs). In this case, infected CSWs transmit the virus through unsafe sex with customers, who then form the "bridge," unknowingly transmitting the infection further into the general population.

It is impossible to accurately establish the extent of commercial sex work, and the prevalence of HIV within the CSW population in Russia, but "snapshot" studies suggest that prevalence is greatly in excess of that in the general population. HIV prevalence rates among female CSWs who also inject drugs have been recorded at 61% in Togliatti and 65% in Kaliningrad (Lowndes et al. 2003), and the need for money to support a drug habit may be a major factor in propelling individuals into sex work. In addition, female drug-using CSWs have been found to be at higher risk of infection through unsafe sex than their non-using counterparts; users are more likely to accept the terms offered by the client, including foregoing the use of condoms, and agreeing to higher-risk sexual practices commonly refused by non-users. CSWs who are non-users also tend to refuse to service drug-using men (where they are identifiable), with the result that these men engage drug-using CSWs almost exclusively. Lower prices offered for sexual services by drug-using CSWs also mean that young boys who are not drug users but who lack money also purchase services from them, leading to exposure of adolescents to infections present in the adult male IDU population (Aral and St Lawrence 2002).

However, the picture painted by these data, of almost inevitable slide from drug injection to HIV infection, and concomitant overlap and interaction between injecting drug users and CSWs, is only partial although such a trajectory may indeed be unavoidable for many individuals. Recent research (Pilkington 2006) amongst young IDUs in Russia challenges this inevitability, presenting evidence derived from ethnographic research into the social norms and circumstances that surround drug use amongst young people. She argues that some users retain strong social ties that help to prevent their slide into the subcultural isolation that normally accompanies drug dependency, of which unsafe communal injecting practices and contact with commercial sex work may be part. Pilkington's fieldwork in the Krasnodar, Samara, and Komi regions of Russia, three diverse localities spanning the far south and far north of the country, found that heroin has penetrated both materially and symbolically into the lives of "ordinary" young people – those in full time education and participating in "normal" mainstream social and cultural institutions. Of the sample group, 13.5% of female and 7.5% of male respondents, who had ever tried any drug, reported the use of heroin, and the high visibility of the drug on the Russian market perhaps contributed to the use of heroin by 14–15 year olds (whereas in the United Kingdom, the mean age of experimentation with heroin is 17).

Pilkington's work challenges the sharp distinction drawn between "recreational" and "problem" drug use within the "normalization thesis", which posits a progression from early recreational drug use to problem heroin use explained by local social exclusionary factors, and provides a different view of the meso and micro-level risk environment for HIV transmission. She argues that within Russian youth cultural practice, a mode of occasional, long-term "safe" heroin use exists, and that this "safe" heroin use is less subculturalized and more embedded within, or alongside, more traditional forms of peer group-based "recreational" drug use. The nature of this heroin use may be an important factor in encouraging but also controlling "recreational" use of heroin (Pilkington 2006, p. 30); "the fact that heroin users continue to maintain strong and diverse friendship and family ties and pursue 'mainstream' goals, status and values may prevent heroin users sliding into the multiple deprivation situations that lead to a downward spiral of social exclusion and compound problem drug use."

This recent study may not be representative of drug use practices throughout Russia, but if it is at least indicative, then it suggests that there are alternative lifestyles for injecting drug users, which do not necessarily include participation in risk behavior, and demonstrates the complex and nuanced nature of the meso and micro-level risk environments.

Tackling IDU and HIV/AIDS: Understanding the Risk Environments

Interventions to tackle the transmission of the HIV virus through IDU have two main points of entry; the prevalence of drug use itself and the unsafe practices that facilitate viral transmission. Given the close links between IDU and HIV, it is essential that interventions address both issues. Unfortunately, in the Russian Federation, this is not always the case; Russian government policy dealing with injection drug use arguably has been acting against initiatives to prevent HIV transmission, and policies aimed at tackling HIV seem to be directed more to the treatment of those infected rather than the prevention of infection.

Considering the first entry point, interventions have taken a hard line against IDU, and until May 2004, possession of even tiny amounts of narcotics was a criminal offence. The relevant law and its enforcement created a climate of fear, as discussed above, with police surveillance of pharmacies where IDUs were known to purchase syringes, detention and incarceration for possession of trace amounts of narcotics, and threat of extortion or arbitrary arrest [Human Rights Watch (HRW) 2004]. Any subsequent imprisonment for narcotics offences only serves to increase IDUs' risk of HIV infection through syringe-sharing in prisons and the poor quality of prison-based HIV-prevention services. The 2004 revision of the legislation saw small-scale drug possession reclassified as an administrative rather than a criminal offence, and the move was perceived to encourage policies grounded in public health and human rights.

As we have already seen, the interaction between police and IDUs in public space is critical to the behavior of drug users, and influences heavily the choices made about sharing injecting equipment. In a recent study, Rhodes, Platt, Sarang, Mikhailova, and Monaghan (2006) found that a sample of Russian police officers interacting frequently with IDUs on the streets of the city of Togliatti, interviewed in 2002, described street policing as a means of maintaining close surveillance of drug users, which may lead to the registration of individuals suspected or proven to be users of drugs. Completion of the registration process further enabled subsequent surveillance, through stop and search procedures, which assisted officers in detaining individuals in contravention of federal legislation. Officers seemed aware of drug users' resultant reluctance to carry injecting equipment linked to their fears of detention or arrest, but the fact that the confiscation of previously used injecting equipment could constitute evidence in relation to drugs possession charges, and that discovery of clean injecting equipment could also be sufficient to raise suspicion meriting further investigation through stop and search or questioning, seemed to outweigh these concerns. These findings suggest that an uneasy relationship existed between street policing and needle and syringe access, in which policing strategies served to undermine needle and syringe accessibility among IDUs. Rhodes et al. (2006) concluded that facilitating partnerships between policing agencies and HIV prevention initiatives is a critical feature of creating environments conducive to risk reduction, but it is as yet unclear whether the 2004 legislation has had an effect on these practices; it would seem reasonable to assume that police cultures and attitudes will take time to adapt.

Where programs to assist IDUs in recovering from addiction are concerned, along with many other countries in Eastern Europe, Russia does not support substitution therapy. Substitution therapy using drugs such as methadone is a treatment approach which assists heroin users to manage withdrawal symptoms and cravings which result when heroin use is reduced or stopped. Taken orally, methadone removes the need to inject, thus significantly reducing users' risk of HIV infection. Although the World Health Organization regards substitution therapy as an essential component of harm reduction programs, the Russian government resists its introduction, citing as justification a responsibility to implement the United Nations drug conventions of 1961 and 1971, Article 38 of the 1961 convention obliges signatories to take all practicable measures to provide treatment for drug dependence, and while there are also provisions in the 1961 and 1971 conventions to control access to methadone, these require prescription rather than outright prohibition. Russia also expresses concerns over the effectiveness of the therapy, and its potential to prolong and even encourage drug addiction. Despite the benefits claimed for the therapy where it is in operation beyond Russia, and widespread outcry over the Russian government's position on this issue, at present there seems little prospect for introduction of methadone substitution therapy for Russia's injecting drug users.

In short, Russian drug treatment services appear to be highly centralized and focussed upon medical approaches, rather than holistic and broad in terms of offering a range of available services to meet a variety of treatment needs. However, a change would require considerable economic resources as well as political commitment at both the federal and the local levels, as well as the commitment of resources to developing the capacity of those working within the treatment system.

Considering the second point of entry, realization of the size and potential impact of the HIV/AIDS epidemic in Russia came late, and although President Putin has recently announced a 20-fold increase in public spending on HIV/AIDS, increasing funds from 150 million rubles in 2005 to 3 billion in 2006 (Novosti, September 27, 2005), the apparent focus is on providing medication to infected individuals rather than focussing on the prevention of infection (Moran, 2005). Even here, due to fears that drug users will not adhere to the medication program, virtually no HIV-infected IDUs in Russia receive treatment with the antiretroviral (ARV) therapy, which can delay the onset of AIDS. As Long et al. (2006) note, only 5,000 individuals in the whole of Russia received ARV therapy in 2005, in the context of a health care system already weakened in the transition from socialism. While there are hopes that treatment programs will increase this figure to over 30,000 during 2007, with increased geographical dispersal of facilities, there is no indication that IDUs will be targeted for this therapy, and the lack of effective assistance to break drugs habits further marginalizes high-risk groups from access to ARV medication.

In terms of the prevention of HIV transmission, education and information dissemination about the risk of infection are not utilized as effectively as might be the case. In the school system, there remains a conservative perception that sex education is controversial, and at present there is no immediate indication that activities will be scaled-up across the country. More practically, the development of harm reduction in Russia *per se* has been hampered by inadequate financing (including a lack of government resources to support projects), and the stigma and marginalization experienced by high-risk groups (DFID 2006).

Conclusion

This chapter has explored the links between drug use and HIV/AIDS in the Russian Federation in the context of what Rhodes et al. (2005) describe as "risk environments." Appreciation of the significance of the geographies of drug use and HIV/AIDS in Russia is central to the design of interventions to tackle these issues; this overview of the macro, meso, and micro-level risk environments demonstrates the complexity of the linkages between drug use and HIV, and the myriad configurations of social situations, structures and places in which risk is produced. Drug use and HIV must clearly be addressed together, and in the context of the Russian Federation, where financial resources are limited and where political commitment to more controversial programs may take time to achieve, interventions must be carefully targeted at the macro level, toward regions which are perhaps most at risk of increase in HIV infections, as well as where prevalence is already high (Moran and Jordaan 2007). At the local level, too, appreciation of the significance of space for social practice, in terms of the operation of policing in public space, and interpersonal interactions in private space is part of a process of decentralization of HIV interventions, in order that local contingencies and contexts can shape appropriate activities.

Chapter 19
Substance Abuse and HIV in China

The Impact of Residence and Residential Mobility

Xiushi Yang

Abstract Using data from a population-based survey conducted in 2003 and employing multilevel modeling, the chapter examines the impact of residential characteristics and mobility on substance abuse and HIV in China. Both individual characteristics and contextual factors are hypothesized to affect individual drug-using behavior and HIV infection. The results suggest that being migrant is associated with significantly less risky drug-using behavior and lower odds of HIV infection. Drug use is also significantly associated with being male, less educated, single, and psychosocial well-being. At the contextual level, drug use is significantly and negatively associated with poverty. HIV infection is significantly correlated with prevalence of drug use in the community. For both drug use and HIV infection, there are significant cross-community variances in the random intercept component, suggesting that the likelihood of substance abuse and HIV infection vary significantly across geographic locations. HIV research and behavioral intervention need to pay particular attention to contextual characteristics.

Introduction

With 650,000 people officially estimated to be living with HIV/AIDS by 2005, AIDS has evolved within two decades from an unheard of disease to an epidemic affecting every population group and geographic location in China (China Ministry of Health, UNAIDS, and WHO 2006). However, significant differences exist in prevalence of HIV across geographic locations (Gong and Shao 2001; Zheng 2001). Although sexual transmission of HIV accounted for 49.8% of new infections in China in 2005, surpassing for the first time that attributable to drug-related transmission (48.6%), drug abuse remains a key source of new HIV infections in the country. In 2004, there were more than one million officially registered drug users in China, of whom more than 75% were active heroin addicts (Tang, Zhao, Zhao, and Cubells 2006). Despite likely

X. Yang
Department of Sociology and Criminal Justice, Old Dominion University, Norfolk, VA 23529
e-mail: xyang@odu.edu

Y.F. Thomas et al. (eds.), *Geography and Drug Addiction*,

serious underreporting, the official statistics make it clear that drug abuse is widespread but varies significantly across geographic locations and different population groups in contemporary China (Fang, Wang, Shi, Liu, and Lu 2006).

While causes of drug abuse and spread of HIV are likely to be complex and multifaceted, increasing migration has been portrayed by the media and implicated in the literature as one of the main catalyst. In fact, residential *immobility* is arguably the key to understanding the absence of drugs, crimes, and commercial sex in pre-reform China (Situ and Liu 1996; Troyer, Clark, and Rojek 1989). Although varied by sources, temporary migrant population, which constitutes the majority of rural–urban migrants in contemporary China, was estimated to have grown from 11 million in 1982 to 79 million in 2000 (Liang and Ma 2004), and estimated 120 million currently (China Ministry of Health et al. 2006). The uprooting and on the move of so many migrant people may create conditions that are conducive to behavioral change and disease transmission. The quick spread of drugs and HIV in China needs to be understood in the context of social and economic changes associated with increasing migration in the country (Smith and Yang 2005; Weniger and Berkley 1996).

Much has been written about the economic causes and consequences of migration. Less studied is the impact of migration on substance abuse and HIV in China. Further, the significant spatial difference in prevalence of both drug abuse and HIV underscores the importance of contextual factors in understanding drug-using behavior and HIV infection. Yet, research on drug abuse and HIV in China has paid little attention to contextual factors that may be conducive to the spread of drugs and consequently HIV. This chapter focuses on the impact of migration and residential contexts on drug abuse and HIV infection. The central hypothesis is that the process of migration renders migrants vulnerable to substance abuse and consequently HIV infection and that causes of drug-using behavior and HIV infection go beyond individual level correlates to also include contextual factors at the residential community level. The results will help better understand the impacts of migration on substance abuse and HIV infection and the contextual underpinnings of individual drug-using behaviors and HIV infection. They may also provide important empirical evidence for the design of behavioral or policy intervention programs that target both individuals and social and residential contexts.

Migration, Drugs, and HIV

Studies in China (Anderson, Qingsi, Hua, and Jianfeng 2003; Li et al. 2004; Smith and Yang 2005) and elsewhere (Hunt 1989; Lansky et al. 2000; Organista and Organista 1997; Skeldon 2000; UNAIDS 2001) have repeatedly identified migration as an important factor leading to the spread of HIV/STDs. From an epidemiological point of view, the spread of infectious diseases such as HIV has always been associated with the movement of people. Migration brings more people into close contact and creates a greater mixing of people at places of destination, which provides the ready environment for viral transmissions. Through the movement of infected persons, migration can in turn offer the convenient vehicle to transport diseases to places where they are previously unknown. However, HIV transmission requires more intimate contacts involving the exchange of body fluids. Migration itself will not spread HIV unless it leads to increases in certain HIV-risk behaviors among migrants.

Indeed, there is a general agreement that migrants are more vulnerable to HIV-risk behaviors such as substance abuse and casual/commercial sex than non-migrants (Anderson et al. 2003; Hu, Liu, Li, Stanton, and Chen 2006; Li et al. 2004; Skeldon 2000; UNAIDS 2001; Yang 2006). A key to understanding migrants' elevated risk behaviors is the potentially reduced social control over individual behavior in the process of migration, which results from migrants' detachment from the usual social and normative control (Yang 2006). Being away from home means a breakaway from family care and supervision and detachment from home community and its associated normative control. This creates some sort of social control vacuum whereby migrants feel less constrained by social norms and values since families and friends back home are unlikely to find out what they do while away from home. The power of social sanction embedded in social control (Gibbs 1982) is thus lost in the process. The transient nature of migrant life and the more anonymous life in places of destination together may render migrants vulnerable to substance abuse.

In addition to lax social control migrants may experience, migrants' peculiar post-migration socioeconomic milieus may also be conducive to risk behaviors (Soskolne and Shtarkshall 2002; Yang 2006). Although not all are alike, many migrants are socially and residentially isolated from the "mainstream" society in the place where they live and work. Once arrived in the city, most migrants are concentrated in the margins of the urban economy (Knight, Song, and Jia 1999; Roberts 1997; Solinger 1999; Wang, Zuo, and Ruan 2002) and live with fellow villagers at the place of work or concentrate in transitional neighborhoods characterized by overcrowding, social disintegration, and lack of social and health services (Ma and Xiang 1998; Zhang 2001). Migrants' social interaction in the city often does not go beyond that with fellow villagers or migrants. Many migrants, particularly temporary migrants, experience little social or cultural assimilation in the place of destination, feel helpless, insecure, discontented, and resentful, and are prone to substance abuse (Anderson et al. 2003).

Migrants' social and residential isolation in cities may further decrease effective normative and formal social controls over their behaviors. On the one hand, the neighborhoods (i.e., the fringe areas of the city) where most migrants live are often characterized by lax law enforcement and poor social integration. Such a living environment is not only conducive to drugs, but also where such socially proscribed and HIV-risk behavior is more acceptable or tolerated. On the other hand, social and economic marginalization and isolation may make migrants indifferent to social sanctions because the very marginal status makes them feel nothing to lose if their behaviors are detected. Consequently, the combination of lax social control and post-migration social isolation may lead to alcohol and/or drug dependence among migrants as a way to escape loneliness, bury anxieties about work and family, and cope with stress and frustration associated with social isolation and marginalization (Jochelson, Mothibeli, and Leger 1991).

Community Contexts, Drugs, and HIV

The importance of social and environmental contexts has long been recognized in studies of public health and health-related behaviors (Diez-Roux 1998; Duncan, Jones, and Moon 1993; Oakes 2004). The literature shows that social and neighborhood conditions are significant factors in explaining the prevalence of HIV risky drug-using behaviors (Crum, Lillie-Blanton, and Anthony 1996; Galea, Ahern, and Vlahov 2003; Latkin, Williams, Wang, and Curry 2005; Wang, Siegal, Falck, and Carlson 1998) and HIV (Wallace, Wallace, Andrews, Fullilove, and Fullilove 1995; Yang 2005). Community or place social and physical characteristics may affect directly the risk of HIV infection through exposure to the virus or indirectly through drug-using behaviors that increase the risk of HIV infection.

Throughout history, the introduction of new infectious diseases to a community has always been closely related to the community's exposure to the outside world. The introduction and subsequent spread of HIV are no exception. Just as it is true for any infections, exposure to HIV virus is a necessary precondition for HIV infection. In fact, if we were completely free from contacts with others and thereby avoiding any potential exposure to HIV, no HIV infection could have occurred. But life is full of interactions with others in both physical and social environments. In the process, characteristics of the community where we live and work can play an important role in determining the extent of our exposure to HIV (Halloran 1998).

First, the proportion of residents in a community who are already infected will determine the extent of potential exposure of its residents to existing pool of transmissible HIV/STDs (Ford and Koetsawang 1991). The intensity of spatial interaction of a community with the outside through migrant network and economic exchange will determine the extent of potential exposure of its residents to new pools of HIV (Wallace et al. 1995; Wood, Chan, Montaner et al. 2000). Second, spatial mobility in and out of a community will create a greater mixing of different at-risk population groups in the community, which in turn increases residents' exposure to HIV and facilitate the spread of the virus. To the extent that migrants are more vulnerable to HIV while away from home, the return of infected migrants will also bring home the AIDS virus and unknowingly pass it on to their sexual partners (Apostolopoulos et al. 2006; Lau and Thomas 2001; Lurie, Williams, Suma et al. 2003).

Indirectly, community social and physical characteristics can affect the risk of HIV through influences over residents' drug-using behaviors. Like any other human behavior, drug-using behaviors are not inborn but learned through context specific socialization (Bandura 1986; Clark 1987). Individuals learn to behave socially by interpreting images or messages they receive in social interactions or in public domains about what is socially acceptable and by observing and imitating the behavior of others they come into direct or indirect contact. In particular, social norms and networks play an important role in influencing drug-using behaviors (Galea et al. 2003; Latkin et al. 2003). For example, the existence of drug-related cultural, social, and physical scenes in a community can lead to more tolerable perceptions about drugs. The size of drug-using population in a community may facilitate the spread of substance abuse by setting real life examples for others to follow. In essence, drug-using behaviors, like any other human behavior, are unlikely under the complete control by individuals and free from the influence of community social and physical environments.

Community social and physical environments can also affect the local opportunity structure that may influence drug-using behaviors (James, Wagner, and Anthony 2002; Rhodes et al. 1999; Yang 2005). Directly, the existence of more drug outlets in a community can lead to easier access to and lower costs of substance abuse (Crum et al. 1996; Galea, Rudenstine, and Vlahov 2005; Weitzman, Folkman, Folkman, and Wechsler 2003). Indirectly, the socioeconomic well-being of the community can determine the extent to which its residents are economically marginalized and socially isolated, which in turn influences behavior and affects the opportunity costs of drug-using behaviors (Brewster, Billy, and Grady 1993; Wilson 1987). Because substance abuse is socially proscribed and incompatible with socially respectable statuses, indulging in it will likely reduce one's chance of achieving the desirable statuses or may lose them if one has already achieved them. However, if opportunities to achieve desirable statuses are few or non-existent, which may particularly be the case in disadvantaged neighborhoods, the opportunity cost of drug-using behaviors will be low, which may be conducive to the spread of drug abuse among residents. Neighborhood disadvantages/disorders are also associated with increased psychosocial stresses, which may lead to greater interpersonal tension and violence and increases in drug abuse as coping and stress reduction mechanisms (Frye et al. 2006; Galea, Freudenberg, and Vlahov 2005).

Data

Data used in the analysis are from a population-based study of the link between migration and the spread of HIV risk drug-using and sexual behaviors in China, which was funded by the National Institute on Drug Abuse. The study covered an entire province in southwestern China (Figure 19.1) and included both a community and an individual sample survey. The community survey took place in 2001 and covered the entire province. The survey used a special questionnaire to compile annual aggregate information at the township level in rural and neighborhood level in urban places on a wide range of socioeconomic indicators, including numbers of registered drug users, crimes reported, and HIV/AIDS cases. All rural townships and urban neighborhoods were included in the survey. Local administrative office or related agencies were sent the special questionnaire and asked to complete the questionnaire with the requested information for the years between 1996 and 2000.

The individual sample survey took place in 2003. Sample selection followed a three-stage sampling procedure. First, tabulations of known HIV/AIDS cases, drug users, and migrants by counties/cities were prepared with data from the provincial public health and public security agencies and the 1995 mini-census. These

Fig. 19.1 Geographic Location of the Research Site, Yunnan Province, in China (See also Plate 18 in the Colour Plate Section)

tabulations were used to rank all counties/independent cities, and from the ranked list of counties/cities, eight were selected, giving priority to places with higher concentration of HIV, drug use, and migrant population and geographically representing the province. Second, all rural townships and urban neighborhoods in each of the eight selected locations were ranked according to estimates of HIV cases, drug users, and temporary migrants, based on existing data from the same government agencies and the 1995 mini-census. From the ranked lists by county/city, five townships and/or neighborhoods were selected from each. Again, the selection was not random but giving priority to places with a combination of high prevalence of HIV, drug users, and temporary migrants and geographically representing the varied parts of the county/city. This resulted in a total of 40 townships and neighborhoods as the primary sampling units (PSUs).

Finally, in each PSU, all individuals 18–55 years of age were listed in one of four categories: HIV positive, drug users, temporary migrants, and non-migrants. They were crosschecked for multiple listings. If an individual appeared in more than one category, the individual was reassigned to only one category according to the following priority order: HIV, drug user, migrant, and non-migrant. For example, a migrant who was also a drug user and HIV positive, that individual was retained in the list of HIV positive persons and removed from the lists of migrants and drug users. Therefore, all individuals would appear in one and only one of the four lists, which were mutually exclusive.

In selecting individuals, disproportionate probability sampling (Bilsborrow, Hugo, Oberai, and Zlotnik 1997) was used to make sure that the resulting sample would contain sufficient numbers of rare populations, e.g., HIV positive and drug users, but not overwhelmed by non-migrants. A target random sample of about 150 individuals from each PSU was planned and distributed as follows: 20 HIV positive, 30 drug users, 40 temporary migrants, and 60 non-migrants. In each category, sample selection started with randomly picking a person from the list and continued selecting at fixed intervals determined by the ratio between the total on the list and the target number for the category. If a list contained fewer than the target number, everyone on the list was selected. Because not every PSU had the target number of subjects in all categories, the actual sample size in a category varied across PSUs.

During the fieldwork, interviewers visited the sampled individuals, explained to them the purpose of the study, their right to refuse, and compensation for their time, and invited them to participate. If the respondent was absent, a second visit was scheduled. If a respondent could not be reached the second time or refused to participate, a replacement was selected randomly from the original sampling list containing the absent or refused respondent unless there was no one left on the list. Participant refusal was low (3.4%). Of the original sample of 5,570, 5,382 individuals consented to participate and completed a face-to-face interview, which took place in private at the respondents' home or if they preferred, a place away from home. All interviews were conducted in Mandarin or the respondent's dialect if the respondent could not communicate in Mandarin.

Methods and Measures

Given the growing consensus that drug-using behaviors and HIV infections are influenced by both individual and contextual factors, many researchers have argued for multilevel analysis of health behavior and outcomes (Duncan, Jones, and Moon 1996; Korff, Koepsell, Curry, and Diehr 1992; Pickett and Pearl 2001). In the analysis, therefore, data from the community and the individual sample surveys are combined to examine through multilevel modeling both individual and PSU level risk factors of drug abuse and HIV infection. Version 9 of the STATA software is used to conduct the multilevel statistical analyses, which will focus on if and to what extent the individual and contextual variables interact and/or jointly explain participants' drug-using behaviors and HIV status.

The dependent variables are a composite drug-using risk index and the odds/probability of being HIV positive. The composite drug-using risk index is based on five dichotomous variables, indicating whether the respondent ever used illicit drugs, ever shared injection needles, started using drugs under 18 years of age, currently uses drugs, and currently injects drugs. Such a composite index is arguably a better measure than any single dichotomous measure alone (Williams et al. 2001). The higher the index, the higher the HIV risk in terms of drug-using behavior. Cronbach's alpha for the composite index with the survey data is 0.84.

The independent variables include individual and PSU-level variables. The key individual variable is migrant status at the time of interview. Migrant is defined as someone who did not possess the official local household registration in the PSU at the time of interview. In addition, a number of individual demographic characteristics and psychosocial well-being indicators are included in the multiple regressions to control for differences between migrants and non-migrants, which may confound the impact of migration on drug using and HIV.

Gender, age, and marital status are self-explanatory. Education is a seven-category ordinal variable, ranging from 1 for illiterate to 7 for four years of college or more education. Being male, young, single, and less educated are all found to be associated with drug abuse in general and risk drug-using behaviors in particular and consequently HIV infection in China (Lai et al. 2000; Tang et al. 2006; Zhou and Li 1999). Ethnicity is a dummy variable coded 1 for the Han majority and 0 for non-Han ethnic minorities. Being ethnic minority has been found a risk factor for both drug abuse and HIV infection (Choi, Cheung, and Jiang 2007; Deng et al. 2007).

For psychosocial well-being, the analysis focuses on the extent of social isolation and lax social control, measured by two composite scales. For the former, a modified version of the UCLA Loneliness Scale (Russell and Cutrona 1988) is used. Respondents reported on a four-point scale how lonely they felt on each of 20 statements (e.g., How often do you feel that you lack companionship? How often do you feel left out? How often do you feel that there are people you can talk to); answers to the 20 statements were summed to form the "loneliness" scale. Lax social control is measured by a modified version of the Attitudes toward Authority Scale (Emler 1999). Respondents reported yes (1) or no (0) on their personal experience with nine events indicating disrespect for laws or use of "deviant" ways to achieve personal ends (e.g., I have carried some kind of weapon in case it was needed in a fight; I have deliberately traveled on a train or a bus without a ticket; I have stolen bicycle(s) from streets). Answers were then summed to create the lax social control scale. For both scales, the higher the score, the more likely the respondent was socially isolated and had behaved in disrespect for laws or deviant ways, indicating lax social control. Cronbach's alphas with the survey data are 0.80 and 0.71 for the loneliness and the lax social control scales, respectively. Both social isolation and lax social control are potential risk factors of drug abuse and HIV infection (Anderson et al. 2003; Deng et al. 2007; Yang 2006).

For community (PSU) characteristics, urban residence is defined as living in neighborhoods in cities and officially established urban towns. Urban living is typically more stressful and associated with greater anonymity, more liberal behavioral norms, increased diversities in population and social networks, and greater exposure and access to drugs (Frye et al. 2006; Galea et al. 2005; Weiss and McMichael 2004). These features of urban living are arguably conducive to the spread of substance abuse and sexually transmitted diseases, including HIV. The other four community (PSU) characteristics are all defined as the means of the respective official annual statistics (1996–2000) from the community survey. Residential mobility is measured as the mean annual total in and out migrants per thousand working age

(15–64 years of age) residents in the PSU, indicative of the extent of population mixing in the PSU and its exposure to and interaction with the outside.

Prevalence of drug use in the PSU is measured by the mean annual number of known drug users per thousand working age residents. The presence of more drug users may facilitate the spread of substance abuse by setting real life examples for others to follow and creating more tolerable perceptions about drugs and in turn the spread of HIV (Yang 2005). PSU level poverty is measured by the mean percent of households living under the government-defined poverty line. It indicates the overall economic conditions and socioeconomic inequalities in the PSU.

Finally, prevalence of HIV is measured by the mean ranking based on reported annual numbers of HIV infections at county/city level. Due to confidentiality concerns, the actual numbers of HIV infection were converted into an ordered rank before the data were released. Two steps were used to convert the raw data into ranking. First, annual reported HIV infections were regrouped into an interval distribution. Second, a numeric value was assigned in ascending order starting with 0 to represent each interval in the distribution. The resulting county/city level prevalence ranking of HIV ranges from 0 to 9. In the analysis, the mean annual county ranking is assigned to PSUs within the same county/city, indicating exposure to existing pool of HIV in the PSU (Ford and Koetsawang 1991).

Results

Table 19.1 presents the bivariate correlation between the dependent and the independent variables. Different from what is expected, being migrant is negatively associated with drug use and HIV. As expected, being male is positively associated while being married is negatively associated with risk drug-using behaviors and HIV. Although the respective correlation coefficient is low, age, education, and ethnicity are all significantly associated with risk drug using and HIV in the expected directions. Both social isolation and lax social control are positively correlated with the two outcome variables. The correlation between lax social control and risk drug using is particularly strong ($r = 0.54$). Lastly, HIV infection is highly correlated with risk drug-using behaviors, as expected.

For community characteristics, with the exception of poverty, all others are correlated with drug-using behaviors and HIV among residents in the expected directions. Individuals living in an urban place and in a community with higher residential mobility and more drug users and people with HIV/AIDS are associated with higher likelihood of acquiring risk drug-using behaviors or HIV. However, all the coefficients are quantitatively small ($r = 0.1$ or lower), suggesting the correlation between these community characteristics and the outcome variables is on the weak side. Different from what is expected, community level poverty is negatively associated with prevalence of risk drug using and HIV. But again, the correlation is quantitatively weak. For a more definitive analysis, we now turn to multivariate and multilevel analysis.

Table 19.1 Bivariate correlation between risk drug-using behavior and HIV infection and individual and community level characteristics

Independent variables	Sample size	Dependent variables	
		Drug-using risk index	HIV/AIDS
Individual level			
Being migrant	5,382	−0.2804**	−0.1537**
Being male	5,355	0.3082**	0.1503**
Age	5,371	−0.0975**	−0.0309*
Being married	5,376	−0.3696**	−0.1806**
Education[a]	5,372	−0.0551**	−0.0273*
Ethnic Han majority	5,351	−0.0360**	−0.0462**
Loneliness	5,382	0.3475**	0.1710**
Lax social control	5,382	0.5390**	0.2925**
Drug-using risk index	5,382	–	0.4556**
Community (PSU) level			
Urban residence	5,382	0.0745**	0.0589**
Prevalence of drug use[b]	4,768	0.0521**	0.1010**
Total migrants[b]	5,382	0.0598**	0.0195
Poverty level[c]	5,382	−0.1138**	−0.0611**
Prevalence of HIV[d]	5,382	0.0315*	0.0692**

[a]Education is an ordinal variable: (1) illiterate or semi-illiterate; (2) elementary school; (3) junior high school; (4) senior high school; (5) vocational school; (6) 2–3 years college; and (7) four years college or more.

[b]Measured as per thousand working age resident population.

[c]Measured as percent of households under government established poverty line.

[d]Composite rank ranges from 0 (low) to 9 (high).

$^{*}p < 0.05$; $^{**}p < 0.01$.

Table 19.2 presents the multilevel analysis of risk drug-using behaviors. When no independent variables are included, results of the random intercept model (Model 1) show highly significant variations (with a standard deviation of 0.255) across PSUs in the estimated constant (intercept), which indicates the mean drug-using risk score in a PSU. This suggests that without considering anything else, the place where one lives (i.e., the physical and social contexts) plays an important role in influencing his or her HIV risk drug-using behaviors. The control of individual level independent variables in Model 2 reduces the cross-PSU variances in average drug-using behaviors. But the variances (0.044) remain statistically significant, and the intra-PSU correlation has actually strengthened (the coefficient increased from 0.039 in Model 1 to 0.049 in Model 2).

Among the individual level variables, age and ethnicity lost their statistical significance in the multiple regression analysis, suggesting that the observed bivariate association between age and ethnicity and risk drug-using behaviors (Table 19.1) may be mediated through other individual level variables. All other individual level variables remained statistically significant. Consistent with the bivariate association, migrants scored significantly lower on the drug-using risk index than comparable non-migrant residents. Being male, single, and with less education were all associated with more risk drug-using behaviors. Both psychosocial well-being indicators

Table 19.2 Multiple linear regression analysis of individual and community risk factors of HIV risk drug-using behaviors[a]

Independent variables[b]	HIV risk drug-using index		
	Model 1	Model 2	Model 3
Individual level			
Migrant		−0.644**	−0.697**
Male		0.440**	0.438**
Age		−0.001	−0.001
Married		−0.590**	−0.618**
Education[c]		−0.092**	−0.106**
Han majority		−0.002	−0.002
Loneliness		0.030**	0.031**
Lax social control		0.339**	0.334**
PSU level			
Urban		–	0.083
Prevalence of drug use[d]		–	0.002
Total migrants[d]		–	0.001
Poverty level[e]		–	−0.007**
Prevalence of HIV[f]		–	–
Random intercept	0.665**	−0.223*	−0.276*
Sample size	5,382	5,249	4,654
Random intercept variances	0.065**	0.044**	0.021**
Intra-PSU correlation	0.039	0.049	0.024

[a]Results are maximum likelihood estimates based on the "xtmixed" model for continuous dependent variable in STATA software.

[b]The reference categories for variables of migrant, male, married, Han majority, and urban are non migrant, female, single, ethnic minority, and rural, respectively.

[c,d,e,f]See notes a, b, c, d, respectively, in Table 19.1.

$^{*}p < 0.05$; $^{**}p < 0.01$.

were significantly and positively associated with drug-using risk, suggesting that people who were socially isolated and had experiences of disrespect for laws or social norms were more likely to also have risk drug-using behaviors.

When individual and community characteristics were examined together (Model 3), the coefficient estimates for all individual level variables hardly changed. This indicates that the associations between individual demographic and social attributes and drug-using behaviors are largely independent of community characteristics. Of the four PSU characteristics, only poverty level remained significant. In contrary to what would be expected, PSU level poverty was associated with a lower prevalence of risk drug-using behaviors. In other words, residents in communities with higher poverty level were significantly less likely to have risk drug-using behaviors.

While both the random intercept variances and the intra-PSU correlation were more than halved in Model 3, they (0.021 and 0.024, respectively) remained statistically significant. It appeared that through whatever mechanisms and in addition to poverty and the other PSU characteristics included in the analysis the social and physical environments of residence in general exerted significant influence over residents' risk drug-using behaviors.

Table 19.3 Multiple logistic regression analysis of individual and community risk factors of the odds of being infected with HIV[a]

Independent variables[b]	Odds of being infected with HIV		
	Model 1	Model 2	Model 3
Individual level			
Migrant		0.246**	0.214**
Male		1.685*	1.530
Age		1.013	1.011
Married		0.772	0.711*
Education[c]		0.991	0.988
Han majority		0.766	0.793
Risk drug-using index		2.508**	2.467**
PSU level			
Urban		–	1.793
Prevalence of drug use[d]		–	1.019*
Total migrants[d]		–	0.996
Poverty level[e]		–	0.986
Prevalence of HIV[f]		–	1.117
Random intercept	0.045**	0.008**	0.004**
Sample size	5,382	5,249	4,654
Random intercept variances	1.043**	0.694**	0.491**
Intra-PSU correlation	0.241	0.174	0.130

[a]Results are based on the "gllamm" model in STATA software and expressed as the odds ratios associated with corresponding one unit change in the independent variables.
[b]The reference categories for variables of migrant, male, married, Han majority, and urban are non-migrant, female, single, ethnic minority, and rural, respectively.
[c,d,e,f]See notes a, b, c, d, respectively, in Table 19.1.
$^{*}p < 0.05$; $^{**}p < 0.01$.

For the risk of HIV, results in Table 19.3 again suggest the importance of residential social and physical environments. For example, when no individual or PSU level correlates were included (Model 1), the average odds of being infected with HIV varied significantly across PSUs. This, along with the sizable intra-PSU correlation (0.241), indicates that like drug-using behaviors the odds of being HIV positive were spatially significantly correlated with where one lived.

Both the cross-PSU variances and the intra-PSU correlation were considerably reduced once the individual level variables were included in the analysis (Model 2). But the average odds of HIV infection remained varied significantly across PSUs. Among the individual level correlates, being migrant was correlated with significantly lower odds of being HIV positive. In fact, for a migrant, his/her odds of being infected with the AIDS virus were less than 25% (OR = 0.246) of that of a comparable non-migrant resident. This was particularly striking in light of the fact that the drug-using risk, a significant risk factor of HIV, had already been controlled.

Consistent with the literature and the earlier bivariate correlation analysis, males had significantly higher odds of being infected with HIV than comparable females. Risk drug-using behavior was the only other individual level variable that remained statistically significant in Model 2. As expected, drug abuse was a significant and powerful risk factor of HIV infection. Other things being equal, each drug-using behavior displayed by respondents (i.e., increase of one unit in the composite drug-using risk index) would more than doubled the odds of HIV infection.

The coefficient estimates for individual level correlates remained mainly unchanged when PSU level characteristics were controlled for in Model 3. Of the five PSU characteristics, only the prevalence of drug use remained statistically a significant risk factor of HIV infection. None of the others, not even the prevalence of HIV in the PSU, was significantly correlated with the odds of being HIV positive. Considering that most of the PSU characteristics were significantly correlated with HIV infection at the bivariate level (Table 19.1), the results suggest that the influence may be mediated by individual level variables, particularly risk drug-using behaviors. The fact that the cross-PSU variances and the intra-PSU correlation remained statistically significant, although further reduced in Model 3, suggests that contextual factors other than those included in the analysis may be important mediators between residential contexts and the likelihood of being infected with HIV.

Discussion and Conclusions

Within two decades, AIDS has evolved from a disease of foreigners to an epidemic affecting every geographic location in China. Drug abuse has been and remains to be a key contributing factor in the AIDS epidemic. In search for answers, both the media and the literature have often blamed the increasing migration for the spread of drugs and HIV in China. While migrants may well be more vulnerable to drug abuse and HIV, little research in China has actually compared migrants with comparable non-migrant residents in the likelihood of drug abuse and/or HIV infection. Further, despite the fact that prevalence of drug abuse and HIV vary significantly across geographic locations, research on drug abuse and HIV in China has rarely incorporated attention to community social and physical contexts. This chapter tries to fill some of the void. Using data from a population-based survey that included both migrants and non-migrants and applying multilevel modeling technique, the chapter focuses on the impact of migration and residential contexts on drug abuse and HIV, emphasizing both individual and contextual risk factors.

The results suggest that being migrant is associated with significantly less HIV risk drug-using behavior and lower odds of being infected with HIV. This appears to contradict the often-negative image about migrants portrayed by the media and in the literature. It is also inconsistent with previous findings using the same data but based on single drug use indicators and without incorporating contextual factors in multilevel modeling (Yang 2006; Yang, Derlega, and Luo 2007). Differences in measurement (single vs. composite drug-using indicator) and model specifications (single vs. multilevel modeling) may help to account for the discrepancy. More importantly, data limitations may have introduced biases that underestimate drug use and HIV among migrants. For example, the survey used official drug user and HIV registries as the sampling frameworks. This would have over sampled non-migrant residents because except for newly reported cases these registries were compiled based on local household registrations. Migrants would not appear on the registries (they would on the registries in their places of origin) because by definition they did not have the local registration at the time of the survey.

Although the survey questionnaire contained detailed questions on drug use and HIV, which would allow migrants to reveal their drug use and HIV status if they had answered the questions accurately, it would unlikely have fully corrected the potential underestimate of drug use and HIV among migrants. Future research on migration and drug use/HIV in China needs to develop innovative sampling techniques that can produce probability samples of drug users and people living with HIV/AIDS among both migrants and non-migrants. This will be methodologically challenging but necessary in order to obtain unbiased samples by migrant status, which are in turn necessary for comparative analysis of drug abuse and risk of HIV by migrant status. Until a link between migration and drug abuse/HIV can be reliably established, it appears fair to conclude that other things being equal migrants are not necessarily or always more vulnerable to drug abuse and HIV as a result of risk drug-using behaviors.

Consistent with previous findings, being male, single, and with less education are all correlated with more risk drug-using behaviors, so are social isolation and lax social control. But being ethnic minority is not significantly associated with risk drug-using behaviors. None of the demographic characteristics appear to be important correlates of HIV infection, of which the only consistent risk factor is drug-using behavior. The lack of significance of ethnicity confirms that both drug and HIV epidemics in China are no longer limited to ethnic minority populations, as they were earlier in the epidemics.

With the only exception of poverty level, the community characteristics as measured have no significant and independent impact on drug-using behaviors in the multiple and multilevel analysis. Different from the common belief in China that drugs are mainly a problem in poor rural areas, drug abuse is found in this study very much an urban problem and is not likely the result of poverty. Similarly, prevalence of drug use in the community is the only community-level risk factor of HIV that remains statistically significant once individual and community characteristics are examined together in the multilevel model. For both drug-using behavior and HIV, the control of community characteristics considerably reduces the random intercept variances and the intra-PSU correlations, suggesting that these PSU characteristics do exert important influence over individual drug-using behaviors and HIV risk. However, the influence of most community characteristics may not be direct but mediated through individual level factors.

The most consistent finding of the study is that individual drug-using behavior and HIV risk are significantly influenced by residential contexts. More research seeking to identify the contextual risk factors is needed for both theoretical understanding of the links between residential contexts and substance abuse and/or HIV and for effective policy prescriptions to moderate the negative impact of social and physical environments. Future studies of drug-using behaviors and HIV risk in China must pay attention to contextual influences and try to understand mechanisms through which community context influence individual behavior. Policy and program interventions to reduce risk drug-using behaviors must address contextual risk factors.

Chapter 20
Placing the Dynamics of Syringe Exchange Programs in the United States

Barbara Tempalski

Abstract Drawing upon the broader health, social, and political geography literature this paper outlines a framework for considering place-based processes through which syringe exchange availability may be understood. It is argued that the geographic distribution of syringe exchange programs (SEPs) in the United States is linked to the social and political conditions of particular localities through three place characteristics: (1) *structural constraints*; (2) *social and spatial distancing of injection drug users*; and (3) *localized action.* Although SEPs remain a controversial issue and face ongoing obstacles from the government, law enforcement and local communities, they continue to operate through the efforts of grassroots organizations and local activists. Action on this issue occurs locally, and the characteristics of place-based factors will affect whether particular areas adopt SEPs.

Introduction

Injection drug use (IDU) has many public health implications, notably its role as a risk factor for infectious diseases such as HIV/AIDS, hepatitis B and C, and endocarditis, (Novick et al., 1997) and, increasingly the co-infection of HIV/AIDS and hepatitis C (Hagan et al., 2005). Preventing the spread of blood-borne pathogens through IDU requires a diversity of approaches, including treatment and detoxification programs, social services, primary health care, outreach, user involvement,

B. Tempalski
National Development and Research Institutes, 71 West 23rd Street, 8th Floor, New York, NY 10010
e-mail: tempalski@ndri.org

This project was supported by National Institute of Drug Abuse grant R01 DA13336 (Community Vulnerability and Response to IDU-Related HIV).

Reprinted from Tempalski, B. (2007). Placing the dynamics of syringe exchange programs in the United States. Health and Place, 13(2), 417–431. With permission from Elsevier.

and access to sterile syringes and injection equipment for people who are injecting (Brooner et al., 1998; Paone et al., 1999; Strathdee et al., 1999). In many parts of the world, syringe exchange programs (SEPs) have been adopted as a critical component of HIV and hepatitis C prevention for IDU. Injectors of heroin and other drugs have ready legal access to sterile syringes throughout much of Western Europe and Australia, as well as in many developing countries (e.g., Brazil and Nepal). In many of these countries, pharmacies sell syringes over the counter, and municipalities or other public bodies run SEPs. In contrast, the United States (U.S.) public health response to blood-borne infections among IDUs has been minimal and geographically uneven (Tempalski, 2005). In most parts of the U.S. legal access to syringes either through pharmacy sales or SEPs remains difficult, and is illegal is many places. Further, since 1998 the federal government has denied funding for SEPs. Understanding the geographic dimensions of SEPs is critically important because these programs have been shown to reduce transmission of HIV and, increasingly, hepatitis B and C, as well as to provide important medical and social services to IDUs (Hurley et al., 1997; MacDonald et al., 2003).

Although SEPs remain controversial and face ongoing obstacles from government and local communities and businesses, they continue to operate and expand through the efforts of grassroots organizations and local activists. Drawing on the broader public health and social and political geography literature, this paper outlines the effects of place characteristics through which geographic variation in SEP availability may be understood. It is hypothesized that the uneven geography of SEPs in the U.S. involves political, socioeconomic, and organizational characteristics of localities that affect service needs, resources, opposition, and localized action.

Place is understood in geography as space endowed with meaning(s) and experience(s) (Cooper et al., 2004; Cresswell, 2004; Kearns and Joseph, 1993), an idea that recognizes that society and space are mutually constituted (Massey, 1997; Soja, 1980; 1997). Central to this concept is that "past and present interactions of social, political, and economic systems give shape to space", and that simultaneously, space configures social relations (Cooper et al., 2005; Kearns and Joseph, 1993; Sibley, 1995). Geographers have been concerned with the role of power embedded in this configuration process, and the interactions between power and sites of resistance and opposition, and human agency (Castells, 1983; Pile, 1997; Smith, 1994). Castells argues that "resistance takes place as a result of demands around three basic structural issues: (1) collective consumption, such as housing, schools, welfare provision and so on; (2) the defense or expression of cultural identities; and (3) the working of the state and/or local government" (taken from Pile, 1997, pp. 9). Thus, economy, culture, and state must be accounted for in the geographical analysis of place.

The points at which power and resistance intersect are areas that shape and give rise to collective action (Martin and Miller, 2003). As Castells argues, Martin (2003) also suggests that resources have been a key to motivating action; this is most evident in the struggle over quality of life issues and the desire of identity groups to be acknowledged within the broader culture. Geographers thus argue that place provides an important mobilizing discourse and identity for collective action (Marston, 2003; Martin, 2003; Miller, 2000). People and groups organized into coalitions actively shape places, and the actions, driven by institutional structures, are never the same from place to place. Thus, people and institutions actively shape place, and simultaneously, place influences the structure of how resources are distributed to individuals and groups (Dear, 1988; Harvey, 1997; Johnston, 1991; Kearns and Joseph, 1993).

Examining the efforts to institute SEP in three U.S. metropolitan areas gives insight into how these forces configure grassroots and institutional response. Bluthenthal (1998) describes an example of collective action to establish SEPs, arguing that government inaction created the political opportunity structure that encouraged harm reduction activists in Oakland, California and elsewhere to set up SEPs. In other situations, strong support by local individuals can lead to wide support for SEPs, as the Tacoma case illustrates. The first publicly-funded SEP in the U.S., in Tacoma, Washington, resulted from the actions of Dave Purchase and others. Recognizing that many IDUs were dying of AIDS and lack of government response, Purchase set up a street-based SEP. Purchase's efforts strongly influenced local program adoption which led to the first publicly funded street-based SEP in the United States.

Yet, in other places, local direct action has been less successful. Attempts to set up SEPs in New Jersey point out political processes producing disconnects between need and services. Injection drug use is the most frequently reported risk behavior among HIV-positive individuals in New Jersey. In 2000, New Jersey reported over 19,000 cumulative IDU-related AIDS cases, and 2.3% of the Jersey City metropolitan area population was injecting drugs (State of New Jersey Department of Health and Senior Services (NJDHSS), 2002). However, the governor and city officials opposed distributing sterile syringes to IDUs, and they used arrests to suppress the state's only publicly visible SEP.

These examples from three places demonstrate different response to the HIV epidemic among IDUs. They illustrate varying philosophies and resource responses within each distinct place. Each one's social, economic and political context has shaped the general characteristics of place, and in turn, given shape and meaning to IDUs' health and access to HIV prevention services.

Thus, in this article, the concept of *place* allows us to engage the set of social and political relationships that create a spatial context in which differential responses to injection drug use–related HIV are structured. As such, this paper offers a context for understanding how place-specific processes—including need for services, local resources, opposition, and grassroots political action—affect the geographic distribution of SEPs in the United States. In this research, it is argued that the geographic distribution of SEPs in the United States is linked to the social and political conditions of particular localities through three place characteristics: (1) *structural constraints*, as they relate to disease vulnerability and service needs; (2) *social and spatial distancing of IDUs* and associated services; and (3) *localized action* concerning disease prevention and health service provision for stigmatized groups. These place-based processes are interrelated: the need for services may lead to activism, which can reduce stigma and ultimately lead to service provision.

This research directly and primarily contributes to geographic knowledge on HIV/AIDS prevention as it relates to health care needs, services, and activism. It also contributes to the broader literature that emphasizes how place-based political, social, and economic processes—such as local economic relations and inequalities, community responses, and grassroots activism—affect the geographies of health care services for people with HIV/AIDS and other stigmatized groups. Thus, this research improves our understanding of the geography of HIV/AIDS prevention services in the hidden and highly stigmatized population of IDUs.

This article expands our knowledge of how place-based political and social processes have led to variations in HIV/AIDS prevention-related services for IDUs, specifically SEPs. At the national level, the U.S. federal government's systematic denial of SEP funding influences the social and political environment of all U.S. localities under consideration. The conceptual framework discussed here shows how the lack of an association between the need for a program and the presence of a program implies that current U.S. political systems are not responding adequately to the important public health problem of infection-related disease. Response to this situation is a matter of local grassroots action, and the characteristics of place-based social and political processes will affect whether particular areas confront this and adopt an intervention.

Activism and Politics: Syringe Exchange Development in the U.S.

Whether the U.S. federal government should support the distribution of sterile syringes for injection drug use has been a highly controversial topic in public policy discourse. In 1988, the U.S. Congress barred federal funding for SEPs, with the proviso that the ban could be lifted by the Surgeon General if he or she determined that such efforts could reduce the risk of HIV infection without encouraging illicit drug use. In 1997 an independent consensus panel convened by the National Institutes of Health concluded that SEPs reduce the spread of injection-related HIV and do not encourage illicit drug use (U.S. General Accounting Office, 1993). Moreover, at least 20 medical, scientific, and other professional associations have endorsed syringe exchange as an effective prevention strategy (Vlahov et al., 2001). Nonetheless, federal funding continues to be withheld.

In contrast, in many parts of the world, SEPs have been accepted as essential components of HIV prevention and other blood-borne diseases. The first SEP was developed in Amsterdam in 1983 as a method of reducing transmission of hepatitis B, another common blood-borne infection among IDUs. After the discovery of HIV infection among IDUs in Amsterdam, the exchange program was expanded and SEPs were implemented in other Dutch cities (Grund et al., 1991). Presently, throughout much of Europe, Australia, Canada, and New Zealand, users of heroin and other injection drugs have ready legal access to sterile syringes.

While syringe exchange was clearly taking on a public health role across European cities, it would remain only a radical idea in the United States for at least another two years. Then, beginning in November 1986, Jon Parker, a recovering IDU and student at Yale University School of Public Health, formed a group called the

National AIDS Brigade and started the first underground needle exchange program[1], distributing and exchanging needles on the streets of New Haven, Connecticut. By the late 1980's, needle exchange programs were operating in over a dozen cities – often in the face of harassment, arrests, and threats of violence. Some programs eventually came to operate with the support of local health authorities using public funds (e.g., Tacoma and New Haven). Other programs were supported and funded through private and charity funds and operated under the consent of the local authorities, as in San Francisco, California; Boulder, Colorado; and Portland, Oregon. Finally, other programs operated under the radar screen of local law enforcement and without official authorization.

By the year 2000, approximately 154 syringe exchange programs were operating in the United States (Beth Israel Medical Center, 2000). Figure 20.1[2] depicts the spatial distribution of SEPs across the largest 96 metropolitan statistical areas in the United States; high geographic clustering of programs is located in the northeast and

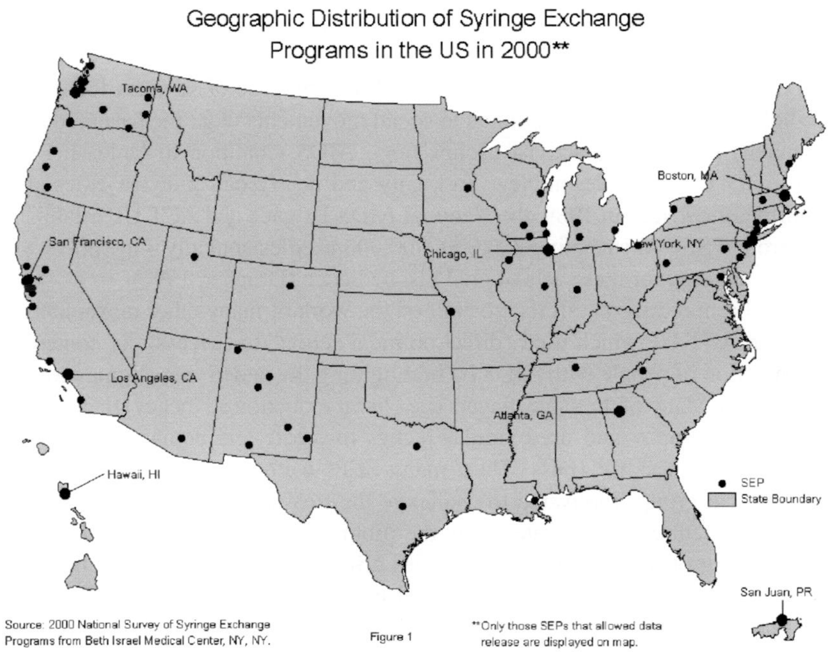

Fig. 20.1 Geographic Distribution of Syringe Exchange Programs in the US in 2000

---

[1] Syringe exchange and needle exchange programs are the same type of program; the terms are used interchangeably throughout the text.

[2] For confidentiality reasons, reporting on the specific names of some cities or programs are prohibited.

northwestern and western regions; while the south and mid-western regions remain low in initiating SEPs.

Despite their proven effectiveness in HIV prevention and despite support from local public health authorities and research institutions, SEPs remain a controversial issue in the United States. They face ongoing obstacles in the form of legal challenges, funding constraints, disapproving attitudes of local residents or business associations, and police harassment of clients and service providers. In the absence of federal support for SEPs, responsibility for implementing this public health measure has fallen to state and local government entities; yet most state and local officials have been reluctant to authorize such programs, and public funding and regulation of syringe acquisition vary by state and locality. Additionally, in many states and localities, the acquisition and exchange of syringes are restricted under prescription and drug paraphernalia laws. Despite this variation, existing SEPs continue to expand and in some areas new programs are beginning; commonly this work is by citizens establishing "facts on the ground" and simply setting up SEPs on their own initiative. More than half the SEPs in the United States are run by nongovernmental organizations and were initiated by activists and local community members (Tempalski et al., 2003a; 2003b).

In the United States, the development and maintenance of services for IDUs—specifically, SEPs—have been linked to social movements (e.g., the harm reduction movement) and specific activist groups (e.g., AIDS Coalition to Unleash Power (ACT UP)). For example, in New York City and nearly half a dozen cities along the northeast corridor in 1989, the National AIDS Brigade and ACT UP established underground SEPs distributing 200,000–300,000 needles annually – members were regularly arrested for these acts of civil disobedience (Drucker, 1990).

These harm reduction efforts are built on the work of many other movements. In particular, ACT UP which used "direct political action" to successfully contest the stigmatization of people with AIDS by highlighting the underlying stigma rooted in homophobia. Many of the same players (i.e., harm reduction advocacy) then adopted a similar philosophy and used similar tactics to create and demand "direct HIV prevention services" for IDUs. Thus, many SEPs were started by AIDS activists from political groups motivated to challenge the treatment and marginalization of IDUs. Harm reduction advocates view the problems of illicit drug use and addiction as arising primarily from drug policies criminalizing the user, and they work toward minimizing the medical and social consequences of drug use. The harm reduction philosophy fosters alternative models to conventional health and human services, as well as drug treatment, while challenging traditional client–provider relationships.

Previous research has shown that political factors and local direct action may substantially influence SEP creation and sustainment (Tempalski et al., 2003b). Using the interpretive framework outlined in this paper, study results indicate that SEP formation, at least in the United States, is based not on relative *need* across localities, but on social and political factors—including active solidarity (ACT UP) and potential solidarity (large numbers of men who have sex with men (MSM) creating greater concern for HIV/AIDS issues)—that may help to form, defend, and possibly sustain SEPs. Hence, AIDS activism supported the harm reduction movement, and local efforts of groups such as ACT UP fostered the harm reduction philosophy and the U.S. expansion of SEPs.

The Relevance of Structural Constraints in IDU-related HIV Vulnerability

Research suggests that health differences across social class and population can be influenced by the social, economic, and political contexts of people's lives (Farmer, 1999; Krieger, 2001; Wallace and Wallace, 1998). For low income or "marginalized" populations, whose lives are narrowly circumscribed by inadequate resources and power, context has critically important effects on health. Prior research documents that social and economic conditions are related to a variety of health and social indicators although the directions of some relationships vary among studies. Ruhm (2000) found that economic good times are associated with worse physical health and with increases in alcohol consumption, though perhaps with better mental health. In a meta-analysis of studies during the 1980s and 1990s, Jin et al. (1997) found that higher unemployment rates are related to higher overall mortality, cardiovascular-related mortality, and suicide.

Krieger (2001) argues that the production and distribution of disease must be understood through constructs of social and economic relations characterized by inequalities and embedded in the larger social, economic, and political structures. Krieger identifies two possible mechanisms through which vulnerability to disease may be understood: (1) that resources and harms are distributed along the lines of social and economic relations; and (2) that these relations condition individual behavior through factors such as relative and absolute deprivation.

The extent of relative deprivation within a society largely determines the degree of harm across particular groups. For example, spatial de-concentration of urban minority communities in New York City (caused by neglect and "planned shrinkage" policies spurred by white governmental fear of urban black revolution) fueled the reemergence of tuberculosis in the 1980's. Deteriorating living conditions, social decay, and reduced fire-fighting capacity contributed to an increased number of fires; the resultant destruction of housing stock dispersed South Bronx communities—and, consequently, multidrug-resistant tuberculosis and AIDS spread within and to adjacent neighborhoods (Wallace et al., 1997).

Although the U.S. HIV/AIDS epidemic has historically affected two groups—MSM and IDUs— in recent years the broadening impact of the epidemic across classes, races, genders, and sexual orientations indicates a growing need for a variety of educational, prevention and treatment services. Resources availability, for example, can vary greatly affecting where and when prevention and treatment services can be implemented. In seeking to understand the factors that shape the HIV/AIDS epidemic among IDUs, and ways in which societies and communities have responded to it, a growing body of research has focused on structural factors that facilitate HIV transmission and its concentration with particular geographic areas and populations (Farmer, 1999; Singer, 1998; Sumartojo, 2000; Taussig et al., 2000). Structural constraints are defined here as social, economic and political environments that shape and constrain individual, community, and societal health outcomes.

Previous research suggests HIV prevalence or incidence among IDUs in a locality is a function of policies about purchasing syringes (Friedman et al., 2001; 2000), syringe exchange (Des Jarlais et al., 1996; Hurley, 1997; MacDonald et al., 2003) urban development (Friedman et al., 1999) or urban "desertification" (Wallace and Wallace, 1998), and indicate that these policies might be altered to reduce HIV transmission. Structural constraints relating to syringe acquisition, which differ by state and locality, can significantly reduce the ability of IDUs to purchase and possess sterile syringes. State and local laws concerning the sale, distribution, and possession of syringes can prohibit any distribution of syringes. Policy processes and regulations governing syringe sales at the state and local levels (e.g., prescription and drug paraphernalia laws) influence the legality of syringe access, the conditions of syringe distribution, and the use of government funds to distribute syringes.

Taussig et al. (2000) look at the legal and regulatory barriers that restrict pharmacy sales of syringes to IDUs and discuss how reducing these barriers can facilitate access to sterile syringes for IDUs and reduce ones' vulnerability to HIV infection. Currently, fourteen states have prescription provisions[3] (Burris, 2005), and thirteen states[4] and the District of Columbia have affirmatively authorized SEPs (Burris, 2002). Additionally, Friedman et al., (2001) research showed that restricting syringe access were associated with HIV transmission. The study indicated that MSAs with anti-over-the-counter-syringe laws had a higher mean HIV prevalence among IDUs than other MSAs (13.8% vs. 6.7%).

Economic and political variables may influence HIV rates among IDUs, and the degree of IDU stigmatization in a locality may be positively associated with HIV transmission. In the case of IDUs, metropolitan-level economic trends affect both their presence and HIV status. Friedman et al. (2004a) found the percent of the population in poverty was positively associated with the population density of injection drug use, HIV prevalence among IDUs, change in HIV prevalence among IDUs, and both black/White and Hispanic/White disparities in IDU population density. Additional research by Friedman et al. (2004b) showed increases in both unemployment and poverty rates between 1990 and 2000 were associated with a higher population density of IDUs in 1998.

The broader sociopolitical system of municipal governance can affect individual IDU behavior and health. For example, policing patterns and arrests can affect drug users' access to services, even in states where syringe access is legal. Cooper, Moore

---

[3] California, Connecticut, Florida, Delaware, Illinois, Indiana, Maine, Massachusetts, New Hampshire, Nevada, New Jersey, New York, Pennsylvania, Virginia and Virgin Islands.
[4] California, Connecticut, District of Columbia, Hawaii, Maine, Massachusetts, Maryland, New Hampshire, New Jersey, New Mexico, New York, Rhode Island, Vermont and Washington.

et al. (2005) found that particular drug-crackdown tactics by police affected IDUs' ability to safely inject. Drug-related arrests were also found to affect IDUs' use of health and treatment services. Research by Bluthenthal et al. (1997) determined that program participation declined following arrests of staff at illegal SEPs. Furthermore, Bluthenthal et al. (1999a) found that IDUs who feared arrest are at least twice as likely as other IDUs to borrow syringes and other injection equipment, perhaps because they hesitate to carry syringes. Bourgois et al. (1997) also found that impoverished or minority IDUs were especially fearful of arrest and therefore particularly reluctant to carry syringes increasing the risk of sharing injection equipment.

The mounting evidence of social and political structuring of race, class, gender and sexuality indicates the need for a variety of approaches to developing services and service provision strategies. Des Jarlais et al. (1995) work in five cities where both HIV prevalence and incidence remained low among IDUs over a minimum of six years suggest three factors contributed to "public health control" over HIV transmission among IDUs: (1) beginning prevention efforts when prevalence was still low; (2) using community outreach to establish trusted communication between health workers and drug users; and (3) providing a ready supply of sterile injection equipment.

However, the design and implementation of HIV prevention strategies to meet the needs of IDUs—such as SEPs or methadone maintenance programs—face challenges to acceptability and feasibility at the local level. For example, while local residents may recognize the pervasiveness of drug use in their community, they may not welcome services for users into their neighborhood (Dear et al., 1997; Wilton, 2000). The reluctance to allow such services is partly attributable to local views of drug users and their behavior as deviant (Daker-White, 1997; Takahashi, 1998). This perception raises important questions relating to the marginalization and exclusion of populations based on gender, sexual orientation, race, and class. These social and political processes rooted in place result in a non-distribution of resources and goods to those in society with the greatest need[s] (e.g., mentally ill, homeless, or HIV-infected persons, prevention and treatment services for IDUs) and thus, largely determine the location of services.

Social and Spatial Distancing of IDUs: Implications for HIV Prevention Services for IDUs in the United States

To explain resistance to the provision of human service facilities social scientists turn to theories of social distancing embedded in concepts of difference: social distancing fuels the stigmatization of individuals, groups and places. Conceptualizations based on difference are rooted in moral beliefs closely related to local cultures; this is, the sense of one's moral responsibility for others decays as one becomes increasingly distanced from others (Smith, 1994). The social and moral distance created by such stigma can carry over to the location and establishment of many programs for services-dependent groups, including HIV-prevention services for IDUs. Syringe exchange programs, for example, can be difficult to establish even in communities hard-hit by injection drug use–related HIV transmission.

Stigma regarding drug use and HIV/AIDS involves not only the devalued individuals and groups but also the locations where such persons receive services; thus, the stigma is extended from person to place. According to Takahashi (1998), the built environment reflect the constructed stigma in society, resulting in a stigmatized landscape in which public and community spaces are viewed as less productive, more dangerous, and less desirable. Within this context, human services facilities can reinforce stigma, because the stigma of places is conditioned partly by the built environment and the types of services provided and clients served. Thus, Takahashi's (1998) research suggests these place-based factors maintain and enforce community boundary definitions, reinforcing current spatial relations of stigma.

During the HIV/AIDS epidemic of the 1980s and 1990s, changing epidemiological patterns and the effects of stigma varied among diverse populations (Farmer, 1992). Although some aspects of this stigma are shared by all HIV-infected persons, others are population-specific. Discrimination against persons with AIDS drew from and compounded preexisting stigmas (especially vis-à-vis MSM and IDUs). As a result, negative attitudes towards drug use enable the institutionalization of prejudices against drug users (Friedman, 1998), and people who have acquired HIV by injecting drugs are doubly stigmatized—for being HIV-positive and for being an "addict."

Stigmatization associated with drug use and HIV/AIDS extends beyond the devaluation or punishment of individuals and groups and often becomes embodied in the location of services used and frequented by clients. This transfers stigma from persons to services and often the places where those services are located. In the United States, this transfer of stigma often embroils harm reduction focused services such as SEPs and can be embodied in community attitudes, law enforcement, and political opposition within various levels of the government (Tempalski, 2005).

The most powerful opposition to establishing SEPs in the United States has come from government authorities or law-enforcement representatives (e.g., district attorneys, police, and politicians) and community leaders (e.g., clergy and business association members). In fact, when these influential groups support SEPs, local resistance to SEPs is greatly weakened. However, the processes that lead to the establishment of SEPs are complex and reflect the social and historical circumstances in each place-based context. In some cases, SEP support was brought on by the direct actions of influential local individuals. For example[5], the establishment of an SEP in Tacoma, Washington resulted from the direct action of a local activist, Dave Purchase, and others. After recognizing that many IDUs were dying of AIDS and that the government was doing nothing to help, Purchase set up a SEP in Tacoma in 1988. Purchase describes the situation in this way:

[5] Interviews provided by the Community Vulnerability and Response to IDU-related HIV Project, Survey of Community Experts (funded by the National Institute on Drug Abuse, Grant# R01 DA13336).

> People were going to die. I had some time on my hands. I had some friends that did help out a lot and never got the credit they deserve. We started in the summer of '88 and every couple of years there's another brouhaha with the same old argument. The fact of the matter is that there have been enough local political people with backbone that have supported us and so has the health department, and so we've weathered attacks. And politics is still a number one problem. AIDS is all politics, it's not science and stuff like that, its all politics (Purchase, 2003).

Purchase's efforts in the late 1980s brought about political support in favor of SEPs and government funding for programs early on in the epidemic. Both are clearly very important for program adoption and sustainability and, in turn, for keeping HIV prevalence low. Efforts of political advocacy groups like ACT UP were just beginning to develop at the time Purchase was working toward gaining local support for a SEP. Thus, his efforts strongly influenced program adoption in Tacoma and led to the first publicly funded street-based SEP in the United States.

In another city, a pro-SEP client believed that the city council's view on syringe exchange was influenced by a very organized community group who managed to vote in city council members who believed that drug use was morally wrong and would oppose syringe exchange to maintain there consituents' support. He had this to say about the locate opposition in his city:

> The Council members are saying it sends the wrong message: one person understands it and sees the health benefits but still feels it sends the wrong message to children, another person is adamantly opposed to anything like that and says all addicts should go into treatment and he actually prefers religiously-based treatment (we didn't even try talking to him he's just so outrageous), the Mayor feels it's a law enforcement issue and therefore you shouldn't declare a state of emergency to support an illegal activity and they can go into recovery and force recovery in the jail system...

Additionally, a different respondent in this city had this to say about trying to get SEP legalized in that city:

> So we are trying to override that [city council vote] because you have basically politicians who are embedded in their conservative morality and that morality is driving politics not public interest or healthcare. So it's pretty scary.

Some AIDS advocacy groups and public health researchers argue that the conservative views of a local community, backed by a Republican majority in local government, can knock down any initiative to establish a legal SEP. In explaining why a SEP did not exist in his city, one local activist described the political environment in the following terms:

> The majority of the city is run by the Republicans, and [the local law enforcement] are ultra-conservative... they can go wherever they want, whenever they want, and clean up what they wish. They make a show of cleaning up prostitutes...

Additionally, a shift in political leadership and SEP support/opposition are not uncommon. In this particular scenario the state legislators who had originally approved syringe exchange were replaced with new officials who did not sympathize with the SEP and turned against it. An SEP supporter described the situation in this way:

> The police were educated about it [syringe exchange] and were very cooperative. This has changed within the last two years, when we've come under public attack and have had more problems with the police.... It's a nightmare. There are new legislators, new cops, everything's changed.... Now there are letters being written against us, we're in the papers and on TV and it's turned into this huge controversy.

Further opposition can result from on-going police harassment of service providers or difficult relations between police and programs stemming from the illegality of drug use and of distributing drug paraphernalia, such as syringes. A local harm reduction worker regarded law enforcement issues relating to drug paraphernalia laws as the biggest obstacle to trying to establish an SEP there:

> There were attempts to set up a public program early on, about 8–10 years ago, but the police would not permit it. No one has made an attempt to do anything publicly in the last five years. It is not illegal to set up a syringe exchange, but it is illegal to carry drug paraphernalia. The police arrested everyone who tried to set up programs.

One SEP worker provided an example of police harassment of SEP employees and clients based the current state and local syringe law in that city:

> If you work for it (the SEP) they arrest you. There's that kind of harassment. There are people being arrested all the time for possession of syringes. Even if they have a card. . .They can show a card and they're told that's no good here. I don't know of any workers going to jail, but I know participants have gone to jail for possession of syringes.

Repressive policies that influence local social and political environments can make program implementation very difficult. Institutional repression based on political policies such as the declaration of the "war on drugs," reinforced by local law enforcement policies, may prevent SEPs from being established. Still other forms of opposition lie in police harassment of service providers and clients, or difficult relations between police and programs created by state drug paraphernalia laws, for distributing or carrying syringes (Bluthenthal, 1999b). Thus, local activists often find it easier to implement underground SEPs than to fight difficult institutionalized policies.

In many localities, prejudice attached to drug use is reinforced by the illegal and covert nature of illicit drug use. Included in this prejudice is a negative perception of injection drug users (e.g., they are junkies and criminals) and IDUs with HIV are poor and non-white (Friedman, 1998). Additionally, in some communities, residents view HIV as a natural, acceptable consequence of deviant behavior (Alonzo and Reynolds, 1995), which intensifies the stigma and creates barriers to initiating HIV prevention services for IDUs. In this case, it is not a negative experience with syringe exchange that leads to opposition to these services, but the overall moral judgment that drug users are evil. For persons holding this view, opposition to SEPs results from an overall equation of HIV with immorality, more so than any personal experience with syringe exchanges. As such, resistance to SEPs may arise even in the absence of adverse events, and such stigma-based resistance remains strong and likely to reoccur.

Localized Action: The Harm Reduction Movement as Disease Prevention

Public and sociopolitical responses prompted by the HIV/AIDS epidemic surpass previous response to past epidemics. Many factors throughout the epidemic (e.g., the "war on drugs," characteristics of affected groups, criminalization of drug treatment, and racial fears) have influenced society's response. These elements of the epidemic have demanded new strategies of public health response to meet the needs of those stigmatized by the disease. Marginalization of specific groups and individuals has been countered by social movements in instances where the state is unwilling or unable to stem the widening health disparities due to HIV/AIDS.

Social movements sparked by grassroots activism have often represented the initial response to a health crises and significantly shaped public health policy (Bluthenthal, 1998; Hoffman, 2003; Szreter, 1988) and health service provision (Petchey et al., 1998). The great sanitary movement in late–18th century Britain, for example, was driven chiefly by local activists appalled by the living and working conditions of the urban poor (Porter, 1999). Other social movements and activism, from the feminist health movement to AIDS activism, have restructured health-related issues, including treatment services, health care reform, AIDS policy, and the destigmatization of particular groups (Banzhaf et al., 1992; Barnett and Barnett, 2003; DeMacro and Johnsen, 2003).

In the HIV/AIDS epidemic's first 20 years, numerous confrontational tactics by advocacy groups resulted in major policy changes. A primary goal of the AIDS movement has been the adoption of a multi-faceted strategy by the U.S. public health system. The AIDS movement grew from the result of the marginalization and exclusion of particular groups based on gender, sexual orientation, race, and class; resulting in widespread social and health disparities among these groups. Early in the epidemic, activists sharply criticized the government and drug companies for their complacency regarding HIV/AIDS. The emergence of political advocacy groups such as ACT UP, which began in 1987 in New York, and then spread to other U.S. cities, is an important factor in lesbian/gay, AIDS, and medical history (Altman, 1994; Stoller, 1998).

Famous for its imaginative street theater, ACT UP, an offshoot from other community-based AIDS organizations, was a magnet for radical young gay men and women. It derived many of its political and cultural practices from a variety of sources as diverse as anarchism, the peace movement, the punk subculture, the feminist health movement, to gay liberation (Epstein, 1996). The success of some local chapters was the result of a natural outgrowth of radical politics (e.g. San Francisco) and/or predominantly gay, white male identity (e.g. New York City) (Epstein, 1997; Stoller, 1998). ACT UP chapters typically had no formal leaders, and in many cities, meetings were consensus based. As a result, direct action tactics of ACT UP chapters were place-specific and highly dependent on local epidemiological needs and local philosophy concerning the epidemic.

ACT UP, and the AIDS movement in general, effected changes in AIDS research, prevention, and treatment methods, ultimately challenging the paternalism of the Western bio-medical model forcing traditional medical authorities to respond to "patient power" (Epstein, 1996, p. 9), and its attendant demands for structural change and treatment advocacy. These achievements are significant because of their impact on an epidemic that public officials had been underemphasizing, and their influence on direct action tactics for other health concerns (e.g., breast cancer and harm reduction services).

The AIDS movement was successful for three reasons: (1) their approach often involved challenging mainstream medical and/or scientific knowledge and practice; (2) activists typically highlighted the embodied experience of people with the disease contesting current biomedical wisdom that the body is a mere vector of infection; and (3) activists involved in the movement typically engage in collaboration with medical and health researchers and professionals to help expand treatment and prevention, and research and funding.

The emergence and mobilization of advocacy for IDU-related HIV services has been built on the work of many other movements, among them: the medical marijuana movement (McQuie, 2003); community services for persons with HIV/AIDS by the Gay Men's Health Crisis[6]; ACT UP and others from the feminist health movement. Accordingly, the harm reduction movement and IDUs advocacy groups are built on the foundations of other movements and borrowed from their particular strengths, specifically the direct action philosophy of ACT UP.

The harm reduction movement, in particular, has successfully convinced clients who are generally put off by other kinds of public health approaches to participate in SEPs. It recognizes that abstinence for drug users is difficult to achieve, and views the problems of illicit drug use and addiction as arising primarily from drug policies that criminalize the user, and works toward minimizing the adverse medical and social consequences of drug use (Clear, 2003). Examples of "actions" associated with harm reduction include syringe exchange, supplying clean injection equipment (e.g. bleach kits, cookers, etc.) overdose training prevention, vein care, wound care, and safe injection space(s). Feasibility of implementing SEPs are subject to a variety of local place-based factors, including the presence of grassroots advocacy; in some cases, underground exchange programs; legal aid for activists who are prosecuted; legal action on behalf of programs working with IDUs; public and private funding for programs; the establishment of programs by public authorities and local community support.

Bluthenthal advocates it was government inaction that created the political opportunity structure that encouraged harm reduction activists in Oakland, California and elsewhere to set up SEPs. The development of an SEP in Oakland, Bluthenthal argues, emerged "through the mobilization of elements of impacted communities and their allies" (Bluthenthal 1998, p. 1151), rather than any rational or sensible

[6] GMHC was the first community-based response to the AIDS epidemic in New York City. It also started the world's first AIDS hotline to handle the public's fears and questions about the epidemic.

cost effective public health strategy designed by experts. Further advocating, it was the diversity of skills of its volunteers and board members embedded in a political environment that encouraged solidarity for the establishment of an SEP. These specific place-based factors contributed to conditions for the emergence of an activist-oriented SEP in Oakland.

In New Haven, Connecticut it would be another six years after Jon Parker's National AIDS Brigade first distributed and exchanged needles on the streets before an IDU in that city would have legal access to sterile syringes. After three years of lobbying from 1987-90, involving the Mayor's Task Force on AIDS and the New Haven Health Department a law was passed in the spring of 1990 amending state statutes that had made syringe purchase without a prescription illegal to allow legal possession of syringes for participants in a demonstration NEP, which was started in New Haven in November of 1990. The NEP officially opened on November 13, 1992 and has continued to this day. The program has since expanded statewide to include other cities such as Danbury, Bridgeport, and Stamford, and at the same time, made syringe purchases in pharmacies available (Kinzly, 2003).

In New Mexico, syringe exchange began on February 2, 1998. After a seven-year legislative struggle, The Harm Reduction Act of 1997 was passed; the Act made syringe exchange legal to programs approved by the New Mexico Department of Health throughout the state. Support came from both Department of Health and community activists – opposition from legislators has gradually waned over the years (McCague, 2004).

The philosophy of harm reduction, developed in Great Britain and the Netherlands, has fostered a global movement. Domestically, it has spread through a network of researchers, providers, and organizations creating health-oriented change at the local community level. National Harm Reduction Coalitions offer a variety of resources, information, and skills training for outreach workers and advocacy groups to work more effectively with drug users and build safer, healthier communities. The movement is broad based and diverse, ranging from grassroots activists, recovering drug users, AIDS activists, lawyers, substance use–HIV researchers, community health educators and service providers, many of whom volunteer at SEPs and have been arrested repeatedly for distributing syringes; it cuts across the various communities hard-hit by substance-abuse and by the HIV/AIDS epidemic, and includes gays and lesbians, IDUs, and many African-American and Latino communities.

The structuring of place processes for the success of implementing SEPs and other harm reduction services concerns the spatial distribution of local organized action including the efforts of individuals, organizations, and networks. In the United States, it is the combined efforts of local organized action that has led to the design and implementation of place-based HIV prevention programs for IDUs. As such, program design based on need is best determined by those braving the crisis, not by government policies based on fear and exaggeration. Grassroots efforts, such as those of the harm reduction movement, defy institutionalized public policies that serve dominant interests and work toward building solidarity for stigmatized groups. U.S. grassroots activists continue to influence HIV prevention efforts to help reduce injection drug use–related harm. Action associated with HIV prevention efforts occurs locally, which in turn affects whether particular areas adopt the intervention.

Discussion and Conclusion

This research suggests that response to IDU-related HIV is responsive to local characteristics as well as to broader factors. This perspective embraces the notion that interactions of social, political, and economic systems give shape to place (Soja, 1980) and, simultaneously, place configures the structure of how resources and harms are distributed. Despite the proven effectiveness of SEPs in preventing HIV transmission, and support for SEPs from local public health and research institutions, these programs remain controversial and face ongoing governmental opposition, inadequate funding, negative attitudes from local residents and business associations, and police harassment. For IDUs, the broader social and political issues attached to sterile syringe access have largely been viewed as issues relating to inequality, marginalization, discrimination, and drug use–associated stigma. These spatially-situated social and political processes have produced large gaps in basic services to vulnerable populations, particularly services to reduce drug-related harm (Tempalski et al, 2003b). An understanding of how these processes influence the availability of services is important because: (1) SEPs reduce HIV transmission and provide important medical and social services to IDUs, and (2) it may clarify the trajectory of U.S. syringe-exchange policy and the limitations of public health reform.

Structural constraints within a particular place are important in facilitating the establishment and operation of SEPs. Embedded in important political issues like the criminalization of drug use is whether the federal government should fund SEPs. Even if the U.S. federal funding ban were lifted, however, major state and local barriers to syringe access or exchange programs probably would remain in place. Public policies, particularly drug control policies, limit access to sterile syringes and injection equipment. State and local laws and regulations can prevent syringe distribution. The political processes and regulations at the state and local levels governing syringe sales (e.g., prescription and drug paraphernalia laws) determine the legality of syringe access, the conditions of syringe distribution, and the use of government funds to distribute syringes. Consequently, in localities with SEPs, the programs are often hampered by statutes outlawing syringe possession, and, as a result, many SEPs remain underground.

The dynamic of social and spatial distancing of IDUs, which is embedded in concepts of difference, is reflected in opposition to and rejection of service – dependent individuals. The processes of stigmatization based on the social production of difference help explain negative attitudes and prejudice toward specific individuals and groups. Examinations of the origins of stigmatization enhanced our understanding of how differences and exclusions are defined and play out at the local level. Several place characteristics are important in explaining the social and spatial distancing of IDUs, including the perceived "dangerousness" of users, the moral (and legal) transgression constituted by drug use, and the belief that people are to blame for their own predicament.

In many localities, these characteristics shape public opinion and heighten opposition to the presence of SEPs in the community, which, in turn, produces large gaps in basic services. Narratives revealed place-related social and political attitudes toward IDUs that might be important in understanding where SEPs are more likely to be established – in particular the attitude of local law enforcement, the political climate at the time of implementing programs and having exceptional political leadership willing to implement controversial programs.

Currently in the U.S., syringe exchange programs exist even though harm reduction activities distinctly contradict current U.S. drug policies and drug laws, which results in SEP activists and staff being arrested and charged with illegal, criminal activity. Pro-SEP harm reduction activists defend themselves as acting out of moral necessity and as being justified in handing out sterile syringes to avert greater harm. It is these places where power and resistance intersect giving rise to collective action and activism; in the case of SEPs – the harm reduction movement. The harm reduction approach, alternatively, emphasizes the dignity and rights of drug users and aims to limit or reduce the adverse health, social, economic, and legal consequences of drug use. Current harm reduction goals exist to minimize drug-related harms by making sure that users have access to sterile syringes and clean injection equipment through the direct services of SEPs.

Finally, the HIV/AIDS epidemic's impact has put new demands on state bureaucratic institutions, demands that have often been met by government inaction. The social and political complexities of designing new public health strategies around the epidemic gave rise to new social movements. These social movements and grassroots activism have influenced the establishment of health care services and provisions for stigmatized groups. For example, ACT UP has brought about political understanding and awareness of AIDS issues and harm reduction actions, such as the creation and expansion of prevention services, while counteracting discrimination and stigmatization of substance users. In the past, ACT UP's direct action campaigns successfully contested the stigmatization of people with AIDS by highlighting the influence of homophobia on popular conceptions of HIV. Many of the same players adopted a similar philosophy and tactics in creating and demanding primary health care and HIV prevention services for IDUs, starting "direct services" as a form of direct action (and often, civil disobedience). The emergence of the harm reduction movement and of organizations that work toward establishing and sustaining direct services demonstrates how health-related social movements innovate in disease prevention.

Continued place-based research on IDU-related HIV services is essential to identify areas in need. This is partly because locality is an important factor in determining whether an IDU are likely to become infected with HIV and where services for IDUs will be readily available. That is, an IDU living in a metropolitan area with high HIV prevalence (e.g., New York City, Jersey City, Baltimore) is more likely to become infected than one living in a low-prevalence area (e.g., Los Angeles, Tucson, or Dayton); and as such, the social and political characteristics of localities are likely to affect HIV prevalence levels and the types of interventions implemented.

To date, the debate in the United States has been driven not by science or public health concerns but by an ideology that sees drug use as "immoral and evil" and SEPs as abetting this antisocial behavior. This view has persisted in Congress, which has barred SEPs from receiving federal funds since 1988, despite clear evidence showing that these programs slow the spread of disease without creating new addicts. In the U.S., syringe exchange programs exist even though harm reduction activities distinctly contradict current U.S. drug policies. This conflict between federal policy and more local public health policies demonstrates why localized action, particularly grassroots action in the form of "direct services" is often the best and only means toward developing HIV prevention efforts and service provision for IDUs.

In this paper, place is seen as a process in which individual efforts by social movement activists and IDUs themselves, together with institutional and structural forces, actively shape the response to and patterns of IDU-related HIV transmission. For health services planners, whether officials or grassroots activists, understanding which place characteristics are related to the presence (and scope) of SEPs can be useful in determining how to facilitate the expansion of SEPs. This understanding of the influence of place-based processes on the development of SEPs and other services for IDUs, may help service providers and advocates overcome structural barriers and capitalize on facilitators to HIV prevention. Further research should follow these attempts to provide information to activists and draw on their experience to contribute to academic and scientific knowledge.

Acknowledgment The research described in this paper is collaboration between Community Vulnerability and Response to IDU-Related HIV project at the National Development and Research Institutes., Inc. (supported by the National Institute of Drug Abuse grant # R01 DA13336) and National Survey of Syringe Exchange Programs (funded by amfAR with support from the Elton John AIDS Foundation and the Irene Diamond Foundation-Tides Foundation) at Beth Israel Medical Center. Special thanks to Dr. Samuel R. Friedman and Dr. Don C. Des Jarlais for their support and use of data from the Community Vulnerability and Response project and Beth Israel Survey of Syringe Exchange Programs respectively; Dr. Sara McLafferty for her un-wilting support and intellectual feedback on the theory framework; Marie Keem and Risa Friedman for transcribing the interviews presented here; and Moriah M. McGrath for her editorial suggestions during the writing of this manuscript.

Chapter 21
The effect of individual, program, and neighborhood variables on continuity of treatment among dually diagnosed individuals

Gerald J. Stahler, Silvana Mazzella, Jeremy Mennis, Sanjoy Chakravorty, George Rengert and Ralph Spiga

Abstract This study reviewed the medical charts of 271 patients diagnosed with comorbid mental health and substance-use disorders who were discharged from a hospital acute inpatient unit to various outpatient treatment programs in Philadelphia. Geographic Information Systems (GIS) technology and logistic regression modeling were employed to investigate the effects of individual, neighborhood, and program-level variables on arrival to the first treatment appointment within 30 days of discharge. Four models are presented. The results of the study suggest that having had three or more treatment episodes prior to inpatient hospitalization, and living in a neighborhood in which temporary or transitional, and presumably, other low income housing is located, increased the likelihood of patients continuing with treatment in the community. Discharge to the preadmission address, a chief complaint of bizarre behavior, close proximity of two or more liquor and/or beer stores, a high density of Narcotics Anonymous (NA) and/or Alcoholics Anonymous (AA) meetings within the neighborhood, an Axis I diagnosis of substance-induced mood disorder, and a urine drug screen positive for heroin reduced the likelihood of attending outpatient treatment. We conclude that geographic and community variables as they relate to substance abuse may add an important dimension to our understanding of patient functioning and well being in the community following inpatient treatment.

Gerald J. Stahler
Department of Geography and Urban Studies, 309 Gladfelter Hall, Temple University (025-27), Philadelphia, PA 19122, United States

Introduction

One of the most robust findings in drug treatment outcome research is that continued abstinence is highly related to length of time in treatment as well as continuity of care (Condelli and Hubbard, 1994; De Leon, 1985; Grella et al., 2000; Hser et al., 2004; Hubbard et al., 1997; Prendergast et al., 2000; Simpson, 1981). We have argued elsewhere that lack of treatment continuity, or client attrition, is one of the greatest problems interfering with treatment effectiveness in substance abuse programs (Stahler et al., 1993). This may be particularly true when clients face transitions in treatment modality, such as when clients are discharged from residential or inpatient care to outpatient treatment in the community. The likelihood of relapse is often a function of how clients deal with the environmental triggers that may be associated with substance use in the community (Lang and Belenko, 2000; Tucker et al., 1990–1991). In recognition of this problem, most treatment programs incorporate some form of relapse prevention involving the identification of environmental triggers as part of the treatment regimen. Indeed, Alcoholics Anonymous (AA) and Narcotics Anonymous (NA) recognize this problem by urging members to avoid the "people, places, and things," which were initially associated with their substance use.

Despite the fact that there is research and clinical evidence to suggest that the environment and environmental cues have such a strong influence on attrition and relapse (e.g., Boardman et al., 2001; Jacobson, 2004; Tucker et al., 1990–1991), very little research has focused on the neighborhood environment, both physical and social, to which recovering individuals are discharged, and how this environment relates to the continuity of care, treatment outcomes, and relapse (Jacobson, 2004). As Jacobson's (2004) excellent review has noted, "... one aspect that remains unexamined by the literature on retention to date is the role of geographic and neighborhood context .... [T]here is good reason to believe that many of the contextual factors that define the treatment client's living and treatment environments and their interactions may play a causal role in both voluntary and involuntary dropout decisions." (Jacobson, 2004, p. 24).

Local geography and the community context of where a patient resides, both in terms of environmental and social attributes, may have an effect on client behaviors that relate to treatment outcomes, such as relapse and attrition (Davis and Tunks, 1990–1991). We believe that this notion that the local environment affects client behaviors applies not only to the American urban environment, but also to virtually any community context in any country. The specific manifestations may vary according to the nation or culture, but the basic principals of the individual client being affected by the local social ecology and environmental contingencies will apply across community contexts. Although linkages between neighborhood environment and substance abuse have been discussed conceptually (Davis and Tunks, 1990–1991; Jacobson, 2004; Joe et al., 1994; Tucker et al., 1990–1991), relatively few studies have examined this relationship empirically. Geographic and neighborhood context variables that have been studied include distance from residence to outpatient treatment facility location (Beardsley et al., 2003; Friedmann et al., 2001; Prue et al.,

1979; Fortney et al., 1995; Saarento et al., 1998; Schmitt et al., 2003; Vaslamatzis et al., 1987); visibility of drug market transactions and other environmental cues which lead to reduced treatment retention (Simpson et al., 1997); neighborhood disadvantage affecting substance use rates or increasing the likelihood of mental health problems (Boardman et al., 2001; Crum et al., 1996; Ennett et al., 1997; Mowbray et al., 1999; Ross and Jang, 2000); drug availability in neighborhoods affecting drug prices and use (Caulkins, 1995); drug market formation affected by geographic factors (Freisthler, LaScala, et al., 2005; Rengert et al., 2005); proximity to self-help and mutual support groups such as AA and NA which increases enrollment (Friedmann et al., 2001; Mankowski et al., 2001; Umbricht-Schneiter et al., 1994); the relationship between density of alcohol outlets and violence (Gruenewald et al., 2006); and living in areas with high housing vacancies or physically deteriorated neighborhoods (Krause, 1996). These factors are not unique to urban environments in the United States of America but also can be observed in other national urban settings. This cross-national generality has been observed in other studies, for example, in the environmental health area. For instance, variations in air pollution within urban settings identified by GIS technology has been associated with pulmonary disease in San Diego and Stockholm (English et al., 1999; Bellander et al., 2001). Furthermore, the effects of the factors we examined very likely have at their foundations fundamental psychological mechanisms which will hold true across nationality in the same way that biologic mechanisms are generalizable (Croner et al., 1996).

Potential environmental risks affecting long-term outcomes for patients who receive inpatient treatment and are then discharged to the community for outpatient treatment are often the social and physical triggers associated with substance use (Lang and Belenko, 2000; Tucker et al., 1990–1991). This is especially true for clinical populations at high risk for relapse, such as patients who have a substance use dependency as well as a co-occurring mental health disorder. These patients have complex treatment needs that complicate their compliance with treatment (Drebing et al., 2002). Rates of relapse and treatment discontinuation varies across different diagnostic constellations, but tend to be particularly high for patients who have co-occurring substance abuse and mental health disorders (Bradizza et al., 2006).

The present study examines the interacting influences of individual, program and neighborhood factors on treatment compliance and continuity for this clinical population. A distinctive feature of this research is the use of Geographic Information System (GIS) software to support the analysis of neighborhood effects on individual behavior. GIS may be considered an extension of conventional database software for spatial data handling and analysis (Longley et al., 2001). It offers the ability to integrate diverse socioeconomic, health, and environmental data so that the spatial relationships among them may be extracted and analyzed. Because of these capabilities, efforts at incorporating spatial analysis into social and health science applications using GIS have been given much attention as of late (Cromley and McLafferty, 2002; Goodchild and Janelle, 2004). Although GIS has long been used to analyze spatially aggregated demographic data, such as U.S. Bureau of the Census data, social and health scientists are now recognizing the utility of GIS for analyzing individual-level survey and administrative records, and linking those records with external data indicating environmental characteristics.

In the health area, this approach has been applied prominently in the context of spatial epidemiology (Elliot et al., 2000) and access to health services (McLafferty, 2003). Only a few researchers have extended this line of research to the domain of substance abuse, where spatial patterns of, and environmental influences on, substance abuse and substance abuse treatment have been explored (e.g., Jacobson, 2004; Latkin et al., 1998; Mason et al., 2004).

In the present research, GIS is used to link individual patients to the characteristics of neighborhoods to which they are released following hospital discharge. This involves combining individual- and neighborhood-level data as well as calculating spatial metrics such as the distance between a patient's residence and certain environmental features we theorize may influence treatment continuity. Following the use of GIS for data integration and the derivation of spatial relationship variables, we use logistic regression for formal hypothesis testing. Here, estimates are obtained for the relative contributions of individual, program, and neighborhood factors to the likelihood of treatment compliance and continuity following discharge from acute inpatient psychiatric hospitalization.

Methods

This study is a retrospective analysis in which the medical charts of 271 psychiatric inpatients with co-morbid mental health and substance-use disorders were reviewed. The psychiatric units used in the study were the Psychiatric Emergency Service, or Crisis Response Center (CRC), and the acute inpatient units at Temple University Hospital-Episcopal Campus in Philadelphia, PA. In the CRC patients are evaluated and stabilized before referral to other levels of care including the acute inpatient units. In this study all patients entered the acute units through the Crisis Response Center (CRC) where they were reevaluated and treated over a stay of approximately 7–8 days. Patients received evaluation and treatment services including rapid intervention with medication, crisis therapy, psychiatric and medical evaluation and stabilization when appropriate, as well as referral to different levels of care. Assessment and final diagnosis were made by staff psychiatrists on the acute units after several days of evaluation and treatment. All patients that were included in the study had no private insurance and, therefore, had been enrolled in Community Behavioral Health (CBH), the behavioral health management organization for all Medicaid recipients in Philadelphia County.

The subject pool for this study from which the final sample was drawn was comprised of all patients referred to the acute inpatient units from the emergency psychiatric facility and admitted to these units between October 7, 2002 and May 30, 2003. Study inclusion criteria were a co-morbid diagnosis (mental health and substance use disorders) and a urine drug screen positive for prototypical illicit drugs of abuse upon entry into the emergency psychiatric center. Subjects were excluded from the study if they had not met the criteria for dual diagnosis and did not have a positive drug screen. Patients were primarily minority, poor, and urban and reflected the general patient population of the hospital and its surrounding community. Table 21.1 shows that the subject population included males (57%) and females (43%), African Americans (62%), Hispanics (24%), and Caucasians (24%), with a median age of 37 for the total sample. About one-quarter (25%) of the patients were homeless, nearly two-thirds (66%) had not completed high school, and three-quarters (75%) were unemployed during the past year. A slight majority of the subjects (55%) had a criminal justice history, and mental health diagnoses included major depression (26%), substance-induced mood disorder (22%), bipolar disorder (15%), and chronic paranoid schizophrenia (13%) among others. Patients tested positive most frequently for cocaine (55%), cannabis (32%), heroin (17%), and alcohol (15%).

Data collection

The procedures of this retrospective chart review had the approval of the Temple University Institutional Review Board. Hospital charts were de-identified by hospital staff prior to chart extraction in accordance with the Health Insurance Portability and Accountability Act guidelines (HIPAA, 1996). Names, addresses, social security numbers, birth dates, and other identifying information that were collected in the chart's face sheet and addressograph were covered during the extraction process. All chart extraction was conducted under the supervision of a hospital employee in a secured space. The chart extraction form did not collect any information relating to name, address, social security number, birth date, or other identifying information. In an effort to be thorough and accurate, several chart documents and forms routinely collected by the hospital were used to extract the information. Chart extraction forms were collected for patients with missing drug screen results, and urinary drug screen information (UDS) was cross-checked with the hospital database. Exclusion of subjects with negative or missing urinary drug screen (UDS) information as well as other missing data resulted in a final subject pool of 271. All analyses were performed on this final sample size.

A 5% random sample of charts (n = 14) was drawn to assess inter-rater reliability on the chart extraction process. A second trained research assistant extracted the same data from the charts on the original sample. The reliability was high, with a 95% inter-rater agreement

After completing the chart review, hospital staff created a database of all patient addresses, and a random number between 0 and 9 was added to each address to preserve address confidentiality, yet allow for a reasonable proximity of location for the analyses. This list of enhanced addresses was added to the research database, along with age, race, and gender.

Table 21.1 Patient characteristics

Characteristic	*n*	%
Gender		
Male	155	57
Female	116	43
Race		
African American	167	62
Hispanic	53	24
Caucasian, other	46	21
Other	5	2
Age		
Range	19–66	
Median	37	
Living situation		
Domiciled	197	73
Homeless/transient	73	27
Completed high school		
Yes	96	35
No	175	65
Employment history		
Employed past year	15	6
Unemployed past year	202	75
Criminal justice history		
Yes	148	55
No	123	45
Chief complaints (can be more than one)		
Suicidal Ideation	91	34
Auditory hallucinations	62	23
Aggressive behavior	43	16
Depression	43	16
Suicide attempt	37	14
Bizarre behavior	34	13
Auditory hallucinations to kill self	27	10
MH diagnosis		
Major depression	71	26
Substance-induced mood disorder	60	22
Bipolar disorder	41	15
Chronic paranoid schizophrenia	34	13
Other	65	24
Positive drug screen (UDS)		
Cocaine	150	55
Cannabis	87	32
Heroin	45	17
Alcohol	40	15
Benzodiazapines	32	12
PCP	24	9
Barbiturates	13	5
Amphetamines	5	2

Variables included

Continuity of treatment, the dependent variable in the analysis, was operationally defined as whether the patient adhered to their treatment regimin by attending their first scheduled post-discharge outpatient appointment within 30 days of discharge. Admittedly, this is a rather simple measure of a more complicated multidimensional process. Indeed, a number of researchers in the mental health and addictions field have developed complex conceptualizations and measures for this construct (e.g., Farrell et al., 1999, 1996; Fortney et al., 2003; Greenberg et al., 2002). However, treatment adherence, a proxy for the more complex conceptualization of continuity of care, has been used in prior research to represent a measure of successful continuity of treatment for dually diagnosed patients (Claus and Kindleberger, 2002; Vendetti et al., 1997). This measure demonstrates construct and face validity. Readmission rates for patients who did not show up for their first outpatient appointment were significantly higher than for patients who did show up for their appointment (Nelson et al., 2000). Furthermore, as Greenberg et al. (2002) noted, in the Health Plan Employer Data and Information Set (HEDIS), which accredits managed care organizations throughout the nation, "... only one measure of the quality of psychiatric care [is used]: the percentage of inpatients hospitalized for a major affective disorder who received an outpatient visit within 30 days of discharge. Selection of this measure ...demonstrates the high face validity ..." (p. 247).

Given the relationship between outpatient therapy attendance and rehospitalization, poor attendance is a significant clinical problem. In a variety of studies examined (Mowbray et al., 1999; Herman et al., 2000; Hansten et al., 2000), follow up rates beyond the first 30 days of inpatient discharge did not differ greatly. That is, the overwhelming majority of patients discharged from inpatient treatment for co-occurring mental health and substance abuse disorders who do not arrive for their first outpatient treatment appointment upon discharge most likely do not make it to outpatient treatment at all. In addition, as time increases between assessment or inpatient hospitalization and follow-up treatment, dropout rates increase (Claus and Kindleberger, 2002).

Explanatory variables used in the regression included individual, program, and neighborhood-level variables. Individual-level variables, gathered from the hospital records, included a number of background characteristics pertaining to the patients, including: race, gender, age, housing status (homeless or domiciled), employment, disability, criminal justice history, high school completion, chief complaint, diagnosis, substances used, prior hospitalizations, prior treatment episodes, and victimization.

Program-level variables also affect continuity of treatment. The initiation of a working alliance with the post-discharge treatment program prior to discharge, supportive services from the post-discharge treatment facility for homeless patients or those with unstable housing, may impact the success or failure of treatment. Finally, accessibility of treatment is critical for those most at risk. Therefore, program-level variables included: pre-discharge visit from the outpatient treatment program, transportation to the outpatient program, housing referral, and distance to treatment.

Neighborhood-level variables were generated to capture the socioeconomic characteristics of the neighborhood to which the clients were discharged. These variables were derived from a variety of publicly available data sources, including U.S. Bureau of the Census tract-level demographic data, as well as directories and data from government and non-profit agencies. The following census data were acquired: individuals living in poverty, households receiving disability and social security disability, households receiving public assistance, individuals over 16 that were unemployed, housing units that were vacant (not abandoned), housing units that were owner occupied, and housing units that were renter occupied. In addition, total numbers of individuals, households, and housing units per census tract were included. The numbers and percentages of those unemployed per tract that were used as variables within this study represented those over 16 who were not at work during the referenced week but had been actively looking for work during the past 4 weeks.

Publicly available lists of meeting locations for double trouble in recovery, dual recovery anonymous, cocaine anonymous, alcoholics anonymous, and narcotics anonymous were obtained by calling local chapters and requesting copies of the 2002 handbooks. A publicly available list of all liquor stores in the city was obtained through the Pennsylvania Liquor Control Board, as was a publicly available list of stores and bars that sell packaged beer to the public. Bars not selling packaged goods but only selling alcohol were not included in this study. A publicly available list of pharmacies was obtained from the Pennsylvania Pharmacy Licensing Bureau.

A publicly available list of shelters was obtained from the city's Office of Emergency Shelter Services, and a publicly available list of single room occupancies and transitional housing addresses was obtained from the city's Office of Housing and Community Development. A list of halfway houses and correctional facilities was obtained from the Pennsylvania Department of Corrections. Finally, a list of free public health clinics, both city-funded and federally qualified, was obtained from the Health Department and the Health Federation of Philadelphia.

The above data were entered into a GIS database for processing. The locations of geographic features, such as shelters and liquor stores, and the post-discharge addresses for patients, were georeferenced using address matching. Georeferencing assigns a coordinate position referenced to the earth surface of a feature. In address matching, the GIS ingests an address encoded as text, matches the address to a particular street segment within a streets database, and returns the geographic coordinate position of the address as a point location.

Following address matching, a number of approaches were used to retrieve the neighborhood characteristics of each patient's post-discharge location. A point-in-polygon overlay operation combined the point-level discharge address data with the polygon-level census data to retrieve the census data associated with each patient. Buffer operations were used to identify the number of liquor stores, shelters, and other geographic features within certain proximities of each patient's post-discharge address. Finally, the distance between the patient's post-discharge address and their respective outpatient treatment program location was calculated using the web mapping application "Mapquest" (Mapquest, Inc.). The calculation obtained by Mapquest represented the fastest route over specific streets and arteries, rather than a point to point calculation. For more information on the details of address matching, overlay, and buffering operations in GIS, which are standard operations included in many commercial-off-the-shelf GIS packages, the reader is referred to Longley et al. (2001).

Data analysis strategy

Ultimately, 101 individual-, program-, and neighborhood-level variables were calculated. These variables were encoded using a variety of data types, including numerical or continuous responses, multi-choice or categorical responses, and yes/no or dichotomous responses. To facilitate the analysis, the numerical and categorical variables were transformed into dichotomous variables. Our motivation for this transformation was to logically reduce the volume of information to a more manageable level. Many of the categorical variables could be logically reclassified into two categories; for example, the categorical variable indicating housing status was reclassified into a dichotomous variable indicating whether a patient was either domiciled or homeless prior to hospital admission. Numeric variables were encoded as being above or below the median.

Following the generation of the dichotomous variables, a variety of approaches were used to reduce and refine the number of variables for inclusion in the logistic regression. Univariate analyses, including t-tests and Chi-squares, were utilized to explore various relationships among the explanatory variables and continuity of treatment. Explanatory variables that did not exhibit significant univariate relationships with the dependent variable were removed. Additionally, correlation matrices and factor analyses were used to address issues of multi-collinearity among explanatory variables. The final set of 14 variables were entered into a series of logistic regressions with continuity of treatment as the dependent variable. These regression models reflect various combinations of variables we hypothesize affect continuity of treatment, while accounting for issues of multicollinearity.

Results

Table 21.2 lists the independent variables used in the regression analyses, and Table 21.3 reports the results of four logistic regression models. Odds ratios of greater than 1.0 indicate an increase in the likelihood that a patient will show up at his or her first treatment appointment after being discharged from an intensive inpatient unit; odds ratios of less than 1.0 indicate a reduced likelihood of attendance. Our analytical strategy involved testing for significance in the influence of, and interactions among, individual-, program-, and neighborhood-level explanators of treatment continuity. For this purpose we ran a number of logistic regression models, beginning with testing for those individual-level factors that proved significant.

Table 21.2 A list and description of independent variables used in the binary logistic regression

Variable	Description
Individual-level	
RET	Whether or not the patient returned home (or to a former address) upon discharge
HIV	Whether or not the patient was diagnosed as being HIV positive
OP	Whether or not the patient had a positive urinary drug screen for heroin
SI	Whether or not the patient had a chief admitting complaint of suicidal ideation
BIZ	Whether or not the patient had a chief admitting complaint of bizarre behavior
NNMH	Whether or not the patient had an Axis I diagnosis of substance-induced mood disorder
MEDNUM	Whether or not the patient had more than one positive urinary drug screen
MEDTX	Whether or not the patient had more than two prior treatment episodes
Program-level	
REF	Whether or not the patient's post-discharge treatment program provided housing referrals
MEDDIST	Whether or not the patient had to travel more than 2.42 miles to his or her post-discharge treatment program
Neighborhood-level	
MEDDPW	Whether or not more than 18.9% of the households in the patient's post-discharge census tract received public assistance
MEDIALC	Whether or not the patient had more than one liquor and/or beer store within 0.4 miles of his or her post-discharge residence
MEDIINST	Whether or not the patient had 1 or more shelters, SRO's, transitional housing facilities, and/or halfway houses within 0.4 miles of his or her post-discharge residence
MEDTNAA	Whether or not the patient had more than six Narcotics Anonymous and/or Alcoholics Anonymous meeting locations within 0.8 miles of his or her post-discharge residence

Following this test, we removed those individual-level variables that were not significant, and reran the model with the program variables added. After again removing those variables that proved to be not significant, we incorporated the neighborhood-level variables into the regression.

Model 1 includes only the individual-level variables as explanatory variables. Likelihood of continuity of treatment is enhanced by the presence of three or more prior treatment episodes (for mental health or substance abuse problems), and is suppressed if the patient returns home following discharge, abuses heroin, enters with a chief complaint of bizarre behavior, and has a substance-induced mood disorder. Diagnosis of HIV, suicidal ideation, and more than one positive urinary drug screen were not significantly related with continuity of treatment.

In Model 2, the significant variables in Model 1 were combined with the program-level variables. The likelihood of continuity of treatment is enhanced if the program

Table 21.3 The results of binary logistic regression of continuity of treatment

Level and variable	Model 1	Model 2	Model 3	Model 4
Individual				
RET	0.36**** (10.22)	0.50** (3.97)	0.38*** (7.44)	0.37**** (9.39)
HIV	0.63 (703)			
OP	0.44** (4.30)	0.49* (3.74)	0.44** (4.63)	0.45** (4.28)
SI	0.63 (2.53)			
BIZ	0.32** (6.24)	0.37** (4.94)	0.39** (4.22)	0.38** (4.59)
NNMH	0.42** (6.39)	0.40*** (7.15)	0.42** (6.23)	0.42** (6.22)
MEDNUM	1.29 (0.72)			
MEDTX	2.13** (5.59)	1.87* (3.62)	2.07** (4.42)	2.27** (6.09)
Program				
REF		1.76* (3.22)	1.20 (0.33)	
MEDDIST		0.62 (2.69)		
Neighborhood				
MEDDPW			0.72 (1.37)	
MEDIALC			0.50** (4.08)	0.48** (4.76)
MEDIINST			2.23*** (7.35)	2.24*** (7.62)
MEDTNAA			0.46*** (6.80)	0.46*** (7.06)
Analysis results				
Constant	2.43* (6.72)	1.77 (2.55)	2.65* (4.95)	2.49** (6.54)
Nagelkerke ($R2$)	0.16	0.16	0.23	0.22
Correct (%)	63	65	69	66

Values indicate the odds ratio; Wald statistic is reported in parentheses.
* Significance < 0.1.
** Significance < 0.05.
*** Significance < 0.01.
**** Significance < 0.005.

offers a housing referral service, though it should be noted that the variable is only significant at a >90% confidence level. Model 3 combines those variables that were significant in Model 2 with the neighborhood-level variables. Model 3 shows an improvement in predictive power over Models 1 and 2, in both Nagelkerke's R^2 and in the percentage of patients correctly classified as attending or not attending their first treatment. This model indicates that the likelihood of continuity of treatment increases with proximity to transitional housing, and decreases with proximity to alcohol sales and AA and NA meetings. The presence of poverty in the patient's discharge location, as indicated by public assistance, was not significantly related to continuity of treatment. Model 4 shows a revised Model 3 with the insignificant variables removed from the equation. Of the explanatory variables, returning to the same home or address (RET) demonstrates the strongest relationship with continuity of treatment, in terms of both the odds ratio and significance. It is an inverse relationship, meaning that returning home was associated with a lower likelihood of attending the outpatient appointment.

Of note is that the program-level variables had only a marginal effect on outpatient treatment attendance beyond that explained by the individual-level characteristics, and that marginal effect was in turn explained by the neighborhood-level variables. The neighborhood factors, on the other hand, substantially increased the predictive power of the model beyond that offered by the individual factors alone. However, the addition of the neighborhood-level variables to the regression equation did not substantially change the magnitude nor significance of the influence of the individual-level variables (with the potential exception of the presence of bizarre behavior), suggesting that individual and neighborhood factors act as separate and distinct influences on treatment continuity.

Another noteworthy result, that is not reported in Table 21.3, concerns the variable 'LIVES.' This variable represents whether the patient was transient or homeless prior to hospital admission. Logistic regression models indicate that this variable is collinear with RET (whether a patient returns to a former address upon discharge). Clearly, a patient can only return home when the patient has a home to which to return. Interestingly, however, LIVES was not significant in predicting continuity of treatment when it was substituted for RET as in Model 4, and Nagelkerke's R^2 also declines markedly to 0.18. This is surprising given the relatively strong relationship of RET with both LIVES and continuity of treatment. Therefore, returning home appears to be the strongest determinant of treatment continuity following discharge.

Discussion

Relatively little research has focused on neighborhood and environmental factors that improve treatment compliance and continuity of treatment. This study is one of the first investigations to examine the influence of these environmental factors on continuity of care. Our results show that the characteristics of the community interact with program and individual variables to influence whether a patient complies with treatment and attends outpatient care. Those patients with three or more treatment episodes for either mental health or substance abuse (or both) disorders prior to inpatient hospitalization, and living in a neighborhood with temporary, transitional, and low income housing, increases the likelihood of patients continuing with treatment in the community. However, those patients who were discharged to their pre-hospitalization address, exhibited bizarre behavior at intake, provided a heroin-positive drug screen at intake, were diagnosed with substance-induced mood disorder, were discharged to a residence with two or more liquor and/or beer stores within 0.4 miles and more than six NA and/or AA meeting locations within 0.8 miles, were less likely to attend the first outpatient treatment appointment.

Although the sample size (n = 271) was relatively small for the analyses performed, the findings of this study converge with our clinical experience and findings in this population. For example, one explanation for why many patients do not continue with treatment as outpatients involves a socio-rational choice model of service participation (Sosin and Grossman, 2003). This model assumes, based on relevant behavioral laboratory, clinical and epidemiological findings, that the hospital, community mental health clinics and community provide the individual with a range of opportunities The individual's valuation of these opportunities are determined by experience with the service provider, program characteristics, program relevance and personal needs. Then the individual weighs the costs and benefits of the alternative opportunities (e.g., hospitalization, continued outpatient treatment) and bases his or her choice on this evaluation. Certainly, psychopathology and other domains of individual functioning may impair the full expression of these skills but patients exercise these within their capacity and the range of opportunities offered by their social and economic environment. Thus, from a socio-rational perspective, it is plausible that patients seeking hospitalization have a complex set of motivations and goals in addition to treatment, such as respite from confiict or drug detoxification. Consequently, those patients who were seeking momentary respite from familial or other psychosocial difficulties might have less interest in continued treatment than patients seeking resolution of symptoms and relief of other long-term chronic problems.

As our previous discussion illustrates, those variables associated with a decreased likelihood of attendance, such as returning to the same pre-hospitalization address, having alcohol and beer outlets nearby, and the individual-level variables of bizarre behavior and substance-induced mood disorders, are therefore quite consistent with clinical experience and expectations. What is less clear is why living in an area with temporary housing would be associated with greater continuity. It may be that these neighborhoods are more affordable to these patients, but it may also be possible that this is an intermediate variable for some other process that affects the main dependent variable.

One variable that was surprising was that a high proximity to AA and NA meetings was associated with a decreased likelihood of continuing treatment. This is in contrast to what others have found that proximity to self-help groups increases enrollment (e.g., Friedmann et al., 2001) and may therefore be presumed to enhance the continuity of treatment. On the other hand, it may be because patients attended meetings *instea*d of continuing in treatment, that these meetings acted as "competitors" or substitutes for treatment, as mentioned above. Another possibility is that the density of self-help meetings are in neighborhoods where there is the greatest demand for these meetings—where there are more current as well as recovering users. In addition, AA and NA meetings are often dependent for financial reasons upon lower-cost space options and donated space. Thus, places where there is the lowest cost real estate or where there are the greatest demand for these meetings, may increase the likelihood of relapse and decrease the likelihood of continuing in treatment. We have no follow-up data to determine whether patients actually attended AA and NA meetings after discharge, and therefore we cannot discern with any certainty the explanation for this finding.

Finally, our results suggest that patients who have had multiple prior treatment episodes manifested a higher rate of continuing treatment than those with less treatment experience. One explanation could be that patients with several prior treatment episodes had developed a relationship with a treatment provider that facilitated contact after discharge. This would be an area for further exploration since we did not examine the data on specific treatment providers. However, we believe it is more likely that this result represents a proxy for a patient's sufficient behavioral organization that enables him or her to attend treatment when necessary; and that there is familial and social support for doing so. Familiarity with the process of treatment may also facilitate a comfort level with treatment compliance.

Given these results, it is important to bear in mind that the study has a number of limitations, and that these findings must be viewed with caution. The study is retrospective in nature, and relies to a large extent on data extracted from hospital charts and census data. In addition, the census tract delineations do not represent actual patient communities. However, these were the only available data that could be used for the purposes of this study. As Ewart and Suchday (2002) note in their study of how urban poverty and violence affect health, "Census boundaries are designed by planners to facilitate the administering of surveys, whereas neighborhood boundaries are defined by the perceptions and social habits of their residents. Census-derived indices of poverty, for example, do not necessarily reflect a neighborhood's degree of exposure to the ill-effects of economic disadvantage because they may include an affiuent city block with a nearby poverty-stricken neighborhood" (Ewart and Suchday, 2002,p. 254). In addition, the dependent variable that was used was attendance at the first post-discharge treatment appointment. While attending the first appointment within 30 days is related to continuity of treatment and ultimately to continued abstinence and stability, it is still a proxy for these outcomes and therefore is a somewhat limited outcome measure. In addition, the neighborhood-level data were obtained from publicly available sources and no attempt was made to verify the accuracy of this information.

Finally, continuity of care, as mentioned earlier, is a far more complex and multidimensional construct than is defined in this study. We conducted the study in a single city in a restricted geographic area which may not be representative of other locations. Although we believe that the general principal that the neighborhood context and local geography will influence patient behavior, especially concerning treatment continuity, we do recognize that context as we use it is narrowly defined. However, we do believe that the notion of environmental contingencies in communities and neighborhoods affecting individual behavior is a robust concept with wide applicability across geographic contexts, but the specific manifestations of this principal may vary widely across different cities, nations, and cultures.

Despite these limitations, there appear to be a number of implications from the results of this study. The research supports the consideration of these geographic and individual-level variables and their interaction when evaluating treatment outcome. In addition, policy needs to incorporate the consideration of geographic factors and their interactions with individual patient level and program-level variables in designing the care of dually diagnosed individuals. In particular, our research points to the potential importance of the effect of returning to one's previous address, residing near alcohol and beer stores, as well as AA/NA meetings postdischarge, as reducing the likelihood of continuing in treatment, whereas living in neighborhoods with low income housing increased the likelihood of attendance.

We believe that the influence of environmental and geographic factors that influence relapse, as well as the epidemiology of substance use represents an important area for future research. Evidence of an increasing recognition of the importance of geographic factors relating to addiction is the National Institute on Drug Abuse's

(NIDA) recent sponsorship of a symposium at the 2006 American Association of Geographers (AAG) annual meeting focused on the geographic aspects of drug addiction. GIS software can facilitate the analysis of environmental influences on substance abuse and substance abuse treatment by both integrating individual-and neighborhood-level data and generating variables that capture the spatial relationships between individuals and geographic features that may influence substance abuse behaviors. We note that a major challenge to the incorporation of GIS methods into substance abuse research is the availability of spatial data on relevant environmental characteristics. While demographic data from the U.S. Census are widely available, the time and effort required for the acquisition and preprocessing of data on the locations of shelters, liquor stores, and the like can be substantial.

Since the results of the study do suggest that geographic features in combination with individual characteristics influence continuity of treatment in the community, we believe that it may be possible to develop a "high risk" patient profile for dually diagnosed patients based on this line of research. Clearly, a great deal more research would be needed to further develop, refine, test, and validate such a profile. We intend to further investigate and refine such an index of risk since this could be of considerable clinical utility in identifying patients who are at the greatest risk for relapse upon discharge, and could be an important aid in discharge planning.

Further research should also focus on additional community variables such as where drug arrests occur and more refined conceptions of neighborhood and neighborhood characteristics. In addition, follow-up data concerning actual relapse and continued abstinence would provide more useful information than the proxy outcome variable that was used in the present study. Nonetheless, it does appear that examining geographic and community variables as they relate to substance abuse may add an important dimension to our understanding of patient functioning and well being in the community following inpatient treatment.

Acknowledgment This project was funded, in part, by a grant from the Pennsylvania Department of Health (#RFA 02-07057123192, Ralph Spiga, Ph.D., Principal Investigator) and by a Research Incentive Funds grant from the College of Liberal Arts, Temple University. The Pennsylvania Department of Health specifically disclaims responsibility for analyses, the interpretations and conclusions. We would also like to express our appreciation for the assistance from L.J. Rassi, D. Haas and S. Maxwell on the project.

Chapter 22
Exploring the Reciprocal Effects of Substance Abuse Treatment Provision and Area Substance Abuse

Matthew E. Archibald

Abstract This study examines the relationship between substance abuse treatment provision and substance abuse. Several questions serve as the basis for testing substance abuse reduction, treatment need-demand, and diffusion hypotheses: Does treatment provision reduce area rates of substance abuse; do area rates of substance abuse (i.e., treatment need-demand) foster increases in treatment provision; and, does spatial and temporal diffusion drive changes in both substance abuse and treatment provision? Analyses of spatial and panel data from 159 counties in Georgia revealed that provision of treatment services did not reduce area rates of substance abuse, in fact, greater density of service providers was related to *increases* in prevalence. Moreover, demand did not foster changes in service provision; only greater service provision at an earlier period. While temporal diffusion of substance abuse and treatment provision took place, there was little evidence of spatial clustering. Strategies to uncover mechanisms underlying these relationships and their implications for treatment policy are discussed.

Introduction

This study investigates the relationship between substance abuse treatment provision and area substance abuse rates. The central questions are whether shifts in treatment services – diagnosis, therapy or referral for rehabilitation of persons with addiction or dependence (US Department of Health and Human Services, Substance Abuse and Mental Health Services Administration 2004) – reduce the prevalence of substance abuse, and whether changes in the pattern of substance abuse influence treatment provision. Moreover, I address the burgeoning interest in spatial relationships and consider the diffusion of substance abuse and treatment provision across geographic domains. Recent studies in the spatial analysis of substance use and abuse

M. E. Archibald
Department of Sociology, Emory University, Atlanta, GA 30322, USA. Telephone: 404-727-5488; e-mail: marchib@emory.edu

Y.F. Thomas et al. (eds.), *Geography and Drug Addiction*,

(e.g., Freisthler, Gruenewald, Johnson, Treno, and LaScala 2005; Gruenewald and Freisthler 2005; Gruenewald, Freisthler, Remer, LaScala, and Treno 2006; Lipton, Gorman, Wieczorek, and Gruenewald 2003) suggest a number of strategies for area-level analysis of addiction and its treatment.

Unfortunately, this relationship remains understudied for two reasons. First, there are still some gaps in the theoretical literature on area health outcomes. In an excellent discussion of the importance of place-based health research, Morenoff and Lynch (2004) note that: whereas the sociological literature on neighborhood effects has taken a "process turn" in recent years and begun to focus more on the mechanisms that explain why neighborhoods matter (Sampson et al. 2002), most research on the neighborhood context of health is still attempting to establish that context matters. This is partly because most health research is framed in a paradigm where individual-level proximal influences – such as behaviors or biomarkers of pathogenic processes – take precedence over contextual factors (p. 407).

Since much of the place-based health research focuses on individual outcomes (Harrison and Sexton 2004), less is known about how organizations, neighborhoods, and communities themselves intersect to produce variation in community-level health. In addition, valid indicators of area substance abuse are not widely available, increasing the difficulty of carrying out this research (McAuliffe, Woodworth, Zhang, and Dunn 2003). Second, there is a dearth of knowledge about geographic variation in substance abuse treatment needs and little is known about "the differential adequacy of treatment resources" (McAuliffe and Dunn 2004, p. 1000). Most studies examining treatment organizations address issues of individuals' geographic and financial access to care, quality of care, and organizational survival.

Two recent trends in the treatment literature can be combined to address these concerns. Jacobson's (2006, 2004a, b) work, for example, links organizations and neighborhood characteristics such as socioeconomic disadvantage, resources, and accessibility, and, McAuliffe and Dunn (2004) compare communities' substance abuse service needs with their level of treatment provision. In order to understand how communities matter in substance abuse treatment and prevention, I borrow from these frameworks to ask how substance abuse treatment organizations and area substance abuse are causally interrelated and then attempt to model that relationship using regression techniques that take into account the spatial processes that influence the reciprocal impact of substance abuse and treatment. Given the assumption that treatment services are targeted to communities with the greatest need-demand based on the degree of abuse (National Health Disparities Report 2006), it is expected that where there is greater provision of services, substance abuse (conceptualized as an indicator of need-demand) will decrease over time, all else being equal. The equality caveat is important since factors such as policing and markets will have a significant influence on use and abuse. It is also expected that greater need-demand for substance abuse treatment will foster an influx of organizations, increasing the density of provider treatment services (Allard 2004).

The spatial hypothesis in these expectations is that geographic proximity leads to the diffusion of substance abuse (see Freisthler, Gruenewald et al. 2005), which can be reversed through the spread of countervailing processes such as substance abuse treatment. The underlying assumption is that treatment provision flows across areas because decision-making in where to locate treatment organizations will respond to demand for services (Greve 2002).

In a broad sense, the theoretical background for this research begs the questions: does the location of public service organizations directly contribute to provision of community goods, in this case, treatment of substance abuse, and what are the factors that determine where (and sometimes, when) new organizations will be established to meet these needs? A key policy issue with regard to the provision of social services is the gap between collective need for services and geographic access to those services (US Department of Health and Human Services, Drug and Alcohol Services Information System Report-DASIS 2001). Institutional and organizational decision-making about location of substance abuse treatment services is anchored in a set of unexamined assumptions about causal relationships between treatment provision, access, and outcomes. By continuing to pursue location decisions based on a tacit theory of relationships, which may or may not be supported by empirical evidence, policy makers run the risk of perpetuating substance abuse rather than ameliorating it.

Research Hypotheses

Substance Abuse Reduction and Treatment Demand

Studies investigating the efficacy of substance abuse treatment, although generally positive (e.g., White 1998), are not without their critics. While Schumacher et al. (2002) and Sindelar and Fiellin's (2001) review of treatment provision efficacy concludes that substance abuse treatment works, and Neubereger et al. (1982) "affirm" that the alcoholic in recovery has a high probability of success, Jacobson (2004a) argues that addiction treatment is characterized by high rates of attrition and relapse. Lee, Reif, Ritter, Levine, and Horgan (2004), for example, found that half of discharged treatment patients on any given day had already been in treatment at some point in their lives and that 30% had been in treatment during the previous year. To make sense of this deficit, some studies explain that treatment for addiction is effective only insofar as supplementary services are available to meet crucial socioeconomic needs (D'Aunno and Vaughn 1995), while others argue that treatment location in disadvantaged communities may be detrimental for individuals' recovery (Jacobson 2006).

The latter implies that, at minimum, geographic access to, and therefore location of, substance abuse treatment, is important for successful treatment of addiction. Research focusing on treatment accessibility, such as the Drug and Alcohol Services Information System Report (8/23/02), suggests that ease of access will foster better treatment outcomes since traveling distances limit those who can take advantage of services. Provision of treatment services in locales that are inaccessible will fail to meet larger demand, limit the pool of those who might benefit from services and differentially influence the prevalence of substance abuse. It is presumed that, as a consequence, substance abuse and mental health providers will tend to cluster in certain areas where indicators of service need, such as poverty and unemployment, are high and gaps in provision seem to warrant the influx of organizations (Allard 2004). Demands for greater accessibility, both geographic and more broadly, strongly imply that substance abuse treatment is efficacious and, if and when conditions are met, areas will benefit from having a treatment provider located nearby. That is, over time, as treatment providers come to occupy a niche in areas that lack services, a decline in substance abuse should be apparent, all else being equal. This reduction hypothesis suggests that: *Greater provision of treatment for substance abuse will be related to decreases in area substance abuse.*

Demand for greater accessibility to treatment services also implies a reciprocal relationship since the prevalence of substance abuse indicates a need for services and drives demand for greater access. Organizations make location decisions for a number of reasons including getting closer to clients and customers and reducing transportation costs (Greve 2000). Treatment organizations may respond to demand for services since their location decisions are likely to be based on the distribution of clients (Jacobson 2004a). For substance abuse treatment, clients and resources may be highly concentrated, resulting in a greater concentration of treatment facilities (Allard 2004). Therefore, the need-demand hypothesis claims that: *Higher levels of area substance abuse will be related to increases in provision of treatment for substance abuse.*

And yet, there are several reasons to consider the null hypothesis with regard to both claims. First, the reduction hypothesis is based on the efficacy of substance abuse treatment which has not been firmly established, so that without successful individual outcomes there are less likely to be area ones. As Beveridge et al. (2000, p. 891) note, small differences in drug use seem to follow no reasonable area patterns. In addition, the direct causal link between clients and treatment, and substance abuse is difficult to sustain. For example, clients may not live in the neighborhoods where they seek treatment although this is the assumption underlying the issue of treatment location and geographic accessibility. Second, factors such as the presence of drug markets and their sanctioning by legal authorities may have a bigger impact on substance abuse than organizations designed to eliminate it. Third, organization theory argues that in human service organizations, performance measures are undeveloped, outcomes are more uncertain, and the bottom line is more ambiguous than in other sectors (Scott 2003). Organizational survival rather than effectiveness may be the ultimate goal toward which resources are directed. Resources will then be divided between survival and operational goals, resulting in a less effective organization. For treatment providers, acquiring resources for survival may substitute for rigorous relapse prevention.

Moreover, mobilization of resources for survival rather than treatment raises the issue whether treatment location decisions are driven by a response to the burgeoning prevalence of substance abuse or for other reasons. Health care markets for example have experienced dramatic changes in the past 10 years and provision of substance abuse treatment has followed suit. Location decisions may have nothing to do with client characteristics and everything to do with profit maximization (Greve 2000), tax shelters, declining property values and seeking institutional support that accrues to agencies with civic missions (Peele 1989). Ozcan, Shukla, and Tyler (1997) found that publicly-funded service organizations that were *less* likely to meet community needs received *more* government funding. Especially during an era of reduced social services and the dismantling of the welfare state, demand for services may not be met with increased provision at all. McAuliffe and Dunn (2004) show that there are large disparities in treatment provision and need (implied by differences in area substance abuse). In short, increases and decreases in treatment provision may not be a function of changes in rates of substance abuse.

Community Disadvantage

Disparities between communities in prevalence of substance abuse and treatment provision are influenced by a variety of social, economic, political, and cultural factors. The influence of these factors is straightforward: "socioeconomic differences in health outcomes have been widely documented for most health conditions in most countries... people who are poorer and who have less education are more likely to suffer from diseases" including chronic conditions associated with addiction (Crimmins, Hayward, and Seeman 2003). Storr, Chen, and Anthony (2004) find that social characteristics of disadvantaged neighborhoods, such as lower levels of income and minority composition, increased access to illegal drugs. Silver, Mulvey, and Swanson (2002) argue that socioeconomic disadvantage is associated with higher rates of substance abuse (among other health outcomes) and Howard, La Bviest, and McCaughrin (1996) show that socio-environmental factors of an area predict treatment outcomes better than race and ethnicity. Goodmand, Siegel, Craig, and Lin (1983) find that socioeconomic disadvantage is related to prevalence of alcoholism. As for the individual-level effects of socioeconomic status, Muntaner, Eaton, Diala, Kessler, and Sorlie (1998) show that financial resources are inversely associated with alcohol and drug disorders. In contrast, some research suggests that substance abuse is not more likely to occur in impoverished neighborhoods. Saxe, Kadushin, Beveridge, Livert, Tighe, Rinskopf, Ford, and Brodsky (2001), for example, argue that the most economically disadvantaged neighborhoods have the most visible drug problems, but that drug use is equally distributed across communities and Freisthler, Gruenewald et al. (2005) show that drug sales and use occur in different places.

Treatment provision and poverty are likely to be related since abuse and treatment provision are a function of resources. Research shows that the distribution of public health services in metropolitan areas is highly associated with levels of poverty and that substance abuse (and mental health) providers tend to be concentrated in these areas (Allard 2004). Other studies such as Allard, Rosen, and Tolmen (2003) argue that central cities (like Detroit) are chronically short of mental health and substance abuse services. Allard's (2004) study suggests that provision tends toward low income areas but it is not known whether need-demand (i.e., higher levels of substance abuse) is high in those areas or not. It is also unknown, as argued above, whether need-demand even drives treatment location decisions. Given some research, it seems likely that areas experiencing greater disadvantage will have higher rates of substance abuse, but it is hard to predict whether treatment provision will be greater or less in those areas.

I tested these hypotheses by focusing on the state of Georgia. Georgia is an ideal case since McAuliffe and Dunn (2004) found that it had the greatest gap between treatment provision and need-demand of any of the 50 states. Consequently, it provides a much stronger test of the hypotheses since the relationships are already skewed in the direction of the null hypothesis. That is, prior research shows that a relationship between treatment provision and substance abuse is very unlikely in the state of Georgia, and the finding of one would strongly suggest that the relationship exists elsewhere.

Data, Measures and Methods

Data

To study area rates of substance abuse and treatment provision, I created an original database of social indicators of community social and economic characteristics, substance abuse treatment provision and substance abuse treatment need for 159 Georgia counties. These counties are the primary geographic unit of analysis. I model substance abuse and treatment provision relationships at the county level and over three time periods, in which complete data were available: 1998, 2000, and 2002. Comparing across the 3 years demonstrates the extent to which the hypothesized relationships are reliable.

Data for these counties (described more fully in the Measures section below) were drawn from a number of publicly accessible data sources: Uniform Crime Reporting Program Data (United States); County Level Arrest and Offenses Data, United States Department of Justice. Federal Bureau of Investigation; Georgia Department of Human Resources, Division of Public Health, Georgia Vital Statistics Report; US Department of Health and Human Services, Substance Abuse and Mental Health Services Administration, Office of Applied Studies, National Survey of Substance Abuse Treatment Services (N-SSATS); Bureau of Labor Statistics Local Area Unemployment, Housing and Household Economic Statistics Division, Small Area Estimates Branch; and the Bureau of Economic Analysis, US Department of Commerce.

Measures

Substance abuse is the excessive and harmful use of alcohol or drugs, leading to clinically significant impairment or distress and often followed by recurrent use

(Diagnostic and Statistical Manual of Mental Disorders 2000). Following McAuliffe and Dunn (2004) and McAuliffe et al. (2003), I use a social indicators approach to measure substance abuse in the aggregate based on several different operational definitions of abuse, such as emergency room visits, arrests for illicit drug sales, and treatment admissions. I constructed a measure of area rates of substance abuse using arrest data for drug possession/sales and alcohol (i.e., drug or liquor law violation) provided by ICPSR's archive of Uniform Crime Reporting Program Data (United States); County Level Arrest and Offenses Data, United States Department of Justice, Federal Bureau of Investigation, and, substance abuse-related mortality data from the Georgia Department of Human Resources, Division of Public Health (Georgia Vital Statistics Report). Relevant mortality rates are those deaths caused by alcoholic liver failure and drug overdoses, and noted as such in coroners' reports. As McAuliffe and Dunn (2004, p. 1000) explain, "it is assumed that a person who died from drug or alcohol dependence, non-dependent abuse, withdrawal or disease... probably has a severe substance use disorder and needed treatment." These authors validated use of this same arrest and morality data, at the state level, by comparing their substance abuse index with data from survey reports. Based on this operationalization, I created a composite index of drug and alcohol arrest rates as well as drug- and alcohol-related mortality. These were standardized as z-scores, summed and then re-scaled so that the resulting index ranges from 0 to 100.[1]

In order to capture longitudinal trends in substance abuse (and treatment provision), data were initially gathered for the years 1997–2003. Unfortunately, data were not available for all of the covariates for each year during this time frame. After exploratory analyses, three time periods at 2-year intervals were selected as the most representative of the data: 1998, 2000, and 2002. A measure of change in the amount of area substance abuse was computed by taking the difference between 2000 and 2002 scores. Since I expected that there will be diffusion of substance abuse over time, I also treated the previous levels of substance abuse as a predictor of changes in substance abuse. To avoid recursive effects, I created a measure of substance abuse lagged 2 years prior (i.e., 1998) to the start of change in the endogenous variable (i.e., 2000).

Substance abuse treatment is an individual or entity that provides alcohol or drug abuse diagnosis, therapy or referral for rehabilitation of persons with addiction or dependence (Brady and Ashley 2005). I measured rate of treatment provision by counting the number of for-profit, non-profit and government treatment providers per 10,000 population using the N-SSATS conducted by US Substance Abuse and Mental Health Services Administration, an agency of the Department of Health and Human Services. Like the substance abuse index, I standardized the measure and re-scaled it so that treatment provision ranges from 0 to 100. Moreover, a treatment provision change score was created to capture longitudinal changes, along with a measure of treatment provision lagged 2 years prior (i.e., 1998) to the start of change in the endogenous variable (i.e., 2000).

Treatment admissions and density of government-owned treatment organizations serve as controls in these analyses since trends in substance abuse may be a function of the volume of treatment services (rather than simply the rate of provision) as well as the degree of state intervention in treatment markets. Annual counts of treatment organization admissions for each county were culled from the N-SSATS reports. Government treatment services were also drawn from N-SSATS and measured by a count of government-owned treatment organizations in each county.

Unemployment, poverty rate and population size. Area disadvantage, delineated by housing stock, unemployment, median income, ethnicity and levels of education, tends to be a factor in differential substance abuse (Storr, Chen, and Anthony 2004) although some research suggests that impoverished neighborhoods are not more drug-dependent than more affluent areas (Saxe et al. 2001). In addition, large metropolitan areas are more likely to encompass disadvantaged areas, although, rural rates of poverty and unemployment also exacerbate conditions expected to influence substance abuse (Van Gundy 2006). As Table 22.1 shows, the size of the adult population across Georgia counties varies widely from a few thousand to over half a million (e.g., Fulton county in Metro Atlanta). I therefore control for unemployment, poverty, and population size. County-level unemployment and poverty rates are defined by the US Census Bureau and published by the Housing and Household Economic Statistics Division, Small Area Estimates Branch. Although the relative estimate of poverty varies by year, example thresholds for

Table 22.1 Descriptive statistics of data used in spatial regression models. Georgia Counties ($n = 159$)

	Mean	Std. dev	Min	Max	Moran's I Y_{it}
Dependent variables					
Substance abuse – change in drug/alcohol arrests and mortality rates					
Period $(t-1)$ to (t)	8.278	18.993	−55.259	85.076	0.0848; $p < 0.040$
Treatment provision – change in drug/alcohol treatment providers					
Period $(t-1)$ to (t)	−1.478	8.019	−55.506	19.120	0.070; $p < 0.075$
Explanatory and control variables					
Substance abuse – rate of drug/alcohol arrests and mortality rates					
Period $(t-2)$	28.240	17.901	0	100	
Period $(t-1)$	27.308	15.637	0	100	
Period (t)	35.585	17.746	0	100	
Treatment provision – rate of drug/alcohol treatment providers					
Period $(t-2)$	10.951	16.882	0	100	
Period $(t-1)$	15.392	17.802	0	100	
Period (t)	13.914	16.365	0	100	
Treatment admissions – yearly substance abuse treatment admissions					
Period $(t\text{-}1)$	461.302	1564.593	0	17325	
Government organizations – # government-owned treatment organizations					
Period $(t\text{-}1)$	0.799	1.054	0	5	
Unemployment – rate of unemployment					
Period $(t\text{-}1)$	4.208	1.098	2.4	8	
Poverty – rate below poverty line					
Period $(t\text{-}1)$	156.405	48.864	38.824	284.182	
Population size – area population size					
Period $(t\text{-}1)$	51761.550	109517.800	2077	816644	

this study were roughly: single = below \$9000; two persons = below \$11,800; three + persons below \$14,000.

Methods

The first law of geography assumes that everything is related, and things that are nearby are more related than things far away. Combine this principle with that of temporal dependence and the chance of creating errors in measurement is great. In both cases, observations will cease to be independent of one another, violating key assumptions of ordinary least squares regression – that the residuals are not correlated and the error variances are homoscedastic. To overcome these problems, spatial regression models are used in the following analyses to model spatial dependence; the issue of autocorrelation is adjusted by inclusion of a lagged dependent variable, and lack of homoscedasticity is adjusted through the use of a special error term, λ. Spatial regression is ideally suited for modeling substance abuse and treatment data since the central units of analysis share contiguous boundaries with one another. It is expected that since substance abuse and treatment provision diffuse across contiguous boundaries, it is necessary to adjust for this bias by explicitly modeling spatial error. The basic model from Anselin (1988) is:

$$\begin{aligned} y &= \rho W y + \varepsilon \\ \varepsilon &= \lambda W \varepsilon + \xi \end{aligned} \tag{22.1}$$

where y is a vector of observations on a dependent variable, Wy is a spatially lagged endogenous variable with a weights matrix W, ρ is the spatial autoregressive parameter for the spatially lagged dependent variable, ε is a vector of error terms, $W\varepsilon$ is a spatially lagged error term with spatial weights matrix W, λ is the spatial autoregressive parameter for the spatially lagged error term, and $\xi \sim \mathrm{N}(0, \Omega)$, where $\Omega_{ii} = h_i(z\alpha)$, when $\alpha = 0, h = \sigma^2$ and the errors are homoscedastic. In this case, the weight matrix is defined by rook contiguity in that the counties expected to influence one another share borders (but no corners of borders). This is the more conservative strategy for assessing contiguity.[2]

The presence of a significant spatial error term indicates correlated errors among neighboring communities. The models also include maximum likelihood estimates of a weighted spatial lag term (ρ), representing the dependent variable of interest. The lag term reflects the extent to which commonalities in the covariates are due to spillovers across neighboring counties. A significant spatial lag parameter is evidence that the original OLS estimate of the impact of the covariates on the outcomes is overstated and must be adjusted (Anselin 1988). Maximum likelihood estimates are used to model the spatially lagged dependent variable and the autoregressive error term.

While the Georgia substance abuse data have a longitudinal component to them, and my model specifies a temporal lag of the dependent variables to obviate any recursion among them, computationally, all variables are treated in the spatial regressions as if the data were cross-sectional.[3] Both GeoDa (GeoDa Software 2004), a program that facilitates analysis of spatially distributed data, and Stata (StataCorp 7.0) were used in the analyses. Spatial regression has been used in a number of healthcare and substance abuse applications (see e.g., Freisthler and Gruenewald 2005; Freisthler, Gruenewald et al. 2005; Mobley, Root Anselin, Lozano, and Koschinsky 2006).

Results

Table 22.1 includes a measure of spatial autocorrelation, Moran's I. This statistic indicates the degree of independence and non-independence of cases based on the rook weight matrix. The statistic uses the rook contiguity weight matrix to estimate the relationship between observations on the change scores for substance abuse and treatment provision. Moran's I for substance abuse and treatment provision is positive, which indicates that there is clustering of similar (rather than dissimilar) change scores among neighboring counties. The Moran's I for treatment provision is not significant, however, and therefore not depicted in Figs. 22.1 and 22.2. The positive significant clustering of cases for substance abuse is narrowly supported by Figs. 22.1 and 22.2. Red and orange-coded counties are those with the highest similar concentrations of substance abuses. These cases are embedded in a system of counties in the northern part of state with the third highest concentration of substance abuse. The two figures (and Moran's I) suggest that substance abuse clusters along geographic dimensions. It is therefore necessary to adjust for any bias this may introduce in the relationship between abuse and service provision.

Table 22.2 presents results of OLS regression analysis of changes in substance abuse on treatment provision, prior rates of substance abuse, organizational variables and area disadvantage. In this Table 22.1 test, the reduction hypothesis by regressing changes in rates of substance abuse at time (t) on provider density in the previous period ($t-1$), and rates of substance abuse ($t-2$), while controlling for the influence of organizations and area disadvantage also lagged one time period prior to changes in rates of substance abuse. Results in Model 1 contradict expectations of the reduction hypothesis: greater provision of substance abuse treatment services does not yield a reduction in substance abuse across Georgia counties. In fact, greater provision of services fosters increases rather than decreases in substance abuse in the later time period, even when level of prior substance abuse is controlled (0.182). Concerning the effects of prior substance abuse themselves; higher rates of substance abuse in the earlier time period tend to result in decreases in substance abuse later on. This relationship, however, cannot be due to the influence of treatment provision, since counties with more services witness *growth* in their rates of substances abuse. Here, policing and drug markets are likely to act as significant mechanisms by which substance abuse either increases or decreases over

A

B

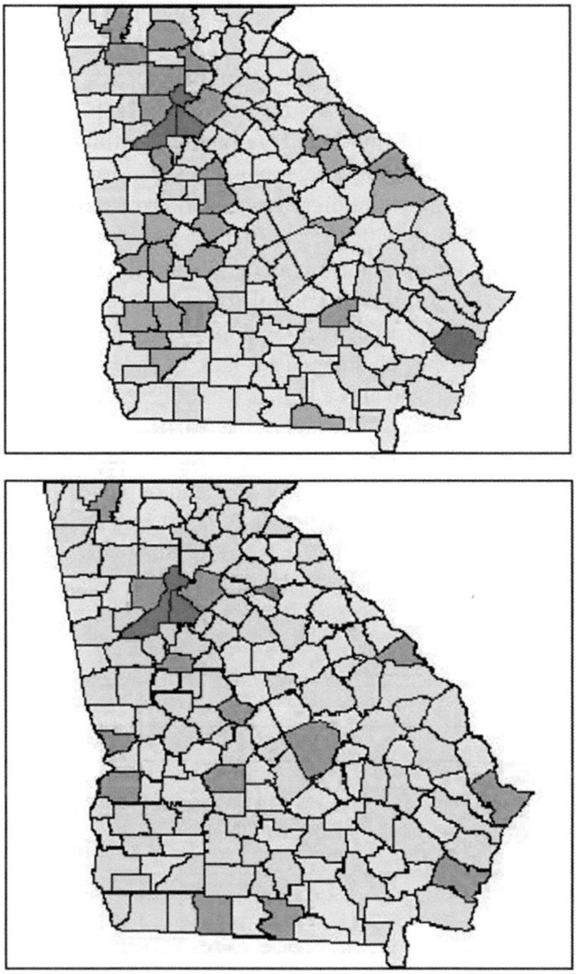

Fig. 22.1 (**a**) Prevalence of substance abuse in Georgia Counties 2002 ($n = 159$): key = red top 1%; orange 90–99%; yellow 50–90%; light blue 10–50%; blue 1–10%; cobalt <1%; (**b**) prevalence of substance abuse in Georgia Counties 2000 ($n = 159$). Key = red >1%; orange 90–99%; yellow 50–90%; light blue 10–50%; blue 1–10%; cobalt <1% (See also Plate 19 in the Colour Plate Section)

time (see e.g., Freisthler, Gruenewald et al. 2005). Interestingly, as unemployment rates increase, substance abuse rates decline, while poverty has the opposite effect, suggesting a complex relationship between work and abuse (negative) and income and abuse (positive). An additional complexity is that government-owned treatment organizations have a negative relationship with substance abuse rates, indicating that areas with government-sponsored treatment are more likely than those with for-profit and non-profit organizations to experience declines in substance abuse.

Note that because there are no causal mechanisms in the theoretical claims, any implied causal links are merely speculative.

Finally, Table 22.2 contains a number of diagnostics for assessing violation of regression assumptions, the most important of which are the tests of spatial dependence based on Moran's index and the Lagrange Multipliers. These tests show no significant autocorrelation in the data. Nonetheless, Models 2 and 3 in Table 22.2 include corrections for spatial autoregression and error to confirm that spatial dependence is not interfering with the substantive hypotheses. As discussed previously, models contain the error term lambda (λ) and the weighted spatially lagged dependent variable rho (ρ), which control for autoregressive error and clustering on the dependent variable, respectively. Neither is significant, so interpretation of results in the models is not challenged by lack of independence of observations. The implication for substance abuse is that while there may be temporal diffusion, from substance abuse at time $t-1$ to substance abuse at time t, there is no significant spatial contagion, as measured by the error terms.

Table 22.3 examines the treatment need-demand hypothesis. In this model, treatment provision is expected to rise as levels of substance abuse, prior levels of provision, organizational characteristics and disadvantage rise. Again, in contrast to expectations, provision of services is not significantly related to changes in sub-

Table 22.2 Regression coefficients and standard errors for spatial analyses of changes in substance abuse: Georgia Counties ($n = 159$)

	Model 2	Model 3	Model 4
Explanatory variables			
Treatment provision *(t-1)*	0.182 (0.093)†	0.176 (0.091)†	0.170 (0.090)†
Substance abuse *(t-2)*	−0.300 (0.093)**	−0.288 (0.090)**	−0.295 (0.090)**
Controls *(t-1)*			
Treatment admissions	−0.001 (0.002)	0.000 (0.001)	−0.001 (0.001)
Government organizations	−4.375 (1.793)*	−4.350 (1.744)*	−4.385 (1.735)*
Unemployment	−5.499 (1.837)**	−5.772 (1.794)**	−5.683 (1.779)**
Poverty	0.133 (0.047)**	0.136 (0.046)**	0.127 (0.046)**
Population size	0.000 (0.000)	0.000 (0.000)	0.000 (0.000)
Lambda		0.108 (0.124)	
ρ			0.143 (0.113)
Constant	18.700 (8.522)*	19.103 (8.507)*	19.198 (8.253)*
Model diagnostics			
Multicollinearity condition number	16.596		
White	21.419		
Moran's I	1.090		
Lagrange multiplier (spatial error)	0.602		
Lagrange multiplier (spatial lag)	1.514		
Model fit			
Log likelihood	−672.80	−672.47	−672.05
AIC	1361.59	1360.93	1362.10

Numbers in parentheses are standard errors
$^{\dagger}p < 0.10$; $^{*}p < 0.05$; $^{**}p < 0.01$; $^{***}p < 0.001$: two-tailed tests

stance abuse, although the coefficient is negative, indicating that growth in treatment services occurs in areas with lower rather than higher levels of substance abuse. The negative relationship has implications for the targeting of services and suggests that treatment location decision-making is less a function of meeting demand for substance abuse services, at least in Georgia, than a function of other causes. These other factors may have more to do with organizational characteristics such as competitive strategy, capacity, institutional affiliation or characteristics of markets such as zoning regulations. The finding of a negative relationship between demand, in the form of levels of substance abuse, and treatment provision, seems contrary to Allard's thesis that social service provision finds its way to areas with greater demand. In fact, neither poverty, itself linked to substance abuse (McAuliffe et al. 2002), nor substance abuse seem to generate greater provision of services, as Table 22.3 demonstrates. Interestingly, the negative direction of the relationship between community disadvantage and treatment, although not significant, indicates that more impoverished counties experience declines in levels of service provision. Overall, the only significant predictor of treatment provision at time t is treatment provision at time $t - 2$. The relationship is negative, indicating that greater service provision early results in declines a few years later. This may be due to competition among service providers, which results in a winnowing out of non-competitive treatment organizations.

Conclusion and Discussion

One implicit assumption of research in the area of substance abuse treatment provision is that increases in availability and geographic access to treatment will serve to eradicate or at least diminish substance abuse in proximity to treatment locations. Studies of treatment outcomes for individuals can now be supplemented with the results of these analyses comparing the effects of services on substance abuse rates in the communities in which they live. And conclusions about problematic efficacy are much the same. Given a reduction hypothesis, it was expected that greater provision of treatment for substance abuse would reduce area substance abuse. Yet, greater provision of services had a *positive* relationship with changes in rates of area substance abuse. Either treatment provision is not effective in reducing substance abuse, or else, assuming treatment services are effective in alleviating individual substance abuse, lack of reduced area substance abuse rates occurs for other reasons. It might mean that residents (those whose reported substance abuse characterizes the community) do not come from the same pool as clients of local treatment programs (Jacobson 2004b), although the underlying assumption of research studies of service provision (e.g., Allard 2004; US Department of Health and Human Services-DASIS Report 2001) suggests that the policy problem is how to provide greater geographic, if not financial, access to services such as treatment (assuming it is efficacious to begin with). It might also mean that policing varies greatly and therefore has an important effect on substance abuse, whether or not treatment is effective. Or, it

Table 22.3 Regression coefficients and standard errors for spatial analyses of changes in treatment provision: Georgia Counties ($n = 159$)

	Model 1	Model 2	Model 3
Explanatory variables			
Substance abuse *(t-1)*	−0.012 (0.047)	−0.008 (0.046)	−0.010 (0.046)
Treatment Provision *(t-2)*	−0.130 (0.040)**	−0.131 (0.038)***	−0.132 (0.039)***
Controls *(t-1)*			
Treatment admissions	0.000 (0.001)	0.000 (0.001)	0.000 (0.001)
Government organizations	−0.668 (0.742)	−0.590 (0.720)	−0.596 (0.721)
Unemployment	0.559 (0.827)	0.490 (0.808)	0.475 (0.804)
Poverty	−0.003 (0.021)	−0.002 (0.021)	−0.001 (0.021)
Population size	0.000 (0.000)	0.000 (0.000)	0.000 (0.000)
Lambda		0.126 (0.123)	
ρ			0.122 (0.120)
Constant	−1.661 (3.764)	−1.741 (3.768)	−1.537 (3.655)
Model diagnostics			
Multicollinearity condition number	16.818		
White	86.434**		
Moran's I	1.298		
Lagrange multiplier (spatial error)	1.004		
Lagrange multiplier (spatial lag)	1.117		
Model fit			
Log likelihood	−546.66	−546.17	−546.15
AIC	1109.33	1108.34	1110.29

Numbers in parentheses are standard errors
$^{\dagger}p < 0.10$; $^{*}p < 0.05$; $^{**}p < 0.01$; $^{***}p < 0.001$: two-tailed tests

might indicate that markets continue to produce substance users at higher rates than treatment cures, so to speak. Finally, the reduction hypothesis may fail even when treatment access and success are excellent with regard to individual outcomes – reducing the likelihood of future substance abuse markedly, without having area effects – because of a temporal lag that fails to alter the ratio of substance abusers and recovered users. The solution to all of these issues may lie with multilevel models that control for treatment outcomes at the individual level while investigating contextual effects at the community and population levels. Having precise, overlapping, information on client and organizational characteristics (such as level of addiction severity, client geographic origins, length of stay, completion rates etc.) and mapping these to community and treatment contexts would enhance our ability to understand the relationship between individual and community-level substance abuse and service provision.

It is also noteworthy to consider that, where higher levels of area substance abuse were thought to increase provision of treatment by increasing demand, this was not the case: higher levels of substance abuse did not foster growth in provision of substance abuse treatment services. This raises the question of organizational and institutional decision-making in provision of services. If the goal is to improve the theoretical understanding and empirical estimation of community-level substance abuse treatment provision as it relates to organizational and area characteristics, it might help to specify the proximate mechanisms of their relationship. For example, what motivates treatment providers to decide to locate in any particular neighborhood, community or region? Are these related to area characteristics such as treatment demand and need, competition from other providers and the structure of markets, or are considerations such as institutional conformity more important (Scott 2003)? Organization theory suggests that some organizations, and indeed organizational sectors, may have lower performance capabilities and might be less competitive but still thrive nonetheless. The key to their persistence lies in their conformity to institutional rules rather than outcomes. To address the issue of how persistently failing organizations thrive, first requires the kind of preliminary analysis undertaken in the study of treatment context.

Another important question is why there is so little spatial dependence in these data. I anticipated that geographic proximity leads to the diffusion of substance abuse (Freisthler, Gruenewald et al. 2005), which will be reversed as treatment organizations respond to demands for services (see Allard's 2004 research on this point). I found, however, that there was no significant spatial diffusion when all explanatory variables are included in the model. It may be that the model has been underspecified and a more sophisticated spatial lag, such as one that measures adjacencies across state boundaries is needed since counties that lie on state borders, particularly city–state boundaries, may have more in common with one another. There may be a more substantive reason for the lack of spatial clustering. McAuliffe and Dunn (2004) show that of the 50 states, Georgia has the largest gap between substance abuse demand and service provision. Their research demonstrates that there is no relationship between the need for services and providers of those services in that state. Consequently, there should be little overlap between the location of providers and little clustering with respect to substance abuse. This is a desirable statistical property (i.e., independence of observations) since it is biased in favor of the null hypothesis. Paradoxically, it may or may not be a socially desirable one, depending on where treatment services are targeted and their affect on community capacity. Future studies might extend county-level analysis of socio-spatial relationships to smaller units such as census tracts and neighborhoods in order to understand whether diffusion operates differently in these contexts (Morenoff and Lynch 2004). The importance of understanding how substance abuse and treatment disperse across geographic boundaries (or not) cannot be emphasized enough.

The purpose of the study was to determine whether treatment provision impacted rates of substance abuse and vice versa. The unexpected direction and significance of effects reveal a rather large gap in our knowledge of the dynamics of this relationship. Pursuing some of these recommendations will permit better specification of the mechanisms underlying the relationship and provide information for addressing how changes in public policy and organizational contingency can enhance substance abuse outcomes. Having more pieces to the puzzle will facilitate decisions concerning treatment provision and direct public policy toward enhancing it.

Acknowledgment The author would like to thank Kendralin J. Freeman and Amy Patterson for their research assistance on this project, and participants at Emory's Sociology Department Seminar for insightful comments on earlier drafts. The author also thanks especially Ivan Cheung for his critical feedback.

Notes

1. Because of the small number of cases, inter-item correlations tended to be low.
2. Weight matrices with queen contiguity were also run, yielding the same results.
3. As noted in the Conclusion section, research on this topic should develop spatial lag models that can handle time-series data as well.

Chapter 23
Using a GIS Framework to Assess Hurricane Recovery Needs of Substance Abuse Center Clients in Katrina- and Rita-Affected Areas

Traci Craig Green and Cynthia Pope

Abstract This study compared profiles of substance abuse (SA) treatment attendees in New Orleans to Houston, Texas, and in hurricane-affected to -unaffected areas within Louisiana (LA); explored the geospatial distribution of points of care for displaced SA clients in areas of projected settlement; and described possible discrepancies between need and available resources. Data from the Federal Emergency Management Agency, US Census Bureau, the Substance Abuse and Mental Health Services Administration, and the Drug Evaluation Network System including Addiction Severity Index composite scores for 5,660 SA clients from 66 LA sites, six ($N = 569$) located in New Orleans and from two sites ($N = 781$) in Houston permitted client comparisons and site-level geospatial analyses. Results suggest the utility of a spatial approach to inform SA policies after a disaster and in hypothesis generation for future research. The results have direct implications on the geography of healthcare for SA in hurricane-affected communities.

Introduction

The impact of the 2005 Atlantic hurricane season and the combined effects of death, injury, destruction, and population displacement from Hurricane Katrina were unprecedented (Centers for Disease Control and Prevention 2006). Hurricane Katrina was the deadliest hurricane since 1928 and the costliest natural disaster in the US history. Together, Hurricanes Katrina and Rita displaced 240,000 households comprised more than 469,000 people from their homes in southern Louisiana (LA) and coastal Mississippi. According to a recent analysis of forwarding zip codes of hurricane evacuees, most people who left Katrina-affected areas relocated to neighboring states and other parishes within LA, places where they either had extended family and friends or that were within a day's drive to their damaged homes. Suburban New Orleans, Houston, TX, and Baton Rouge, LA were the most common places

T. C. Green
67 Marshall St., Medford, MA 02155, USA. Tel.: 617 909 9919; Fax: 781 395 2125
e-mail: traci.c.green@yale.edu

Y.F. Thomas et al. (eds.), *Geography and Drug Addiction*,

of relocation (Tizon and Smith 2005). Nearly two years after the hurricanes, about 100,000 displaced people are still living in Houston (Associated Press 2007).

For those in treatment or in need of treatment for drug and alcohol addictions, the relocation forced an additional hardship: to seek substance abuse (SA) and mental health (MH) services in unfamiliar communities. In addition, other evacuees are likely to be at risk for SA and addiction or relapse which is likely also to co-occur with post-traumatic stress disorder (PTSD) and/or depression resulting from the immense challenges and loss imposed by the hurricanes (Cottler, Compton, Mager, Spitznagel, and Janca 1992; Kessler, Sonnega, Bromet, Hughes, and Nelson 1995; Vlahov et al. 2002; Volkow 2005; Weiss et al. 2002). Increases in both frequency and prevalence of SA occur among disaster survivors in the short and long terms, often accompanied by a concomitant rise in SA treatment demand (Deren, Shedlin, Hamilton, and Hagan 2002; Factor et al. 2002; Joseph, Yule, Williams, and Hodgkinson 1993; Movaghar et al. 2005; ; Vlahov et al. 2002; Vlahov et al. 2004a; Vlahov, Galea, Ahern, Resnick, and Kilpatrick 2004b; Weiss et al. 2002). Indeed, physicians treating hurricane-affected patients detected increased SA and MH disturbances early on as people responded to their losses (Graham 2006; Mehren 2006). Relapses among people who had recently been in treatment and who had lost support networks have been reported (Bergeron 2006; Curley 2005) and are expected to continue during the process of individual and community recovery (US Department of Health and Human Services 2005). Lessons learned about trauma and SA after September 11 of initiation of substance use, relapse, and an increased intensity and need for SA prevention and intervention services were noted by the Substance Abuse and Mental Health Services Administration (SAMHSA) (McDuff and Ford 2005). Approximately 25–30% of people exposed to severe trauma and 5–10% of people exposed to moderate trauma will develop substance use problems (Foa, Stein, and McFarlane 2006). People affected directly by disaster are at greatest risk for PTSD and depression (Galea et al. 2002a,b). The natural trajectory of MH and SA problems after a disaster indicates that these problems tend to be greatest in the initial months following the disaster, then generally decline in the affected population (Norris 2005). However, many key questions about this trajectory persist, for example, it is unclear if the decline is linear or not (Norris 2005) and how this trajectory may differ for affected populations where pre-disaster prevalence of MH and SA problems was high. There were 500,000 people directly impacted by the Hurricane Katrina; up to 200,000 of whom could develop need for SA treatment (National Mental Health Association 2006). This situation has prompted one group in LA to frame the problem as another familiar disaster metaphor: "We are on the leading edge of an emotional and psychological tsunami that threatens the health and recovery of our community" (MultiQuest 2006).

Drug and alcohol addiction etiology reveals an apt fit of the chronic disease model. From longitudinal studies, it is known that abuse of drugs and alcohol is usually chronic over a lifetime (Hser, Anglin, and Powers 1993; Hser, Hoffman, Grella, and Anglin 2001; Vaillant 1983). Thus, drug and alcohol addictions, like all chronic diseases, require adherence to treatment regimens, access to prescription medications, and management through regular healthcare provider interaction

(McLellan, Lewis, O'Brien, and Kleber 2000; McLellan 2002; Vest and Valadez 2006). These key elements of proper chronic disease management are important to instill and maintain in times of both stress and calm, and they fit within the paradigm of primary healthcare practice.

In addition to being chronic, SA may also act as a coping mechanism in response to trauma. This, in turn, can lead to dependency or addiction as well as other psychiatric disorders (Austin and Godleski 1999). Large-scale environmental devastation coupled with large-scale trauma and endemic socioeconomic distress in the Hurricanes Katrina and Rita areas may have an unprecedented effect on the panoply of psychiatric responses in the affected population.

Studies of other recent natural disasters reveal that class and socioeconomic status influence the efficacy of evacuation and influence subsequent physical and MH outcomes of the affected population (Fussell 2005). For example, following Hurricane Andrew in 1992, people who trusted information sources urging evacuation, those who had family in nearby safer cities, and those with the financial means to move into a hotel were more likely to evacuate, whereas households that were lower income, minority, or supported elderly or disabled persons were less likely to evacuate (Gladwin and Peacock 1997). In the case of Hurricane Katrina, many of those who received government assistance to relocate were of a precarious socioeconomic status, compared to New Orleans and LA residents. A survey of Houston Red Cross shelter residents conducted soon after relocation found low employment rates, low household incomes, and low rates of home ownership prior to the hurricanes (Brodie, Weltzien, Altman, Blendon, and Benson 2006). Evacuees were also disproportionately African American, had low levels of education, and tended to lack health insurance coverage, relying primarily on public healthcare provided by the New Orleans Charity hospital system. Two in five evacuees also suffered from a chronic condition, a proportion exceeding that of LA and other New Orleans residents. Two-thirds of the respondents who reported an intention to relocate cited a preference for the Houston area. Another study conducted within Austin, TX shelter residents on September 8–9, 2005 found demographic and socioeconomic constraints similar to the Houston shelter residents, high self-reported prevalence of chronic conditions, and notable reports of mental illness and past year use of an illegal substance (Vest et al. 2006).

One way of examining the potential needs of substance abusers displaced by the hurricanes is to study the profiles of clients who have since entered treatment. From September, 2005 through July, 2006, the Texas Department of Health Services enumerated 567 (711 with duplicates) hurricane-related admissions to SA treatment, 66% of whom were from out of state and most of those had been born in New Orleans (71%) (Texas Department of Health Services 2006). In addition to the low socioeconomic status and education levels evident among the Katrina and Rita evacuees, the new clients tended to be opioid users, likely due to the rush to accommodate opioid-dependent clients in hosting communities (Maxwell 2006).While these early reports of treatment recipients are revealing, a multitude of factors influences the dynamics of treatment access, and in times of crisis, such as a natural disaster, those who receive treatment may not reflect the full spectrum of need.

Another important source of complementary information in a rapid assessment of the needs of substance abusers displaced by the hurricanes is the status of the pre-disaster populations. Using data from sources that predate the events of September 2005, we may be able to anticipate both the critical and unique areas of a public health response. For example, the interplay of the lower socioeconomic status of the evacuees and existing or subsequent SA problems could have numerous possible implications for the host community's ability to treat SA. Psychological sequelae of disasters are known to be more frequent and severe among vulnerable populations, such as substance abusers (Galea et al. 2002a; Schuster et al. 2001). Consequently, if there was a higher baseline prevalence of SA problems in the evacuee population compared to pre-hurricane New Orleans residents, such a discrepancy may lead to an underestimate of the response needs. In addition, evacuees with SA problems may require more extensive social services than would be expected in the substance abusing population of the host community. Thoroughly examining pre-existing data sources as well as collecting timely, precise new information are part of a comprehensive approach to assessing and responding to health crises. Moreover, new tools can help better understand nuances in these data. Geographic Information Systems (GIS) is being used increasingly in public health and can be a powerful tool in disaster response and management (ESRI 2006); (Cromley and McLafferty 2002; Meade and Earickson, 2000).

While the size of the Hurricane Katrina and Rita evacuee population who already suffer from or who are at risk of SA is unknown, the endemicity of SA in LA and the sheer magnitude of the natural disaster give public health officials cause for concern. A primary question is whether or not the existing SA and MH services in the communities hosting evacuees have the capacity to care for them. In this instance, capacity refers to the surge capacity, or "the ability of the health care system to expand care capabilities in response to a sudden, unexpected patient influx, whether this is a smaller 'daily' event, such as the arrival of 10 trauma patients into the emergency department from a bus crash, or a larger-scale or catastrophic event, such as a major earthquake." (Jenkins, O'Connor, and Cone 2006). For this chapter, the term capacity hereafter refers to surge capacity, and more explicitly, the ability to accommodate the latter, catastrophic surge.

Houston received the largest number of people displaced by the hurricanes, primarily from New Orleans, so we focused this study on comparisons of data from New Orleans vs. Houston and, within LA, hurricane-affected vs. -unaffected areas. We used GIS to conduct statistical and geospatial analyses to: (1) compare profiles of SA center attendees in New Orleans to those in the primary areas of displacement: Houston, TX, and the unaffected areas within LA; (2) explore the geospatial distribution of points of care for hurricane victims with SA and MH needs in the areas of projected settlement; and (3) assist recovery planning efforts by describing these available resources, possible discrepancies, and barriers in Houston and hurricane unaffected areas in LA with respect to SA and MH care. The overarching aims of this analysis were to generate hypotheses for further research and to illustrate the use of GIS in response capacity planning.

Methods

Data Sources

Using data from the 2000 US Census, the Federal Emergency Management Agency (FEMA) Mapping and Analysis Center, SAMHSA, Geographic Names Information System (GNIS), the Texas Department of Health Services treatment client data,[1] and the Drug Evaluation Network System (DENS) database dated September 2005,[2] analyses at the individual and SA treatment site levels were conducted, as detailed below. We calculated a density of hurricane-related admissions per treatment slot available in Houston. Data for the numerator were obtained from the Texas Department of Health Services for all SA treatment sites within Houston zip codes reporting hurricane-related treatment admissions from September 2005 and July 2006. The denominator was a count of all possible treatment slots available at sites within the respective hosting Houston zip codes. Sites were contacted directly by telephone to obtain the number of possible treatment slots (inpatient and/or outpatient) at their facility if this information was unavailable from publicly accessible sources. In this study, only two dimensions of surge capacity and capacity to care for SA clients were examined: treatment sites and treatment slots. There are many elements of such capacity, including system integrity, space (size and quality), staff (numbers and skill), and supplies (quality and volume) (Kelen and McCarthy 2006b).

The DENS data provided scores from the seven Addiction Severity Index (ASI) (McLellan, Luborsky, Woody, and O'Brien 1980) composites. The ASI is a standardized instrument administered as a semi-structured interview that was designed for use on admission to a drug and alcohol program. The ASI is used extensively to assess severity of addiction and the need for treatment. Its administration upon treatment entry is required by 42 states and the Veteran's Administration. More specifically, the ASI measures severity of problem areas that are typically associated with alcohol and drug abuse: medical status, employment and support status, drug use, alcohol use, legal status, family and social relationships, and psychiatric status (McLellan et al. 1985, 1992). Select items from each domain are combined and weighted using a scoring algorithm to create composite scores. The composite scores range from 0 to 1; higher scores indicate greater need for treatment in a given area. Composite scores can be compared across domains, to better understand the current needs of the patient. The ASI has demonstrated excellent reliability and validity across diverse patient populations (Hodgins and El Guebaly 1992; Joyner, Wright, and Devine 1996; McLellan et al. 1985).

---

[1] Counts of clients entering substance abuse treatment to all publicly funded treatment sites in Texas by zip code were kindly supplied by Jane C. Maxwell, PhD, from the Gulf Center for the Study of Addictions in Austin, Texas, as part of data collected by the Texas Department of Health Services.

[2] DENS database September 2005 is available from Treatment Research Institute, 600 Public Ledger Building, 150 South Independence Mall West, Philadelphia, PA 19106–3475.

Other variables selected for this analysis from the DENS database were relevant client characteristics (e.g., gender, race/ethnicity, employment, and medical history). Data from 5660 SA center clients from 66 sites in LA, six of which ($N = 569$) were located in New Orleans, comprised the LA data set. Data from 781 client scores at two Houston sites were also obtained. ASI scores were collected from 2002 to 2004 and site, the only geographic locator, was identified by zip code. We used mean, median, mode and proportions to summarize the site-level data.

Analysis

Geocoding (i.e., the process of relating an address to a geographic area) was performed on the DENS data sites, all available SA and MH service sites (SAMHSA data), and all hospitals in LA and Houston (from GNIS). In this way, it was possible to differentiate sites located in hurricane-affected and -unaffected areas. Analysis of individual-level data, i.e., client demographics and ASI composite scores, were conducted first. T-tests, chi-square tests, and their non-parametric equivalent (i.e., Fisher's exact, Kolmogorov–Smirov Z tests) compared data from clients in New Orleans to Houston and, once identified as an affected site by geocoding, data from clients in hurricane-affected to those from unaffected areas within LA. Next, site-level analyses were conducted. For reasons of privacy, the DENS does not release individual client addresses or geographic locators more specific than the zip code of a client's treatment site for analysis, geospatial analyses were conducted in the aggregate at the treatment site level. Proximity to care, one aspect of the broader concepts of access to care, surge capacity, and geographic access, was analyzed in this study at the site level. We did not endeavor to include other aspects of the broader concept of access to care such as staff number, languages spoken, training levels, etc., nor of other elements of geographic access such as transportation or mobility factors, due to limitations of available data sources. Proximity, or distance, is a key factor in the uptake of SA treatment services (Jacobson, Robinson, and Bluthenthal 2006; Schmitt, Phibbs, and Piette 2003). Thus the site-level analyses involved descriptive statistics, buffer analysis for proximity of care sites, and, for the LA data, creation of a spatial surface of the mean ASI composite site scores, mapped by zip code tabulated area (ZCTA). The ZCTA is a specially designed statistical entity built from US Census blocks that allows for translation of zip codes to US Census area units (e.g., blocks, tracts), thereby facilitating area-level analysis. ZCTA-level data from the 2000 US Census on the proportion of vacant housing units were mapped to reflect areas of potential resettlement for hurricane victims in Houston.

Two of the analytic methods employed in this study warrant further detailed explanation: proximity analysis using buffers and spatial surface creation using kriging. A spatial analysis allows for the exploration of the inter-relationships of features of a map. Proximity analysis, one of the tools of spatial analysis, draws from the concept of placing a buffer around an object to determine what features are within a certain distance of another feature (Theobald 2005). In this study, we used proximity analysis to examine proximity to healthcare and SA treatment facilities for areas with more available vacant housing, i.e., areas of likely resettlement for hurricane-effected populations in Houston. Using GIS, it is possible to query the maps to see what proportion of communities live within closer proximity to these health services. Alternative analytic methods to describe geographic access to care points could not be performed (see, for example, Lin 2004) due to the difficulties in estimating hidden populations from publicly available data (i.e., substance abusers are not enumerated in the Census) and lack of information on where hurricane evacuees have relocated within Houston. Buffer analysis permits a visual depiction of the proximity of probable residence to key locations of healthcare and SA treatment and an estimate of the social distribution of geographic accessibility to care.

Further explanation of kriging and its use in this study is also warranted. We can treat the sample of ASI composite scores in LA collected by the DENS as geostatistical data describing a SA environment, with its own spatial distribution. The treatment sites describe the longitudinal and latitudinal locations (X and Y axis) of the SA environment while the median ASI composite scores measured there correspond to the severity (Z axis) of SA. Together, the X, Y, and Z values describe a three-dimensional space which can be "smoothed" to create a surface. However, because the DENS data represent a sample rather than a census of ASI composite scores across LA treatment sites, interpolation of non-sampled sites was undertaken. Drawing from an underlying assumption of geography that things closer together tend to be more alike than things farther away, there are several possible interpolation methods. The method ultimately selected for this study was ordinary kriging, which has a long and successful history in geostatistics (Altman 2000; Cressie 2000). Kriging interpolates values at unmeasured points by weighting measured points that are closest more than points that are farther away but also incorporates patterns of autocorrelation. Different from deterministic techniques for interpolating values which control either the extent of similarity (i.e., inverse distance weighting) of the values or the degree of smoothing in the surface (e.g., radial basis functions), geostatistical methods rely upon a random spatial process (Cressie 2000). That is, there is an assumption that at least some of the spatial variation of natural phenomena can be modeled by random processes with spatial autocorrelation. Kriging methods first quantify the spatial structure of the data using semivariogram modeling, which fits a curve through the observed spatial variation between points that have been grouped into classes of distance. The semivariogram quantifies the spatial dependence or autocorrelation in the data. Then, using the fitted model from the semivariogram, the spatial data configuration, and the values of the measured sample ASI composite score points around the prediction locations, kriging makes predictions for the unknown values at the specific locations in LA. Kriging model comparisons can be made to determine the best fitting model, using the criterion of mininimizing the root mean squared prediction error (Cressie 2000; Milillo and Gardella 2006), as was performed in this study.

The outcomes of interest for the geospatial analysis focus on the ASI medical, drug, and alcohol composites in the interest of space. Analyses were conducted in ArcGIS version 9.1, SAS version 9.1 and SPSS version 12 at the alpha = 0.05 level.

Results

Geocoding of SA and MH sites was conducted with 80% sensitivity, a minimum candidate score of 10 and a minimum match score of 60. Iterative matching of addresses was also attempted. Fifty-one of the 65 (79%) Houston sites and 15 of the 16 (94%) New Orleans sites were matched. Hospitals in Houston and LA were added with latitude and longitude coordinates. We were unable to locate the treatment sites for two zip codes reporting hurricane-related treatment admissions, representing 10 of the 229 treatment entries (<5%). Density values for the two zip codes were therefore set to missing.

Zip codes from the treatment centers receiving hurricane-related admissions from September 2005 to July 2006 numbered 107, dispersed throughout the state but clustered in Houston, Texarcana, Austin, and San Antonio (Fig. 23.1). The bulk of admissions occurred in the Houston zip codes (44.2%, 249/563) for Katrina and Rita evacuees. For Katrina-related admissions only, Houston area zip codes alone received the majority of admissions: 62.7% (229/365).

Demographics: New Orleans and Houston

Table 23.1 compares the New Orleans to Houston area clients and presents detailed results of the bivariate statistical tests. To summarize these comparisons, there were

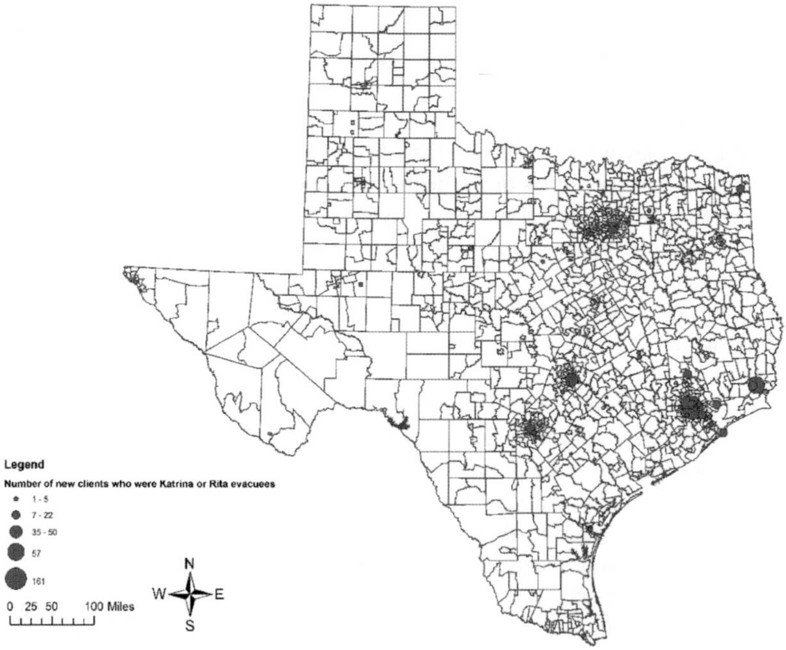

Fig. 23.1 Distribution of Hurricanes Katrina- and Rita-related treatment admissions, September 2005 to July 2006, Texas (See also Plate 20 in the Colour Plate Section)

Table 23.1 Comparison of DENS substance abuse center client characteristics: New Orleans vs. Houston and hurricane-affected vs. hurricane-unaffected areas in Louisiana

	New Orleans, LA ($N = 569$)	Houston, TX ($N = 781$)	Statistic & p-value	Affected ($N = 863$)	Unaffected ($N = 4,797$)	Statistic & p-value
Age: mean (SD)	36.5 (11.2)	35.3 (9.8)	2.08, $p < 0.05$	36.1 (10.9)	33.7 (10.2)	−5.82, $p < 0.0001$
Female (%, N)	27.8 (158)	1.0 (8)	218.3, $p < 0.0001$	27.2 (235)	30.2 (1,451)	3.18, NS
Race non-white (%)	76.1	55.7	57.14, $p < 0.0001$	59.6	37.8	142.51, $p < 0.0001$
Ethnicity Hispanic (%)	5.0	14.1	29.4, $p < 0.0001$	5.1	2.2	25.2, $p < 0.0001$
Years of education completed: mean (SD)	11.8 (2.4)	12.0 (2.2)	5.296, $p < 0.0001$	11.77 (2.4)	11.55 (2.2)	−2.61, $p < 0.01$
Address owned by client/family (%)	34.8	3.3	96.6, $p < 0.0001$	43.1	62.7	111.29, $p < 0.0001$
Median time at this address in months (IQR)	24 (81)	6 (60)	4.83, $p < 0.0001$*	24 (92)	24 (93.7)	0.63, NS *
No. of people dependent on client: mean (SD)	0.77 (1.3)	0.51 (1.1)	3.7, $p < 0.0001$	0.79 (1.3)	0.94 (1.4)	2.96, $p < 0.01$
Spend most of the time with			16.99, $p < 0.0001$			12.75, $p < 0.01$
Family	46.1	34.4		45.7	50.4	
Friends	24.0	28.7		23.6	24.7	
Alone	30.0	36.9		30.7	24.9	
ASI-composite score: mean (SD)						
Medical	0.1805 (0.31)	0.1414 (0.28)	2.27, $p < 0.05$	0.2109 (0.34)	0.1615 (0.29)	−3.94, $p < 0.0001$
Drug	0.1255 (0.12)	0.1746 (0.13)	−6.77, $p < 0.0001$	0.1306 (0.12)	0.1361 (0.14)	1.14, NS
Legal	0.1510 (0.18)	0.1949 (0.19)	−4.04, $p < 0.0001$	0.1625 (0.19)	0.1618 (0.20)	−0.099, NS

Table 23.1 (continued)

	New Orleans, LA ($N = 569$)	Houston, TX ($N = 781$)	Statistic & p-value	Affected ($N = 863$)	Unaffected ($N = 4,797$)	Statistic & p-value
Family/social	0.1874 (0.24)	0.2819 (0.26)	−6.43, $p < 0.0001$	0.1947 (0.24)	0.1679 (0.22)	−2.94, $p < 0.01$
Employment	0.7008 (0.30)	0.7094 (0.27)	−0.525, NS	0.6844 (0.30)	0.6238 (0.31)	−5.39, $p < 0.0001$
Psychological functioning	0.2002 (0.24)	0.2370 (0.22)	−0.275, $p < 0.0001$	0.2414 (0.26)	0.2236 (0.25)	−1.84, NS
Alcohol	0.1683 (0.24)	0.2925 (0.28)	−8.38, $p < 0.0001$	0.1787 (0.24)	0.2172 (0.28)	4.08, $p < 0.0001$
Drug that is biggest problem for client						
Heroin	4.7	3.0	1.4, NS	4.3	0.3	123.43, $p < 0.0001$
Cocaine	25.8	15.5	10.98, $p < 0.001$	21.6	11.9	55.27, $p < 0.0001$
% smoking	74.9	61.5	16.05, $p < 0.0001$	61.6	59.1	1.0, NS
% snorting	17.5	27.3	10.47, $p < 0.001$	26.8	28.9	0.79, NS
% injecting	6.3	8.0	0.88, NS	9.8	10.7	0.36, NS
Cannabis	20.0	3.0	42.58, $p < 0.0001$	17.2	11.9	17.28, $p < 0.0001$
Alcohol	8.1	3.7	5.65, $p < 0.05$	10.9	11.7	0.38, NS
Alcohol to intoxication	4.9	4.4	0.09, NS	6.3	8.2	3.43, NS
Alcohol and one or more drugs	25.4	66.1	124.32, $p < 0.0001$	25.3	35.9	34.02, $p < 0.0001$
No alcohol, more than one drug	4.7	1.5	5.38, $p < 0.05$	5.2	7.8	7.06, $p < 0.01$
In controlled environment, past 30 days**	31.1	60.2	111.48, $p < 0.0001$	38.9	29.3	31.65, $p < 0.0001$
Has chronic medical problem(s)	31.5	17.5	33.1, $p < 0.0001$	33.0	26.9	12.9, $p < 0.0001$

Table 23.1 (continued)

	New Orleans, LA ($N = 569$)	Houston, TX ($N = 781$)	Statistic & p-value	Affected ($N = 863$)	Unaffected ($N = 4,797$)	Statistic & p-value
Days experienced medical problem in past 30: mean (SD)	5.1 (10.4)	3.4 (8.4)	2.97, $p < 0.01$	6.0 (11.1)	4.5 (9.4)	−3.64, $p < 0.0001$
Taking medication for psychiatric problems in past 30 days	15.4	10.0	8.03, $p < 0.01$	19.4	16.7	3.68, NS
Emotionally abused in past 30 days	12.5	9.2	1.96, NS	12.5	16.7	8.92, $p < 0.01$
Has a profession, trade, or skill	54.6	73.5	27.0, $p < 0.0001$	58.1	57.8	0.033, NS
Someone contributes to support (non-institutional, cash, food, housing)	40.2	25.7	16.5, $p < 0.0001$	44.4	49.2	6.55, $p < 0.01$
Constitutes majority of their support	74.5	43.8	63.7, $p < 0.0001$	73.5	70.1	2.08, NS
Days paid for working in past 30: mean (SD)	6.4 (9.5)	4.5 (8.5)	3.62, $p < 0.0001$	5.9 (9.4)	7.5 (10.3)	4.49, $p < 0.0001$

Unless otherwise noted, the test statistic reported comparing continuous values (e.g., age) refers to t-test values and for comparing dichotomous or categorical values (e.g., race, ethnicity) refers to chi-squared test values.

*Kolmogorov–Smirov Z test

**Controlled environment includes alcohol/drug treatment, medical treatment, psychiatric treatment, jail/prison, other.

IQR = Interquartile range, NS = not statistically significant, SD = standard deviation.

numerous notable differences between clients in the two cities. New Orleans SA treatment sites treated proportionately more female clients, and the treatment communities differed in terms of modal substances abused (i.e., crack and marijuana in New Orleans vs. alcohol in Houston). Racial and ethnic differences between the settings stood out: 76.1% (429/564) of clients in New Orleans were of non-white race, 93.2% (400/429) of whom were Black; 5% (28/562) were Hispanic. In Houston, 55.7% (393/706) of clients were of non-white race, 74.6% (293/393) of whom were Black and 14.1% (108/765) were Hispanic. New Orleans clients were less likely to have a profession, trade, or skill, tended to live and spend more time with family, and were less likely to have recently been in a controlled environment, compared to Houston area clients.

ASI Composite Scores: New Orleans and Houston

Contrasting the ASI composite scores reveals differences in the degree of difficulty in functioning in many domains. Clients in both settings reported high degrees of distress on the employment scale. Aside from this domain, the similarities in rank of areas in which clients reported distress departed: New Orleans clients had greater distress on the psychological disturbances, family/social, and medical problems composites whereas the Houston clients were troubled by distress measured in the composites, in descending order, of alcohol, family/social and psychological disturbances. For Houston clients, medical problems caused the least distress, of the domains assessed by the ASI. While New Orleans SA treatment clients had significantly higher mean ASI medical scores, Houston clients had higher ASI scores on all other composites, save for similar ASI employment scores. The higher ASI medical scores in New Orleans were reflected by the higher proportion of clients reporting chronic health problems, recent hospitalizations, and use of psychiatric medications. Figure 23.2 illustrates the distribution of ASI-Medical scores by sites in ZCTAs of New Orleans and their proximity to SA treatment sites and hospitals, indicating the high density of services that were available to clients in New Orleans. In addition, evident is the immense damage Hurricane Katrina caused to these core services, affecting most of the Orleans Parish SA treatment sites and several hospitals. In particular, ZCTAs where SA treatment clients had higher ASI-Medical scores (i.e., darker colors) but also closer proximity to SA and hospital services suffered hurricane damage and evacuation.

Proximity to Resources: Houston

The DENS site-level data for Houston and New Orleans and the available housing characteristics from the Census were joined to the US Census's Topologically Integrated Geographic Encoding and Referencing system (TIGER) file street map at the ZCTA level. If we assume that people displaced by Katrina will be renting

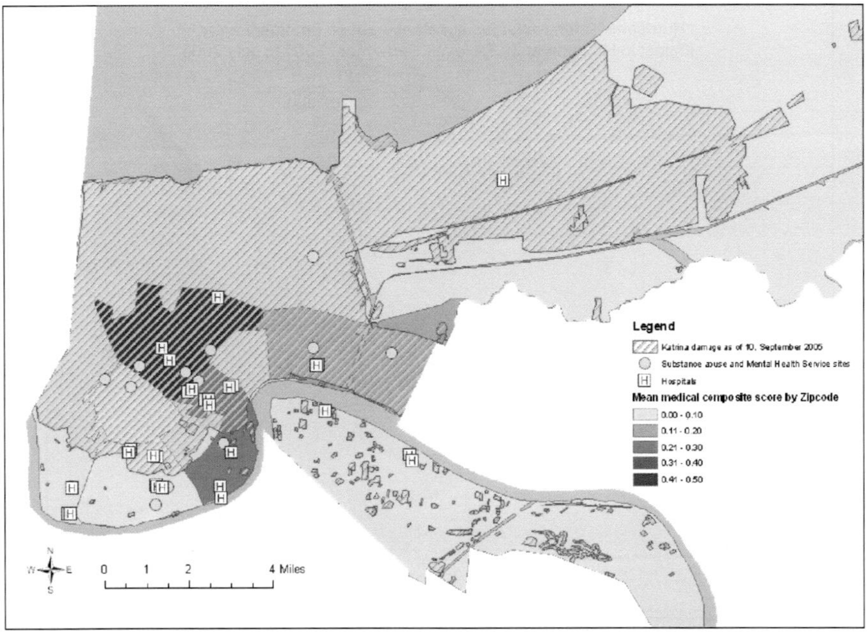

Fig. 23.2 Hurricane Katrina-affected areas of New Orleans, Louisiana: ASI Medical Composite Scores by zip code and proximity to medical services (See also Plate 21 in the Colour Plate Section)

apartments or houses in the Houston area (rather than buying homes), we can anticipate that areas with proportionately more available vacant housing would be more likely to host evacuees in the short to medium term. Using Census data for the number of vacant housing units as a proportion of all available housing units for each Houston-area ZCTA, buffers of 2-mile radii were drawn around both the hospitals and SA treatment sites to look at proximity to medical and SA treatment center sites for areas of likely residence. A 2-mile radius around the care sites was considered accessible by foot or public transportation for this population, selected after review of relevant healthcare accessibility and utilization literature (Marcus, Fortney, Olfson, and Ryan 1997; Marcus, Olfson, Fortney, and Ryan 1997; Fortney, Booth, Blow, Bunn, and Cook 1995a,b; Fortney, Lancaster, Owen, and Zhang 1998; Meade et al. 2000; Mooney, Zwanziger, Phibbs, and Schmitt 2000). A basic underlying assumption of the proximity analyses is that people who live closer to health resources are more likely to use them (Meade et al. 2000); they are not necessarily more likely to have higher health status or better treatment outcomes, however, than those who do not live nearby the health resources. While the number of hospitals ($N = 75$ in Harris county, where Houston is located) and SA/MH service-providing centers is high in the Houston area ($N = 72$), proximity to these services from where the displaced clients are likely settling is not consistent or easily accessible (i.e., distances from service centers >2 miles; Figs 23.3 and 23.4). In fact, there are many ZCTAs with a high proportion of vacant housing units that appear isolated

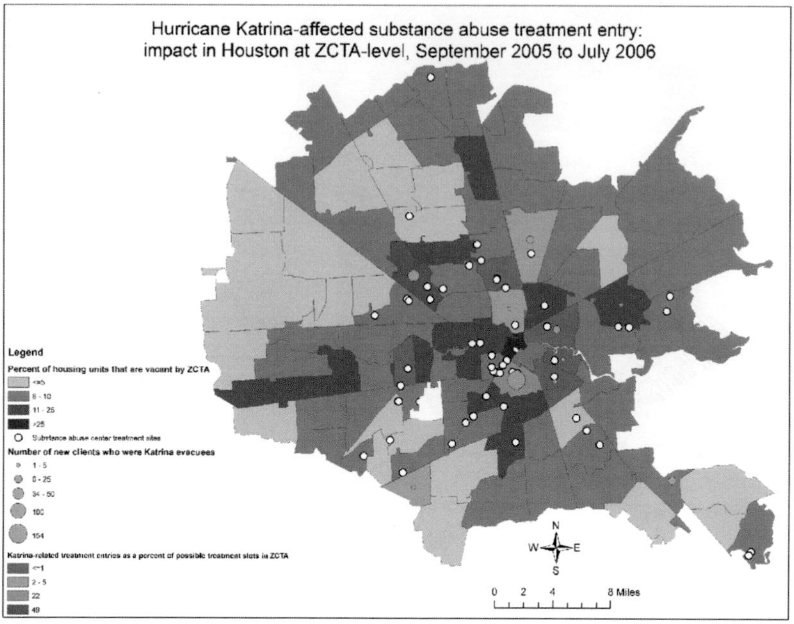

Fig. 23.3 Hurriane katrina-affected substance abuse treatment entry: Impact in Houston, Texas at ZCTA-level, September 2005 to July 2006 (See also Plate 22 in the Colour Plate Section)

with respect to hospitals and SA treatment services. While 77% of Houston's ZCTAs lie within a 2-mile radius of one or more hospitals, only 20.8% of them have more than 10% vacant housing. Similarly, 68% and 57% of ZCTAs in Houston lie within a 2-mile radius of one or more SA treatment sites or treatment sites for women, respectively, yet only 20.5% and 22.8% (women-specific treatment) of the ZCTAs have more than 10% vacant housing available. These data suggest that geographic access to hospitals and SA treatment sites may be limited for people living in those areas, such as those displaced by the hurricanes. Treatment options for women with SA problems in Houston were noticeably limited, with only 29.7% of the 72 sites treating women, as compared to 44% of the 16 sites in New Orleans (Fig. 23.5).

Figures 23.3–23.5 also display the count and density of the hurricane-related treatment center admissions through July 2006 by ZCTA (admissions per treatment slot available). This aspect of Fig. 23.3 reveals that hurricane-related treatment entries clustered in Houston's downtown area ZCTAs. Most Katrina-related admissions comprised 5% or less of a ZCTA's possible treatment slots, but two ZCTAs carried a larger burden of treatment entries. The overlap of higher treatment admission densities in areas with greater proportions of vacant housing units suggests that evacuees who are seeking and receiving treatment for their SA problems are living close to treatment sites. Furthermore, the observed overlap supports the use of the ZCTA proportion of vacant housing units as a proxy for the location of the evacuees' resettlement. Figures 23.4 and 23.5 show that, for those receiving treatment,

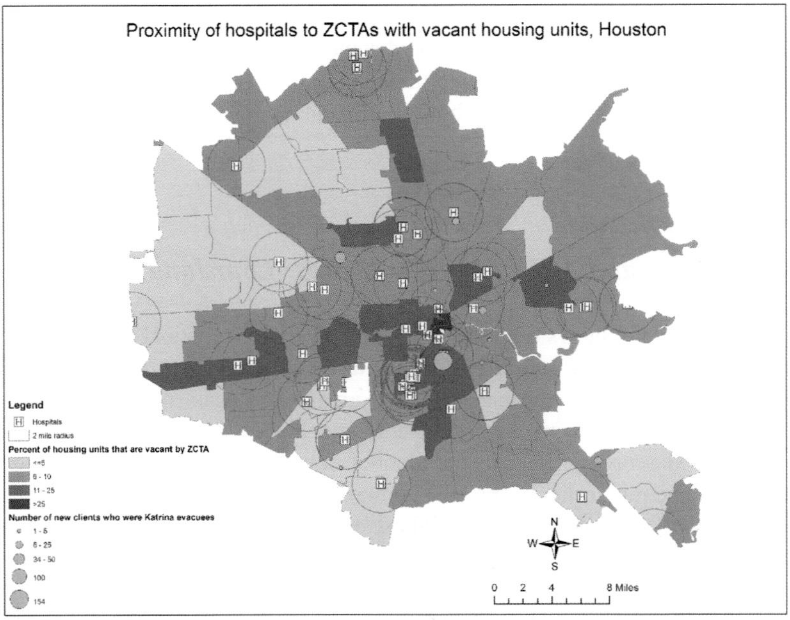

Fig. 23.4 Proximity of hospitals to ZCTAs with vacant housing units, Houston, Texas (See also Plate 23 in the Colour Plate Section)

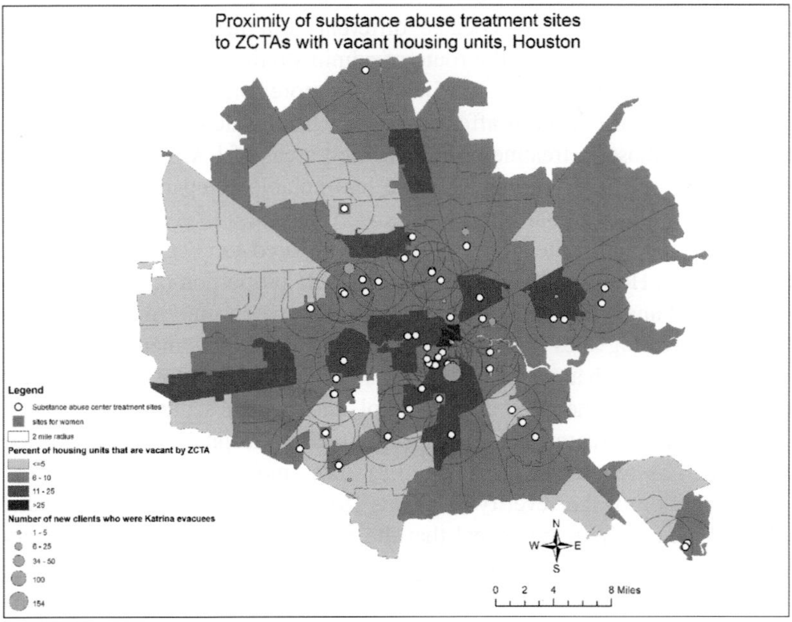

Fig. 23.5 Proximity of substance abuse treatment sites and sites with women drug user services to ZCTAs with vacant having units, Houstan, Texas (See also Plate 24 in the Colour Plate Section)

access to Houston hospitals and SA treatment sites with options for women with SA problems does not appear consistent, and distances to these resources are great. The scales of Fig. 23.2 vs. Figs 23.3–23.5 underscore how much larger a city Houston is compared to New Orleans and how the distances to resources differ for their denizens.

Demographics and ASI Composite Scores: Louisiana

Table 23.1 (shaded) shows detailed results of bivariate statistical tests comparing characteristics of clients within LA from hurricane-affected to those from unaffected areas. Within LA, hurricane-affected areas were again more likely to have clients with more severe medical problems and higher ASI medical scores, and had higher ASI scores on all but the alcohol composite, compared to unaffected areas in LA. The rank order of the ASI composites for clients in hurricane-affected and -unaffected areas revealed the highest scores were on the ASI employment subscale, followed by the psychological functioning subscale. Thereafter, medical problems remained a main source of distress for hurricane-affected area clients, compared to alcohol for clients in unaffected areas. Racial and ethnic differences in SA treatment clientele indicated that hurricane-affected areas had significantly greater diversity than unaffected areas, though the disparities were less pronounced compared to the Houston clientele contrasts. While proportionately more clients in hurricane-affected areas abused cocaine, different from the Houston clientele, there were no notable differences in the route of administration (i.e., snorting, smoking, injecting) reported by LA clients. Marijuana was more commonly noted as the most problematic drug for clients in affected areas, whereas alcohol was slightly more problematic for those in treatment in unaffected areas of LA. Differences in drug abuse patterns for clients within LA were less notable compared to the Houston client contrasts. Hurricane-affected clients were also less likely to live in a home that they or someone in their family owned compared to clients from hurricane-unaffected areas. However, there was a comparable proportion of female clients in the affected and unaffected areas and more than adequate women drug user services ($N = 90$ SA treating sites in unaffected areas, 40 treating women, 44.4%). Using kriging techniques with an exponential fit that minimized the root mean square prediction error, the smoothed ASI drug, alcohol, and medical scores are overlayed on ZCTAs with ≥ 1 SA treatment center, locations of all hospitals, and the hurricane-affected areas, as relevant (Figs 23.6–23.8). The smoothed surface indicates the spatial distribution of the severity of clients' SA problems across the state and in the affected areas. The maps reveal that the SA treatment sites appear to service most of the areas of need. The ASI drug scores are higher in the northwestern and southeastern parts of LA while the ASI alcohol scores are distinctly higher outside of the hurricane-affected areas. The ASI medical score smoothed rates are highest in hurricane-affected areas of LA but the distribution of hospitals in nearby affected and unaffected areas appears adequate.

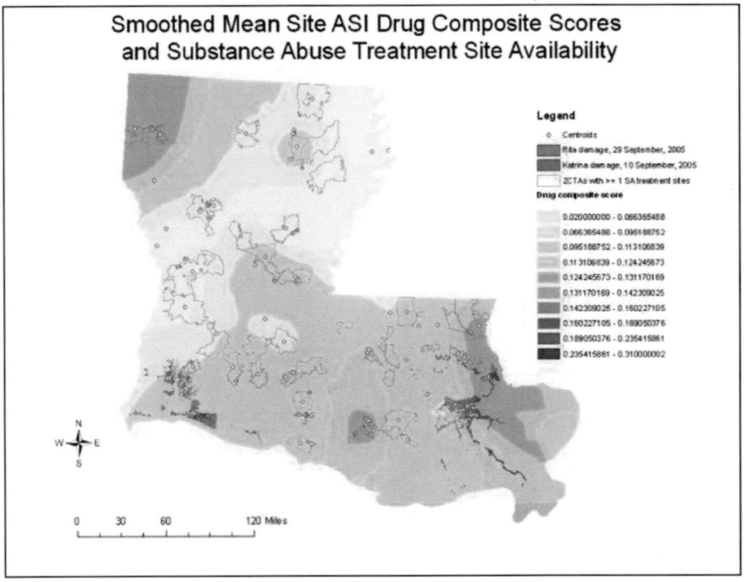

Fig. 23.6 Smoothed mean site ASI drug composite scores and substance abuse treatment site availability, LA (See also Plate 25 in the Colour Plate Section)

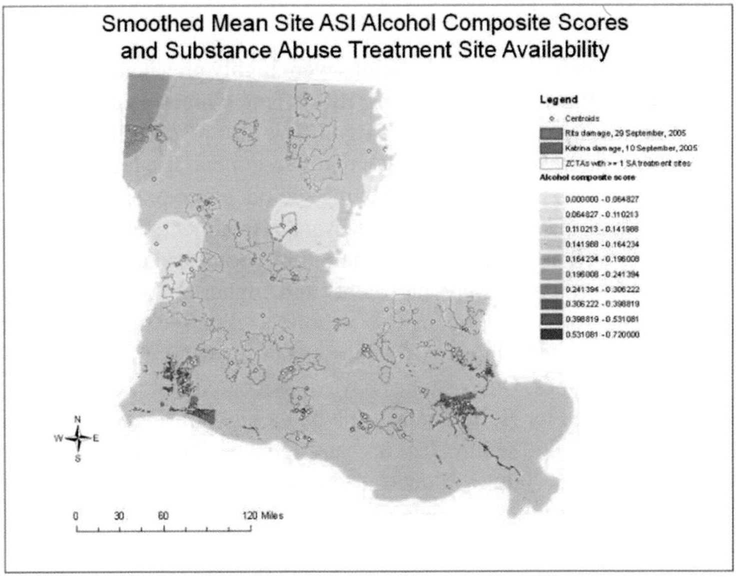

Fig. 23.7 Smoothed mean site ASI alcohol composite scores and substance abuse treatment site availability, LA (See also Plate 26 in the Colour Plate Section)

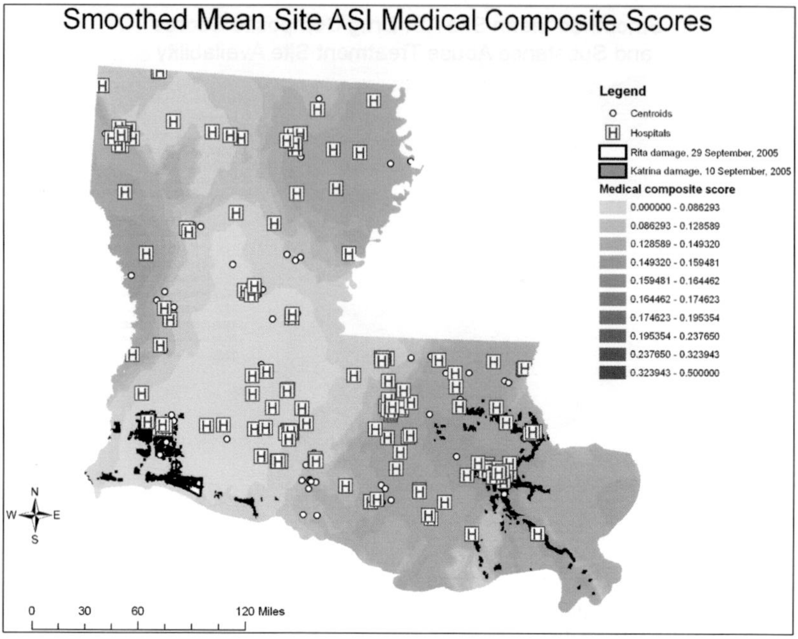

Fig. 23.8 Smoothed mean site ASI medical composite scores and locations of hospitals, LA (See also Plate 27 in the Colour Plate Section)

Discussion

This analysis found that comorbid conditions were highly prevalent among SA treatment clients from Hurricane Katrina- and Rita-affected areas as compared to their counterparts in hosting communities. Using GIS, it also revealed apparent discrepancies in the type and distribution of women's treatment services and healthcare locales. These needs, as well as the contrasting sociodemographics of the hurricane-affected populations, may be particularly relevant to treatment service providers and policy makers, as well as to health researchers for hypothesis generation.

Chronic health problems were prevalent, and higher ASI medical composite scores were found for New Orleans SA treatment clients compared to Houston SA treatment clients and within LA for hurricane-affected compared to those unaffected by the hurricanes. Other studies have documented that the population from New Orleans displaced by Hurricane Katrina had a high prevalence of chronic diseases (Brodie et al. 2006; Millin, Jenkins, and Kirsch 2006; Perry, Dulio, Artiga, Shartzer, and Rousseau 2006; Rudowitz, Rowland, and Shartzer 2006; Vest et al. 2006); the population with SA problems is no different. The Centers for Disease Control and Prevention's morbidity surveillance data revealed that similar proportions of hurricane evacuees sought care at healthcare facilities for mental illness (which included alcohol and drug abuse, withdrawal) and chronic illnesses in the wake of Hurricane Katrina, while visits to evacuation centers more often involved treating infectious and acute problems (Centers for Disease Control and Prevention 2006b). The importance of integrating primary care and the management of chronic conditions, including screening for SA problems and referring people to treatment, following a disaster is underscored in all of these studies.

The health science of disaster response and management encompasses attending to patients and communities experiencing acute crises as well as caring for and reinstating disrupted management channels for chronic medical conditions such as SA. Yet, SA is often divorced from disaster response and it is infrequently considered in the surge capacity and response capability literature (Jenkins et al. 2006; Kelen et al. 2006a,b; McCarthy, Aronsky, and Kelen 2006; Rothman, Hsu, Kahn, and Kelen 2006). Unique to this health problem, the time horizon for accommodating surges in clients needing SA treatment may differ from that typically considered in surge responses [i.e., 48–72 hours (Kelen et al. 2006a)]. Immediate needs of some SA patients, e.g., patients on substitution therapy, may follow a short timeframe; but the comparatively longer latency of incident or recurring SA problems resulting from depression and coping with the disaster-related trauma, suggest a redefinition of surge capacity and response capability planning to encompass a longer time horizon for chronic health conditions like SA. The reality for hosting communities is that Hurricane Katrina evacuees who have chronic health problems are not likely to return to New Orleans and other hurricane-affected areas until infrastructure to support their health conditions such as health services, jobs, transportation, and affordable housing improve (Perry et al. 2006). Thus, longer term response capability planning now is necessary.

Efforts should be made to ensure that hurricane-displaced clients have healthcare coverage and access to hospitals and primary healthcare to optimize their SA treatment and prevent excess morbidity and mortality. Lack of health insurance is an established predictor of not completing SA treatment (Garcia, McGeary, Shultz, and McCoy 1999). Yet even those evacuees with Medicaid had difficulties transferring coverage to their state of new residence (Perry et al. 2006). Other solutions put in place for evacuees are not reaching those in need. For example, to augment healthcare options for the uninsured and underinsured in the Harris County Hospital District, where Houston is located, a discount (Gold Card) for primary healthcare services is available. Receipt of the discount is a function of household income (those who fall below 250% of the federal poverty line) and lack of eligibility for federal assistance programs (e.g., Medicaid). The Gold Card was extended to hurricane evacuees, waiving the six-month county residency requirement. While the program is crucial, it is meeting less than two-thirds (62.6%) of the demand for primary care in low income uninsured households of Harris County (Center for Health Services Research 2006). It is not clear if or where there are gaps in supply and demand specifically among the hurricane-displaced in Harris County. Beyond coverage, physical access to healthcare and SA treatment services is also a determinant of their uptake (Jacobson et al. 2006; Schmitt et al. 2003). Our maps suggest that distances to care are one barrier to use of the Houston-area services for hurricane evacuees, a finding echoed in "Voices of the storm: Health experiences of low income Katrina survivors" (Perry et al. 2006).

Our analysis suggested that womens' SA treatment services and the special challenges of women drug users may be in need of attention in Houston to appropriately care for hurricane-displaced female clients. Women tend to carry additional burdens during the aftermath of a disaster and are at greater risk of PTSD and depression than men (Galea et al. 2002a; North et al. 1999). Sixty-eight percent of female caregivers in one recent study of LA FEMA recipients reported a high prevalence of MH disability due to symptoms of depression, anxiety, or other psychiatric disorders, and they were 2.5 times more likely to have a child with MH problems (Abramson, Garfield, and National Center for Disaster Preparedness 2006; Abramason, Garfield, and National Center for Disaster Preparedness 2006). Even when medically indicated, a woman's SA treatment may be foregone, delayed, or cut short to provide and care for her family. Social problems, such as homelessness, domestic violence, involvement of child protection services, and psychiatric diagnosis, are identified as hurdles to SA treatment completion among women (Kelly, Blacksin, and Mason 2001). Treatment options that give women the opportunity to maintain an integral role as caregivers with sufficient social support are crucial to their recovery success and their family's resettlement. An alternative explanation for our findings may be that women in Houston are receiving treatment for MH and SA needs in different environments such as primary care offices that were not analyzed in this study. Future studies should explore some of these hypotheses, as well as how demographics in Houston, New Orleans, and LA more generally have changed since the hurricanes.

Similar to our findings, Maxwell, Podus, and Walsh (2006) reported profiles of clients entering treatment in Texas who were Katrina or Rita evacuees. Sociodemographically, they found an increase in clients who were African American, of lower socioeconomic status, and had lower education levels, characteristics comparable to our New Orleans and hurricane-affected clients in LA. Though their study sample had a high proportion of methadone-maintained clients (due to the rush to accommodate this subpopulation of Katrina evacuees), the commonalities in treatment populations are notable. SA treatment clients who were evacuees of Hurricane Katrina had lower treatment completion compared to evacuees of Hurricane Rita (Maxwell, Podus, and Walsh 2006). Reasons for lower completion may include differences in race, gender, drug of abuse, degree of trauma experienced, and/or the availability of services to attend to these differences. Our analysis revealed differing patterns of difficulties in functioning across the ASI domains between the New Orleans and Houston clients as well as between the hurricane-affected and -unaffected clients in LA. Within the state, anecdotal reports echo what our data revealed: several sociodemographics of displaced clients and the dynamics of drug use in hurricane-affected vs. -unaffected hosting communities differ. Reports of backlash by treatment programs who would try to keep out the 'undesirables' from New Orleans are emerging (Bigg, Buanauro, Gutenson, and Maxwell 2006). Comprehension of client differences and patterns of distress may assist practitioners to determine needs for a variety of SA treatment modalities and support services, interventions tailored to real conditions, and timely contact with culturally competent care.

To be clear, problems in the Texas and Louisiana MH and SA systems predated the hurricanes of 2005. The hurricane-displaced population arrived in Texas to a MH system that was already stretched, attempting to serve more than 125,000 patients on a state budget that ranked 48th in the country for per capita spending on MH (National Mental Health Association 2006). Moreover, publicly funded methadone maintenance treatment slots were limited in Houston before the evacuees' arrival (Maxwell, Pullum, and Tannert 2005). In the SAMHSA treatment planning region where New Orleans is located, adult unmet drug treatment needs are nearly twice that of adults in the region where Houston is located (SAMHSA, Office of Applied Studies 2006). The sizable influx of adults into the Houston area, many of whom were already in need of SA treatment, suggests that the drug treatment gap there will likely widen.

In LA, before the 2005 hurricanes, SAMHSA estimated that one in five Louisianans experienced a mental disorder in any given year, which translates to roughly 650,000 adults and 245,000 children. Directly resulting from the hurricanes, an additional 30% of Louisianans can be expected to develop symptoms and show signs of a mental disorder (National Mental Health Association 2006). MH care in pre-hurricane LA reached less than half of those in need, and consisted of a patchwork of outpatient and care services (Seeman 2005). With respect to SA treatment, demand far exceeded supply in LA. Estimates by the Louisiana State Office of Addictive Disorders suggest that 600,000 state residents met the criteria for alcohol or drug dependence. Waiting lists for treatment numbered 1,200 to 1,800 people every day (Finan 2005). LA had only 32 detoxification beds, 20 of which were at Charity Hospital in New Orleans (now closed) and 416 inpatient treatment beds statewide prior to the hurricanes (Curley 2005). The state lost 19 outpatient treatment programs and 25 prevention programs statewide, or approximately one-third of SA services to the 2005 hurricanes (Curley 2005). Since nearly all SA treatment services in LA are state or federally funded, appropriate response to the health crises of Hurricanes Katrina and Rita depends on the availability of new mechanisms and alternative sources of funding. Private and non-profit organizations have shown remarkable ingenuity, flexibility, and commitment to continuity of care (see, for example, www.hopenetworks.org and Toriello, Morse, Morse, Kissinger, and Pedersen-Wasson 2007), but cannot be the sole source of support in the long term for the complex and costly undertaking of SA treatment, MH care, and hurricane recovery.

The main piece of legislation enabling federal assistance to disaster victims is the 1974 Robert T. Stafford Disaster Relief and Emergency Assistance Act (P.L. 100–707) (Williams 2005). In the fall of 2006, efforts to amend the Stafford Act to explicitly include SA in addition to MH assistance and training in Presidentially declared disaster areas succeeded in the Senate but failed in the House of Representatives. Thus, for states responding to and recovering from disasters, there are still no adequate resources for preventing or relieving SA problems in people at high risk of initiation and relapse, nor funds for long-term or traditional SA services, nor disaster worker training to better meet this special population's needs in future disasters. Even after Hurricanes Katrina and Rita, we are no better prepared to respond to the MH and SA problems that inevitably will result from future disasters. Data from this analysis suggest that people with SA problems displaced by Hurricanes Katrina and

Rita have distinctly different sociodemographics and treatment needs which may not be met through current care provisions in Houston and LA. SA-specific funding for states affected by a disaster to diagnose and recommend programs tailored to the needs of disaster-displaced treatment clients and populations at risk of is one potential policy tool that could be considered in future disaster response.

In this chapter, we used GIS to help compare and better understand the profiles of SA center attendees in New Orleans to those in the primary areas of displacement of Houston, Texas, and unaffected areas within LA as well as to iteratively explore and generate hypotheses for future research. Building on results from conventional statistical tests conducted at the individual-level, GIS allowed for exploration of spatial patterns of medical, drug and alcohol addiction severity across treatment sites, depicting the contours of a unique risk environment in the hurricane damaged areas of New Orleans and in the areas of displacement. In this way, we can better understand the range and likely SA treatment needs of patients displaced by the hurricanes. The individual-level analyses revealed important differences in chronic health and medical problems among clients from hurricane-affected sites, which generated hypotheses that could be explored using GIS. By conceptualizing adequate access as minimal distance to medical care and substance treatment, GIS helped to explore proximity to and availability of these resources across Houston, highlighting areas likely inhabited by those displaced by the hurricanes. In terms of response capacity, using GIS to geocode points of care and SA treatment revealed important discrepancies, such as differences in women's treatment services, which might otherwise not have been detected. The kriging results generated in GIS suggest spatial relationships that could be tested in future research on the impact of Hurricanes Katrina and Rita on actual healthcare and SA treatment resource use across LA. This chapter illustrates only a few of the many research tools available to navigate the evolving field of SA geography.

Effective use of GIS requires timely data collected at a level of scale and specificity that can inform the problems at hand. Presently, the application of GIS to the study of SA and addictions is limited by a lack of geographic-specific data, inadequate sharing of available data, poor communication of data, and duplication of data.[3] One key aspect of the present analysis was geocoding, yet this technique

[3] Two examples are offered. The first author encountered numerous difficulties when placing telephone calls to the substance abuse treatment sites listed in the SAMHSA Treatment Locator files for Houston that potentially received hurricane-displaced clients. Several of the phone numbers were incorrect, disconnected, or, when reached, reluctant, or unwilling to give information on the facility's treatment slot capacity, often over concerns of privacy. As a second example, for use of the DENS data set, the street locations of the treatment sites themselves were kept confidential, again due to privacy concerns, preventing against place-identified, site-specific data analysis. To determine whether the Louisiana DENS sites were located in hurricane-affected or flooded areas, we asked the Treatment Research Institute staff to match the DENS site addresses to the affected areas and zip codes we provided, thereby blinding us to the site's identity. A special code was generated indicating whether the site was located in a hurricane-affected or flooded section. The zip code of individual clients' home addresses would have been more useful for our analyses, but their access is entirely forbidden.

requires access to geographic locator information about which individuals and/or sites may have privacy concerns, as well as accurate address and Census street files to map addresses with a degree of certainty. Balancing privacy and confidentiality protection with the need for improvements in the availability, collection, flow, and accessibility of geographic-level data on drug and alcohol use and misuse is integral to advancing this area of research.

Study Limitations

We made several assumptions that may not be appropriate for answering the research questions. For example, it was assumed that the clients attending SA treatment from 2002 to 2004 were similar to attendees in 2005. To explore the impact of these assumptions, bivariate Pearson correlations (ρ) comparing across time were conducted on the mean ASI scores, client age, and the number of clients at each site. The analysis revealed a strong and significant correlation pattern ($\rho = 0.35–0.84$, $p < 0.01$) over time for all but the ASI legal composite scores ($\rho = 0.24$, $p = 0.01$), thereby giving credibility to the assumption of client characteristic stability in this analysis. Lacking client-level location data, it was assumed that clients lived near the SA treatment sites they attended so that meaningful conclusions could still be drawn about geospatial patterns from the site-level data. Moreover, we assumed that people did not relocate from the Houston and LA areas after their initial displacement due to the disaster. While there is no evidence to the contrary for the Houston evacuees, new data on those displaced within LA revealed that families moved on average 2.5 times in the year following the hurricanes, but stayed within LA (Abramson et al. 2006). It is unclear how this degree of instability may further affect response capabilities and people's access to care within LA. In addition, it is not clear to what extent the SA treatment sites in Houston, New Orleans, and LA more generally are representative of all SA treatment sites in those cities and states, so selection bias is possible.

Data from SAMHSA collected during the same time period as the DENS validate many of the state-specific trends observed, for example the substances most commonly abused in LA and New Orleans (Duffy 2004). Nevertheless, the SAMHSA SA treatment locator database used for geocoding is not comprehensive. It includes facilities that: are included on SAMHSA's Substance Abuse Treatment Services (I-SATS); are approved by the State Alcohol and Drug Abuse authority as a SA treatment facility; and have responded to the most recent annual National Survey of Substance Abuse Treatment Services (conducted annually). We opted to use the SAMHSA Treatment Locator database because it is publicly available and is the resource to which the Texas Department of Health and Human Services website links for finding SA treatment services in one's neighborhood. The 2005 hurricanes also devastated coastal Mississippi, but due to limitations on data availability, we restricted the analysis to the hurricane-affected populations of LA. Methadone maintenance programs were not sampled by the DENS and therefore provided no data for this segment of the treatment population. It is a further limitation of the DENS database that only two SA treatment sites contributed data for the Houston metropolitan area. Moreover, it would have been helpful to obtain more specific information on the types of chronic health conditions from which SA clients suffer; however, these data are not available in the version of the ASI released by the DENS. It should be noted that we did not incorporate other important aspects of surge capacity and response capabilities of Houston and LA, such as staffing and expertise, system integrity, supplies (e.g., availability of naltrexone, buprenorphine), or transportation to care sites. There were also other aspects of barriers to treatment and healthcare access that were not included in the present analysis, such as language services, handicap accessibility, and transportation times, due to data limitations. Finally, it would be more accurate to have US Census block level data on housing rather than ZCTA-level data for our buffer analysis. We chose to hold constant the unit of analysis across the maps and analyses to communicate our findings.

Conclusion

Options for SA treatment and healthcare in Houston and hurricane-unaffected areas in LA are available, though greater distances to resources, sociodemographic differences, and preexisting unmet treatment needs should be considered in disaster response capability and recovery planning. More generally, our analysis suggests that efforts be made to ensure access to hospitals and primary healthcare for SA treatment clients affected by the hurricanes, including the possibility of extending public healthcare coverage to those displaced by the hurricanes regardless of eligibility status. Whether or not Houston shelter residents fulfilled Medicaid eligibility, the healthcare needs of the evacuees in Brodie and colleague's study were comparable, suggesting the need for broader criteria in response to massive relocation after natural disasters (Brodie et al. 2006).

Women's treatment services and the special challenges of women drug users may need attention in Houston; further study of this discrepancy is warranted. Better health outcomes in disaster recovery involve adequate access to a nexus of three healthcare components: primary healthcare, MH care, and SA treatment. Our findings endorse the integration of SA treatment into the primary care practice model and the need to interweave MH and care of other chronic health problems into SA treatment settings.

It is often difficult to rapidly obtain data on the health needs of people affected by disaster, especially concerning sensitive health concerns like their behavioral and MH needs. In this situation, reliance upon data sources that predate the disaster may serve as a proxy starting point for appropriate responses. This study has illustrated how statistical and technological applications such as GIS can be applied to existing data sources for hypothesis generation and disaster response capability and recovery.

Acknowledgment We thank Drs. Theodore Holford, Robert Heimer, Lauretta Grau, Kim Blankenship, and Jeannette Ickovics for their comments on earlier drafts of this chapter. The comments and suggestions from colleagues David Green, Erika Martin, TJ Ghose, Eric Poolman, Sarah Bray, Enrique Pouget, and Kavita Misra at the Center for Intedisciplinary Research on AIDS at Yale University are greatly appreciated.

Chapter 24
Using GIS to Identify Drug Markets and Reduce Drug-Related Violence[1]

A Data-Driven Strategy to Implement a Focused Deterrence Model and Understand the Elements of Drug Markets

Eleazer D. Hunt, Marty Sumner, Thomas J. Scholten and James M. Frabutt

Abstract A new drug enforcement model was implemented at the High Point Police Department to reduce violence related to overt street dealing. There is a relationship between overt drug dealing and violent crime. The model targets street-level dealers to cease dealing and offer community assistance. The result is a sustained decrease in drug offences and violent crime of 31% and 37%, respectively. This chapter outlines the use of a geographic information system (GIS) in implementing the model and producing maps and statistics to assess the outcome of this model. GIS illustrates the structure of drug dealing, how it is spatially organized, shows the spatial relationship between dealing locations and crime, and assesses the model using temporal and location change maps and statistics.

Introduction

Beginning in the early 1990s, the primary drug enforcement method consisted of police crackdowns (Caulkins 1993:848). Local media throughout the country routinely reported large-scale drug sweeps conducted by one or more law enforcement agencies. Stories tell of dozens of people arrested, multiple houses searched, jails and courts clogged due to processing of the increasing numbers of offenders. Unfortunately, there was often a return to the same level of drug dealing and related violence after a few days of the police sweeps (Becker 2006). Local citizens who were stopped and questioned during these operations viewed the police as an occupying army (Scott 2002). Traditional law enforcement techniques have not been successful

E.D. Hunt
High Point Police Department, High Point, North Carolina, USA

Y.F. Thomas et al. (eds.), *Geography and Drug Addiction*,

or sustainable in closing drug markets and reducing violence (Shepard and Blackley 2005). Too often, these traditional methods have been counterproductive because citizens become as suspicious of law enforcement as of the drug dealers and the solution may be no better than the problem. Intertwined with this situation are racial and ethnic elements that all too often lead to a schism and level of distrust between law enforcement and the community (Kraska and Cublellis 1997).

It is clear that overt street-level drug dealing is toxic to a neighborhood because of the direct nexus between drug dealing and violence. To overcome this relationship, a new strategy to reduce drug-related crime and violence in neighborhoods was developed, implemented, and assessed – based on a geographic data-driven methodology. The strategy – a focused deterrence model – eliminates all of the usual arguments from both law enforcement and the community, regarding how law enforcement has responded to the drug problem. It includes dealer locations, the distribution of dealers within the market, and the relationship between dealer locations and crimes. This strategy is data-driven using a geographic information system (GIS) as part of the methodology to identify the target enforcement areas and to assess the results of the strategy.[2] From a geographic perspective, the use of this model revealed a detailed understanding of the spatial structure of walk-up, drive-through, open-air drug markets. Point data are based on addresses, which provided locations of dealer sales, reported crimes, and 911 calls.

The Focused Deterrence Model

The data-driven focused deterrence strategy used to close drug markets and reduce drug-related violence grew out of the focused deterrence or "pulling levers" framework (Gladwell 2002; Kennedy 1997, 1998; Kleiman 1993). High Point became one of the first cities to replicate Boston's focused deterrence violence prevention strategy (Kennedy, Braga, and Piehl 2001; Kennedy, Piehl, and Braga 1996; Veen 1997), by launching the interagency Violent Crime Task Force (VCTF) in 1997 to reduce guns and gun-related violence[3] (US Department of Justice 1999, 2001). The VCTF model was modified to use in targeting overt drug markets within High Point.

Drug dealers have learned from the criminal justice system that there are no substantial consequences for dealing drugs. Kennedy argues that dealers engage in dangerous criminal activity but are not irrational; if law enforcement officials and the community can craft a message and develop a new way of responding to the dealers, which includes accountability from them and community support, dealers will respond positively and modify their behavior and modus operandi. If police, community members, and family join together to send a unified message that drug dealing is not acceptable, it can have a powerful impact.

To apply this model, an operational plan was developed, which systematically addressed drug sales, identified one specific location in the city, and engaged drug dealers and their families in stopping drug dealing.[4] The plan also included clear and predictable sanctions, and provided a range of social services and community help. Finally – and most importantly – the model mobilized the community in setting standards of acceptable behavior. The initial stage of the model is simple: Select a target area, direct police to focus on street-level dealers, make undercover purchases to use later as evidence against selected offenders, and – with community endorsement – put drug dealers on notice that there is zero tolerance for drug dealing, while detailing the alternatives for ending their dealing activities. The model is designed for the non-violent individual who generally does not, yet, have a felony conviction.

The model involves several elements and follows a process flow (Fig. 24.1[5]). The model consists of three phases; identification, notification, and resource delivery/community support. The identification phase selects an area within the city to target and generates a list of all the dealers operating in the area that will be targeted for the initiative. The notification phase informs the targeted dealers and significant others/family that they have been selected to participate in a public meeting. The public and law enforcement inform the dealers that overt dealing will no longer be tolerated and it will stop immediately. Resource delivery and community support is a major part of the follow-up and monitoring phase. Police check for any dealing and the community helps provide resources and assistance to the dealers. GIS is utilized in the first step to identify areas that may be suitable for implementing the model and is used in the last step to extract and generate data for a range of longitudinal and statistical analysis.

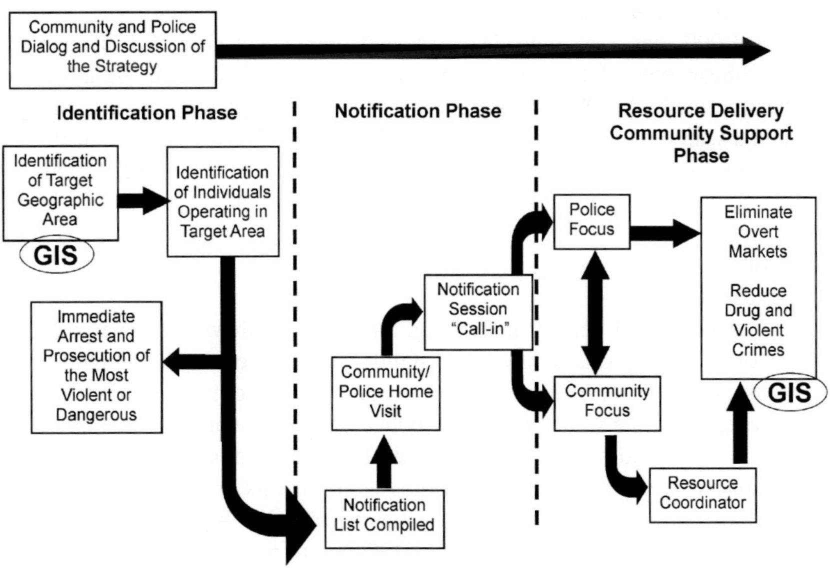

Fig. 24.1 Strategy flow model showing the process in three phases: identification, notification, and resource delivery community support

The Role of GIS in Data-Driven Methodology

The first step in selecting the target area is based on a series of GIS-generated density maps, using one calendar year as a baseline. The research team did not approach the process by asking the question, "Where are there drug markets?" Rather, the team asked, "Where are there densities of violent, sex or weapons crimes that may be spatially concurrent with drug sales?" Based on years of collecting crime data for this community, it was determined that person crimes (murder, robbery, and assault), theft or robbery, sex and prostitution, and weapon offenses are generally related to an area of active drug dealing.

Without a definitive set of crimes that are associated with or indicative of drug activity, our approach of selecting the crime categories is inclusive versus restrictive, arguing that a narrower reselection could be conducted later. A four-step process was used. The first step involved identifying the data needed for analysis: 911 calls, drug arrests, field contacts, and a category of grouped crimes labeled serious, consisting of: murder, rape, robbery, aggravated assault, weapons, sex, and prostitution. Second, the relevant attributes were extracted from the records database including date and time, address, nature or offense, and XY coordinates. Third, density maps of each layer were generated. These maps were analyzed individually and in different merged combinations, such as 911 calls and field contacts. Where densities were generated, the underlying crime and arrest reports were carefully analyzed – euphemistically called "unpacking" – to determine if the incident was related to drugs.

The density maps were created using a kernel type density algorithm, a search radius of 1,000 feet, area units in acres, and output cell size of 100 feet. Display is bilinear interpolation using nine symbology classes and an equal interval classification method. The resulting maps consisted of high resolution, small grid cells, and a search radius (K) that clusters only nearby offenses. These density map specifications follow recommendations by Eck et al. (2005:28) on developing and structuring hot spots using kernel densities, "...for an application that requires output for focused police patrolling..." as compared to other forms of spatial clustering techniques or approaches (see Poulsen and Kennedy 2004).

The four individual density maps (Fig. 24.2) do not show the same pattern or distribution of densities, and they differ significantly. For example, the places with the highest density of 911 calls do not correspond to field contacts or drug arrests. It did, however, illustrate some patterns of police operations and false hot spots. The police department and county court and jail create false hot spots (denoted by "A" and "B," respectively) due to default addresses for arrests or incident reports. The 911 calls have a greater proportion of densities based on public housing (denoted by "C") or economically disadvantaged areas – there are no hot spots in the northern portion of the city, where new subdivisions are being built. Drug paraphernalia and possession offenses frequently clustered near a community homeless shelter. Field contacts and sex crimes, when analyzed more carefully, increased in frequency in close proximity to areas of known drug activity.

Fig. 24.2 Four density maps showing the differences between 911 calls, drug arrests, field contacts, and serious crime (See also Plate 28 in the Colour Plate Section)

Given the diversity of each map, the four base maps were merged to form a combined or synthesized density map (Fig. 24.3). The result was a map with several densities that could potentially be associated with drug activity. The densities vary in size and shape. It is important to stress that at this point no clear delineations of where drug markets are located has been determined; the densities are simply pointing to where violent and serious crimes have occurred and may be associated with drug activity.

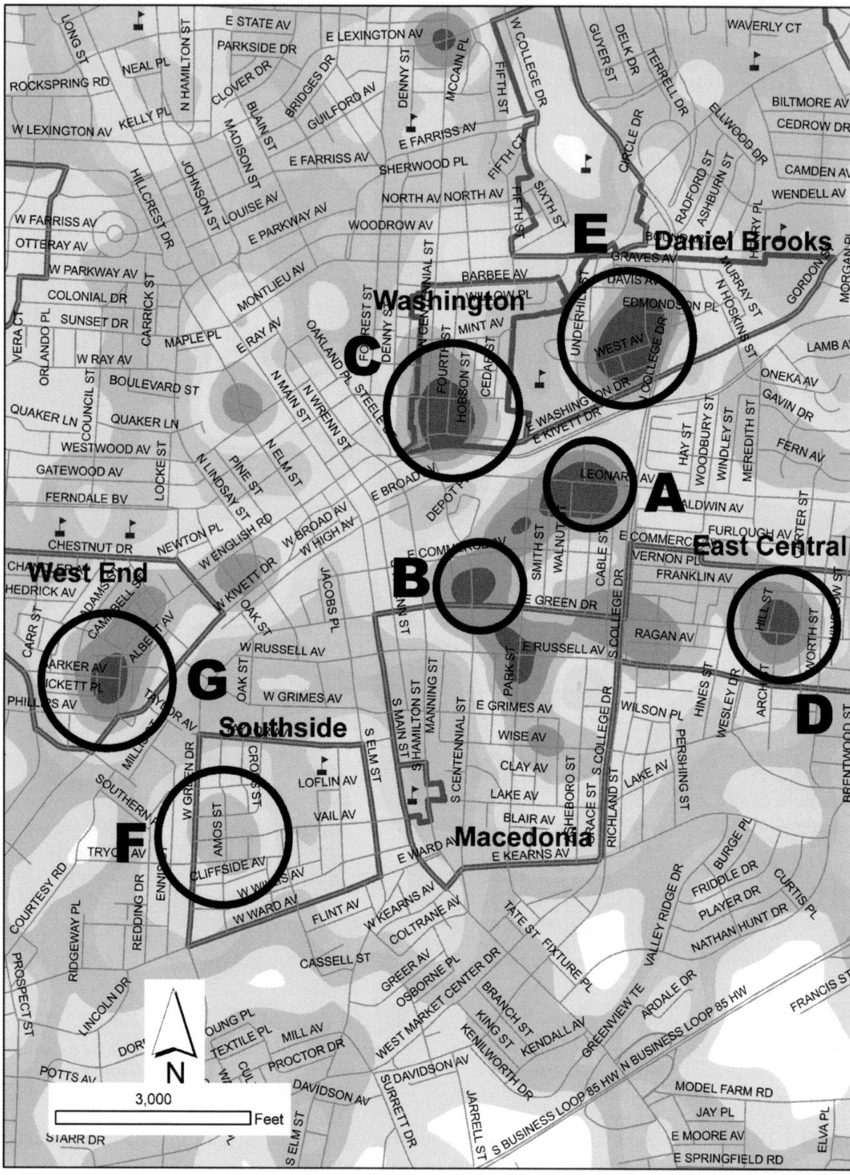

Fig. 24.3 2003 baseline density map and process of elimination, illustrating the identification and selection of a focus area (See also Plate 29 in the Colour Plate Section)

The identification of possible drug markets starts with understanding the conditions that constitute each density and unpacking all the police reports related to each density – classifying each report into violence with drug incidents, violence with no drug incidents, no violence with drug incidents, and no violence no drug incidents. Figure 24.3 visually suggests that there are seven or eight hot spots that are potential locations of drug market activity. A simple process to select one of the densities was based on the attributes of each density; Fig. 24.3 illustrates the selection process. The map shows two densities (A and B), that are artificial and are the result of reports and calls that default to the police department or county court and jail. Two areas were excluded (C and D) because recent large-scale narcotics operations and arrests disrupted crime patterns, and closer examination of the crime reports revealed that most of the serious crime and arrests was not associated with drug activity. Three densities remain (E, F, and G), each within a traditionally defined neighborhood – Daniel Brooks, Southside, and West End. Further review of data and the hot spot analysis eliminated the Southside neighborhood (F) because the count of serious and drug-related crime was not as high in frequency as the others. The Daniel Brooks neighborhood (E) was excluded because the core of the neighborhood consisted of a large public housing complex that introduced several variables (transient population, high turn over, skewed economics and race statistics) that for the initial project area would make it more difficult to understand the true crime patterns and drug sales as they related to the community. It was clear from reviewing the data that the West End (G) had an active drug market associated with violent crimes, including the most recent drug-related homicide. The target area for implementing the model was a five or six block area within the highest density polygon.

For the purpose of this research, established neighborhood polygons were used to define the area within which the data would be collected and comparative data from a second neighborhood would be used as a control. The idea is to use a polygon that contains the market activity and an adjacent area that may be influenced by the market. This could be police response areas or beats, census boundaries, arbitrary grids, or some form of calculated catchment area[6], perhaps based on the calculated densities. Because market boundaries cannot be definitively drawn, the above examples would generate a drug activity area or sphere of influence. It is suggested that distinct boundaries of a drug market will not be generated, relying on a dynamic boundary that will change temporally.

The Spatial Structure of Drug Markets

What does a drug market look like? There have been a number of studies that have examined the general organization of drug markets and related crime and disorder (Ratcliffe 2002; Rengert, Ratcliffe, and Chakravorty 2005:5–13; Weisburd and Mazerolle 2000) and several that focus on hot spots related to identifying the spatial distribution of a drug market (Eck 1995; Harocopos and Hough 2005; Jacobson 1999; Lupton et al. 2002; Weisburd and Green 1995). These studies look at the general patterning or location of a drug market using specific crime categories or boundaries based on kernel densities; the resolution to look at the relationship between the location of street-level dealing and crime patterns and police activity to a specific address or XY coordinate had been unavailable.[7]

The GIS provides the means to analyze multiple layers of tabular and geographic data, allowing for spatial and numeric statistical analysis and increased accuracy and detail. The GIS-based research for this initiative provides the resolution to look at how the market is geographically arranged. GIS provided a way to map the individual houses and street corners where drugs were being sold, locations of each crime, field contacts and 911 calls that occurred in the area, and then analyze the pattern of crime and identify a spatial distribution of the market. Based on the analysis of the data, it is apparent that the drug market caters to walk-up, drive-through curbside drug transactions. In theory, the market would be defined as a "small local market in equilibrium" (Rengert et al. 2005:41). Mass transit, such as city buses, does not appear to be a factor; auto theft and recovery patterns suggest a portion of customers travel in and out of the area from other parts of the city. Most walk-up sales are to individuals within and around the neighborhood, a portion of them being prostitutes. There is a lack of complete data concerning the remaining drive-up sales and where they originate – from the data analysis, it is inferred that these individuals are from the southwest portion of the city or county. Based on the police records and vice unit activity, this market is pernicious, perennial, with a history of related violent crime. For a period of 10 years prior to the start of this project, the area has produced one homicide per year.

Figure 24.4 shows a large-scale map of the West End neighborhood with the density map overlaid by streets and buildings. The West End neighborhood includes data counts from the entire neighborhood as part of the analysis. It covers 212 acres, and based on 2000 Census data, has a population of 1,754 (669 white, 447 black, 342 Hispanic, and 296 other) with the highest number of persons (289) between the ages of 5 and 17, median age of 33, 552 households with an average family size of 3.26, and 613 housing units (61 vacant, 162 owner occupied, and 390 renter occupied).

Based on undercover investigations – step two in the identification phase of the model – a list of known street-level dealers was compiled and used to conduct surveillance and undercover drug purchases. Controlled drug purchases were made at 16 locations as signified by the black crosses and one street corner signified by the black square (Fig. 24.4). What is clear is the small number of individuals who comprise and maintain an active drug market – only 11. Several patterns are evident:

1. The even distribution of the sale locations and associated houses.
2. The elliptical pattern of the houses more-or-less evenly distributed (interior ellipse).
3. The density of serious crime, drug arrests, field contacts, and 911 calls that occur within the ring of houses – thus generating the highest density within the ring of drug sale locations.

This pattern suggests that when generating a density related to drug dealing, the dealers may not be operating with in the calculated density, rather outside or ringing the density. Various categories of crime also followed a similar pattern. Figure 24.4 also shows burglaries (white circles) ringing around the outside of the drug houses in a similar, larger elliptical orientation. There is a relationship between the need for

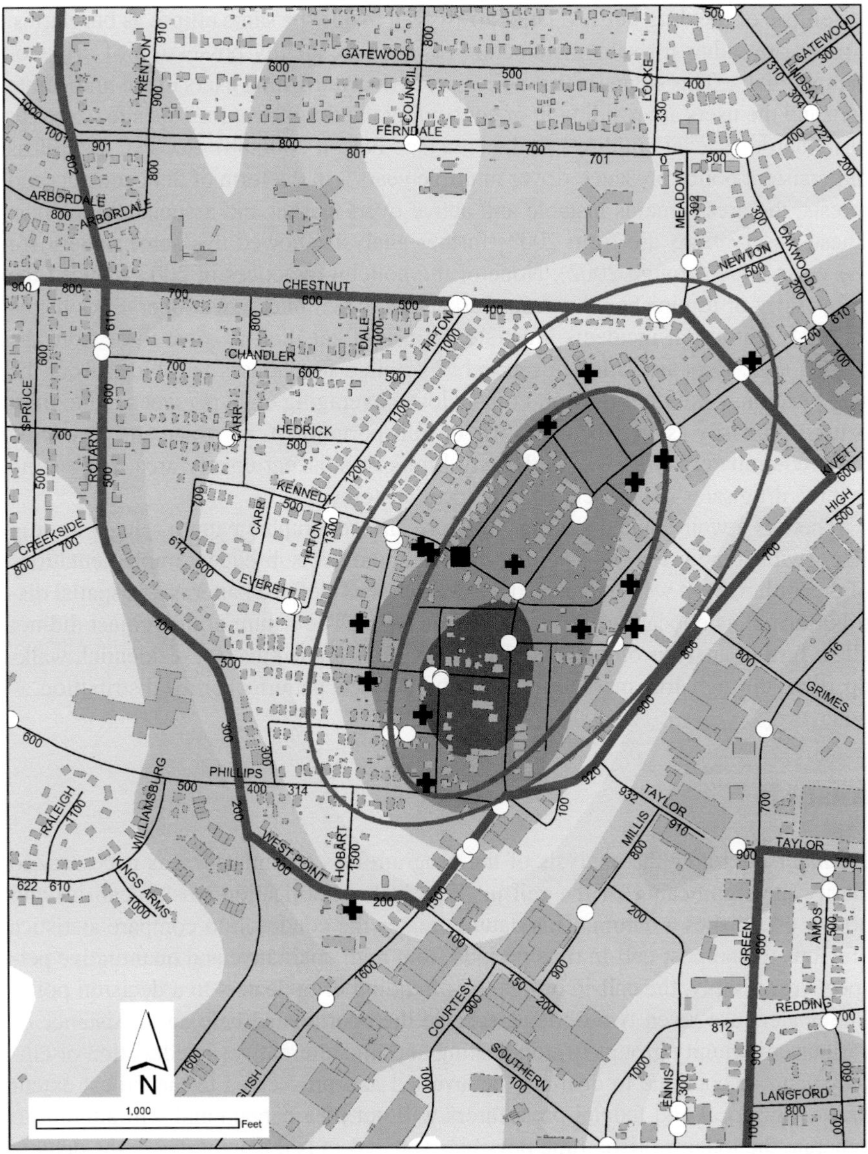

Fig. 24.4 West End neighborhood showing density ringed with drug sale and burglary locations (See also Plate 30 in the Colour Plate Section)

money or items used to barter for drugs and it is apparent that the local, walk-up customer would use the surrounding neighborhood as a resource.

Field contacts (not shown) are initiated by officers and are generated from interviewing persons on the street, in a car, or at a house. Analysis showed that field contacts are conducted throughout the neighborhood with increased frequency within the ellipse of the drug houses. Drug arrests fall inside the same ellipse as burglaries. Arrests are either for sale/manufacture or more frequently possession of drugs or paraphernalia. This suggests that police are actively arresting users in the immediate area and the users are residing in or consuming the drugs near the sale site – a result of numerous vacant buildings in the area (see Felson 2006:88). Second, regardless of persistent police presence – over multiple years – in the form of field contacts and arrests, the area remains a stable and active overt market and associated violence. These observations are from 2003 (further analysis showed this same pattern for the years 1999 through 2002). None of the policing practices in 2003 reduced the drugs sales or diminished the drug market area when undercover operations began in 2004. Auto theft incidents illustrate the use of car theft to drive to or from the market area. Cars are stolen or recovered within a few houses or a block from the sale location. Auto theft in High Point is low – averaging eight per week. Auto theft is used primarily as an expedient form of transportation. Cars stolen and recovered are predominately from within the city limits or adjacent county area, suggesting that the drug sales are associated with local customers.

There is a well-defined internal structure and geographic pattern related to drug sales (with some variation, a similar pattern in the two subsequent implementations of the initiative). It was unexpected to see such well-defined patterns and spatial distributions of sale locations to various crime types. The nature of the project did not allow further analysis of the area of study, but it is argued that for residential, walk-up, drive-through drug markets, this may be a typical pattern of crime distribution.

Analysis

To understand how the analysis of the program was formulated, it is necessary to clarify the significance of the call-in (Fig. 24.1 – notification session/call-in) and how it establishes a datum. This datum creates the condition to compare statistics previous to and post-call-in time periods from both qualitative and quantitative perspectives. Simply, the call-in brings the offending drug dealers to a decision point. The dealers are given two options: receive the coordinated help and assistance of various community groups to stop dealing and find alternatives in education or employment, or continue to deal and be arrested and prosecuted to the fullest extent of the law; local and federal prosecutors will not plea bargain and will attempt to generate the longest prison time possible. The call-in is the point where the dealers need to make a rational choice between the two alternatives.

Qualitative

The power of the call-in was immediate. The next morning, five of nine offenders called a resource coordinator and asked for assistance, mostly for employment opportunities. The resource coordinator received one call from a man who said that he had been dealing drugs in the West End neighborhood, heard about the call-in, and wanted help in finding a job so that he could quit.

The West End drug market vanished overnight. Dealers and prostitutes were no longer present in the area. Street corner and drug house activity, drive-up buyers, and prostitution activity were no longer visible. The character of the neighborhood changed immediately; residents ventured outside again, children played at playgrounds, people cared for their property, and many other signs of transformation occurred. Street and narcotics officers soon picked up a clear sense from offenders throughout High Point that the West End had become a "no go" area for drug dealing. The market was genuinely closed. Vice/narcotics detectives, freed from time consuming but essentially pointless street enforcement, have been freed to pursue more serious drug traffickers. They are making far more productive cases, and drug seizures are up by a factor of ten (2006), making progress on the core drug problem in High Point. From the in-depth analysis of the West End neighborhood, the vice/narcotic unit now had a complete ethnography of the dealers, associates, families, and structure of the neighborhood, providing in-hand knowledge to conduct future drug enforcement.

The absence of street dealers and prostitutes was obvious based on direct observation, but how much had really changed? In the weeks following, the call-in vice/narcotics detectives sent two informants to 16 locations and attempted to make drug purchases. None of the dealers at those locations would sell to the informants. Informants now spot-check these neighborhoods once a month. The overt drug market in the West End has now been closed for nearly three and a half years. The same pattern is evolving in two other subsequent initiative areas.

Quantitative

The use of GIS was central in the analysis phase. A variety of spatial analyses were conducted in order to assess change in crime counts and spatial distribution. A comparative analysis of pre – and post – call-in data in the form of numerical counts supports the success of the West End drug initiative and the positive changes in the community. The date of the call-in (May 18, 2004) was used as a datum. Table 24.1 illustrates how the data was collected and organized, showing the intervals in days

Table 24.1 Analysis method to collect crime data. May 18, 2004 is the datum point with values extracted in equal intervals of 100 and 50 days

Etc	300 Days	250 Days	200 Days	150 Days	100 Days	May 18 2004	100 Days	150 Days	200 Days	250 Days	300 Days	Etc
	29	26	25	20	18		12	12	16	17	22	
PRE CALL-IN							POST CALL-IN					

for the aggregation and collection of data. Analysis began 100 days after the call-in, and counts for serious and drug crimes were calculated in 50-day intervals afterward. Data for the entire city, the West End neighborhood, and a control neighborhood were collected. Crime incidents were extracted from the records management database and added to the geographic data layers. The neighborhood polygons were used to select out only the data from the two neighborhoods.

Analysis suggests that drug sales have not been displaced and that no other hot spots emerged, rather, drug sales and violent crimes have declined overall after the first and each subsequent call-in. Nor is there the effort to establish markets in less economically desirable areas as Robinson and Rengert (2006) suggest. Figure 24.5 shows the change in serious crimes. The first map shows the baseline data from 2003. The remaining maps show temporal sequences of 300, 565, and 730 days after the call-in and Fig. 24.6 shows the change for sale/manufacture of drugs based on choropleth maps generated from census blocks. The first map shows the condition of sale and manufacture in the first quarter of 2004, before the initiative began. The second map shows the second quarter of 2004 when the initiative began with the call-in. The third map shows the third quarter of 2004 following the call-in; a reduction in sales and manufacturing of drugs is evident. The fourth map is the first quarter of 2006 – 2 years after the call-in – no reported sales or manufacturing of drugs is reported within the West End neighborhood. Each of the targeted offenders is tracked, and all but one remained in the West End neighborhood and did not continue overt selling.

The significant changes in serious and in the sale/manufacture of drugs within the West End neighborhood and the areas surrounding it, particularly to the south and east, are an example of diffusion of benefits (Clarke and Eck 2005:13 and 51). Diffusion of benefits operates in the following manner: *Crime prevented HERE is usually not shifted THERE. Moreover something extra happens: When preventing crime HERE, it goes DOWN nearby* (Felson 2007). Thus, drug sales have been reduced well beyond the target area and our expectations. Figures 24.5 and 24.6 only compare two periods of time, but an analysis of change using GIS-generated density/choropleth maps has been completed quarterly from 2003 to 2007, showing the same results.

In the West End neighborhood, violent and drug crime numbers dropped dramatically (Fig. 24.7). Small absolute numbers make for large percentage shifts, particularly for short comparison periods, but more than 3 years after the call-in, the reduction in violent crime appears to have stabilized at a one-third decrease – an average of 36.7% (Fig. 24.7a). The violent offenses that have occurred namely robbery and assault, when reviewed are not related to drug activity. Most important, there has not been a homicide and two rapes and two weapon violations were reported in the West End since the intervention. Drug offenses have a similar pattern with an average decrease of 30.85% (Fig. 24.7b).

The density maps illustrate the change and intensity in hot spots across the city and indicate a lower density per acre. In 2003, there were 29–36 incidents per acre and in 2004 and 2005, the density per acre declined to 20–23. Violent crime citywide has decreased 20% over the 2 years following the start of the initiative.[8]

Fig. 24.5 Count of serious crime by census blocks comparing 2003 baseline counts with 300, 565, and 730 day counts after the call-in (See also Plate 31 in the Colour Plate Section)

Fig. 24.6 Sales and manufacturing counts by census blocks comparing a 2004 yearly quarter before the call-in, the quarter the call-in occurred, the following quarter, and the first quarter of 2006 (See also Plate 32 in the Colour Plate Section)

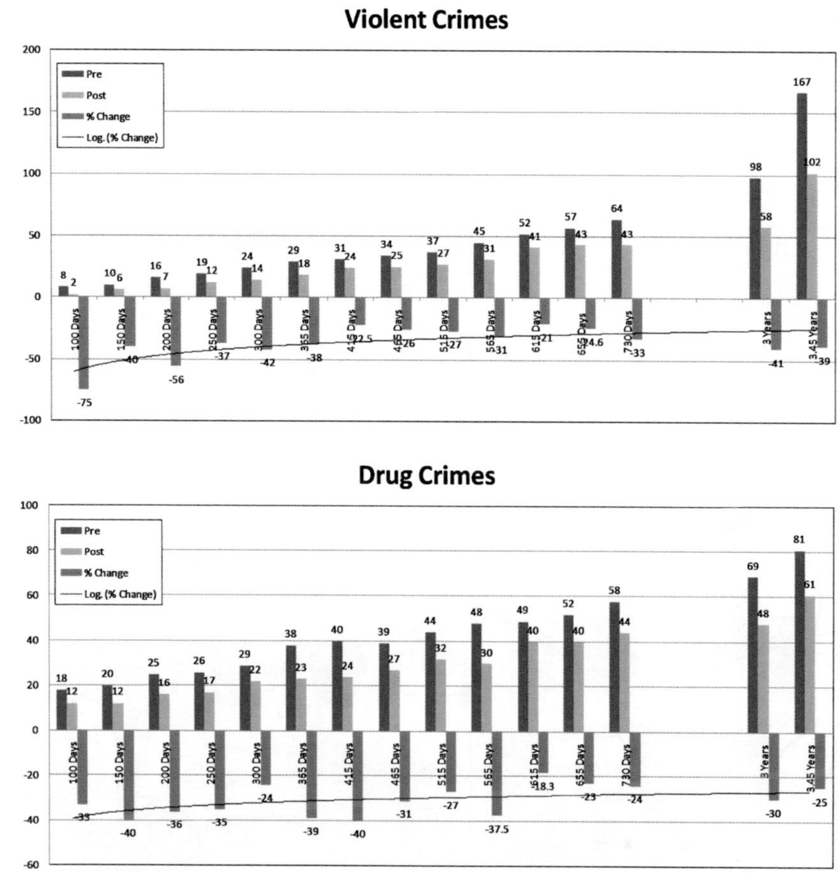

Fig. 24.7 Change in the counts of violent offenses 3.45 years before and after the West End call-in (**a**) and changes in the counts of drug offenses 3.45 years before and after the West End call-in (**b**) (See also Plate 33 in the Colour Plate Section)

An 18 month comparative change map showing densities of manufacture and sale of drugs prior to and after the call-in indicates significant changes in the location of this activity (Fig. 24.8). The change map is based on geographic spatial statistics using high/low clustering – informally referred to as hot/cold density change maps – specifically the Getis-Ord general G algorithm. The change map encompasses a time period that includes two initiative areas: West End and Daniel Brooks. Figure 24.8 shows significant decreases in both of these areas and the surrounding area. Areas that show a relative increase do not reflect the formation of a new drug market. Some of these increases are the result of increased policing at high schools and arrests for drug paraphernalia around a homeless shelter. Several of the areas that show an increase – when reviewed – did have one or more dealers making sales but inside a house; in this case, the high/low clustering has some ability to identify developing drug sales locations. Overall sales and manufacturing decreased, and new markets are not apparent.

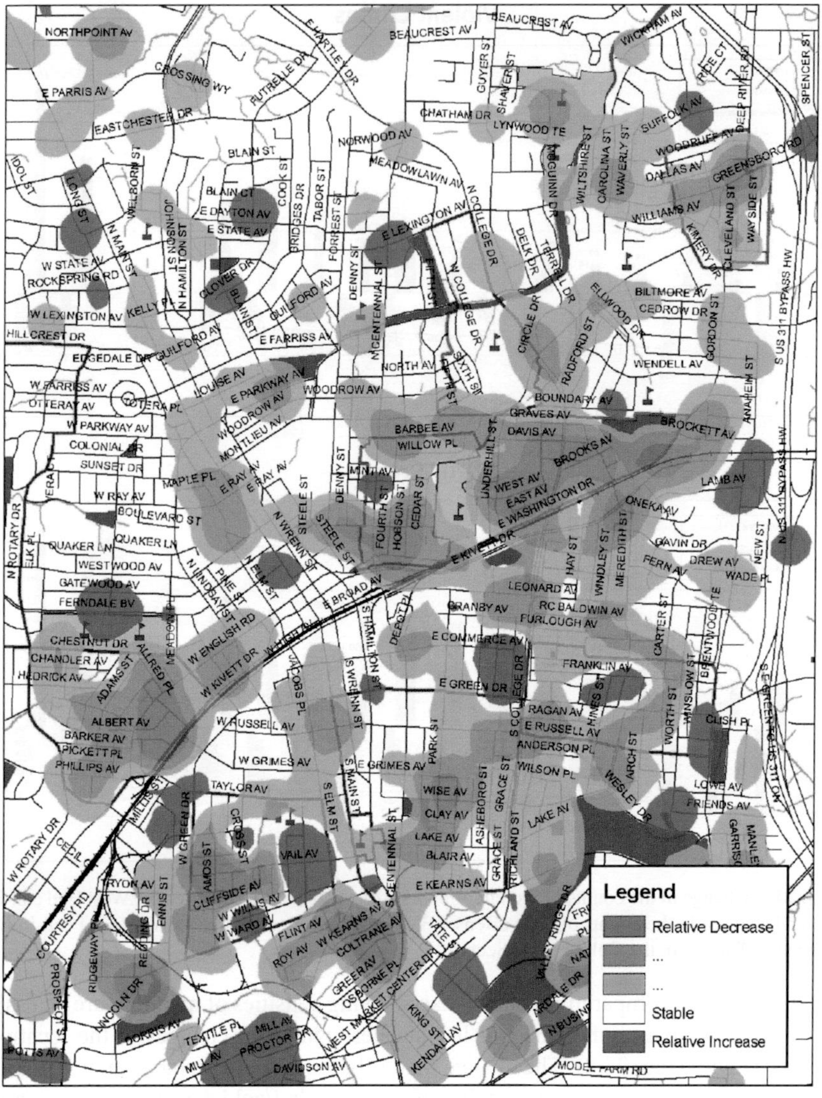

Fig. 24.8 Change in sale/manufacturing January 2003–June 2004 compared to July 2004 to December 2005 (18-month comparison using the Getis-Ord General G algorithm) (See also Plate 34 in the Colour Plate Section)

As part of the post call-in, citizens were encouraged to call 911 for any reason. Citizen calls for service in the West End neighborhood increased 10% within the first 150 days from the call-in. Increase in calls averaged 5% for the 2 years following the call-in. The nature of the calls shifted those that dealt with drug, suspicious activity, and shots fired in the first 6 months following the call-in to quality of life calls such as noise problems, juvenile problems, assistance, and intoxicated persons.

Conclusion

Community residents joined with police to deliver the message that drug dealing, drug-related crime, and violence would no longer be tolerated in their neighborhood. The consequences of continued drug dealing were clearly explained to offenders, and police presented a deadline to offenders and watched for signs of continued drug-related activity. Those who ignored the warning were swiftly prosecuted and used as examples to other offenders. Those who accepted the offer of help worked through a resource coordinator to take advantage of community services and resources. The community was extremely supportive of this concept because the warning and the offer of help were fair.

Residents reported an improved quality of life because street drug dealing abated following the notification. The partnership helped the citizens voice their outrage toward open-air drug dealing, and they became vigilant in calling police to report illicit activity. Police redeemed the citizens' faith in their ability to bring positive change to their neighborhoods, and narrowed the schism between law enforcement and community. Residents from other areas of the city are asking for the street drug initiative in their neighborhood. The officers assigned to the West End beat met and developed their post initiative strategy. The officers have taken responsibility for the neighborhood – even off-duty officers attend West End Community meetings. While community members worked closely with police to take back their neighborhood, they also learned how to keep it. Strong community participation at neighborhood meetings and the increase in calls to police are evidence that they remain vigilant and committed. Every violent crime that occurs is reviewed in an effort to determine common traits – based on GIS spatial or temporal relationships – or any other connections to individuals or other activities.

Geographic information system was an important tool used in the research methodology to map drug and serious offenses and to identify the highest priority drug-related hot spot in the city. The methodology used for this project has been applied to two subsequent areas in the city; it has been integrated into the crime analysis process. Changes in the type of data used to build the density maps allow for it to be used for similar efforts (e.g. domestic violence, prostitution).

In generating the densities, one area where the methodology can be modified is 911 data. Using all categories of dispatched data, densities that are larger in area and overwhelm the drug arrests, serious crime, and field contacts will be generated. The 911 data needs to either be used selectively – calls related to drugs, guns, and persons crimes – or it cannot be used at all. A combination of field contacts, drug arrests, and serious crime will generate similar patterns and locations of drug-related densities.

The concept of a focused deterrence strategy, while successful, also provided the resolution to look at the structure of a walk-up, drive-through overt drug market. It is suggested that while generating hot spot maps – based on a combination of crime types – that identify the general market area, it is the address or parcel level mapping of the dealer locations and various crimes that define the sales market, while the violent crime occurs inside the dealing area and the property crimes occur outside the dealing area. The methodology allows easy extraction of data for analysis. Selecting date ranges and using polygons to "cookie-cut" the data needed for analysis is easily acquired. While the initial use of GIS was not to drill further into the spatial relationships between dealing and crime, there is still fertile ground for analyzing geographic patterns at the micro-level and running new types of geospatial statistics that are now available. The spatial relationships of property crimes, prostitution, and auto theft can be analyzed to the scale of individual addresses. Analysis of home and work addresses of drug buyers can identify the spatial extent of the drug market – how far is an individual traveling to obtain drugs? Routes of travel, journey to purchases, and the geographic structure of the associated property crimes (burglary, larceny, theft) can be developed.

The reduction in violent crime and drug-related crimes has been sustained with no evidence of displacement. The strategy has been implemented in two other areas of the city with the same results. The drug markets closed down overnight.

Due to the program's success and sustainability, members of the High Point Police Department view this drug strategy as a philosophy rather than a project. It has become institutionalized within the police department. This process has profoundly changed operations within the police department. The department now sees itself as "data-driven" relying on GIS-based crime analysis to identify focus areas and hot spots. A focus area is a situation that may need a coordinated effort of several units of the department and may last several weeks or months; for example, serial residential burglaries or gang activity. In contrast, a hot spot is a short-duration project where a brief, short-term enforcement in a small area is needed; for example, prostitution or speeding. The use of GIS to monitor crime patterns is a daily activity; web-based GIS crime analysis tools are available to the entire police department.

The drug reduction strategy is based on applied theoretical models that have been developed over the past 20 years. It has worked in High Point and is being implemented in several cities around the country (Nashville, TN; Winston-Salem, NC; Raleigh, NC; Newburg, NY; Providence, RI; Rockford, IL) and in 2009 the US Department of Justice, Bureau of Justice Assistance is sponsoring ten cities across the US to replicate the strategy. Because the initiative is logically structured and given the similarity of patterns seen in overt drug markets, the model can be replicated in any size community, it is manageable, and will work if implemented correctly. With the resources of GIS, over time, the resolution and spatial understanding of drug markets will be refined. As such, associated property and person crime patterns, and drug arrests may predict an emerging drug market. Further, dealers in these areas will more readily be identified for future initiatives.

Notes

1. This program was one of four finalists selected for the 2006 Herman Goldstein Award for Problem Oriented Policing (http://www.popcenter.org/about-conference-papers.htm) and received the Harvard University, John F. Kennedy School of Government, 2007 Innovations in American Government Award.

2. For a brief review of the progression of GIS being integrated into crime analysis, review the following references. Gottlieb et al. (1994) illustrates crime analysis using simple coordinate graphing. Block et al. (1995) illustrates the application of early GIS software and spatial modeling. Harries (1999) and Eck et al. (2005) illustrate more advanced GIS techniques. These references do not include the area of geographic or environmental criminology that is now making great use of GIS.
3. For a complete discussion of the model and implementation see: High Point Police Department (1999) and (2000).
4. For context, the city of High Point has a population of just over 94,000 persons and comprises 54 square miles. The population of High Point has grown 25% in the last 16 years. Economically, the City reflects the rustbelt cities of the great lakes – the loss of manufacturing jobs and idled blue-collar workforce.
5. Source: Dr. James M. Frabutt et al., at The Center for Youth, Family, and Community Partnerships at the University of North Carolina at Greensboro serving as the Project Safe Neighborhoods Research Partner for the US Attorney's Office, Middle District of North Carolina. Supported by PSN funding (Award #2002-GP-CX-0220) through the US Department of Justice, Office of Justice Programs.
6. See Hunt (1992) for an early example of integrating GIS with catchment area analysis.
7. Epidemiology has incorporated GIS both as a research methodology and for the spatial accuracy that can be achieved. See Clarke et al. (1996).
8. Based on published UCR reports, the decrease from 2003 through 2005 is 21.5%: http://www.ncsbi.gov/crimestatistics/crimestatistics.jsp.

Chapter 25
Modeling the Spatial Patterns of Substance and Drug Abuse in the US

Sucharita Gopal, Matt Adams, Mark Vanelli

Abstract This chapter focuses on how spatial analysis might be used to improve the identification and treatment of substance use disorders (SUDs) and co-existing mental disorders in clinical practice. Using data from the National Survey on Drug Use and Health, this chapter uses Local Indicators of Spatial Association methodology to map regional differences in drug use at the sub-state level for the period 1999–2001 and from 2002 to 2004. It also maps the geographic correlation between co-morbid mental illness and substance use since psychological illness and drug addiction often co-exist. What results are maps that identify how areas of high and low substance abuse prevalence migrate from established foci over time and occasionally emerge as "outbreaks" unrelated to prior patterns of substance abuse. Such maps suggest substance abuse may be seen as a slow-moving chronic illness epidemic appropriate for public health surveillance. In future, clinicians and public health administrators might use maps that identify substance abuse hot spots specific to a region in order to improve case finding and treatment of SUDs. This chapter is based on an ongoing and active collaboration between a practicing psychiatrist and a geographer and discusses the relevance of spatial analysis from the perspective of its potential ability to improve patient care.

Background – The Cost of Drug and Substance Addiction

Drug addiction or substance dependence is a chemical or psychological dependency on a substance. Prolonged addiction may lead to degeneration of cognitive abilities and both general mental and physical health. Drug addiction may not only render negative consequences on the individual but also place undue emotional and financial stress on the family unit. Consequently, the sum of many individual addictions imposes unwarranted financial stress on society. A recent study published by the Office of National Drug Control Policy stated that the overall cost of drug abuse was

S. Gopal
Boston University, Boston, MA, USA

Y.F. Thomas et al. (eds.), *Geography and Drug Addiction*,

$143.4 billion in 1998, up 5.9% annually since 1992 (ONDCP 2001). Expenditure is focused on adverse effects to the health-care system, the welfare system due to losses in productivity, as well as strains on law enforcement and the judicial system.

Substance Use Disorders and Co-existing Mental Illness

Drug addiction or dependence may exist alone or may co-exist with a mental illness. When a mental disorder co-exists with a substance use disorders (SUDs), an individual is said to have a dual diagnosis. Typically, the course of serious mental illness (SMI) is worsened by co-existing SUDs. This is well-illustrated in the case of bipolar disorder where the lifetime prevalence of co-morbid alcohol or drug abuse and dependence range from 48 to 61% (Vornick and Brown 2006). Among individual with bipolar disorder current alcohol or drug abuse or dependence may increase the risk of treatment resistance and ongoing symptoms in the face of otherwise effective treatment; elevate the risk of suicide, aggression and violence; increase the likelihood of hospitalization; and contribute to worse function and quality of life (Buckley 2006). Prior cannabis use among those with bipolar disorder is linked with the duration of manic symptoms while the duration of alcohol abuse may help predict the duration of depressive symptoms (Strakowski et al. 2000).

Substance use disorders contribute to worse clinical outcomes through both direct effects on brain chemistry and function, and indirectly, by increasing the likelihood of medication discontinuation (Keck et al. 1988; Sajatovic et al. 2006). Repeated illness episodes are in turn more likely to be severe, frequent, and less medication responsive (Post et al. 1986; Perles et al. 2006; Kendler et al. 2000). Conversely, when both SUD and SMI receive treatment, individuals often succeed in achieving remission, independent living, competitive employment, and an improved quality of life (Weiss et al. 2005; Drake et al. 2004).

A delayed or missed diagnosis is the norm for patients with SUDs. Nearly half of all patients with a SUD never make contact with the health-care system and when contact is made, it is often decades after the onset of the problem. In this case, "contact" refers to a health-care provider being told of the existence of the problem, not subsequent treatment of the disorder. The time it takes until 50% of those with a SUD disorder make treatment contact is over 10 years for those alcohol dependence; over 20 years for those with drug abuse; and over 30 years for those with alcohol abuse. Only with substance dependence do 50% of cases make contact in less than a 10-year period (Kessler et al. 1998, 2005).

When both a SUD and mental disorder co-exist, treatment for both disorders occurs in only a minority of cases. Data from the 2004 National Survey on Drug Use and Health (NSDUH) estimate that roughly 4.6 million Americans have both a mental health and substance abuse disorder. Among these individuals, 47% reported receiving treatment for a mental health disorder and 11% for a SUD. Only 6% reported for both substance use and a mental health disorder.

The Geography of Addiction

Geographic information system (GIS) and spatial analysis provide powerful tools for analyzing addiction and epidemiology (Riner et al. 2004; Wieczorek and Hanson 1997). The use of GIS for addiction and SUD can be categorized into four areas: GIS is used for mapping SUDs (e.g., Richards et al. 1999), for resource allocation and service planning (e.g., Field and Beale 2004), epidemiological research such as exploring the spatial association between locations in the environment and risk for SUD (e.g., Wilson and Dufour 2000; Millar and Gruenewald 1997), including crime (Gorman et al. 2005) and for prediction (e.g., Beale 2004; Kraus et al. 2003; Gruenewald and Treno 2000).

Addiction research is further enhanced by developments in the field of spatial statistics. In this context, three recent trends are worth noting. First, small area studies are becoming increasingly used with the ability to properly account for high variance of estimates in small geographic areas and characterize overall geographic trends and patterns (Elliot et al. 1993; Wennberg et al. 1973). Second, Bayesian methods are becoming popular tools for disease mapping. Besag et al. (1991) describes a Bayesian approach, which separated spatial effects from heterogeneity. Waller et al. (2004) uses a Bayesian hierarchical model that accommodated covariates and spatial structure that change over time. Third, the widespread use of GIS and links to statistical packages has further encouraged spatial data analysis (Lawson 2001; Waller et al. 2004; Banerjee et al. 2004).

The National Survey on Drug Use and Health

Due to the illegal nature of the use, sale, and distribution of illicit substances, it is difficult to determine the extent of drug addiction in the United States. Perhaps, the most complete and geographically precise source of data on substance use is provided by The NSDUH, an anonymous household survey of illicit drugs, alcohol, and tobacco abuse of people over the age of 12 living among the non-institutionalized civilian population in the United States. The Substance Abuse and Mental Health Administration conduct the survey, which annually interviews roughly 67,000 people. The definitions of drug abuse and dependence used in the NSDUH survey are based on definitions found in the 4th edition of the *Diagnostic and Statistical Manual of Mental Disorders* or DSM-IV (APA 1994).

The NSDUH provides prevalence estimates of substance abuse or dependence in past month among persons aged 12 or older in a given set of years. As the NSDUH captures data at both the state and sub-state levels, it consequently provides data with a higher degree of spatial resolution than survey data based on state level data. Sub-state level refers to an intermediate level of census geography between state and county level in the census hierarchy and consists of an aggregation of counties in each state. For the time period 1999–2001, NSDUH data is available in aggregate and at sub-state level for individual ages 12–17, 18–25, and 26 and older. For the time period 2002–2004, data is available for the same sub-state geographies in aggregate but not for individual age cohorts.

Estimates in the NSDUH survey are based on a survey-weighted hierarchical Bayes (SWHB) estimation approach, and the 95% prediction (credible) intervals are generated by Markov Chain Monte Carlo techniques. Thus, the NSDUH data is not drawn directly from the survey data but a SWHB-estimation. Since the NSDUH is well-sampled, geographically distributed survey, the SWHB technique borrows strength from ancillary higher resolution data to better estimate data in surrounding small geographic units, to then aggregate the data to the sub-state or state level. This technique is generally referred to as small area estimation and has been well-developed over the past few decades (Ghosh and Rao 1994; Folsom et al. 1999).

Drug Abuse Definition

A respondent in the NSDUH survey was defined as having abused that substance if he or she met one or more of the following four abuse criteria and was determined not to be dependent on the respective substance in the past year.

1. Serious problems at home, work, or school caused by the substance, such as neglecting your children, missing work or school, doing a poor job at work or school, or losing a job or dropping out of school.
2. Used the substance regularly and then did something that might have put you in physical danger.
3. Use of the substance caused you to do things that repeatedly got you in trouble with the law.
4. Had problems with family or friends that were probably caused by using the substance and continued to use the substance even though you thought the substance use caused these problems.

Drug Dependence Definition

A respondent was defined as having drug dependence in the survey if he or she met three or more of the following six dependence criteria.

1. Spent a great deal of time over a period of a month getting, using, or getting over the effects of the substance.
2. Used the substance more often than intended or was unable to keep set limits on the substance use.
3. Needed to use the substance more than before to get desired effects or noticed that same amount of substance use had less effect than before.
4. Inability to cut down or stop using the substance every time tried or wanted to.
5. Continued to use the substance even though it was causing problems with emotions, nerves, mental health, or physical problems.

6. The substance use reduced or eliminated involvement or participation in important activities.

Patients are considered to have a problem with substance abuse or dependence, not both.

In addition to the six listed above, a seventh criteria for withdrawal was also included, Withdrawal was said to have occurred if a respondent had experienced a certain number of substance-specific withdrawal symptoms (e.g., having trouble sleeping, cramps, hands tremble). For further details, see http://www.drugabusestatistics.samhsa. gov/NSDUH/2k4NSDUH/2k4results/appb.htm#SPD):

Serious Psychological Distress Definition

Serious psychological distress (SPD) is a non-specific measure of psychological distress NSDUH survey based on a 6-question (K6) scale that asked about the 1 month in the past year that an individual was at their worst emotionally. Responses to these questions produced a score from 0 to 24; a score of 13 or greater identified the individual as having SPD. This threshold was based on research suggesting that scores above this threshold meet the criteria for a SMI. SPD was referred to as SMI in prior NSDUH reports. Criteria for a SMI is established by having a DSM-IV or equivalent ICD-9-CM equivalent diagnosis in the past year that results in impairments that interfere with or limit one or more major life activities.

Other Data Sources

Other data that could be included for a more complete analysis of addiction is patient hospital visits and arrest records that are annually available for the United States. The disadvantage of using non-survey institutional data is that hospital inpatient or outpatient services rely upon access to hospitals and health insurance, which is unequally distributed throughout the country. Similarly, drug related arrests are subject to the differing law enforcement practices and legal standards of individual states. Due to difficulties in normalizing the data, both hospital visits and arrest records were visually evaluated and discounted for use in this study. Any future study on addiction will benefit from inclusion of such data.

Mapping the Geography of Substance Use Disorders Using Local Indicators of Spatial Association

The primary objective of the research is to describe the spatial patterns of SUDs in United States and to identify statistically significant areas of high and low drug use. The most suitable statistic for this purpose is a measure of spatial autocorrelation that tests to see if there is any systematic pattern in the spatial distribution of a variable. In this context, *positive spatial autocorrelation* describes patterns in which neighboring areas are similar to each other. *Negative autocorrelation* describes patterns in which neighboring areas are dissimilar to each other. One of the oldest measures of spatial autocorrelation is Moran's I statistic (Moran 1950) that is still a *defacto* standard for determining the level of spatial autocorrelation. The Moran's scatterplot is helpful in the visualization of four types of spatial autocorrelation. Quadrant HH refers to an observation with high value surrounded by observations with high values while quadrant LL refers to an observation with low value surrounded by observations with low values. Quadrant LH (HL) refers to an observation with low (high) value surrounded by observation with high (low) values. Quadrants HH and LL signify clusters while HL and LH are spatial outliers.

While global autocorrelation is useful in some contexts, it may not pick out local patterns. Anselin (1995) introduced Local Indicators of Spatial Association (LISA) to assess the significance of spatial autocorrelation at a local scale. LISA represents the local version of Moran's I measure. A positive value for LISA indicates spatial clustering of similar values (can be high or low) while a negative value indicates spatial clustering of dissimilar values between an observation and those in its neighborhood.

In the present research, GeoDa (Anselin 2006) software is used to analyze spatial patterns in SUDs, since it can easily assign spatial weights and offers robust options for statistical inference. Contiguity of spatial weights is the most important criteria in conducting LISA, as different weights produce different results. Two choices for constructing weights are available; selection of either depends on the type of geographic data involved. For points, various distance decay measures are available. For contiguous polygons, as with the sub-state geographies used in this research, contiguity based on adjacent neighbors is appropriate. There are essentially two types of contiguity, referred to as the queen or rook contiguity; the queen's contiguity assigns spatial weights based on neighbors in all directions, while the rook's assigns only in the four cardinal directions. In this research, a queen's contiguity was chosen due to the absence of specific knowledge that the spatial nature of substance abuse is inherently restricted to the four cardinal directions. It was assumed that the neighboring drug addiction occurred regardless of these directional constraints.

Both global and local Moran's I was estimated for this research. The global Moran's I addressed the question what is the extent of clustering in the total area as well as if this clustering is significantly different from a random spatial distribution. The Moran's scatterplot is used to evaluate global spatial autocorrelation. A spatially random permutation procedure was performed to re-compute the statistic 999 times to arrive at a stable reference distribution at 99% confidence interval. In the presence of global spatial autocorrelation, local Moran's I was then used to determine the local component. Local Moran's I identifies the local clusters of SUDs (high–high or low–low). Again, randomization of the spatial distribution was performed 999 times. Clusters relating positive spatial autocorrelation significant at the 95th and 99th percentiles are included in the results. This is an important distinction, as clusters at this confidence level remained stable over multiple random permutations. Although still significant, less confidence should be placed for those at the 95th percentile (Anselin et al. 2006). Negative spatial autocorrelation (spatial outliers) is not considered in the analysis.

The final results of LISA analysis showing clusters of high (red-orange) and low values (dark and light blue) are visualized as choropleth maps. In addition, means for all high clusters (hot spots), low clusters (cold spots), and no clusters were computed and are displayed to relate the magnitude of the addictions on all maps. The maps show SUDs at the sub-state level, for *any illicit drug, illicit drugs other than marijuana, marijuana, cocaine, binge drinking, and cigarettes.* In addition, each LISA map for the second time period is compared with the results presented in the NSDUH 2004 report (Source) that provides the rates of SUDs in different regions of the US.

Figure 25.1 shows LISA maps for illicit drug use for the period 1999–2001 that exhibits the following trends. Illicit drug use in youths 12–17 is more likely to occur in the Western states of California, Colorado, Oregon, Montana, New Mexico, and Arizona and in the Northeastern states of Vermont and New Hampshire. Further, among adults 26 and older, illicit drug use is less prevalent in the Midwestern (North and South Dakota, Nebraska, Iowa, and Minnesota) and Mid-Atlantic states (Pennsylvania, West Virginia, Virginia, Maryland, and Kentucky). In addition, hot spots of illicit drug use shrinks as youths become adults in New England but increase as youths become adults in California, Oregon, Western Washington state, and Nevada.

Figure 25.2 shows that for the period 2002–2004 relative to the period 1999–2001, "outbreaks" of illicit drug use in northern Montana and New Mexico not present in the earlier period. There is a migration into New York and Maine of drug abuse hot spots from established foci in Vermont, Western Mass, and Southern New Hampshire as well as a migration of illicit drug use from Northern to Southern California. There is a greater geographic resolution and specificity with regards hot spots of high illicit drug use prevalence using sub-state data.

The NSDUH survey date for 2004 shows that marijuana is dominant amongst all SUD since marijuana use is 56.8% compared with drugs other than marijuana (around 23%). Figure 25.4 shows the LISA map for marijuana in the US for the 2002-2004 for all (above 12 years of age). Compared with the first time period, hot spots of marijuana use have expanded in Oregon, New Hampshire and Maine in the second time period.

Figure 25.3 shows LISA maps for marijuana abuse or dependence for the period 1999–2001. Among adults, hot spots of marijuana abuse and dependence are largely coastal phenomena, confined to the New England States and to California, Oregon, and Western Washington. An isolated hot spot of marijuana is observed in central Colorado. Two large, contiguous bands of low marijuana abuse and dependence can be identified. One runs through the Midwest from North and South Dakota straight down to Texas and Louisiana while the other occupies the Appalachian and Mid-Atlantic states of Virginia, West Virginia, Pennsylvania, Kentucky, and Ohio. Low youth marijuana use is also concentrated in the southeastern states.

Figure 25.4 shows that in the period 2002–2004 relative to 1999–2000, marijuana abuse or dependence migrates into New York state, New Hampshire, and Maine from established foci in Vermont, Western Mass, and Southern New Hampshire. Marijuana abuse or dependence also migrates from Northern to Southern California.

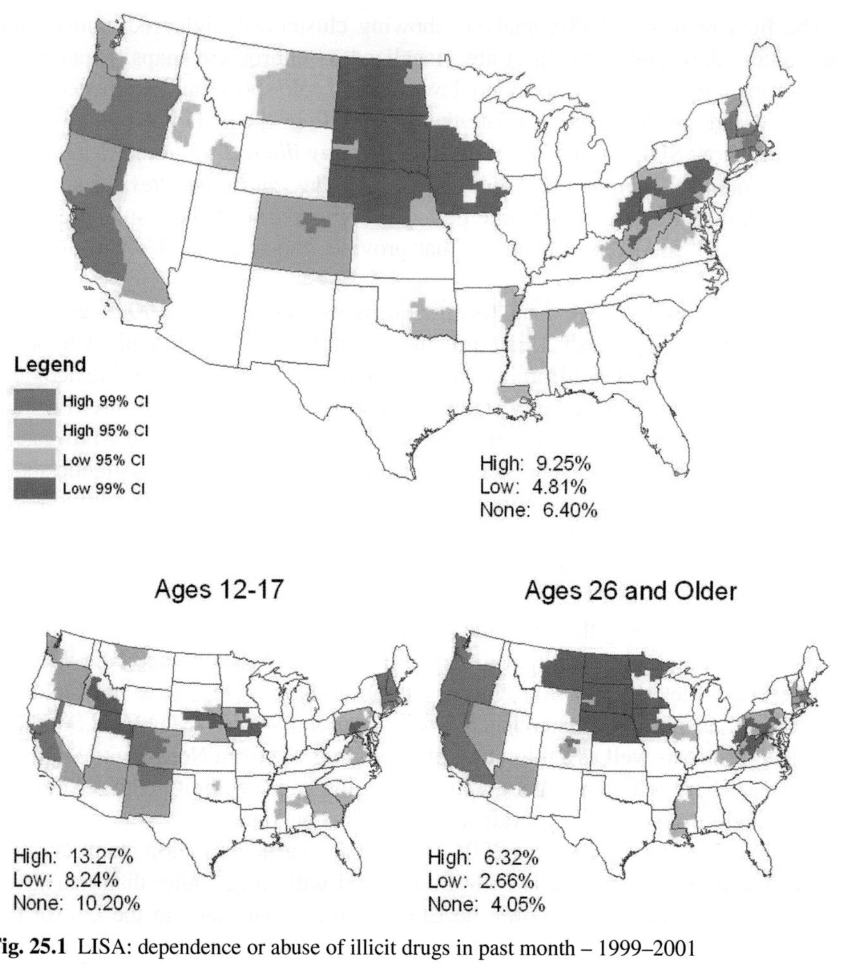

Fig. 25.1 LISA: dependence or abuse of illicit drugs in past month – 1999–2001
Source: NSDUH household survey data, 2002–2004. Illicit drug use is defined as: any use of marijuana/hashish, cocaine (including crack), heroin, hallucinogens, inhalants, or prescription psychotherapeutic agents used non-medically (pain relievers, tranquilizers, stimulants, and sedatives). "Estacy" (MDMA), LSD, PCP, peyote, mescaline, mushrooms are considered hallucinogens. Methamphetamine is considered a stimulant. Use of alcohol or cigarettes, which is illegal for youths, is not included. (See also Plate 35 in the Colour Plate Section)

In addition, marijuana abuse or dependence disappears from its hot spot in Colorado. There is an "outbreak" of marijuana abuse and dependence emerge in Northern Montana and eastern Wisconsin.

Figure 25.5 shows LISA maps for Illicit Drug Use Other Than Marijuana ("hard drugs") for the period 1999–2001 that show the same general hot and cold spots locations as when marijuana was included. However, hard drug use is more extensive in the west in Nevada, northern Arizona and in Oklahoma, where it is a problem most evident in adults. In Colorado, hard drug use hot spots occur until the age of 26

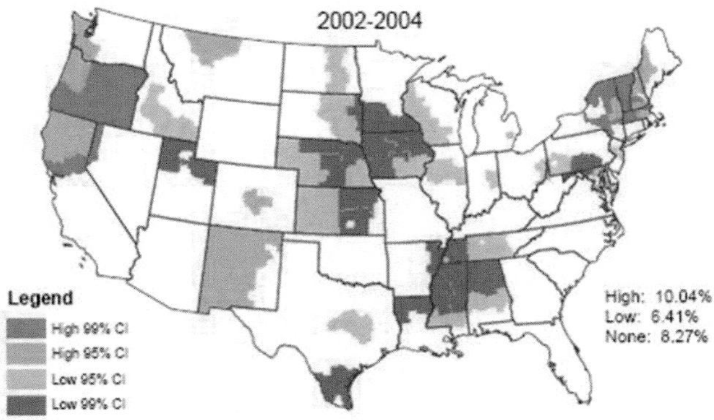

Fig. 25.2 Dependence or abuse of illicit drugs in past month, 2002–2004

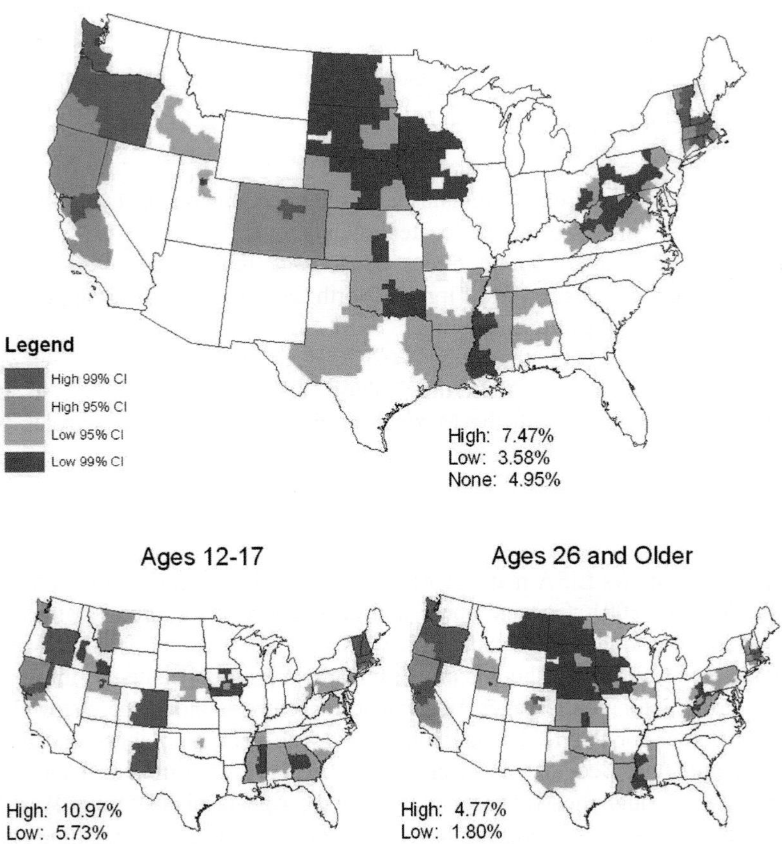

Fig. 25.3 LISA: dependence or abuse of marijuana in past month 1999–2001.
Source: NSDUH household survey data 1999–2001. Marijuana use includes use of both marijuana and hashish. (See also Plate 36 in the Colour Plate Section)

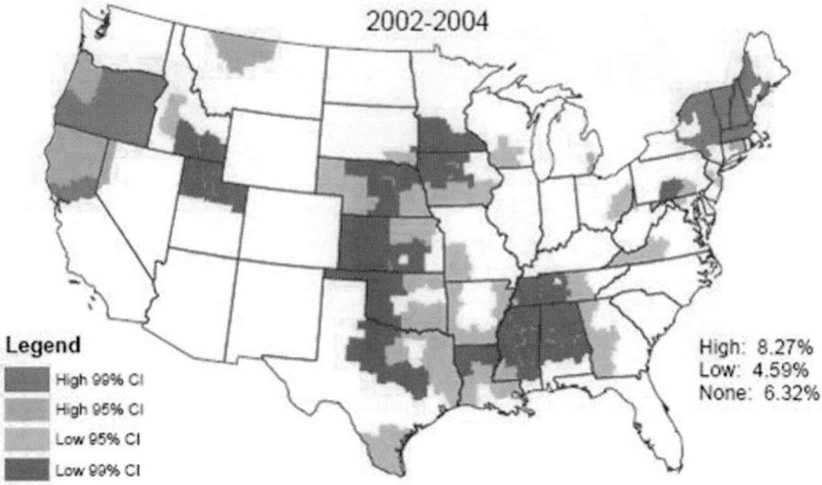

Fig. 25.4 LISA: dependence or abuse of marijuana in past month – 2002–2004
Source: NSDUH household survey data 2002–2004. Marijuana use includes use of both marijuana and hashish. (See also Plate 37 in the Colour Plate Section)

and then fall off. Hard drug use remains a coastal phenomenon, as it was with marijuana, with an additional hot spot located in Oklahoma. High rates of "hard drug" use for youth 12–17 are observed in most of Montana; in a number of Southeastern states (Louisiana, Arkansas, West Virginia, North Carolina, and Mississippi); and in parts of Colorado and California.

Figure 25.6 shows that in the period 2002–2004 relative to the period 1999–2001. A dramatic intensification of "hard drug" abuse and dependence migrates to states bordering or contiguous with the Mexican border, such as Texas, Arizona, New Mexico, and Okalahoma. There are "Outbreaks" of "hard drug" abuse and dependence in North Carolina and Kentucky while there is the disappearance of "hard drug" hot spots in Oregon, California, and Massachusetts relative to the previous time period.

Figure 25.7 shows LISA maps of Abuse or Dependence on Cocaine in the past Month for 1999–2001. Cocaine abuse and dependence is elevated in a swath of states contiguous to the US-Mexican border: Texas, New Mexico, Arizona, Nevada, Colorado, and California. There is a widespread pattern of cocaine use throughout the western US for youth's ages 12–17 that concentrates along the states on the Mexican border for the 18–25 age cohort. As with marijuana abuse and dependence, an area of low cocaine abuse and dependence extends from North and South Dakota down through the Midwest and through the Appalachian states of West Virginia, Pennsylvania and Ohio. Similar patterns persist in Fig. 25.8 for time period 2001–2004.

Figure 25.9 shows LISA data for Binge Drinking for the period 1999–2001. Binge alcohol consumption rates for youth ages 12–17 are highest in the contiguous northern Midwest states of Montana, Wyoming, North and South Dakota,

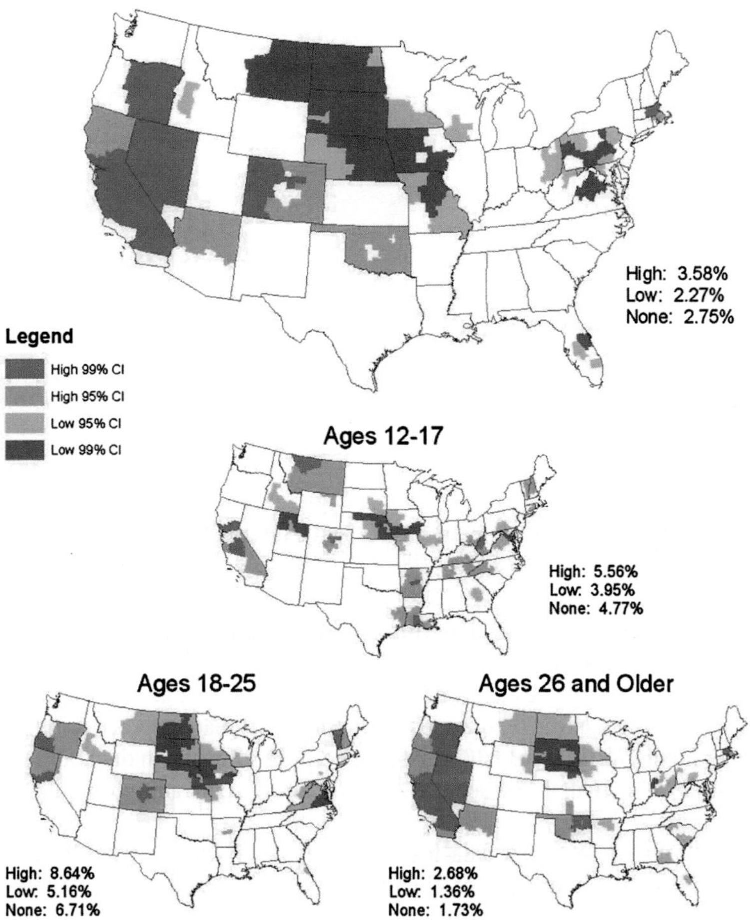

Fig. 25.5 LISA: dependence or abuse of illicit drug use other than marijuana in past month – 1999–2001

Source: NSDUH household survey data 1999–2001. Illicit drug use of drugs other than marijuana/hashish is defined as: any use of, cocaine (including crack), heroin, hallucinogens, inhalants, or prescription psychotherapeutic agents used non-medically (pain relievers, tranquilizers, stimulants, and sedatives). "Estacy" (MDMA), LSD, PCP, peyote, mescaline, mushrooms are considered hallucinogens. Methamphetamine is considered a stimulant. Use of alcohol or cigarettes, which illegal for youths, is not included. (See also Plate 38 in the Colour Plate Section)

Minnesota, Nebraska, and Iowa and extend down into eastern Colorado and New Mexico. Binge alcohol rates for youth 12–17 are also high in eastern Massachusetts. Rates of binge alcohol drinking are lowest for youth 12–17 years in the southeastern states of Virginia, Tennessee, North Carolina, South Carolina, Georgia, and Mississippi. Areas of elevated binge drinking for adults 26 year and older are concentrated in a similar but slightly smaller geography than for youth 12–17 years.

Figure 25.10 shows that in the period 2002–2004, relative to the period 1999–2001, there is a southern migration of binge drinking from northern Midwest

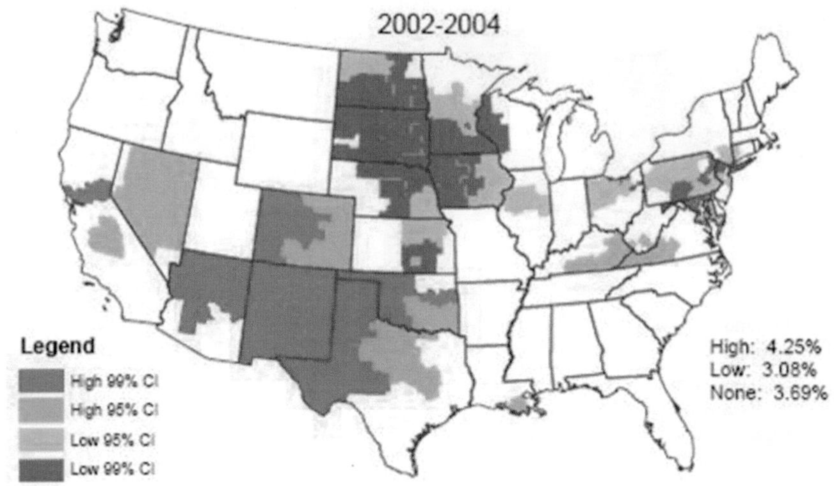

Fig. 25.6 LISA: dependence or abuse of illicit drug use other than marijuana in past month – 2002–2004 (See also Plate 39 in the Colour Plate Section)

states (North and South Dakota and Minnesota) into Illinois and Iowa and westward migration into Wisconsin.

There is a seemingly westward migration of binge drinking from Massachusetts into New York state. NSDUH survey data simply notes that underage drinking rates are higher in the Northeast (32.3%) and Midwest (31.4%) than in the South (26.2%) and West (27.3%). Our LISA analysis shows the hotspots and problem areas for binge drinking at the sub-state level.

Figure 25.11 shows LISA data for Cigarette Use in the period 1999–2001 exhibiting a pattern of high rates of cigarettes use in the southeastern ("bible belt") states of Kentucky, Tennessee, Arkansas, Louisiana, North Carolina, South Carolina, Ohio, and Virginia. There are high rates of cigarette use in the Midwestern sates of Missouri, Arkansas, and Oklahoma. Many of the same states that had low rates of binge drinking but have high rates of cigarette use. High rates of smoking in youth ages 12–17 in the northern Midwest states of North Dakota, South Dakota, Minnesota, eastern Montana, and Wyoming.

Figure 25.12 shows that for the period 2002–2004 relative to the period 1999–2001, new cold spots of cigarette use in New Mexico, Wisconsin, and Southern California. NSDHU survey data reports that cigarette smoking rates are lowest in the West (19.7%), and highest in the Midwest (27.8%) and South (26.8%).

Modifiable Areal Unit Problem Related to Scaling

The Modifiable Areal Unit Problem (MAUP) is a potential source of error when spatial data is aggregated (Unwin 1996). Each aggregation scheme is likely to produce different spatial variations (Openshaw 1984). During the last two decades, several

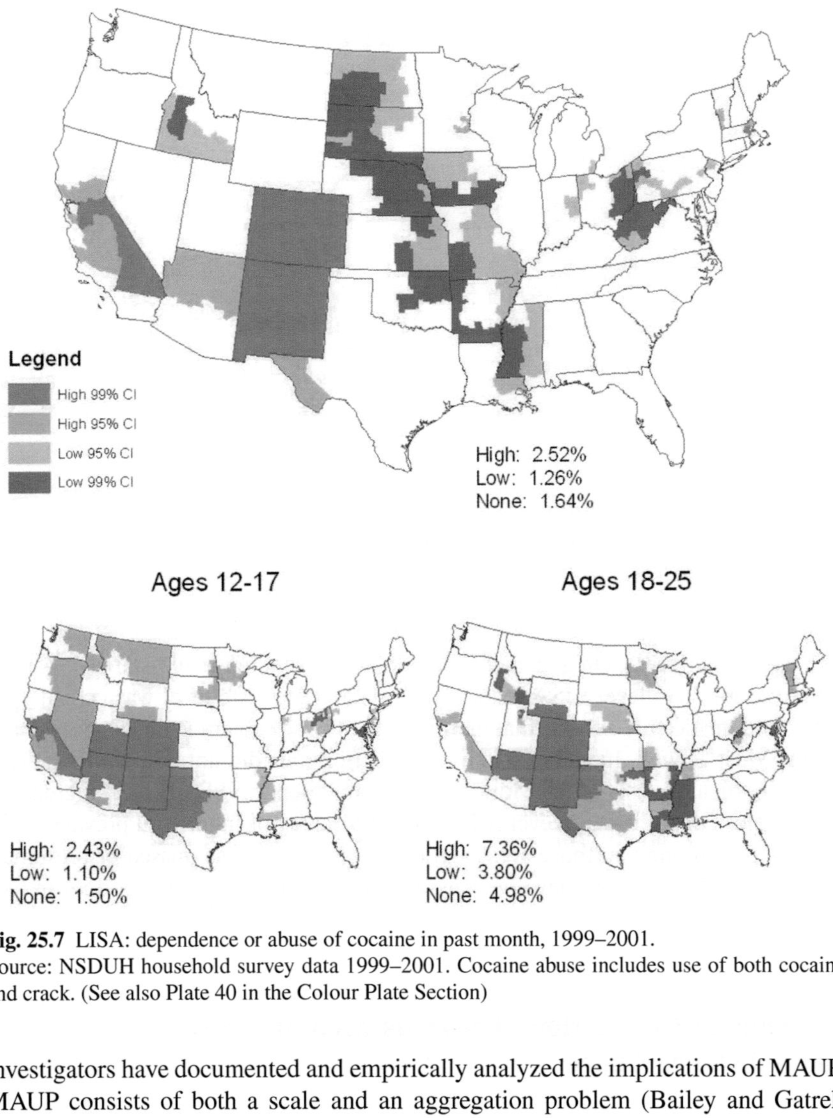

Fig. 25.7 LISA: dependence or abuse of cocaine in past month, 1999–2001.
Source: NSDUH household survey data 1999–2001. Cocaine abuse includes use of both cocaine and crack. (See also Plate 40 in the Colour Plate Section)

investigators have documented and empirically analyzed the implications of MAUP. MAUP consists of both a scale and an aggregation problem (Bailey and Gatrell 1995). The scale problem is relatively well-known and produces different statistics from the same set of data when the information is grouped at varying levels of spatial resolution (e.g., census blocks, tracts, counties, states, regions). The zoning effect is the variability in statistical results obtained within a set of modifiable areal units as a function of the various ways these units can be grouped at a given scale, and not as a result of the variation in the size of those areas.

Figure 25.13 shows the well-documented MAUP related to scaling using the NSDUH dataset. Important geographic variations in substance abuse patterns are lost when the analysis is not carried out at the sub-state level. This is illustrated with the comparison of the state and sub-state maps for marijuana use. Large areas of

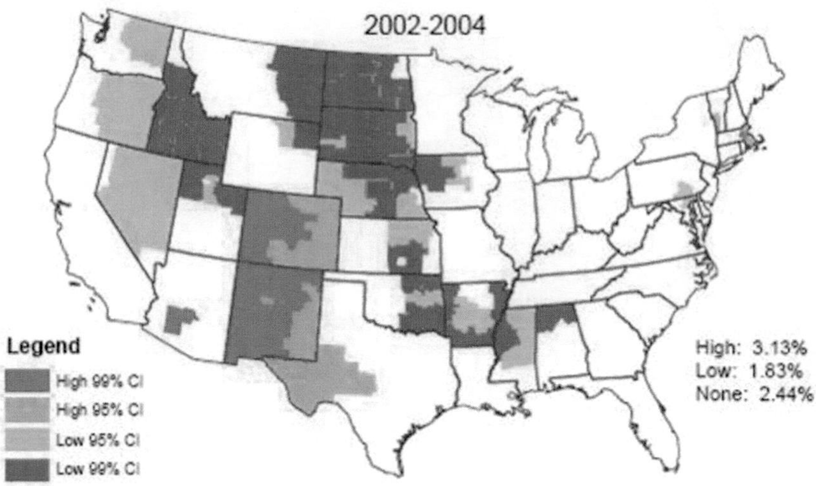

Fig. 25.8 LISA: dependence or abuse of cocaine in past month, 2002–2004 (See also Plate 41 in the Colour Plate Section)

high use in California, Oregon, and Colorado are present in the sub-state LISA map but not at the state level LISA map. If only data at the state level was considered, one would mistakenly assume that marijuana is only a problem in the Northeast and not the West as well. In addition, the state distributions did not hold up to the rigorous significance tests of the Moran's scatterplot. Hence NSDUH data aggregated at scales less than the state level help to identify areas of high SUD prevalence of value to practitioners in those areas and that would otherwise missed or presumed to not exist. (Risk of false negative – assuming no SUD when hot spot does in fact exist).

Geographic Co-location of Serious Mental Illness and Substance Abuse

SMI Hot Spots

Before tackling the issue of co-location, it is important to analyze patterns of SMI alone.

Local Indicators of Spatial Association analysis for SMI alone at 95% confidence over 999 permutations revealed hot spots of especially high SMI in 3 pockets – North Dakota (Lake Region and South Central), New Mexico (Region 2) and Oregon (region 4) with LISA indices of 0.01, 0.07, and 0.01 respectively. However, only North Dakota and New Mexico continue to be hot spots even at 99% confidence.

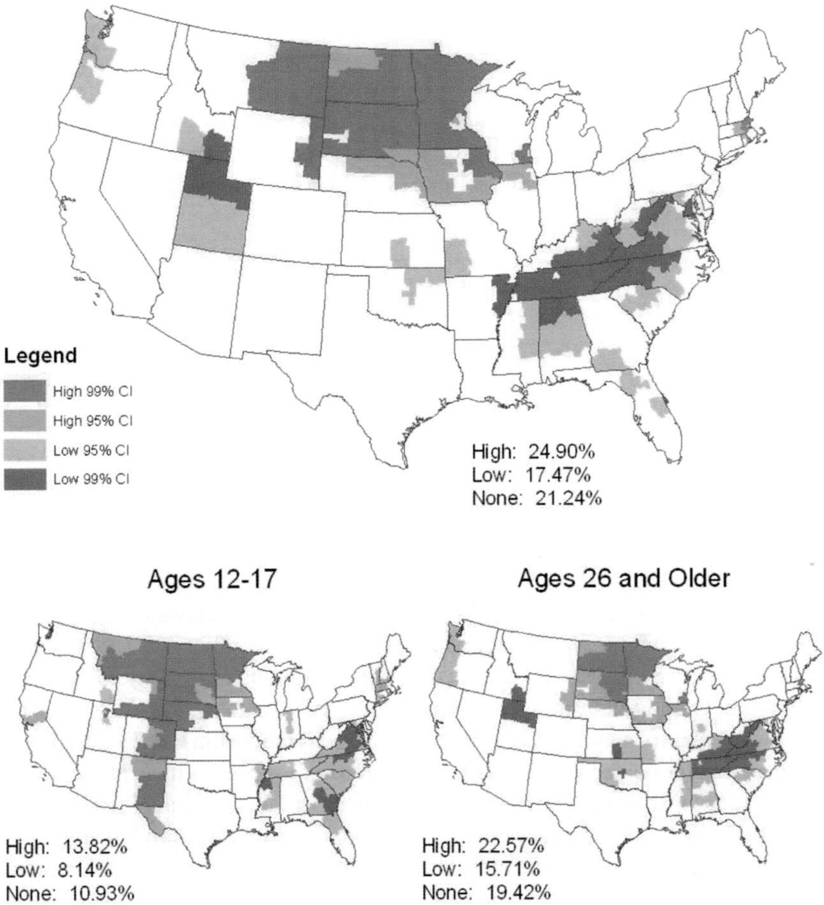

Fig. 25.9 LISA: binge alcohol use in past month, 1999–2001
Source: NSDUH household survey data 2002–2004. Binge alcohol use defined as five or more drinks at the same time or within a couple of hours of one another on the same occasion at least once in the past 30 days. A "drink" is defined as a can or bottle of beer, a glass of wine or wine cooler, a shot of liquor, or a mixed drink (See also Plate 42 in the Colour Plate Section)

An important objective of this research is to evaluate the co-occurrence of mental health and SUDs. Co-morbidity is analyzed using SMI and drug use data. Pearson correlations are used for variable selection for all NSDUH and NESARC drug variables. While correlating data across surveys is not ideal, the NSDUH does not have state level data for all individual substances. For consistency with the NSDUH, data from the NESARC was aggregated to state geographies and converted to percentages per 100,000 persons. Two drug variables were correlated at 99% confidence, one from each survey. From the NSDUH, illicit drug use other than marijuana in the past month had a 0.512 correlation with SMI. Amphetamine use in the past year from the NESARC had a 0.418 correlation with SMI. Multivariate LISA analysis

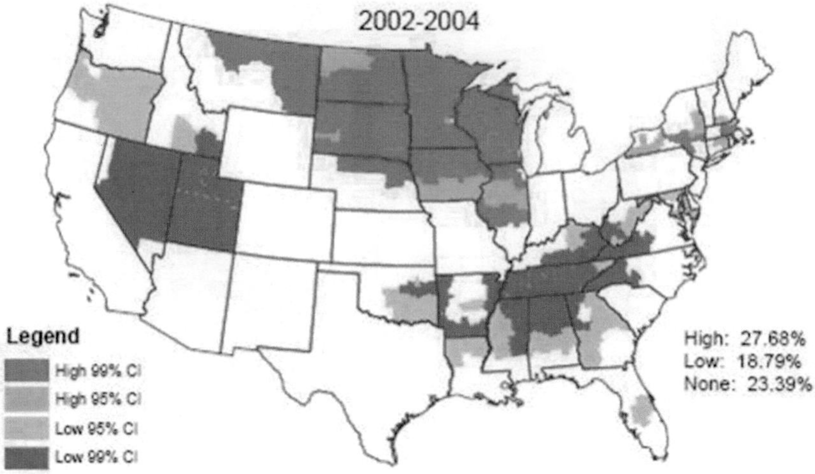

Fig. 25.10 LISA: binge alcohol use in past month, 2002–2004 (See also Plate 43 in the Colour Plate Section)

at 95% confidence of SMI and Illicit Drugs in Past Month shows clusters in parts of Washington, Oregon, Wyoming, and Rhode Island. Multivariate LISA analysis at 95% confidence of SMI and Marijuana in Past Month shows clusters in parts of Washington, Oregon, Wyoming, and Massachusetts.

While correlations and LISA analysis show the relationship between SUDs and SMI, they cannot be used for prediction. Therefore, we used a multiple regression model that used SMI as the dependent (response) variable, and amphetamines use and "other" drugs as the independent (predictor) variables. Although this model implies causality between drug use and SMI, it is primarily used for hypothesis testing. This is essential to establish so that the co-occurrence of addiction and mental health may then be modeled in space. While substance abuse can contribute to mental illness, it is also known that illicit drugs are often used to self-medicate uncomfortable emotional states associated with mental health.

The statistical results indicated a good fit for the model, with an overall R^2 of 0.423, a F-statistic of 17.56, a p-value of 0.0000, and an error variance of 0.534. The resulting equation for the model may be written as follows:

$$Y = 4.6367 + 1.1068 \times X_1 + 0.2569 \times X_2,$$

Where Y = SMI, X_1 = other drugs, and X_2 = amphetamines

Figure 25.14 shows the model residuals mapped to view spatial discrepancies. The model residuals are displayed as a series of warm and cool colors. In standard practice, the residuals are computed as the observed value minus the predicted value from the linear model. The red, orange, and yellow colors indicate areas where SMI is greater than would be predicted by the linear model. Conversely, the blue colors indicate where SMI is less than would be predicted by the linear model. In general, there is a correspondence between residual values and drug use; states with high

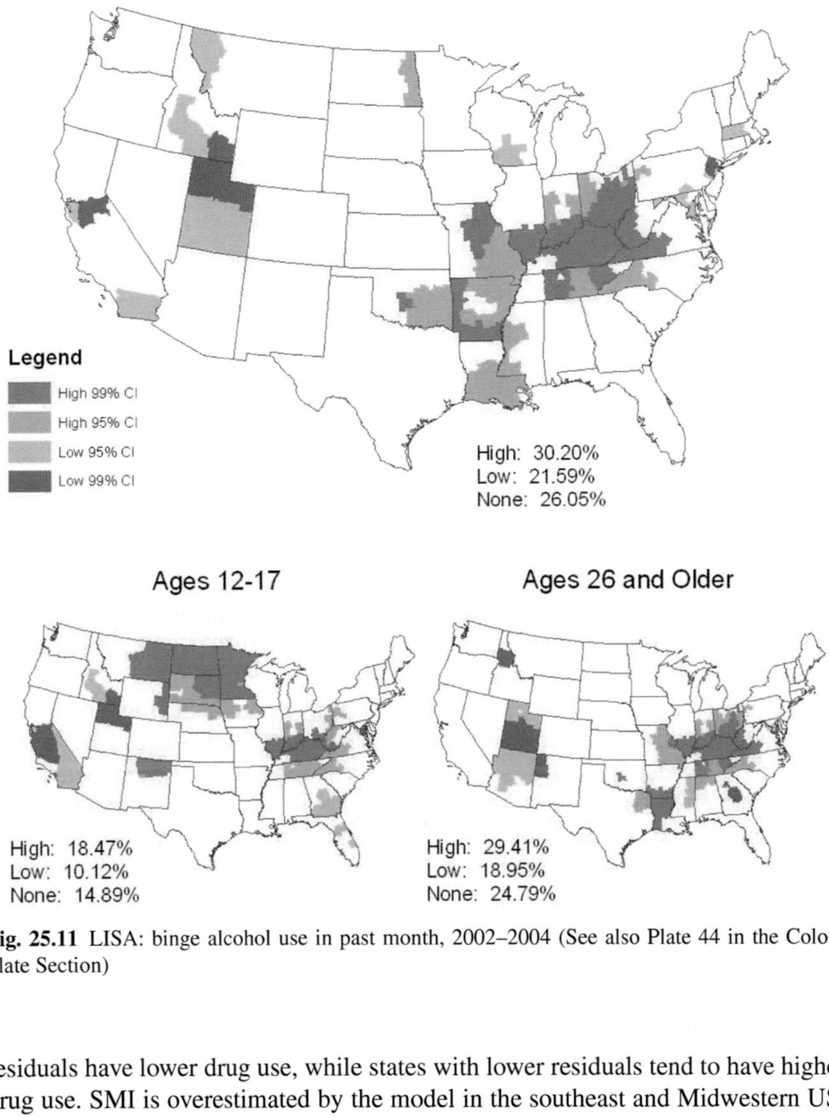

Fig. 25.11 LISA: binge alcohol use in past month, 2002–2004 (See also Plate 44 in the Colour Plate Section)

residuals have lower drug use, while states with lower residuals tend to have higher drug use. SMI is overestimated by the model in the southeast and Midwestern US, while it is underestimated in the west, southwest, and northeast (Fig. 25.14). These visual trends more accurately describe the co-occurrence of SMI and substance abuse in space, but further exploration to capture non-linear trends is warranted before any further conclusions are drawn.

To guard against a spurious relationship, two tests were performed. In the first, additional variables that are correlated with SMI were added to the multiple regression model to test for stable coefficients in the proposed model. In accordance with prior variable selection, additional Pearson correlations were run on a variety of basic demographic variables obtained from the US Census 2000, describing race, gender, age, income, and percent urban. In addition, all variables pertaining to alcohol and tobacco use from the NSDUH were included. Four variables were

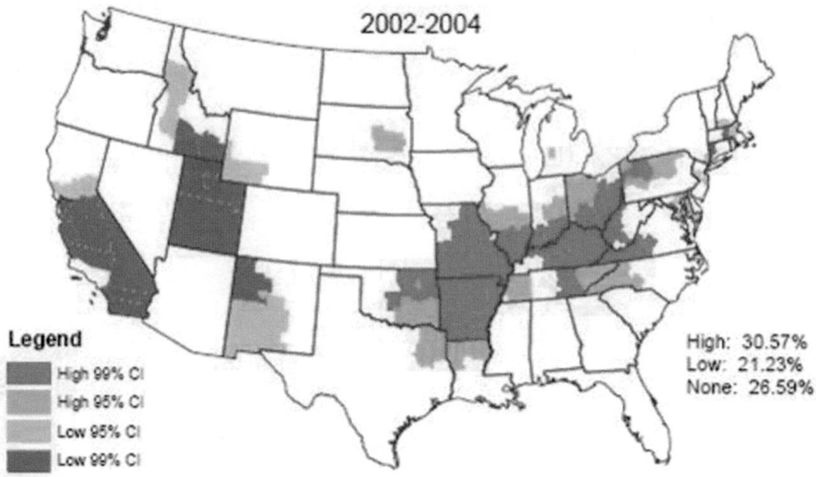

Fig. 25.12 LISA: cigarette use in past year, by age: 2002–2004 (See also Plate 45 in the Colour Plate Section)

significant at the 99% confidence level – alcohol use in past month (−0.381), illicit drug use other than marijuana in past month, ages 18-25 (0.392), median household income (−0.464), and per capita income (−0.465).

The results of the stepwise addition of each variable to the multiple regression model indicated that the parameters in the proposed model were indeed stable after evaluation of model numerical outputs. The second test was to guard against the intermediary effects of another variable, such as age, to explain the proposed model relationship as well as one of the independent variables. Breakdowns by age were available for the other drug variable. As such, each age variable was substituted in the model for the other drug variable. Significant reductions in the R^2 as well as other model indictors signified that age was not a factor in examining co-morbidity.

A Doctor's Perspective

Addictive and mental disorders have their foothold in youth and young adults (Kessler 2005). Seventy-five percent of those with SUDs experience the onset of their illness by age 27. Narrow interquartile ranges help further define the onset of most SUDs as occurring between the late teenage years and the late 20s. Statistically speaking, we know who is at risk, but not *where and for what*.

Epidemics are typically characterized by person, time, and *place*. Spatial analysis has the potential to help clinicians improve case finding by knowing which substances use disorders are most prevalent in the area in which they live and practice. Such knowledge may help clinicians ask better screening questions about substance abuse and help increase their index of suspicion regarding the presence of SUD when behavioral, cognitive, or emotional changes occur.

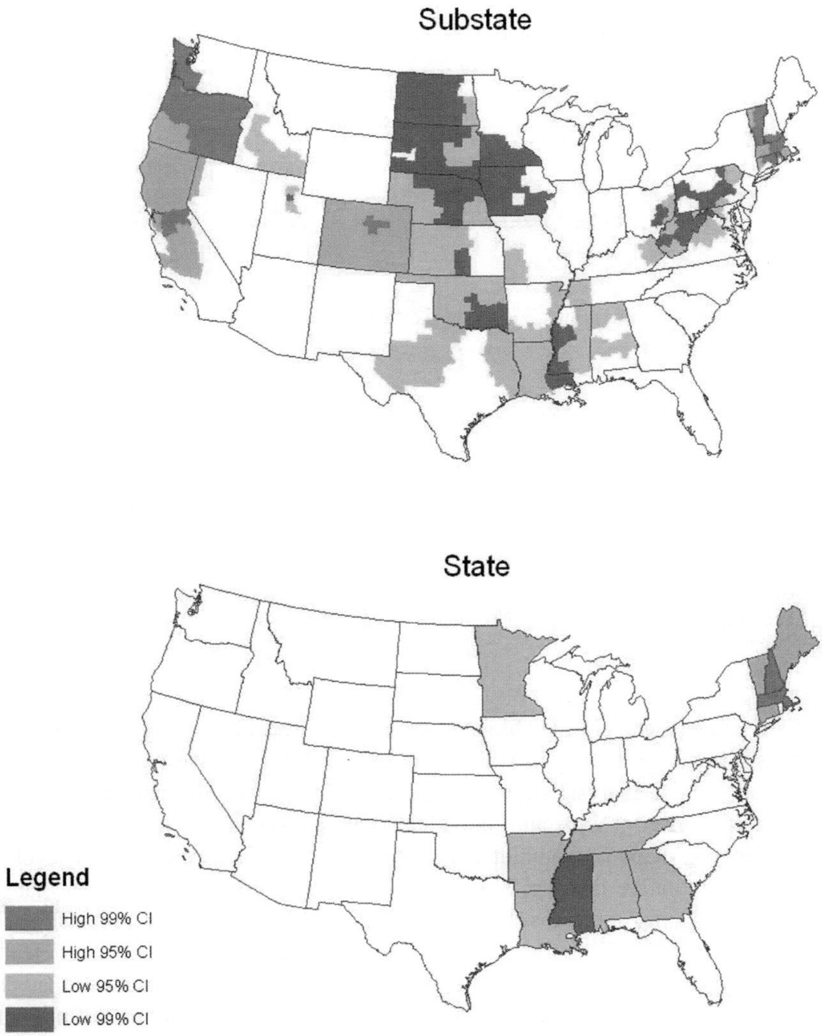

Fig. 25.13 LISA: state vs. sub-state for marijuana use (See also Plate 46 in the Colour Plate Section)

What might be the value of spatial analysis in clinical practice?

First, spatial analysis and resulting maps are of practical value in their ability to cut through a blizzard of medical information so that one can quickly see the important patterns, trends, and anomalies relevant to the area where one lives and practices. Absent such data, physicians may acknowledge that SUDs are a problem, but perhaps not a problem for their patients or in their geography. Spatial data helps make otherwise abstract data real by tying it to a place. For example, professional training

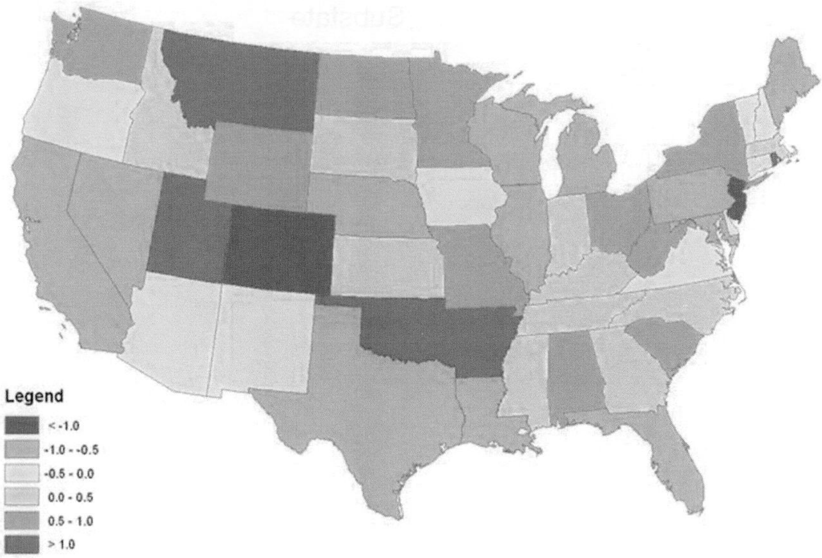

Fig. 25.14 Map of residuals of the linear regression model of SMI (See also Plate 47 in the Colour Plate Section)

helps prepare you for the possibility of amphetamine abuse in a patient with aggressive behavior, anxiety, and elevated blood pressure, but it would not tell you, as maps do, this possibility is most likely in Maine, Wisconsin, Idaho, and West Virginia.

Second, spatial analysis allows for the creation of risk information relevant to teens, multicultural groups, and individuals who possess a low level of health literacy. Because maps communicate visually, they do not require a high level of verbal or quantitative literacy. This is an increasingly critical point given that an estimated 37% of Americans have a level of health literacy that prevents them from understanding all but the most basic information (National Center for Health Statistics 2003). These maps might be used as educational and diagnostic aids as maps charting substance abuse hot spots help to identify and define the problem without pointing a finger at an individual. Such maps might be used to help initiate discussions about substance abuse problems that are locally prevalent.

Third, maps might also be used to help departments of public health and employers more selectively to target the substance abuse issues most relevant to their work force. For example, while employers in states bordering Mexico are more likely to face elevated rates of cocaine abuse relative to other states, employers in the Midwest are more likely to encounter elevated rates of binge drinking.

Conclusion

This chapter illustrates the relevance of GIS and spatial analysis in describing and analyzing a serious health problem in the US. Both global and local Moran's I was estimated for this research. While global Moran's I addressed the question what is the extent and significance of clustering in the total area, local Moran's I was used to determine the local component. Local Moran's I identifies the local clusters of SUDs (high–high or low–low). Sensitivity analysis was used to assess the stability of clusters dependent on both the number of permutations and the significance level. The resulting maps show SUDs at the sub-state level, for *any illicit drug, illicit drugs other than marijuana, marijuana, cocaine, binge drinking, and cigarettes*. In addition, LISA maps for two time periods, 1999–2000 and 2002–2004, are compared to understand the changes in the spatio-temporal pattern of SUDs in the United States, especially, movement of clusters as well as appearance and disappearance of clusters. For example, marijuana abuse or dependence migrates into New York state, New Hampshire, and Maine in 2002–2004 period from Vermont, Western Mass, and Southern New Hampshire in 1999–2000. Similarly, marijuana abuse or dependence also migrates from Northern to Southern California. While an "outbreak" of marijuana abuse emerges in Northern Montana and eastern Wisconsin, the previous hot spot located in Colorado disappears. Such spatio-temporal trends are useful in understanding the phenomena and patterns of drug abuse in the US as well as making predictive models of where the hot spots would emerge.

This chapter presents results of a shared research agenda between a physician and a geographer and describes spatial analysis relevant to mapping substance abuse disorders and SMI as well patterns of co-morbidity between the two. GIS and spatial analysis potentially offers health-care practitioners new tools that can be used to identify SUDs that are most prevalent in their region. From a practical perspective, this research illustrates that spatial analysis and resulting maps are of practical value to physicians and health-care workers in simplifying and presenting medical data to identify important patterns, trends, and anomalies relevant to the area where one lives and practices. Maps are also useful tools for communicating important risk information to individuals who possess a low level of health literacy. Such research may also be useful for formulating and implementing public health planning and policies. In addition, the research highlights the need for data integration by state and federal agencies of sub-state or finer level spatial data when available to address MAUP problems related to scale. Future research might consider other data relevant to provider and patient decision-making. This might include the spatial distribution on medication adherence or diagnostic data. Such data integration and further analysis may provide health-care practitioners better analytical tools to target and improve care in areas of risk.

Chapter 26
Reconceptualizing Sociogeographic Context for the Study of Drug Use, Abuse, and Addiction

Mei-Po Kwan, Ruth D. Peterson, Christopher R. Browning, Lori A. Burrington, Catherine A. Calder and Lauren J. Krivo

Abstract Problem drug use, including abuse and addiction, are public health concerns that have wide-ranging social consequences. Among the many social factors identified as relevant for illicit drug use and abuse, community context remains relatively understudied by comparison with individual, family, and peer risk factors. Yet, the etiology of drug abuse points to characteristics of individuals that tend to cluster within disadvantaged neighborhood contexts (e.g., poverty, single-parent families, and early childhood behavioral problems). In this chapter, we propose a new conceptualization of sociogeographic context for analyzing the potentially complex relationships between contextual risk factors and drug use, abuse, and addiction. Our conceptualization goes beyond the conventional notion of local context as comprised of static neighborhood conditions to encompass dynamic patterns of movement of local residents and non-residents across time and space that affect individual behaviors in significant ways. We suggest that some types of individual and community spatio-temporal use patterns may contribute to problematic drug behaviors because they generate higher levels of social isolation.

Introduction

Problem drug use, including abuse and addiction, are public health concerns that have wide-ranging social consequences (ONDCP 2001; Robert Wood Johnson Foundation 2001). Among persons age 12 and over in 2005, 19.7 million (8.1 percent of the US population in this age range) had recently used an illicit drug (SAMHSA 2006). Marijuana is the most widely used such drug, but significant numbers engage in the use of drugs such as cocaine (2.4 million), hallucinogens (1.1 million), and non-medical psychotherapeutic drugs (6.4 million). A great deal of research seeks to account for the social sources of such illicit drug use emphasizing a range of risk and protective factors. These generally fall into two broad

M.-P. Kwan
Department of Geography, The Ohio State University, Columbus, OH 43210-1361, USA. Tel.: 614-292-9465; Fax: 614-292-6213
e-mail: kwan.8@osu.edu

Y.F. Thomas et al. (eds.), *Geography and Drug Addiction*,

categories, those that lie within individuals and their interpersonal environments (e.g., attitudes, peer groups, family), and those that reflect societal contextual conditions (e.g., laws and norms, economic deprivation, neighborhood disorganization; see Hawkins, Catalano, and Miller 1992; NIDA 2004, 2007).

Among the many social factors identified as relevant for illicit drug use and abuse, community context remains relatively understudied by comparison with individual, family, and peer risk factors. Yet, the etiology of drug abuse points to characteristics of individuals that tend to cluster within disadvantaged neighborhood contexts (e.g., poverty, single-parent families, and early childhood behavioral problems). In this chapter, we seek to contribute to research on drug use by proposing a new conceptualization of sociogeographic context for analyzing the potentially complex relationships between contextual risk factors and drug use, abuse, and addiction. However, our conceptualization goes beyond the conventional notion of local context as comprised of static neighborhood conditions to encompass dynamic patterns of movement of local residents and non-residents across time and space that affect individual behaviors in significant ways. We suggest that some types of individual and community spatio-temporal use patterns may contribute to problematic drug behaviors because they generate higher levels of social isolation.

A New Concept of Sociogeographic Context

Although not a common approach in studies of drug activity, considerable research has sought to examine the influence of social and geographic environments on other problem behaviors (e.g., crime, adolescent sexual activity). In conducting such work, scholars have focused on the neighborhood of residence as the most relevant context affecting behavioral outcomes. The typical analytic approach examines residential neighborhood characteristics such as poverty, assuming that neighborhood effects operate through connections that exist among those residing in the same area. This approach has major limitations for producing a thorough understanding of how the dynamics of context operate to encourage or limit problem behaviors. It assumes that neighborhood *geographic* contexts are equivalent to the significant *social* contexts affecting people without assessing where individuals actually spend time while engaged in daily activities. Specifically, this approach ignores more fluid and dynamic aspects of the impact of sociogeographic contexts including: (1) how much time people actually spend in their residential communities; (2) where else people go, how much time they spend there, and what activities individuals are involved in when they travel outside of their neighborhoods; (3) what types of areas other residents or peers travel to and how prevalent and time-extensive these extra-community activities are; and (4) what types of non-residents regularly spend time within the borders of a given local area and what activities they are engaged in while there.

All of these aspects of people's uses of and movement across space and time comprise the dynamic social character of geographically-based communities – dynamics whose patterns and consequences have been undertheorized and rarely examined empirically. We argue that human social action does not take place at one time point and wholly within neighborhoods of residence; nor are neighborhood environments stable and comprised only of residents. They also include those who regularly visit for work, child care, health care, and/or informal socializing. Further, the movement and activities of residents and non-residents change over time. Our new conceptualization of sociogeographic context incorporates these dynamics to expand our understanding of how individuals and communities are situated in space and time in ways that affect the likelihood of problem drug behaviors. This approach facilitates research that integrates sophisticated theoretical and methodological tools from sociology, geography, and statistics to deepen the understanding of contextual influences on drug use, abuse, and addiction.

Neighborhoods as Social Contexts: Limitations of Past Research

A large body of research investigates the sources of problematic drug behaviors among adolescents and adults (see reviews by Hawkins et al. 1992; Galea, Nandi, and Vlahov 2004). This literature has long focused on how individuals' own characteristics and those of their families and peers affect substance use outcomes. Recently, studies have begun to emphasize the additional impact of the local context, arguing that neighborhoods possess properties that shape residents' options and thereby contribute to more or less problematic behavior (Galea, Rudenstine, and Vlahov 2005; Hawkins, Van Horn, and Arthur 2004; Sampson, Morenoff, and Gannon-Rowley 2002). While extant drug research provides important insights into the individual and structural sources of various types of problem drug use, the research discussed below does not fully capture the dynamic ways in which people are connected across space and time to diverse types of contexts that may have implications for drug use, abuse, and addiction.

To date, the body of literature examining neighborhoods as social contexts for illicit drug use and related behaviors is relatively small. Moreover, this work fails to demonstrate a consistent effect of community characteristics on illicit drug use. Rather, the findings are contingent on: (1) the aspect of drug use under consideration (e.g., actual drug use versus perceived drug use; marijuana use versus other types of illicit drugs); (2) the particular neighborhood characteristic being examined (e.g., census-defined measures of neighborhood disadvantage, survey-based measures of neighborhood features, or investigator-defined community types); and (3) the specific nature of the effect being proposed (e.g., direct versus indirect). Several studies indicate that differences in neighborhood disadvantage (variously defined in terms of one or more census characteristics such as poverty status, levels of joblessness, or as an index capturing multiple community features) provide a part of the explanation for higher levels of illicit drug use among adolescents and adults (Abdelrahman et al. 1998; Boardman et al. 2001; Crum, Lillie-Blanton, and Anthony 1996; Hoffman 2002). In contrast, Allison et al. (1999) and Esbensen and Huizinga (1990) found that such neighborhood disadvantage measures are not significantly associated with adolescent substance use, and Hoffman's (2002) research indicates that high poverty areas have less adolescent drug use. Relatedly, Simons et al. (1996) found that community disadvantage does not directly affect substance abuse among boys, but this factor has an indirect effect on such behavior through its association with the quality of parenting and deviant peers. Similarly, Brook, Nomura, and Cohen (1989) demonstrated that neighborhood influences on the frequency, duration, and severity of drug use are mediated by family and peer factors. Some analysts have used census-based measures of residential instability and population composition (percent Black, percent Latino, percent foreign born) to assess the effects of neighborhood context on drug use. Frank,Cerda, and Rendon (2007) found that Black concentration affects current and frequent drug use, but such use is not affected by Latino or immigrant concentration. Beyond census-based neighborhood measures, analysts have assessed the effects of local social organization such as perceived social cohesion and neighborhood disorder on problematic drug behaviors. Adolescent drug use is negatively associated with perceived neighborhood cohesion (Duncan, Duncan and Strycker 2002) and positively associated with perceived neighborhood disorder (Jang and Johnson 2001), although the latter effect is partially mediated by social learning and bonding variables.

A larger body of work has focused on neighborhood influences on crime and problem behaviors other than drug use. These studies should be informative for the investigation of the sources of illicit drug activities. This literature demonstrates a consistent effect of community disadvantages, including poverty, in increasing individual criminal victimization and offending (Morenoff, Sampson, and Raudenbush 2001), adolescent delinquency (Sampson, Morenoff, and Raudenbush 2005), dropping out of school (Crane 1991), low birth weight (Morenoff 2003), early sexual initiation (Browning, Leventhal, and Brooks-Gunn 2004), and teen pregnancy (Harding 2003). Further, some studies show that differences in neighborhood disadvantage provide much of the explanation for higher levels of problem behaviors among minorities as compared to Whites (Browning, Leventhal, and Brooks-Gunn 2004; McNulty and Bellair 2003; Sampson et al. 2005). Aspects of social disorganization also have effects on various outcomes. Violent victimization and early sex, for instance, are more likely in areas that have lower levels of collective efficacy (e.g., Browning, Leventhal, and Brooks-Gunn 2005; Sampson, Raudenbush, and Earls 1997). Research examining the influence of neighborhood characteristics on aggregate crime shows strong positive effects of disadvantage on criminal involvement overall, and within racial/ethnic groups. Such findings are robust to variation in the measurement of disadvantage and crime, and to the choice of the local unit of analysis (Peterson and Krivo 2005; Sampson and Bean 2006).

Although the neighborhood literature on drug use, crime, and a variety of other problem behaviors has demonstrated the contextual embeddedness of such outcomes, as noted previously, the knowledge base suffers from significant limitations emanating from the narrow, static resident-based approach that pervades current research. In this approach, the influence of local context is considered only as the effect of social characteristics of residential neighborhoods (e.g., disadvantage or perceived disorder) on adult and teen problem outcomes. This ignores the more complex ways in which individuals move across urban space over time, and through a variety of contexts that might constrain or encourage their chances of using or abusing illicit drugs, or otherwise getting into trouble.

To address this limitation, spatial regression models that incorporate the influence of other nearby communities have become more widely applied (Anselin et al. 2000). However, these have mainly been used to examine criminal outcomes, and we are aware of no studies that have taken this approach in analyzing drug use or other related health outcomes. Moreover, spatial analytic techniques conceptualize and measure proximate community influence as a relatively simple function of spatial distance from the focal neighborhood (Browning, Feinberg, and Dietz 2004; Morenoff et al. 2001). This is unfortunate because the effects of neighborhoods are more likely to result from a combination of the aggregate social and economic character of area residents, the relative number and types of people who visit the area on a regular basis, the purpose for which outsiders visit, the extent to which individuals leave their local residential areas, the social and economic character of the communities residents visit, the purposes for which people visit other areas, and the relative amount of time spent in their own and other neighborhoods. Capturing this fluid reality requires moving to a dynamic approach for conceptualizing and measuring local contextual effects on behavior.

Research in geography and transportation science has attempted to characterize people's activity-travel patterns that are generated as people move across space and time in urban environments. Doing so reorients research from a focus on static residential environments alone to a more dynamic way of considering the contexts in which people operate. However, this literature has generally simply measured spatial patterns and examined their relationships with individual characteristics (e.g., age, gender, income, etc.) (e.g., Lu and Pas 1999; Saxena and Mokhtarian 1997), rather than linked these patterns to behavioral outcomes (e.g., Golob 2000; Pendyala et al. 1991). Further, with few exceptions (e.g., Kwan 1999a,b), neither the temporal patterns of activity that are associated with individual activity space, nor the relationship of space–time use with contextual factors (e.g., characteristics of the neighborhood) is addressed.

We propose that the study of drug use, abuse, and addiction would be enhanced by analyses that move beyond the static, resident-based understanding of the role of neighborhoods. In the following section, we discuss important elements of a more fluid and dynamic conception of community contextual effects, one that explicitly acknowledges that residents spend different amounts of time in their neighborhoods of residence, and have access to resources in areas in which they do not reside but spend considerable time. The most significant element of this contextual approach rests upon the way in which varying space-time use patterns reflect differential levels of social isolation.

Social Isolation and Problem Drug Behaviors

Social isolation is central to theoretical discussions describing the mechanisms linking neighborhood disadvantage to a host of problem behaviors that include illicit drug use. According to this argument, areas with high concentrations of the most disadvantaged segments of the population foster problem behaviors among individuals and in the aggregate because of the substantial prevalence of persons who lack sustained contact with mainstream society (Wilson 1987). The idea underlying this notion of social isolation is that, relative to members of other groups, individuals in disadvantaged neighborhoods have less contact with resource-rich individuals and settings either inside or outside their communities. We draw on this notion but extend it to conceptualize social isolation as the degree and nature of connections to other places that people experience through their routine daily activities (e.g., going to and from work and school, caring for children, shopping). This conceptualization treats social isolation as a variable. Further, it recognizes the dynamic quality and multidimensionality of social isolation as embedded in the ways people move across space and time.

We consider social isolation as representing the extent and character of connections between a focal neighborhood and other communities as residents undertake their regular activities. Our concept of the *extent* of social isolation refers to the frequency with which residents of a focal area venture outside their own neighborhood, and correspondingly the prevalence of visitors to this focal area. By the *character* of social isolation, we mean the degree of compositional similarity among residents of a focal neighborhood, those who enter the area, and those in the areas that residents visit. This distinction makes clear that the degree to which people and areas are socially isolated is contingent upon how often people leave their own community, how much others come to their area, and the types of places to which they go in ways that are measurable and separable in their potential effects. For example, a neighborhood would be socially isolated in *extent* if on a regular basis few non-residents visit the area to work, shop, and go to school, and few residents leave to accomplish these activities in outside neighborhoods. In contrast, another neighborhood may have many exchanges of residents and visitors going in and out, but these activity flows may involve people and areas that share similar characteristics, so that residents are socially isolated in *character*. This dimension of isolation applies to both disadvantaged and advantaged contexts. That is, social isolation may be among those who are similarly disadvantaged, potentially compounding its impact, or among those who are similarly affluent with potential implications for further advantaging individuals and neighborhoods.

This specification allows us to consider social isolation as having both individual and neighborhood expressions that are theoretically and empirically distinct. On the one hand, individuals exhibit space–time activity patterns as they move through their lives in local environments. We can think about how much an individual leaves his/her neighborhood on a regular basis and whether he/she visits areas that are more or less similar to his/her neighborhood of residence when going to work, shop, or visit the doctor. On the other hand, individual activity patterns accumulate to provide aggregate contexts that are more or less connected with more or less resourced outside environments yielding a neighborhood level of social isolation. Thus, irrespective of a particular individual's patterns of activity, and hence his/her social isolation, he/she may live in an area where very few residents venture out of the community, only a few visit the area, and the visitors all have similar characteristics and come from areas that are similar to those in the neighborhood. Extending this logic, it is clear that a variety of complex combinations of individual and neighborhood social isolation in extent and character is possible and that researchers can examine their separable effects on involvement in drug use, abuse, and addiction.

For instance, focusing on aggregate levels of social isolation, we suggest three mechanisms through which neighborhood social isolation works to influence drug abuse and addiction in a particular sociogeographic context. First, social isolation may affect the strength of informal social control norms. Normative orientations regarding expectations for behavior and intervention to control the misuse of public space within communities have been shown to be important determinants of crime in urban contexts (Sampson et al. 1997), and may have implications for the prevalence of drug use and abuse. Shared community behavioral standards may lead both to reinforcement of socialization practices that discourage drug use and intervention to control overt manifestations of drug activity (e.g., quasi-public or "open air" drug markets). Second, the extent of movement into and through neighborhoods results in distinct social ecologies that have implications for the occurrence of a variety of problem behaviors, including drug use and related activities. Communities that maintain active streets with pedestrians engaged in conventionally-oriented activity will also provide fewer opportunities for the emergence of drug markets and associated drug abuse and addiction. Third, cultural transmission processes disseminate behavioral orientations within communities through social learning of action patterns. Exposure to behaviors and their perceived rewards and costs is likely to be significantly influenced by the extent and character of social isolation. For instance, exposure to non-residential neighborhood contexts where drugs are prevalent or tolerated may increase the likelihood of drug use.

Drawing on the logic of these mechanisms, hypotheses about how social isolation affects problematic drug use within urban neighborhoods can be generated and refined to reflect the complexities of contemporary society. First, an extremely economically disadvantaged neighborhood that is highly socially isolated in extent as reflected in the fact that its residents rarely leave the area captures a key aspect of Wilson's (1987) discussion of social isolation. Such areas, already compromised in their capacity to effect informal social control, do not reap the benefits of extra-neighborhood resources and exposure to conventional sources of influence outside of the neighborhood. Thus, informal social control norms and their implementation are likely to be imperiled in disadvantaged communities where the extent of social isolation is marked by significant residential confinement. In turn, the aggregate and individual likelihood of illicit drug activity will increase.

Additionally, the conceptualization provided here suggests potential effects on drug activities of the extent of social isolation evident in extra-neighborhood residents' patterns of travel into a focal neighborhood. Here our reference is to the social ecological and routine activity consequences of travel patterns *into* a neighborhood. Active streets traversed for conventional purposes yield deterrent effects on the misuse of public space (including public substance abuse and other drug activities) above and beyond the influence of residents' normative orientations. Dense social ecologies fostered by the incorporation of a disadvantaged neighborhood into the activity spaces of residents of other areas may add a source of guardianship not captured by resident-based informal norms. However, travel into the neighborhood may yield a social control benefit only beyond a certain threshold. Infrequent travel into a focal area may produce only sparse social ecologies and limited additional monitoring of public space.

Residents of disadvantaged communities may also be highly socially isolated in character: residents spend time in other neighborhoods and receive visitors from other areas, but these other communities are demographically similar in character to the focal neighborhood. In this case, the social characteristics of origin and destination neighborhoods may combine to influence behavioral tendencies. For example, similarly disadvantaged neighborhoods may help foster illicit drug use among residents of each area through introducing a wider network of opportunities for acquiring drugs and by enhancing the cultural transmission of attitudes favorable to their use. Mears and Bhati (2006) make this same point regarding the spread of criminal involvement.

Disadvantaged neighborhoods characterized by low levels of social isolation may experience benefits that help buffer the impact of internal structural deficits. These neighborhoods will have residents who move freely to other areas that are demographically different in character from their own, and may receive middle class visitors for such activities as work and church attendance, partially addressing the deficit in contact with mainstream society. If so, we would expect residents of the focal neighborhood to experience less unconventional behavior, and to exhibit more positive individual outcomes. Having additional people in the area should increase the number of "guardians" on the street making it more difficult for drug exchanges to take place. This may, however, be tempered by the extent to which visitors have a vested interest in the focal neighborhood. For example, those who work in the neighborhood may have greater interest in informal and formal social control than those who visit to shop. Adult residents of the focal neighborhood who spend time in settings with higher levels of employment or where residents have access to more resources, may be able to forge social or political connections that could be called upon to benefit the focal neighborhood and its residents (Coleman 1990; Sampson 2003). They may bring opportunities that reduce reliance on drug trafficking as a source of income and, in turn, reduce drug use, abuse, and addiction. Also, residents may learn about drug prevention and treatment facilities or develop the means to bring such services to their neighborhood. The result would be relatively fewer drug problems than in disadvantaged areas that are more socially isolated in extent or character.

There is less theorizing about the role of social isolation for residents of economically advantaged communities. There are a variety of possibilities. For example, it is reasonable to anticipate that social isolation (in character) would benefit advantaged communities through access to resources, reinforcing positive normative values (anti-drug use orientations), and heightened levels of social organization of other resource-rich communities. Yet, it is also possible that in combination with resources for the purchase of drugs, such isolation will enhance the likelihood that drug problems remain in the private sphere, going undetected, and thereby unaddressed, escalating the extent of the drug problem in the community. Alternatively, contact of residents of advantaged communities with disadvantaged communities could mean exposure of the former to drugs and associated vagaries that might not be readily available in their communities.

Finally, the potentially differential effects of social isolation at different stages of the life course needs to be acknowledged. First, age may condition the magnitude of the effect of social isolation. Socialization and learning, mechanisms through which social isolation is hypothesized to influence behavior, may be more consequential for adolescents than for adults, resulting in stronger effects of isolation for the former group. However, among youth, very young adolescents (ages between 11 and 12) may spend less time in activity spaces beyond the home as a function of their generally lower levels of autonomy and more frequent supervision by parents and caretakers. Their more restricted activity patterns suggest more limited effects of social isolation on the youngest adolescents. Second, age may condition the type of outcome with which social isolation is associated. Older adults may be less likely to experience social isolation effects on behaviors such as drug use and arrest. Evidence suggests that adults at older ages who exhibit continuity in problem behavior may be expressing "life-course persistent" inclinations that are relatively less elastic to context (Moffitt 1993).

Conclusion

Extant research on drug abuse and addiction has largely neglected the etiological significance of community context. We suggest that a focus on contextual factors may add significant explanatory power to current models of drug abuse and addiction. However, the prevailing approach to neighborhood effects research remains hampered by a static, resident-based orientation that assumes constant levels of exposure to neighborhood environments across individuals and communities.

We have described in this chapter a new conception of sociogeographic context, elaborating the concept of social isolation as a useful tool for understanding contextual influences on drug abuse and addiction. With this new notion of context, we seek to move to a dynamic perspective on the influence of social and geographic space, acknowledging the fluid and complex nature of spatial influence. Individuals spend varying amounts of time in their neighborhoods of residence, resulting in both individual and, crucially, aggregate (neighborhood level) mobility patterns with potentially significant consequences for drug-related behavior. For instance, individuals who live in structurally disadvantaged *and* socially isolated neighborhoods may be significantly more likely to experience drug abuse and addiction than individuals in similarly disadvantaged communities with substantial access to external resources.

Recent neighborhood-based data collection efforts provide an unprecedented opportunity to operationalize the concept of social isolation (e.g., using geocoded information on activity patterns at both the individual and neighborhood levels). In combination with sophisticated innovations in GIS-based tools for examining complex patterns of space–time use, this perspective has the potential to significantly advance our understanding of crime, drug abuse, and other problem behaviors

Acknowledgment This work was supported by a grant from the National Science Foundation's Sociology and Law and Social Science Programs (SES-0528232). We would like to thank Reginald A. Byron and Jae-Yong Lee for their capable assistance.

Chapter 27
Spatial Analytic Approaches to Explaining the Trends and Patterns of Drug Overdose Deaths

Charlie DiMaggio, Angela Bucciarelli, Kenneth J. Tardiff, David Vlahov and Sandro Galea

Abstract To effectively utilize and interpret spatial analyses, substance use researchers, public health practitioners and policy makers should be familiar with some of the available data analytic techniques, each of which comes with advantages and drawbacks. In this chapter we first discuss three cluster detection tools and their associated software applications. We then present a Bayesian hierarchical approach, briefly reviewing its theoretical underpinnings, commonly used models, and how inferences may be drawn from a sample-based posterior distribution. We demonstrate the use of each approach on a set of substance abuse mortality data, comparing the results across the four tools. Our empiric illustration considers the role of neighborhood-level socioeconomic status (SES) in explaining opiate-related overdose deaths in New York City. We end with a discussion of the implications of the choice of technique and software on interpreting spatial analyses of substance abuse and conclude that the choice of a method will be driven by the question to be answered, data and software availability and the intended audience or context in which the research is being conducted.

Introduction

Mapping techniques and spatial analysis have been used in a number of studies seeking to describe and analyze substance abuse. Spatial analytic studies have demonstrated the correlation of drug use to deprivation indices (Squires, Bleeching, Schlecht, and Ruben 1995); the role social networks play in urban adolescent substance abuse (Mason, Cheung, and Walker 2004); the effect of ecologic level

S. Galea
Department of Epidemiology, University of Michigan, 1214 S. University, Room 243, Ann Arbor, MI 48104, USA
e-mail: sgalea@umich.edu

Y.F. Thomas et al. (eds.), *Geography and Drug Addiction*,

variables, such as legal prohibitions against alcohol sales (Schulte, Aultman-Hall, McCourt, and Stamatiadis 2003); and whether frequency and type of drug use are geographically located independent of neighborhood characteristics (Latkin, Glass, and Duncan 1998).

There are a number of spatial analytic tools available to epidemiologists, each having advantages as well as drawbacks. To effectively utilize and interpret spatial analyses of substance abuse, researchers, public health practitioners, and policy makers should be familiar with some of the available data analytic techniques.

In this chapter, we first discuss three cluster detection tools and their associated software applications: nearest neighbor index (NNI; ESRI 2005), Ripley's K-function (Levine 2004), and a space–time and time permutation scan statistic (Kuldorf 2005). We briefly describe these techniques and then demonstrate their use on a set of substance abuse mortality data, comparing the results across the three tools. We then introduce hierarchical spatial modeling (Imperial College and Medical Research Council 2004). We will discuss the advantages and disadvantages of a Bayesian approach, some commonly used models, and how to draw inferences from the sampled posterior distribution. We will demonstrate this approach on our data set and compare the results to those we obtained with cluster detection tools. As an empiric illustration, we consider the role of neighborhood-level socioeconomic status (SES) in explaining opiate-related overdose deaths in New York City (NYC). We conclude with a discussion of the implications of the choice of software and techniques on interpreting spatial analyses of substance abuse.

Data and Variable Definitions

We manually reviewed medical files at the Office of the Chief Medical Examiner (OCME) of NYC and identified all cases of fatal accidental drug overdose occurring in the city between 2000 and 2004. The OCME is responsible for assessing all deaths of persons believed to have occurred in an unnatural manner in NYC. Therefore, all overdose deaths in NYC would have been reviewed by the OCME and included in this chart abstraction.

The OCME medico-legal investigators use the decedent's medical history, the circumstances and environment of the fatality, autopsy findings, and laboratory data in attributing the cause of death and other criteria to each case being reviewed. The variables we abstracted for our analysis included the decedents' gender, age, address of residence, and location of injury. We geocoded residential and injury locations using ArcGIS, version 9.1 (ESRI 2005). For analysis purposes, place of injury (location of death) was used. In the analysis, only decedents who were successfully assigned an address of injury were included. Overdose deaths may include more than one drug being present; to increase the reliability of our measures across the different analytic techniques, we restricted our analyses to cases in which opiates were the only (in the case of scan statistics) or primary cause of death.

Table 27.1 Demographic characteristics of successfully geocoded opiate-related drug overdose deaths, New York city 2000–2004

	No.	%
Total	2426	
Sex		
Male	1883	77.6
Female	543	22.4
Age		
15–24 yrs	146	6.0
25–34 yrs	431	17.8
35–44 yrs	898	37.0
45–54 yrs	762	31.4
55–64 yrs	162	6.7
65–74 yrs	25	1.0
Over 74 yrs	1	0.0
Race/ethnicity		
White	1069	44.1
Black/not Hispanic	560	23.1
Hispanic	776	32.0
Asian	11	0.5
Other	10	0.4
Year of death		
2000	543	22.4
2001	455	18.8
2002	484	20.0
2003	506	20.9
2004	438	18.1

Descriptive Epidemiology

From 2000 through 2004, the OCME reported 3982 fatal overdose deaths within NYC. Of these, 3777 occurred among NYC residents, in which 2516 were determined to have opiate toxicity as the primary cause of death. Together, 96.4% (2426 out of 2516) cases were successfully geocoded. These cases constituted the study base for our subsequent analyses. Their demographic characteristics are presented in Table 27.1.

Cluster Detection Techniques

We conducted cluster analyses for all opiate-related deaths. We first described cross-sectional spatial distribution of all fatal opiate-related deaths in NYC that occurred between 2000 and 2004 using an average NNI statistic. We then utilized an L function transformation (derived from Ripley's K-function) to produce graphs to assess at what distance the clustering (if present) was observed to be the greatest for each year. We then used a space–time permutation model to assess both the spatial and temporal clustering opiate-related overdose deaths. This approach uses count (case) data only, and assesses not only spatial clustering characteristics but also the role of time as a variable over the 5-year period. In this way, we determined if any clusters were statistically significant when adjusting for the year in which the deaths occurred. Finally, we applied a space–time scan statistic that differs from the permutation model in that it also adjusts for the underlying population at the census tract level.

First-order Clustering Technique: Average Nearest Neighbor Index

Description

The global presence or absence of clustered overdose incidences can be assessed using the average NNI. This index is a measure of how similar the mean distance of all cases is to the expected mean distance for a hypothetical random distribution (Mitchell 2005). The equation for calculating the average NNI is expressed as (Mitchell 2005):

$$d = \left(\frac{\sum_i Ci}{n}\right) - \left(\frac{0.05}{\sqrt{n/A}}\right)$$

where the average NNI (d) is equal to the summed distance to each feature nearest neighbor ($\sum_i Ci$) divided by the number of features (n) or the 'observed distribution' of mean features minus the product of 0.05 divided by the square root of the number of features (n) divided by the study area (A) or the 'expected mean distance for a random distribution' (Clark and Evans 1954). Clustering is suggested when the observed average distance is greater than the mean random distance ($d < 1$). An index value close to 1 indicates randomness, while a value greater than 1 indicates dispersion of cases. Within ArcGIS, version 9.1, tests of significance (a z-score and p-value) are included with the NNI output. If the z-score is negative, this suggests the cases are clustered. Conversely, if the z-score is positive, this suggests the cases are dispersed, while a value close to zero indicates the random nature of cases.

Application

Table 27.2 displays the average NNI results of all opiate-related drug overdose deaths. The results suggest that the greatest clustering of any overdose occurring in NYC was in 2000 (NNI = 0.7636; z-score $= -10.5474$). Given the statistically significant negative z-score values of all 5 years, individually, opiate-related drug overdose demonstrates significant clustering for all years between 2000 and 2005.

Table 27.2 Average nearest neighbor analysis of opiate-related drug overdose deaths, New York city, 2000–2004*

Year of Death	Average Nearest Neighbor Ratio	z-score (Standard Deviations)	p-value
2000–2004	0.7023	−27.8189	<0.0001
2000	0.7636	−10.5474	<0.0001
2001	0.7216	−11.2869	<0.0001
2002	0.7659	−9.9835	<0.0001
2003	0.8235	−7.1310	<0.0001

*Based on weighted counts of injury location and direct distance measurement.

Moreover, clustering became less dense toward the later years of the study period as displayed by the gradually increasing z-scores.

Second-order Clustering Technique: Ripley K-Function (L-Transformation) Statistic

Description

While the average NNI considers only the distance between one case and its nearest other case, the Ripley's K-function statistic is a second-order statistic that considers the complete distribution of all distances in the point pattern of cases (Levine 2004). It tests the cumulative distribution function of the entire set of inter-point distances among the point data. When K statistics are transformed into a square root function, the result is called a L function transformation (L(d)). The square-root transformation results in a linear function. This statistic can be very useful when exploring the nature, in terms of distance, of the case clustering within the entire study area. L function equation is expressed as (Levine 2004):

$$L(d) = \sqrt{\frac{A \sum_{i \neq j} \sum I_{ij}\, d_{ij}}{\pi n(n-1)}}$$

The numerator is the Ripley's K-function, where the distance (d) is measured between case (i) and every other case (j); then each distance is multiplied by the weight for the case paring (I_{ij}), and all the values are summed ($i \neq j$ indicates the distance between cases) are not included in the sum (Levine 2004). Finally, the result is multiplied by the study area (A) and divided by the number of cases (n) squared. The denominator is π multiplied by the number of possible case pairings (represented as $n-1$). The square root of the product is then taken. At any given distance (represented by the x-axis of the result graph), if the line of observed L values is above that of the expected values [L = 0 or complete spatial randomness (CSR)], then the cases are more clustered than expected for a random distribution (with the peak of the graph representing the greatest clustering detected at a specified distance) (Levine 2004). Once the curve falls below the CSR line, cases at that point become dispersed at a given distance.

To test the null hypothesis of global spatial randomness of opiate-related overdose for the period of interest, we computed a 95% CI (referred to as the envelope) of the L-function (L(*d*)) using a Monte Carlo method of specifying 100 simulated random patterns (Levine 2004). At a given distance (represented on the *x*-axis), a value of L(*d*) (represented on the *y*-axis) outside the confidence interval (CI) envelope is interpreted as a significant departure from CSR toward clustering or dispersion. When the function peaks at the largest, most positive value and remains outside the CI envelope, this is considered to be the distance at which cases tend to be the most clustered.

Application

Figure 27.1 presents a graph of the L-function statistic for the entire 2000–2004 time period, and suggests that most clusters of opiate-related overdose are fairly compact and that the greatest clustering occurs at distance of approximately 12,007.34 feet or 2.27 miles [distance at which L(*d*) peaks in the output]. This suggests that most clusters occur with a radius of approximately 2.25 miles. Thereafter, the cases become more dispersed. The L(*d*) curve also remains outside the95% confidence envelope and, therefore, remains statistically significant. Additional curves for individual

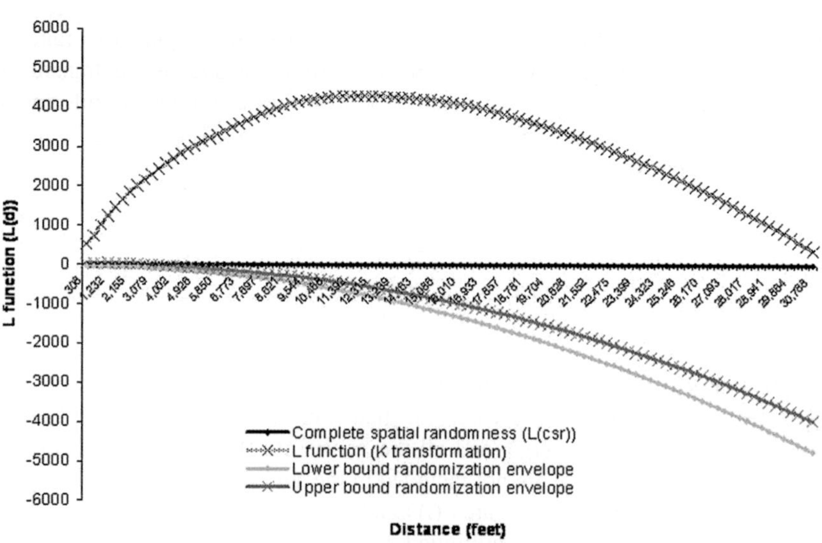

Fig. 27.1 Graph of L-function statistic. New York City opiate-related overdose deaths, 2000–2004

years (not presented) are also clustered at approximately 2–3 miles, with the tightest clusters observed to be in 2000 (10,775.82 feet or 2.27 miles).

Space–Time Modeling Techniques: Space–Time Permutation Statistic and Space–Time Scan Statistic (SaTScan)

Description

Pure spatial analyses, e.g., NNI and Ripley's K, are useful when exploring cross-sectional health outcomes. When the variable of 'time' (in units of hours, days, months, years, etc.) is of interest, we will need a model that assesses the trend of the outcome over both space and time. We are interested in whether the same areas experience clustering year after year, asking: Are the cases clustered and, if so, do they continue to cluster over time given the nature of the study area?

The space–time permutation scan statistic model uses only case data. There is no requirement for specifying the underlying population data. It makes minimal assumptions about the time, geographic location, or size of the potential case clustering. The model adjusts for what is termed as 'purely' spatial and temporal variation in the case data for a given area (Kulldorff, Heffernan, Hartman, Assuncao, and Mostashari 2005). Using a Poisson-based probability model, a series of overlapping scanning windows (cylindrical in form) move across the spatial plane (the base of the cylinder) while also scanning the point data for temporal clusters (the height of the cylinder). The circular base represents the geographical area or the study area while the height of the cylinder scans for time (in days, months, or years) clustering. For each location, the scanning window calculates the number of observed and expected cases. The statistical significance of an observed 'cluster' is then evaluated taking into account the multiple testing methods (0, 9, or 999 Monte Carlo replications). For each center and radius of the cylinder base, the method iterates over all possible temporal cylinder lengths. Cylinders can be geographically large and temporally short (forming a flat disc), or can be geographically small and temporally long (forming a pole), or any combination in between. The number of observed cases is divided by the calculated expected number of cases for each cylinder to the power of the observed inside the cylinder, and then multiplied by the observed, divided by the expected to the power of the observed outside the cylinder. The approximation, a Poisson generalized likelihood ratio, is expressed as (Kulldorff et al. 2005):

$$\left(\frac{C_A}{\mu_A}\right)^{C_A}\left(\frac{C-C_A}{C-\mu_A}\right)^{(C-C_A)}$$

where C is the total number of observed cases, μ_A represents the mean number of expected number of cases within the cylinder, and C_A represents the observed number of cases within the cylinder.

The space–time scan statistic is also based on Poisson modeling, just as the space–time permutation model, but allows for scanning of purely spatial, purely temporal, and special temporal clusters.

These models are most readily applied using the SaTScan software package, which is available for download (Kuldorf 2005) after registration and can be translated in ArcGIS for viewing of the cluster statistics.

Application

The space–time permutation scan statistic of opiate-related overdose was mapped to give a visual display of the model output (see Fig. 27.2a). In terms of spatiotemporal clustering for opiate-related overdose citywide, seven clusters were detected. The primary cluster was detected in the northwestern portion of Manhattan and the southern region of Bronx in 2001, while the other six secondary clusters were located in various parts of the city. It is notable that none of the clusters reached a level of statistical significance.

The majority of opiate-related drug clusters seem to have occurred in 2000, although the primary cluster in northern Manhattan/South Bronx was detected in 2001. It is suggested that opiate-related fatal overdose is not only changing incidence pattern, but also such cases are decreasing.

In our comparison space–time scan statistic model, we attempted to see how clusters may change when adjusting for the underlying population counts. In this analysis, we used population counts at the census tract level for the entire NYC area for a finer and more exact population adjustment. Using a small neighborhood unit of population count adjustments allows for a finer resolution of cluster detection.

The space–time scan statistic detected fewer clusters, all of which were statistically significant (Fig. 27.2b). What remains as the primary cluster (just as with the space–time permutation model cluster map) is the one located in the northern region of Manhattan and the southern region of Bronx. Compared to the space–time permutation cluster map, the central radius point of the cluster is shifted slightly west (encompassing more of Manhattan than Bronx), the radius of the cluster is approximately one mile larger (radius = 3.596 miles as compared to 2.664 miles), and the year in which cases were significantly detected was 2002–2003 as compared to 2001 for the space–time permutation model.

There are additional differences between the two approaches. A cluster in lower Manhattan detected by both methods is somewhat larger in the space–time scan, which additionally detected higher than expected cases in both 2000 and 2001 as compared to only 2000 in the space–time permutation scan.

Bayesian Hierarchical Models

Bayesian hierarchical modeling is frequently used in spatial epidemiological analyses, but may be unfamiliar to some substance abuse researchers. Multilevel spatial modeling, though, has been used to capture context in community studies

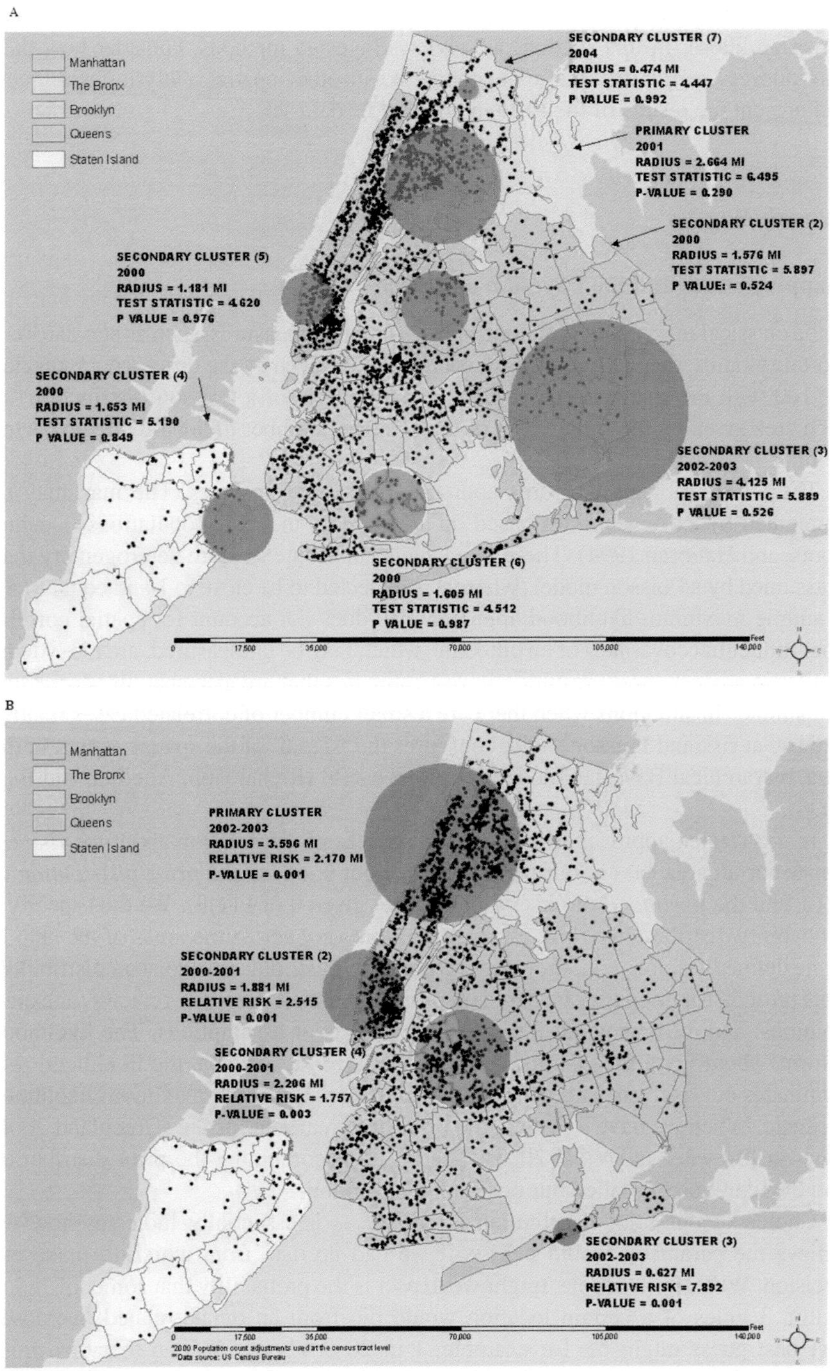

Fig. 27.2 Opiate-related deaths, New York City, 200–2004. (a) Space-time permutation scan indicating 7 clusters (b) Controlling for underlying population count (See also Plate 48 in the Colour Plate Section)

of substance abuse (Luke 2005) and in studies of drug-related crime (Law and Bayesian 2004). In this section, we review Bayesian methods, consider how they may address certain difficulties encountered in other spatial analytic techniques, and present the results of their use in our sample data set.

Description

Mapping issues and the Bayesian approach

In the classical maximum likelihood approach to risk measures, such as standardized mortality ratios, the risk estimate for each area j is given by the observed j/expected j * 100 with the standard error under an assumption of a Poisson distribution for each area given by the square root of the observed number divided by the expected number.

There are problems with this approach for spatial analyses. The map may be dominated by extreme values based on a few cases in small populations (Devine, Louis, and Halloran 1994). These rare events contribute to more heterogeneity than is assumed by a Poisson model (where μ is expected to be close to 1 and equal to S). A simple maximum likelihood approach also does not account for spatial correlation. Influential covariates of an outcome, which may be unmeasured, are likely to be similar in adjacent areas resulting in risk estimates that are also spatially correlated and similar. In situations when there are a small number of correlated cases relative to those at risk and Poisson 'noise' obscures the 'signal' of the spatial pattern in the data, hierarchical Bayesian modeling can be useful (Richardson, Abellan, and Best 2006).

In a Bayesian approach, our two main sources of information about the risk estimate for an area (θ) are our prior beliefs about θ, called the *prior distribution* or $p(\theta)$, and the *likelihood* of observing our data given θ or $L(y|\theta)$. We thus specify a probability distribution of risk estimates (θ) that vary across the areas of the map in some defined fashion, e.g., they may be normally distributed or Poisson distributed. This prior distribution may be based on previous studies, literature reviews, or expert opinions, and informs about θ through our beliefs or assumptions. The likelihood informs about θ via the data itself. When we have lots of data, the likelihood predominates our analysis, and our results will essentially be the maximum likelihood estimate. When we have less data, the prior has greater influence (Greenland 2006; Lawson, Browne, and Vidal 2003). The result of combining the prior distribution and the likelihood is called our posterior distribution.

Choice of our prior distribution is critical as it essentially indicates how we believe the parameter would behave if we had no data from which to make our decision. What, for example, might we expect is the probability that someone living within 3 miles of a certain location would die from an opiate-related overdose? Our best guess might be, for example, 1 in 20 or about 5%, and that this probability varies around this point estimate in a normal fashion with a variance of, say, 0.01 or 1%. This estimate may be based on previous studies, law enforcement data, clinical experience, or a combination of sources. What if we conduct a study that indicates the risk of an opiate overdose within 3 miles of the location is 45%? How likely is our observed data given our postulated prior probability? Our posterior distribution combines our expectation with what we actually observe. In a very common sense way, it tells us, for example, that if the results of our study differ markedly from our best existing information, we should perhaps be somewhat skeptical.

The approach is hierarchical or mixed because we specify a distribution of *hyperparameters* (λ) for our risk parameter, θ allowing it to vary across each area. One could, for example, say that y_i is the empirical (observed) rate of substance abuse-related deaths in zip code i, θ is the true underlying rate, and λ how that true rate varies (Banerjee, Carlin, and Gelfand 2004).

As noted, the posterior distribution ($\Pr[\theta|y]$) is based on our prior assumptions and our observed data. It follows Bayes' theorem and is proportional to the likelihood times the prior: (Greenland 2006)

$$\Pr[\theta|\mathrm{y}]\alpha\Pr[\mathrm{y}|\theta] * \Pr[\theta],$$

As described by Richardson, Abellon, and Best (2006), hierarchical Bayesian spatial models describe observed cases in an area as Poisson distributed with a mean equal to the expected number of cases times the risk for that area:

$$\mathrm{O}_i \sim \mathrm{Poisson}(\rho_i E_i).$$

At the second level of the model, the risk for each area (ρ) is transformed to a log scale (making relationships additive rather than multiplicative) and is described as an intercept term (a) and two random effects, one spatial (θ), the other non-spatial ($\boldsymbol{\Psi}$):

$$\log \rho_i = a_i + \theta_i + \psi_i$$

The spatially structured component is described as a conditional autoregressive (CAR) Gaussian process [$\theta \sim$ CAR normal (W, τ_θ)] where the conditional distribution of each θ_i, given all the other θ_i's, is normal with μ = the average θ of its neighbors and a precision (τ_θ) proportional to the number of neighbors. W represents the matrix of neighbors that defines the neighborhood structure. The non-spatial component of the model (ψ_i) is defined as normally distributed with $\mu = 0$ and precision (τ_ψ). The model is completed by assigning hyper-priors to the precision terms τ_θ and τ_ψ.

The Poisson–gamma model

This hierarchical Bayesian approach most frequently described in the mapping literature is the Poisson–gamma model. In this formulation, the risk (θ) is described as a set of parameters that may include any number of explanatory variables (Lawson et al. 2003). The prior distribution of the observed outcome y is described as $y|\theta \sim \text{Po}(\theta E)$, and the hyper-prior distribution of risk is $\theta|\acute{\alpha}, \beta \sim$ gamma $(\acute{\alpha}, \beta)$, with $\mu = \acute{\alpha}/\beta$ and $\sigma^2 = \acute{\alpha}/\beta^2$ (Banerjee et al. 2004; Lawson et al. 2003). We could further specify $\acute{\alpha}$ and β, but we assume that beyond a certain point, further model specification will have little practical effect on our results.

We commonly choose a non-informative (proper) or arbitrarily vague-prior that is uniform or 'flat' to allow the data to predominate and lead us to a posterior distribution that is dominated by the likelihood. A gamma (0.5, 0.0005) has been suggested as reasonable (Law and Bayesian 2004).

For simple models for which there is a closed form (i.e., they behave as true distributions and integrate to 1), we can estimate the posterior distribution directly via the maximum likelihood estimate, and a Bayesian approach is unnecessary. But for most reasonably realistic models, we will not be able to find a closed form and will need sample-based approaches.

Empirical Bayes methods approximate the posterior distribution (Devine et al. 1994). Full Bayes methods base inferences on a sample of the full posterior distribution. The results from such a sample are not as informative as the closed form itself, but are usually sufficient for inference. We increase the precision of our estimates by increasing the sample size (Banerjee et al. 2004).

One way to construct a sample from the posterior distribution is through Markov Chain Monte Carlo (MCMC) methods. Like a 'random walk' seen in time series analysis, the resulting series has no 'memory'. Subsequent values depend only on the current value, and the series converges to a stationary distribution assumed to be the posterior. Unlike traditional Monte Carlo methods, MCMC methods produce correlated samples, because they base subsequent values on current values. Methods such as thinning every other value may help decrease this correlation (Imperial College and Medical Research Council 2004).

Transition probabilities for selection into the series are typically determined through the use of the so-called Gibb's sampler. A special case of the Metropolis–Hastings algorithm, the Gibb's sampler generates conditional probability distributions of a parameter given all other parameters, and transition probabilities are generated that result in a proposal value that accepts or rejects the value with a probability of 1 or 0 (Lawson et al. 2003). The algorithm is useful in the context of Markov random fields where the joint posterior distribution is complicated but the full posterior prior distributions have simple forms.

As noted, the spatial Poisson process consists of two components: uncorrelated global heterogeneity ($\boldsymbol{\Psi}$), usually due to unmeasured confounders or effects throughout the data, and correlated or specific heterogeneity, due to spatial correlation or local effects (θ) (Lawson et al. 2003; Richardson et al. 2006. To capture both spatial variation and non-spatial random effects in an additive fashion, we model the natural log of risk as the sum of these two components (da Silva and Melo 2004).

Setting the correlational structure

Since we want to model the spatial components so that geographically close areas present similar risks, we use information from other areas in the region to reduce random variation unrelated to the risk represented by our risk estimate. This should take spatial correlation into account and result in smoother informative maps. Here, we see the advantage of a Bayesian approach. Modeling θ as a random variable rather than a fixed variable allows us to set a spatial correlational structure.

We can describe this structure via Markov random fields where each θ, given all the other θ's, depends only on its neighborhood. A Markov random field is a locally specified joint distribution that can be determined by its full conditionals. Given a joint distribution, $\Pr[y_i \ldots y_n]$, the set of full conditional distributions, $\Pr[y_i | y_j]$, that we can create from it are uniquely determined. Brook's lemma tells us that we can go in the opposite direction. If we have a set of full conditional distributions, we can get the unique joint distribution from which they arose (Banerjee et al. 2004).

A locally determined, weighted structure can be represented by a CAR Gaussian model where the conditional distribution of each θ is given by:

$$\theta i | \theta j \sim \mathrm{N}_1(\Sigma w_{ij}\theta_j / \Sigma \mathrm{w_{ij}},\ 1/\tau \Sigma w_{ij})$$

where j is not equal to i and is an element of d, the set of the neighbors of i (da Silva and Melo 2004).

The simplest and most commonly used definition of a neighborhood is the existence of a common border between areas. In this case, the weights are specified as $w_{ij} = 1$ if j is in d, and $w_{ij} = 0$ if j is not in d. In this case, the Σw_{ij} is simply the number of neighbors of area i. So, the conditional prior mean of θ is given by the arithmetic average of the spatial effects of its neighbors, and the conditional prior variance is proportional to the number of neighbors.

This structure has been used in a number of disease mapping studies (da Silva and Melo 2004), but other approaches are possible. We could set up a proximity matrix of weights, w_{ij}'s, based on the distance between or other relationships between the spatial units, e.g., a set of first order w_{ij}'s ($w_{ij}[1]$), if an areal unit is, say, less than some predefined distance (Banerjee et al. 2004). We could also adopt more complicated relationships between spatial units to represent, for example, the movement of commuters from one area to another. The more the neighbors we include, the more the smoothing we achieve. The limit would, of course, be averaging over all the spatial units, which would be the overall average and would not be very informative at the local level.

Interpreting results

An important consideration in MCMC methods is diagnosing convergence to the stationary Markov Chain. A commonly accepted approach is to run and dynamically monitor a specific number, e.g. 3, parallel chains and examine the trace plots for when they start to overlap as an indication of convergence. We then discard the burn-in period samples and base inference on the stationary Markov Chain. The Gellman Rubin statistic is useful in diagnosing convergence. It compares variation within chains to those between chains for evidence of scale reduction. When the scale reduction factor reaches 1, there is evidence of convergence. Other convergence statistics are based on examining individual chains (Lawson et al. 2003).

Once the posterior distribution has been sampled, a Bayesian CI has a straightforward interpretation. In a 95% CI, we are 95% certain that the true value lies within it. It is most easily obtained by chopping off the $\acute{\alpha}/2$ tails of the posterior probability distribution.

Being a sample-based approach, in MCMC methods, no two analysts will end up with exactly the same results. This makes variance assessment crucial. Recall that MCMC produces correlated samples. This will lead to underestimates of the variance. One may 'thin' the samples to decrease correlation, but this would result in throwing out information. Variance estimates based on an effective sample size (ESS), though, are available (Banerjee et al. 2004).

Application

We calculated standardized mortality ratios (SMR) for opiate-related deaths in NYC for the years 2000–2004 using the expected number of overdose deaths in a zip code tabulation area based on the mean number of such deaths in NYC throughout the 5-year study period. We were interested, in this example, in drawing inference about the potential role of SES as an explanatory variable for opiate-related overdose deaths in NYC. We used zip code level median household income (MHI) as a proxy for neighborhood-level SES throughout this example.

In the model, the likelihood of the observed values in the standardized morbidity ratio is modeled as a Poisson distribution. The log of the observed value is a function of the log of the expected value, an intercept, and the coefficient for a normally transformed median household income measure. Random effects are represented by a Gaussian intrinsic CAR model with the weights for adjacent neighbors set at 1. Non-informative prior distributions are placed on the intercept, the coefficients, and on tau, the precision term for the CAR prior distribution for random effects.

Our interest is in mapping the zip code level risk estimates while accounting for the potential instability and autocorrelation of those rates and controlling for MHI.

For this empiric illustration, we used WinBUGS (Imperial College and Medical Research Council 2004) software to run three parallel MCMSs with over-dispersed initial values for 120,000 iterations. The first 60,000 iterations were discarded as a burn-in, and our inferences were based on the second 60,000 iterations. We assessed convergence by examining trace histories for parallel chains, and we used R (20) software to conduct the Brooks, Gelman and Rubin, and the Geweke convergence diagnostics as well as the Heidleberger and Welch stationarity tests. We present our results as median values for the coefficients with their associated 95% equal-tailed

Bayesian CIs, histograms (kernel density graphs) of the sampled distributions, and maps comparing raw and smooth risk estimates.

As indicated in Fig. 27.3a, the histogram for β1 (MHI coefficient) is smooth and normally distributed. The Gellman Rubin statistic was calculated to be 1 for most of the runs (Fig. 27.3b), and the trace history for β1 appears to reasonably overlap. Taken together, this information gives us confidence that the model appropriately converged to the posterior distribution and that our inferences based on this posterior distribution are valid.

The median value for the MHI coefficient was −0.3782 (95% CI 0.5681, −0.1855). The interpretation is not straightforward because the dependent variable in the model is the natural log of the SMR, and the MHI variable itself has been normalized to achieve appropriate convergence in WinBUGS. It does, though, indicate that SES, as measured by MHI, is strongly and significantly inversely related to the number of heroin overdose deaths in a zip code area. Essentially, the number of opiate-related deaths increases in a linear fashion as MHI declines. That this is as expected indicates, to a certain extent, the validity of the model. Also, and perhaps more importantly, the subsequent fitted SMR values now control for this important potentially confounding variable.

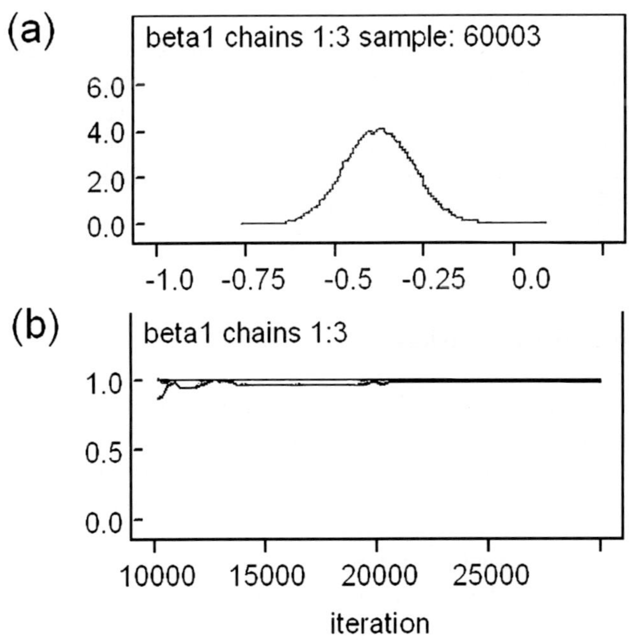

Fig. 27.3 Results from Markov Chain Monte Carlo run, association of median household income with opiate overdose standardized mortality ratios, New York City zip code tabulation areas, 2000–2004. (a) Histogram for Median Household Income beta coefficient. (b) Trace history for Median Household Income beta coefficient

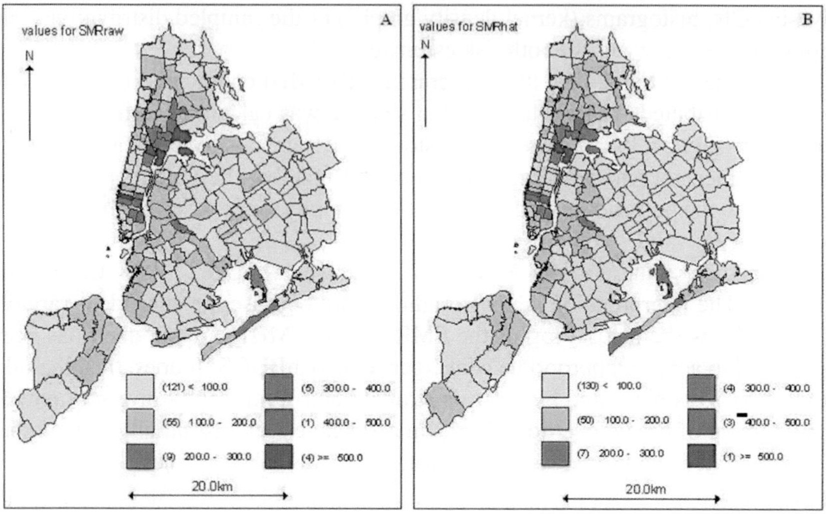

Fig. 27.4 Opiate-related standardized mortality ratios, New York City zip code tabulation areas, 2000–2004. (a) Unadjusted (b) Smoothed estimated adjusted for median household income and autocorrelation with Bayesian Hierarchical Modeling

Figure 27.4 presents the raw and fitted zip code level SMRs. It appears that if we had simply mapped the raw SMRs, we would infer greater than expected rates in such areas as South Bronx and Northern Queens. Looking at the fitted values, these potential clusters seem to become less severe when we take the underlying distribution of the population and its SES characteristics into account. The most evident area of continuing concern is Harlem in northern Manhattan which, despite controlling for MHI, continues to display much greater than expected numbers of heroin overdose deaths.

Discussion and Conclusions

The methods presented in this chapter have much to offer the substance use researcher. They can be viewed as offering incrementally more information and detail as one progresses from first-order cluster detection methods, such as the NNI through scan statistics, to more explanatory analytic techniques, such as hierarchical modeling.

While we did not spend much time on it, an essential first step in any spatial analysis is to describe the data in terms of summary statistics and simple plots. Not only does this provide key descriptive information, it also allows the researcher to assess whether the data meet the assumptions underlying subsequent tests, e.g., Poisson distributions of the outcome of interest.

First-order clustering methods offer a relatively straightforward and easily interpretable global assessment of whether clustering exists and how tight the clustering appears to have been. These are important considerations, but most analysts will want to know exactly where and when the clustering occurred and whether any apparent clustering was simply due to chance. Scan statistics are an important public health tool in this regard. They are fairly straightforward in their application and interpretation, and allow the incorporation of a time variable.

When a population denominator is not available or appropriate, the space–time permutation model is very useful. Although none of the clusters detected through the use of space–time permutation model (Fig. 27.2a) were statistically significant, the results suggest that certain areas had higher counts than expected and that there were temporal changes in opiate-related drug overdose. Having a measure to simply detect higher than expected case counts is particularly worthwhile in public health research. In a setting where timely results based on possibly incomplete count data is a prime consideration, such as in syndromic surveillance (Heffernan et al. 2004), the scan statistic may be one's first choice.

When population estimates are available, adjusting for areas that are more highly populated is appropriate, and the space-time scan statistic is a better, more precise measurement of cluster points. In our example (Fig. 27.2b), we were able to more accurately describe the location and statistical importance of clusters detected by the space–time permutation model. Again, when public health concerns are uppermost, this method may be particularly useful.

Bayesian methods may be most appropriate when potential explanatory variables are available and one's interest is in assessing the *determinants* of health outcomes on a spatial level. It is important to appreciate that this is a smoothing method. When cluster detection is of uppermost concern, caution must be exercised that potential clusters are not smoothed away. While its appropriate utilization requires knowledge of MCMC and sample-based methods, full Bayesian analysis, as presented in our example, is not always necessary. Good empirical approximations are available and obviate the need to learn and use new complex statistical software (Devine et al. 1994; Greenland 2006). But, when data are sparse and highly correlated and there is concern over noise obscuring spatial signals, a full Bayesian approach can help describe both the determinants and the patterns of the outcome of interest at a finer level.

It should be noted that we do not, in this discussion, dwell extensively on the implications of the results of the specific example we have used here to illustrate the material being described in this chapter. However, we considered the role of neighborhood-level SES in explaining rates of heroin-related overdose in the largest US urban area. Conceptual frameworks that consider the complex etiology of substance use and its consequences (Galea, Rudenstine, and Vlahov 2005) have long suggested that a full consideration of the determination of substance use requires that we consider a range of individual- and group-level factors to understand population patterns of substance use. The hierarchical approach introduced here, suitably expanded, can be applied to test specific hypotheses and to pursue spatial etiologic questions, incorporating determinants at group and individual levels as necessary.

Ultimately, we note that the limitations of available data, including, for example, the use of zip codes as a neighborhood proxy, have been well discussed elsewhere in the literature about the multilevel determination of population health (Osypuk and Galea 2007), and pertain just as much to spatial analyses as they do to all other types of epidemiologic analyses. Future work that makes use of the methods introduced here to address specific substance use-related etiologic hypotheses may benefit from application of these methods at different group levels of inference.

In conclusion, an appreciation for where and when health outcomes occur adds much to substance abuse epidemiology. A number of spatial and temporal methods are available, each with its own advantages and disadvantages in terms of complexity, data requirements, and underlying assumptions. The choice of a method will be driven by the question to be answered, data and software availability, and the intended audience or context in which the research is being conducted.

Acknowledgment Dr. DiMaggio's efforts were supported by the US National Institute on Drug Abuse, grant number 1R03DA023431-01 and Centers for Disease Control and Prevention Health Protection Research Initiative Grant Number 1 K01CE000494-01. The efforts of Drs. Tardiff, Vlahov and Galea were supported in part by US National Institutes on Drug Abuse grants numbers DA06534 and DA017642.

References

Abdala, N., Grund, J.-P.C., Tolstov, Y., Kozlov, A.P., & Heimer, R., (2006). Can home-made injectable opiates contribute to the HIV epidemic among injection drug users in the countries of the former Soviet Union? Addiction, 101, 731–737.

Abdelrahman, A.I., Rodriguez, G., Ryan, J.A., French, J.F., & Weinbaum, D. (1998). The epidemiology of substance use among middle school students: the impact of school, familial, community and individual risk factors. Journal of Child and Adolescent Substance Abuse, 8(1), 55–75.

Abler, R., Adams, J., & Gould, P. (1971). Spatial Organization: The Geographer's View of the World. Englewood Cliffs, NJ: Prentice-Hall.

Abramason, D., Garfield, R., & National Center for Disaster Preparedness, O.A. (2006). On the Edge: Children and Families Displaced by Hurricanes Katrina and Rita Face a Looming Medical and Mental Health Crisis. Available at: http://www.ncdp.mailman.columbia.edu/files/marshall_plan.pdf. [On-line].

Acevedo-Garcia, D. (2000). Residential segregation and the epidemiology of infectious diseases. Social Science and Medicine, 51, 1143–1161.

Acevedo-Garcia, D. (2001). Zip code-level risk factors for tuberculosis: neighborhood environment and residential segregation in New Jersey, 1985–1992. American Journal of Public Health, 91, 734–741.

Acevedo-Garcia, D., Lochner, K., Osypuk, T.L., & Subramanian, S. (2003). Future directions in residential segregation and health research: a multilevel approach. American Journal of Public Health, 93, 215–221.

Ackerman, W. (1998). Socioeconomic correlates of increasing crime rates in smaller communities. The Professional Geographer, 50(3), 372–387.

ADAM. (1998). Annual report on drug use among adult and juvenile arrestees 1998 Washington, DC: National Institute of Justice, U.S. Department of Justice.

Adelekan, M.L., & Stimson, G.V. (1997). Problems and prospects of implementing harm reduction for HIV and injecting drug use in high risk sub-Saharan African countries. Journal of Drug Issues, 27(1), 97–116.

Agency for Healthcare Research and Quality (2006). National Healthcare Disparities Report: Summary. Rockville, MD: Agency for Healthcare Research and Quality. Available at: http://www.ahrq.gov/qual /nhdr03/nhdrsum03.htm. [On-line]

Agnew, J. (1987). The United States in the World Economy. Cambridge: Cambridge University Press.

Agudelo, I. (1999). El Rápido Tránsito: Imágenes de la Adolescencia y la Juventud en Nicaragua (67). Sistema de las Naciones Unidas en Nicaragua, Documento de Trabajo, Managua, Nicaragua.

Ahern, J., Pickett, K.E., Selvin, S., & Abrams, B. (2003). Preterm birth among African American and white women: a multilevel analysis of socioeconomic characteristics and cigarette smoking. Journal of Epidemiology and Community Health, 57, 606–611.

Aitken, C., Moore, D., Higgs, D., Kelsall, J., & Kerger, M. (2002). The impact of a police crack-down on a street drug scene: evidence from the street. The International Journal of Drug Policy, 13, 189–198.

Akbik, H., Butler, S.F., Budman, S.H., Fernandez, K., Katz, N.P., & Jamison, R.N. (2006). Validation and clinical application of the Screener and Opioid Assessment for Patients with Pain (SOAPP). Journal of Pain and Symptom Management, 32(3), 287–293.

Alan Guttmacher Institute (AGI). (1995). Teenage Pregnancy and the Welfare Reform Debate. New York, NY: Alan Guttmacher Institute.

Alaniz, M.L., Cartmill, R.S., & Parker, R.N. (1998). Immigrants and violence: the importance of neighborhood context. Hispanic Journal of Behavioural Sciences, 20, 155–174.

Albert, D.P., Gesler, W.M., & Levergood, B. (2000). Spatial Analysis, GIS and Remote Sensing: Applications in the Health Sciences (Hardcover). Sleeping Bear Press, Farmington Hills, MI.

Alcabes, P., Beniowski, M., & Grand, J.-P. (1999). Needle and syringe exchange in Poland and the former Soviet Union: a new approach to community-impact studies. Journal of Drug Issues, 29, 697–703.

Aldstadt, J., & Getis, A. (2006). Using AMOEBA to create a spatial weights matrix and identify spatial clusters. (A Multidirectional Optimum Ecotope-Based Algorithm) Geographical Analysis, 38, 327–44.

Allard, S., Tolman, R., & Rosen, D. (2003). Proximity to service providers and service utilization among welfare recipients: the interaction of place and race. Journal of Policy Analysis and Management, 22(4), 599–613.

Allard, S.W. (2004). Access to Social Services: The Changing Urban Geography of Poverty and Service Provision. Washington, DC: The Brookings Institution, Center on Urban and Metropolitan Policy, Survey Series.

Allard, S.W., Rosen, D., & Tolman, R. (2003). Access to mental health and substance abuse services among women receiving welfare in Detroit. Urban Affairs Review, 38(6), 787–807.

Allison, K.W., Crawford, I., Leone, P.E., Trickett, E., Perez-Febles, A., Burton, L.M., & Le Blanc, R. (1999). Adolescent substance use: preliminary examinations of school and neighborhood context. American Journal of Community Psychology, 27(2), 111–141.

Almog, M., Curtis, S., Copeland, A., & Congdon, P. (2004). Geographical variation in acute psychiatric admissions within New York City 1990–2000: growing inequalities in service use? Social Science and Medicine, 59, 361–376.

Alonzo, A.A., & Reynolds, N.R. (1995). Stigma, HIV, and AIDS: an exploration and elaboration of a stigma trajectory. Social Science and Medicine, 41(3), 303–315.

Altman, D. (1994). Power and Community: Organizational and Cultural Responses to AIDS. London: Taylor & Francis Ltd.

Altman, N. (2000). Krige, smooth, both or neither? (with discussion). Australian and New Zealand Journal of Statistics, 42, 441–461.

Amaro, H., Whitaker, R., Coffman, G., & Heeren, T. (1990). Acculturation and marijuana and cocaine use: findings from HHANES 1982–1984. American Journal of Public Health, 80, 54–60.

American Psychiatric Association. (1994). Diagnostic and Statistical Manual of Mental Disorders DSM-IV-TR, 4th Edition. Arlington, VA: American Psychiatric Association.

American Psychiatric Association. (2000). Diagnostic and Statistical Manual of Mental Disorders, 4th Edition. Washington, DC: American Psychiatric Association.

Anderson, A., Qingsi, Z., Hua, X., & Jianfeng, B. (2003). China's floating population and the potential for HIV transmission: a social-behavioural perspective. AIDS Care, 15(2), 177–185.

Anderson, N. (1923). The juvenile and the tramp. Journal of Criminal Law, Criminology, and Police Science, 14, 290–312.

Andresen, N.A. (2006). Crime measures and the spatial analysis of criminal activity. The British Journal of Criminology, 46(2), 258–285.

Aneshensel, C.S., & Sucoff, C.A. (1996). The neighborhood context of adolescent mental health. Journal of Health and Social Behavior, 37, 293–311.

Anselin, L. (1988). Spatial Econometrics, Methods and Models. Dordrecht: Kluwer.

Anselin, L. (1995). LISA: local indicators of spatial autocorrelation. Geographical Analysis, 27, 93–115.

Anselin, L. (2003). GEODA 0.9 Users Guide. Urbana-Champaign, IL: Spatial Analysis Laboratory, University of Illinois. Available at: http://sal.agecon.uiuc.edu/geoda.

Anselin, L., Cohen, J., Cook, D., Gorr, W., & Tita, G. (2000). Spatial analyses of crime. In: D. Duffee (ed.), Measurement and Analysis of Crime and Justice; Criminal Justice (pp. 213–262). Washington, DC: National Criminal Justice Reference Service.

Anselin, L., Ibnu, S., & Youngihn, K. (2006). GeoDa: an introduction to spatial data analysis. Geographical Analysis, 38(1), 5–22.

Apostolopoulos, Y., Sonmez, S., Kronenfeld, J., Castillo, E., McLendon, L., & Smith, D. (2006). STI/HIV risks for Mexican migrant laborers: exploratory ethnographies. Journal of Immigrant and Minority Health, 8(3), 291–302.

Appel, P.W., Ellison, A.A., Jansky, H.K., & Oldak, R. (2004). Barriers to enrollment in drug abuse treatment and suggestions for reducing them: opinions of drug injecting street outreach clients and other system stakeholders. The American Journal of Drug and Alcohol Abuse, 30(1), 129–153.

Aral, S.O., Padian, N.S., & Holmes, K.K. (2005). Advances in multilevel approaches to understanding the epidemiology and prevention of sexually transmitted infections and HIV: an overview. The Journal of Infectious Diseases, 191(Suppl 1), S1–S6.

Aral, S.O., & St Lawrence, J.S. (2002). The ecology of sex work and drug use in Saratov Oblast, Russia. Sexually Transmitted Diseases, 29, 798–805.

Arauz, R., Ortells, P., Morales, A., Guevara, M., & Shedlin, M.G. (1997). Sexo Inseguro. Investigación Cualitativa Sobre Comportamiento Sexual de Alto Riesgo Respecto al SIDA en Nicaragua. Managua, Nicaragua: Fundación Nimehuatzin.

Associated Press. (2007). Many Houston Katrina evacuees unemployed 2 years later. Available at: http://www.news8austin.com/content/headlines/?ArID=185163 [On-line].

Austin, L.S., & Godleski, L.S. (1999). Therapeutic approaches for survivors of disaster. The Psychiatric Clinics of North America, 22, 897–910.

Avruskin, G., Jacquez, G., Meliker, J., Slotnick, M., Kaufmann, A., & Nriagu, J. (2004). Visualization and exploratory analysis of epidemiologic data using a novel space time information system. International Journal of Health Geographics, 3(1), 26.

Baca, C.T., & Grant, K.J. (2005). Take-home naloxone to reduce heroin death. Addiction, 100(12), 1823–1831.

Bailey, T., & Gatrell, A. (1995). Interactive Spatial Data Analysis. Harlow, England: Longmann Scienctific and Technical.

Bailey, T.C., & Gatrell, A.C. (1995). Interactive Spatial Data Analysis, 2nd Edition. London: Longman.

Bakken, R. (2006). Answered Prayers. Grand Forks Herald. pp. 1–2A, Grand Forks, ND.

Ball, J.C., & Chambers, C.D. (1970). Overview of the problem. In: J.C. Ball, & C.D. Chambers (eds.), The Epidemiology of Opiate Addiction in the United States. (pp. 5–21). Springflied, IL: Charles C. Thomas.

Ballesteros, M.F., Budnitz, D.S., Sanford, C.P., Gilchrist, J., Agyekum, G.A., & Butts, J. (2003). Increase in deaths due to methadone in North Carolina. JAMA, 290(1), 40.

Bandura, A. (1986). Social Foundations of Thought and Action: A Social Cognitive Theory. Englewood Cliffs, NJ: Prentice-Hall.

Banerjee, S., Carlin, B.P., & Gelfand, A.E. (2004). Hierarchical modeling and analysis for spatial data. New York: Chapman & Hall/CRC

Banzhaf, M., Morgan, T., & Ramspacher, K. (1992). Reproductive rights and AIDS: the connections. In: M. Banzhaf, et al. (eds.), Women, AIDS, and Activism (pp. 199–209). Boston: South End Press.

Barnett, E., & Casper, M. (2001). A definition of social environment. American Journal of Public Health, 91(3), 465.

Barnett, R., & Barnett, P. (2003). If you want to sit on your butts you'll get nothing!: community activism in response to threats of rural hospital closure in southern New Zealand. Health and Place, 9(2), 59–71.

Bastos, F.I., Strathdee, S.A., Derrico, M., & Pina, M.D.F. (1999). Drug use and the spread of HIV/AIDS in South America and the Caribbean. Drugs, Education, Prevention and Policy, 6(1), 29–49.

Beardsley, B.A., Wish, E.D., Fitzelle, D.B., O'Grady, K., & Arria, A.M. (2003). Distance traveled to outpatient drug treatment and client retention. Journal of Substance Abuse Treatment, 25, 279–285.

Becker, G.S., Grossman, M., & Murphy, K.M. (1994). An empirical analysis of cigarette addiction. The American Economic Review, 84, 396–418.

Becker, M. (2006). Focus: drug raids. How effective is the drug war? Buffalo News, p.B1, Buffalo, NY.

Beletsky, L., Ruthazer, R., Macalino, G.E., Rich, J.D., Tan, L., & Burris, S. (2007). Physicians' knowledge of and willingness to prescribe naloxone to reverse accidental opiate overdose: challenges and opportunities. Journal of Urban Health, 84(1), 126–136.

Bell, D.C., Carlson, J.W., & Richard, A.J. (1998). The social ecology of drug use: a factor analysis of an urban environment. Substance Use and Misuse, 33, 2201–2217.

Bell, W. (1954). A probability model for the measurement of ecological segregation. Social Forces, 32, 357–364.

Bellander, T., Berglind, N., Gustavsen, P., Jonson, T., Nyberg, F., Persagen, G., & Jarup, L. (2001). Using geographic information systems to assess historical exposure to from traffic and house heating in Stockholm. Environmental Health Perspectives, 109, 633–639.

Bellis, D.J. (2003). Hotel Ritz – Comparing Mexican and U.S. Street Prostitutes: Factors in HIV/AIDS Transmission (pp. 41–68). New York: Haworth Press.

Bembry, J.X., & Norris, D.F. (2005). An exploratory study of neighborhood choices among moving to opportunity participants in Baltimore, Maryland: the influence of housing search assistance. Journal of Sociology and Social Welfare, 32(4), 93–107.

Bengal, Department of Agriculture. (1920). Season and Crop Report. Calcutta: Department of Agriculture.

Bennet, G.C., Yaus Wolin, K., Robinson, E.L., Fowler, S., & Edwards, C.L. (2005). Perceived racial/ethnic harassment and tobacco use among African–American young adults. American Journal of Public Health, 95, 238–240.

Bergeron, K. (2006). Katrina Continues to Take Toll on Mental, Physical Health. The Sun Herald. Biloxi, MS.

Berkman, L., & Kawachi, I. (2000). A historical framework for social epidemiology. In: L. Berkman, I. Kawachi (eds.), Social Epidemiology (pp. 3–12). New York: Oxford University Press.

Bernardinelli, L., Clayton, D., & Montomoli, C. (1995). Bayesian estimates of disease maps: how important are priors? Statistics in Medicine, 14, 2411–2431.

Besag, J., York, J., & Mollie, A. (1991). Bayesian image restoration, with two applications in spatial statistics. Annals of the Institute of Statistical Mathematics, 43, 1–59.

Beth Israel Medical Center. (2000). Beth Israel National Survey of Syringe Exchange Programs. New York, N.Y.

Beveridge, A.A., Kadushin, C., Saxe, L., Rindskopf, D., & Livert, D. (2000). Survey estimates of drug-use trends in urban communities: general principles and cautionary examples. Substance Use and Misuse, 35, 891–923.

Beyrer, C. (1998). War in the Blood: Sex, Politics and AIDS in Southeast Asia. London: Zed Books.

Beyrer, C. (2001). Shan women and girls and the sex industry in South-east Asia: political causes and human rights implications. Social Science and Medicine, 53, 543–550.

Beyrer, C., et al. (2000). Patterns of HIV spread associated with drug trafficking. Proceedings of the Global Research Network, Third Annual Meeting, Durban, August 2000.

Beyrer, C., Razak, M.H., Lisam, K., Chen, J., Lui, W., & Yu, X.F. (2000). Overland heroin trafficking routes and HIV-1 spread in south and south-east Asia. AIDS, 14(1), 75–83.

Biernacki, P., & Waldorf, D. (1981). Snowball sampling: problems and techniques of chain referral sampling. Sociological Methods and Research, 10, 141–163.

Bigg, D., Buanauro, E., Gutenson, J., & Maxwell, S. (2006). An attempt at mobile care for drug users affected by Hurricane Katrina. California Society of Addiction Medicine Newsletter, 8–9.

Billy, J.O.G., & Moore, D.E. (1992). A multilevel analysis of marital and nonmarital fertility in the U.S. Social Forces, 70, 977–1011.

Bilsborrow, R.E., Hugo, G.J., Oberai, A.S., & Zlotnik, H. (1997). International Migration Statistics: Guidelines for the Improvement of Data Collection Systems. Geneva: International Labor Office.

Blakely, T., Lochner, K., & Kawachi, I. (2002). Metropolitan area income inequality and self-rated health – a multi-level study. Social Science and Medicine, 54, 65–77.

Block, C.R., Dadoub, M., & Fregly, S. (1995). Crime Analysis Through Computer Mapping. Washington, DC: Police Executive Research Forum.

Bluthenthal, R.N. (1998). Syringe exchange as a social movement: a case study of harm reduction in Oakland, California. Substance Use and Misuse, 33(5), 1147–1171.

Bluthenthal, R.N., Kral, A.H., Erringer, E.A., & Edlin, B.R. (1999a) Drug paraphernalia laws and injection-related infectious disease risk among drug injectors. Journal of Drug Issues, 22, 1–16.

Bluthenthal, R.N., Kral, A.H., Erringer, E.A., & Edlin, B.R. (1999b). Drug paraphernalia laws and injection-related infectious disease risk among drug injectors. Drug Issues, 22, 1–16.

Bluthenthal, R.N., Kral, A.H., Lorvick, J. and Watters, J.K., (1997). Impact of law enforcement on syringe exchange programs: a look at Oakland and San Francisco. Medical Anthropology, 18, pp. 61–83.

Bluthenthal, R.N., Lorvick, J., Kral, A.H., Erringer, E.A., & Kahn, J.G. (1999) Collateral damage in the war on drugs: HIV risk behaviors among injection drug users. The International Journal of Drug Policy, 10, 25–38.

Blumenthal, W.J., Springer, K.W., Jones, T.S., & Sterk, C.E. (2002). Pharmacy student knowledge, attitudes, and beliefs about selling syringes to injection drug users. Journal of the American Pharmaceutical Association (Washington), 42(6 Suppl 2), S34–S39.

Boardman, J.D., Finch, B.K., Ellison, C.G., Williams, D.R., & Jackson, J.S. (2001). Neighborhood disadvantage, stress, and drug use among adults. Journal of Health and Social Behavior, 42, 151–165.

Bobkov, A., et al. (1998). A sudden epidemic of HIV type 1 among injecting drug users in the former Soviet Union: identification of subtype A, subtype B, and novel gagA/envB recombinants. AIDS Research and Human Retroviruses, 14(8), 669–676.

Boerl, M.W., Sterk, C., & Elifson, K. (2006). Baby boomer drug users: career phases, social control, and social learning theory. Sociological Inquiry, 76(2), 264–291.

Bonilla-Silva, E. (2001). White Supremacy and Racism in the Post-Civil Rights Era. Boulder, CO: Lynne Riener Publishers.

Booth, B.M., & Curran, G.M. (2006). Variations in drinking patterns in the rural South: joint effects of race, gender, and rural residence. The American Journal of Drug and Alcohol Abuse, 32(4), 561–568.

Boulos, M., & Phillipps, G.P. (2004). Is NHS dentistry in crisis? Traffic lights maps of dentists distribution in England and Wales. International Journal of Health Geographics, 3(1), 10.

Bourgois, P., Lettiere, M., & Quesada, J. (1997). Social misery and the sanctions of substance abuse: confronting HIV risk among homeless heroin addicts in San Francisco. Social Problems, 44, 155–173.

Bowdle, T.A. (1998). Adverse effects of opioid agonists and agonist-antagonists in anaesthesia. Drug Safety, 19(3), 173–189.

Bower, A. (2003). Sharing family values – troubled parents are getting a second chance: foster care for them, along with their kids. Time, Feb. 17 161(7): 62–3.

Bowser, D. (2003). West of the Creek: Murder, Mayhem, & Vice in Old San Antonio. San Antonio: Maverick Publishing Company.

Boyle, K. (2004). Arc of Justice: A Saga of Race, Civil Rights, and Murder, in the Jazz Age. New York: Henry Holt and Company, LLC.

Bradizza, C.M., Stasiewicz, P.R., & Paas, N.D. (2006). Relapse to alcohol and drug use among individuals diagnosed with co-occurring mental health and substance use disorders: a review. Clinical Psychology Review, 26, 162–178.

Bradshaw, M.J., & Vartapetov, K. (2003). A new perspective on regional inequalities in Russia. Eurasian Geography and Economics, 46, 403–429.

Brady, T.M., & Ashley, O.S. (eds.). (2005). Women in substance abuse treatment: Results from the Alcohol and Drug Services Study (ADSS) (DHHS Publication No. SMA 04-3968, Analytic Series A-26). Rockville, MD: Substance Abuse and Mental Health Services Administration, Office of Applied Studies.

Brantingham, P.L., & Brantingham, P.J. (1993). Nodes, paths and edges: considerations on the complexity of crime and the physical environment. Journal of Environmental Psychology, 13, 3–28.

Brantingham, P.L., & Brantingham, P.J. (1999). Theoretical Model of Crime Hot Spot Generation. Studies on Crime and Crime Prevention, 8, 7–26.

Braveman, P.A., Cubbin, C., Egerter, S., Chideya, S., Marchi, K.S., Metzler, M., & Posner, S. (2005). Socioeconimc status in health research: one size does not fit all. JAMA, 294, 2879–2888.

Breuer, B., Pappagallo, M., Tai, J.Y., & Portenoy, R.K. (2007). U.S. board-certified pain physician practices: uniformity and census data of their locations. The Journal of Pain, 8(3), 244–250.

Brewster, K.L., Billy, J.O., & Grady, W.R. (1993). Social context and adolescent behavior: the impact of community on the transition to sexual activity. Social Forces, 71, 713–740.

Britt, H.R., Carlin, B.P., Toomey, T.L., & Wagenaar, A.C. (2005). Neighborhood level spatial analysis of the relationship between alcohol outlet density and criminal behavior. Environmental and Ecological Statistics, 12, 411–426.

Brockerhoff, M., & Biddlecom, A.E. (1999). Migration, sexual behavior and the risk of HIV in Kenya. The International Migration Review, 33, 836–856.

Brodie, M., Weltzien, E., Altman, D., Blendon, R.J., & Benson, J.M. (2006). Experiences of Hurricane Katrina evacuees in Houston shelters: implications for future planning. American Journal of Public Health, 96, 1402–1408.

Bronfenbrenner, U. (1979). The Ecology of Human Development. Cambridge: Harvard University.

Bronfenbrenner, U. (2005). Making Human Beings Human. Thousand Oaks, CA: Sage.

Brook, J.S., Nomura, C., & Cohen, P. (1989). A network of influences on adolescent drug involvement: neighborhood, school, peer, and family. Genetic, Social, and General Psychology Monographs, 115, 125–145.

Brooner, R., Kidorf, M., King, V., Bielenson, P., Svikis, D., & Vlahov, D. (1998). A drug abuse treatment success among needle exchange participants. Public Health Reports, 113(1), 130–139.

Brouwer, K.C., Case, P., Ramos, R., Magis-Rodríguez, C., Bucardo, J., Patterson, T.L., & Strathdee, S.A. (2006). Trends in production, trafficking, and consumption of methamphetamine and cocaine in Mexico. Substance Use and Misuse, 41(5), 707–727.

Brouwer, K.C., Firestone, M., Lozada, R., Magis-Rodríguez, C., & Strathdee, S.A. (2005). Migration Trends and Drug Treatment Needs among Injection Drug Users in the Mexico/U.S. Border City of Tijuana, Mexico. Paper presented at the 4th International Conference on Urban Health, Toronto, Canada.

Brouwer, K.C., Strathdee, S.A., Magis-Rodriguez, C., Bravo-Garcia, E., Gayet, C., Patterson, T.L., et al. (2006). Estimated numbers of men and women infected with HIV/AIDS in Tijuana, Mexico. Journal of Urban Health, 83(2), 299–307.

Brown, A. (1982). Modelling the spatial distribution of suburban crime. Economic Geography, 58(3), 247–261.

Browning, C. (2002). The span of collective efficacy: extending social disorganization theory to partner violence. Journal of Marriage and the Family, 64(4), 833–850.

Browning, C.R., Feinberg, S.L., & Dietz, R. (2004). The paradox of social organization: Networks, collective efficacy, and violent crime in urban neighborhoods. Social Forces, 83, 503–534.

Browning, C.R., Leventhal, T., & Brooks-Gunn, J. (2005). Sexual initiation during early adolescence: the nexus of parental and community control. American Sociological Review, 70(5), 758–778.

Broyles-Gonzalez, Y. (2001). Lydia Mendoza's Life in Music/La Historia de Lydia Mendoza. New York, NY: Oxford University Press.

Brugal, M.T., Domingo-Salvany, A., Maguire, A., Cayla, J.A., Villalbi, J.R., & Hartnoll, R. (1993). A small area analysis estimating the prevalence of addiction to opioids in Barcelona, 1993. Journal of Epidemiology and Community Health, 53, 488–494.

Bucardo, J., Brouwer, K.C., Magis-Rodriguez, C., Ramos, R., Fraga, M., Perez, S.G., et al. (2005). Historical trends in the production and consumption of illicit drugs in Mexico: implications for the prevention of blood borne infections. Drug and Alcohol Dependence, 79(3), 281–293.

Buchanan, D., Shaw, S., Teng, W., Hiser, P., & Singer, M. (2003). Neighborhood differences in patterns of syringe access, use, and discard among injection drug users: implications for HIV outreach and prevention education. Journal of Urban Health, 80(3), 438–454.

Buckley, P.E. (2006). Prevalence and consequences of the dual diagnosis of substance abuse and severe mental illness. The Journal of Clinical Psychiatry, 67(Suppl), 5–9.

Bullington, B. (1977). Heroin Use in the Barrio. Lexington, MA: Lexington Books.

Buonanno, P., & Montolio, D. (2008). Identifying the socio-economic and demographic determinants of crime across Spanish provinces. International Review of Law and Economics. Retrieved 1 August, 2007, from http://www.paolobuonanno.it/doc/IRLE_revision.pdf.

Burdon, W., Messina, N., & Prendergast, M. (2004). The California treatment expansion initiative: aftercare participation, recidivism, and predictors of outcomes. The Prison Journal, 84(1), 61–80.

Burris, S. (2005). Personal communication. James E. Beasley Professor of Law. Temple University Beasley School of Law.

Burris, S., Blankenship, K.M., Donoghoe, M., Sherman, S., Vernick, J.S., Case, P., et al. (2004). Addressing the "risk environment" for injection drug users: the mysterious case of the missing cop. The Milbank Quarterly, 82(1), 125–156.

Burris S, Vernick J, Ditzle A, & Strathdee S. (2002). The legality of selling or giving syringes to injection drug users. In: TS Jones & PO Coffin (eds.), Preventing blood-borne infections through pharmacy syringe sales and safe community syringe disposal. Journal of the American Pharmaceutical Association, 42 S13–17.

Burrows, D. (2001). Estimating the costs of implementing widespread needle and syringe exchange provision: examples from the Russian Federation and Ukraine. Paper presented at the Twelfth International Conference on the Reduction of Drug Related Harm, New Delhi, 1–5 April 2001.

Butler, B. (2000). Mapping in action: Illinois methamphetamine risk-model. Crime Mapping News, 2(4), 8–9.

Cadora, E. (2002). Criminal justice and health and human services: an exploration of overlapping needs, resources, and interests in Brooklyn neighborhoods. Paper Prepared for the From Prison to Home: The Effect of Incarceration and Reentry on Children, Families, and Communities National Policy Conference Convened by the U.S. Department of Justice and the Urban Institute, Washington, DC, January 30–31. Available at: http://www.urban.org/UploadedPDF/ 410633_CriminalJustice.pdf.

Cahill, M., & Mulligan, G. (2003). Crime determinants in Tucson, AZ. Urban Geography, 24(7), 582–610.

Caldwell, C.H., Sellers, R.M., Bernat, D., & Zimmerman, M.A. (2004). Racial identity, parental support, and alcohol use in a sample of academically at-risk African–American high school students. American Journal of Community Psychology, 34, 71–82.

Cameron, L., & Williams, J. (2001). Substitutes or complements? Alcohol, cannabis and tobacco. The Economic Record, 77, 19–34.

Carrieri, M.P., Amass, L., Lucas, G.M., Vlahov, D., Wodak, A., & Woody, G.E. (2006). Buprenorphine use: the international experience. Clinical Infectious Diseases, 43(Suppl 4), S197–S215.

Casavantes, E.J. (1976). El Tecato: Social and Cultural Factors Affecting Drug use Among Chicanos, Revised Edition. Washington, DC: National Coalition of Spanish Speaking Mental Health Organizations.

Castells, M. (1983). The City and the Grassroots. Berkeley, CA; University of California Press.

Caulkins, J., & Pacula, R.L. (2006). Marijuana markets: inferences from reports by the household population. Journal of Drug Issues, 36(1), 173–200.

Caulkins, J.P. (1993). Local drug markets' response to focused police enforcement. Operations Research, 41(5), 848–863.

Caulkins, J.P. (1995). Domestic geographic variation in illicit drug prices. Journal of Urban Economics, 37, 38–56.

CDC. (1983). Leads from the MMWR. Heroin-related deaths – District of Columbia, 1980–1982. JAMA, 250(4), 463–464.

CDC. (2000). Unintentional opiate overdose deaths – King County, Washington, 1990–1999. MMWR Morbidity and Mortality Weekly Report, 49(28), 636–640.

CDC. (2004). Unintentional and undetermined poisoning deaths – 11 states, 1990–2001. MMWR Morbidity and Mortality Weekly Report, 53(11), 233–238.

CDC. (2005). Atypical reactions associated with heroin use – five states, January–April 2005. MMWR Morbidity and Mortality Weekly Report, 54(32), 793–796.

Census and Statistics Department. (2000). Population and Household Statistics Analysed by District Council District. Hong Kong: General Household Survey Section (1), Census and Statistics Department.

Census and Statistics Department. (2001). Population and Household Statistics Analyzed by District Council District. Hong Kong: General Household Survey Section (1), Census and Statistics Department.

Census and Statistics Department. (2002a). Demographic Trends in Hong Kong, 1981–2001. Hong Kong: Demographic Statistics Section, Census and Statistics Department.

Census and Statistics Department. (2002b). Population and Household Statistics Analyzed by District Council District. Hong Kong: General Household Survey Section (1), Census and Statistics Department.

Census and Statistics Department. (2003). Population and Household Statistics Analyzed by District Council District. Hong Kong: General Household Survey Section (1), Census and Statistics Department.

Census and Statistics Department. (2004). Population and household statistics analyzed by District Council District. Hong Kong: General Household Survey Section (1), Census and Statistics Department.

Census and Statistics Department. (2005). Population and Household Statistics Analyzed by District Council District. Hong Kong: General Household Survey Section (1), Census and Statistics Department.

Census and Statistics Department. (2006a). Census and Statistics Department, Statistics by Subject. Retrieved 1 November, 2006 from HKSAR, Census and Statistics Department Web site: http://www.censtatd.gov.hk/hong_kong_statistics/statistics_by_subject/index.jsp?subjectID=2&charsetID=2&displayMode=T.

Census and Statistics Department. (2006b). Hong Kong Annual Digest of Statistics 2006. Hong Kong: Government Printer.

Center for Communication Programs. (2005). HIV/AIDS Communication Strategies. Retrieved 30 November, 2006, from http://www.jhuccp.org/topics/hiv/la.shtml.

Center for Epidemiological Work Group. (1999). Epidemiologic Trends in Drug Abuse: vol. I: Highlights and Executive Summary. Bethesda, MD: National Institute on Drug Abuse.

Center for Health Services Research. (2006). Shortage of Primary Care for the Uninsured. Houston Health Services Research Collaborative, The University of Texas School of Public Health, Fall Newsletter Available at: http://www.sph.uth.tmc.edu/hsrc [On-line].

Center for Strategic and International Studies [CSIS]. (2005). Russia and HIV/AIDS, Opportunities for Leadership and Cooperation, A report of the CSIS task force on HIV/AIDS. Washington, DC: Brookings Institution/CSIS Joint Delegation to Russia.

Centers for Disease Control and Prevention. (1999). HIV Counseling and Testing in Publicly Funded Sites. Annual report 1997–1998. Atlanta, GA: Centers for Disease Control and Prevention.

Centers for Disease Control and Prevention. (2001). HIV Prevalence Trends in Selected Populations in the United States: Results from National Serosurveillance, 1993–1997. Atlanta: Centers for Disease Control and Prevention.

Centers for Disease Control and Prevention. (2002). HIV/AIDS Surveillance Report 2001. Centers for Disease Control and Prevention. Atlanta, GA.

Centers for Disease Control and Prevention. (2006a). Morbidity Surveillance After Hurricane Katrina – Arkansas, Louisiana, Mississippi, and Texas, September 2005. Morbidity and Mortality Weekly Report, 55, 727–731.

Centers for Disease Control and Prevention. (2006b). Public Health Response to Hurricanes Katrina and Rita – United States, 2005. Morbidity and Mortality Weekly Report, 55, 229–231.

Central Registry of Drug Abuse (CRDA). (2006). Fifty-Fifth Report 1996–2005. Hong Kong: Narcotics Division, Security Bureau.

Centre for Health Protection, Department of Health. (2006). HIV Surveillance Report, 2004 Update. Hong Kong: Centre for Health Protection, Department of Health.

Centro Nicaragüense de Derechos Humanos. (2004). Human Rights in Nicaragua. Annual Report 2003. Managua, Nicaragua: CENIDH.

Chainey, S., & Ratcliffe, J. (2005). GIS and Crime Mapping. Chichester: Wiley.

Chaloupka, F.J. (1991). Rational addictive behavior and cigarette smoking. The Journal of Political Economy, 99, 722–742.

Chaloupka, F.J., Grossman, M., & Tauras, J.A. (1999). The demand for cocaine and marijuana by youth. In: F.J. Chaloupka, M. Grossman, W.K. Bickel, & H. Saffer (eds.), The Economic Analysis of Substance Use and Abuse (pp. 133–156). Chicago: The University of Chicago Press.

Chamarro, C.F., Iriás, N., & Gutiérrez, J.C. (1996). Los Nuevos Votantes: Política, Sexo y Drogas (99). Managua, Nicaragua: Centro de Investigaciones de la Comunicación, CINCO.

Chambers, C.D., Cuskey, W.R., & Moffett, A.D. (1970). Mexican–American opiate addicts. In: J.C. Ball, & C.D. Chambers (eds.), The Epidemiology of Opiate Addiction in the United States (pp. 202–221). Springfield, IL: Charles C. Thomas.

Chan, G.M., Stajic, M., Marker, E.K., Hoffman, R.S., & Nelson, L.S. (2006). Testing positive for methadone and either a tricyclic antidepressant or a benzodiazepine is associated with an accidental overdose death: analysis of medical examiner data. Academic Emergency Medicine, 13(5), 543–547.

Chan M.K. & Lee S.S. (2004). Can the low HIV prevalence in Hong Kong be maintained? AIDS Education and Prevention, 16(3 Suppl A),18-26.

Chandra, S. (2000). What the numbers really tell us about the decline of the opium regie. Indonesia, 70, 101–123.

Chavira, R. (2003). Mexico Seeing Surge in Drug use. Miami Herald, p. from http://www.mapinc.org/drugnews/v2003/n1197/a2006.html. (Accessed 27 July, 2007)

Chein, I., Gerard, D.L., Lee, R.S., & Rosenfeld, E. (1964). The Road to H. New York: Basic.

Chelala, C., & Beyrer, C. (1999). Drug use and HIV/AIDS in Burma. Lancet, 25(354), 1119.

Chen, L., Walker, S., & Tong, S. (2002). The impact of the variation in death certification and coding practices on trends in mortality from ischaemic heart disease. Australian Health Review, 25(4), 189–197.

Cherpitel, C.J. (1993). Alcohol and violence-related injuries: an emergency room study. Addiction, 88, 79–88.

China Ministry of Health, UNAIDS, and WHO. (2006). 2005 Update on the HIV/AIDS Epidemic and Response in China. Beijing: China Ministry of Health.

Choi, S.Y.P., Cheung, Y.W., & Jiang, Z.Q. (2007). Ethnicity and risk factors in needle sharing among intravenous drug users in Sichuan Province, China. AIDS Care, 19(1), 1–8.

Chuang, Y.-C., Ennett, S.T., Bauman, K.E., & Foshee, V.A. (2005). Neighborhood influences on adolescent cigarette and alcohol use: mediating effects through parent and peer behaviors. Journal of Health and Social Behavior, 46(2), 187–204.

Ciccarone, D., & Bourgois, P. (2003). Explaining the geographical variation of HIV among injection drug users in the United States. Substance Use Misuse, 38(14), 2049–2063.

Cicero, T.J., Dart, R.C., Inciardi, J.A., Woody, G.E., Schnoll, S., & Munoz, A. (2007). The development of a comprehensive risk-management program for prescription opioid analgesics: researched abuse, diversion and addiction-related surveillance (RADARS). Pain Medicine, 8(2), 157–170.

Cicero, T. J., Inciardi, J. A., & Munoz, A. (2005). Trends in abuse of Oxycontin and other opioid analgesics in the United States: 2002-2004. J Pain, 6(10), 662-672.

Cisneros Santamaría, N.P. (1995). Conocimiento, Percepción del Riesgo y Estrategias de Prevención Sobre VIH/SIDA que Tienen los Estudiantes del Primer Ingreso en Tres Universidades de Managua. Managua, Nicaragua: UCA, Psicología, monografía (58).

Clark, N.M. (1987). Social learning theory in current health education practice. Advances in Health Education and Promotion, 2, 251–275.

Clark, P., & Evans, F. (1954). Distance to nearest neighbor as a measure of spatial relationships in populations. Ecology, 35, 445–453.

Clark, W. A. V. (1984). Housing consumption and residential mobility. Annals of the Association of American Geographers, 74, 29-43.

Clark, W.A.V., & Ledwith, V. (2006). Mobility, housing stress, and neighborhood contexts: evidence from Los Angeles. Environment and Planning A, 38(6), 1077–1093.

Clatts, M.C., & Davis, W.R. (1999). A demographic and behavioral profile of homeless youth in New York City: implications for AIDS outreach and prevention. Medical Anthropology Quarterly, 13(3), 365–374.

Clatts, M.C., Davis, W.R., & Atillasoy, A. (1995). Hitting a moving target: the use of ethnographic methods in the evaluation of AIDS outreach programs for homeless youth in NYC. Qualitative Methods in Drug Abuse and HIV Research. NIDA Research Monograph, 157, 117–135.

Claus, R.E., & Kindleberger, L.R. (2002). Engaging substance abusers after centralized assessment: predictors of treatment entry and dropout. Journal of Psychoactive Drugs, 34, 25–31.

Clear, A. (2003). Personal communication. Executive Director, Harm Reduction Coalition, New York City, New York.

Cliff, A., & Haggett, P. (2004). Time, travel and infection. British Medical Journal, 69, 87–99.

Cockings, S., & Martin, D. (2005). Zone design for environment and health studies using pre-aggregated data. Social Science and Medicine, 60(12), 2729–2742.

Cohen, D., Ghash-Dastidar, B., Scribner, R., Miu, A., Scott, M., Robinson, P., Farley, T.A., Bluthenthal, R.N., & Brown-Taylor, D. (2006). Alcohol outlets, gonorrhea, and the Los Angeles civil unrest: a longitudinal analysis. Social Science and Medicine, 62, 3062–3071.

Cohen, D., Spear, S., Scribner, R., Kissinger, P., Mason, K., & Wildgen, J. (2000). "Broken windows" and the risk of gonorrhea. Am J Public Health, 90(2), 230–236.

Cohen, D.A., Farley, T.A., & Mason, K. (2003). Why is poverty unhealthy? Social and physical mediators. Social Science and Medicine, 57(9), 1631–1641.

Coleman, J.S. (1990). Foundations of Social Theory. Cambridge, MA: The Belknap Press of Harvard University.

Collins, C., & Williams, D.R. (1999). Segregation and mortality: the deadly effects of racism? Sociological Forum, 14, 495–523.

Colorado Department of Public Health and Environment. (2003). Cleanup of clandestine methamphetamine labs guidance document. Available at: http://www.cdphe.state.co.us/hm/methlab.pdf. [On-line]

Condelli, W.S., & Hubbard, R.L. (1994). Relationship between time spent in treatment and client outcomes from therapeutic communities. Journal of Substance Abuse Treatment, 11, 25–33.

Cone, E.J., Fant, R.V., Rohay, J.M., Caplan, Y.H., Ballina, M., Reder, R.F., et al. (2004). Oxycodone involvement in drug abuse deaths. II. Evidence for toxic multiple drug–drug interactions. Journal of Analytical Toxicology, 28(4), 217–225.

Conroy, M.S. (1990). Abuse of drugs other than alcohol and tobacco in the Soviet Union. Soviet Studies, 42, 447–480.

Cooper, H., Friedman, S.R., Tempalski, B., Friedman, R., & Keem, M. (2005). Racial/ethnic disparities in injection drug use in 94 U.S. metropolitan statistical areas. Annals of Epidemiology, 15, 326–334.

Cooper, H., Moore, L., Gruskin, S., & Krieger, N. (2004). Characterizing perceived police violence: implications for public health. American Journal of Public Health, 94, 1109–1118.

Cooper, H., Moore, L., Gruskin, S., & Krieger, N. (2005) The impact of a police drug crackdown on drug injectors' ability to practice harm reduction: a qualitative study. Social Science and Medicine, 2005a, 61, 673–684.

Corey, S.D., Burris, S., Kraut-Becher, J., Lynch, K.G., & Metzger, D. (2005). Effects of an intensive street-level police intervention on syringe exchange program use in Philadelphia, PA. American Journal of Public Health, 95, 233–236.

Cottler, L.B., Compton, W.M. III, Mager, D., Spitznagel, E.L., & Janca, A. (1992). Posttraumatic stress disorder among substance users from the general population. The American Journal of Psychiatry, 149, 664–670.

County-Level Detailed Arrest and Offense Data. U.S. Dept. of Justice, Federal Bureau of Investigation. Uniform Crime Reporting Program, 1990–2002 [Computer file]. 3rd ICPSR ed. Ann Arbor, MI: Inter-University Consortium for Political and Social Research [producer and distributor].

Cox, S., & Shipley, M. (1997). Counting the uncatchable? An epidemiological method for counting drug misusers. Social Psychiatry and Psychiatric Epidemiology, 32, 19–23.

Crane, J. (1991). The epidemic theory of ghettos and neighborhood effects on dropping out and teenage childbearing. American Journal of Sociology, 96, 1226–1259.

Cressie, N. (2000). Geostatistical methods for mapping environmental exposures. In: P. Elliott, J.C. Wakefield, N.G. Best, & D.J. Briggs (eds.), Spatial Epidemiology: Methods and Applications (pp. 185–204). Oxford, England: Oxford University Press.

Cresswell, T. (2004). Place: A Short Introduction. Wiley Blackwell, Hoboken, NJ: Blackwell Publishing.

Crew, R.E., & Davis, B.C. (2003). Assessing the effects of substance abuse among applicants for TANF benefits: the outcome of a demonstration project in Florida. Journal of Health and Social Policy, 17(1), 39–53.

Crimmins, E.M., Hayward, M.D., & Seeman, T.E. (2004). Race/ethnicity, socioeconomic status, and health. In: N.B. Anderson, R.A. Bulatao, & B. Cohen (eds.), Critical Perspectives on Racial and Ethnic Differences in Health in Late Life (pp. 310–352). Washington, DC: The National Academies Press, National Academy of the Sciences.

Crofts, N., Louie, R., Rosenthal, D., & Jolley, D. (1996). The first hit: circumstances surrounding initiation into injecting. Addiction, 91, 1187–1196.

Crofts, N., Reid, G., & Deany, P. (1998). Injecting drug use and HIV infection in Asia: the Asian Harm Reduction Network. AIDS, 12(Suppl B), S69–S78.

Cromley, E.K., & McLafferty, S.L. (2002). GIS and Public Health. New York: Guilford Press.

Croner, C.M., Sperling, J., & Broome, F.R. (1996). Geographical information systems (GIS): new perspectives in understanding human health and environment relationships. Statistics in Medicine, 15, 1961–1977.

Crum, R., Lillie-Blanton, M., & Anthony, J.C. (1996). Neighborhood environment and opportunity to use cocaine and other drugs in late childhood and early adolescence. Drug and Alcohol Dependence, 43, 155–161.

Cunningham, J., & Liu, L. (2003). Impacts of federal ephedrine and pseudoephedrine regulations on methamphetamine-related hospital admissions. Addiction, 98(9), 1229–1237.

Curley, B. (2005). Katrina Causes Wave of Addiction Problems. Join Together Available at: http://www.jointogether.org/news/features/2005/katrina-causes-wave-of.html [On-line].

Curtis, D. (2004). Looking for strengths in response to AIDS: individual, group and public authority roles in strategy. Public Administration and Development, 24, 51–59.

Curtis, A., Mills, J., & Leitner, M. (2007). Katrina and vulnerability: the geography of stress. Journal of Health Care for the Poor and Underserved, 18, 315–330.

D'Aunno, T., & Vaughn, T.E. (1995). An organizational analysis of service patterns in outpatient drug abuse treatment units. Journal of Substance Abuse, 7(1), 27–42.

da Silva, S.A., Melo, L.M., & Ricardo, E. (2004). Spatial Analysis of Incidence Rates: a Bayesian Approach. Sao Paolo, Brazil: Universdade Federal do Parana.

Daker-White, G. (1997). Drug user's access to community-based services. Health and Place, 3(4), 217–228.

Dasgupta, N., Kramer, E.D., Zalman, M.A., Carino, S. Jr, Smith, M.Y., Haddox, J.D., et al. (2006). Association between non-medical and prescriptive usage of opioids. Drug and Alcohol Dependence, 82(2), 135–142.

Davis, J.R., & Tunks, E. (1990). Environments and addiction: a proposed taxonomy. The International Journal of the Addictions, 25, 805–826. De Leon, G. (1985). The therapeutic community: status and evolution. The International Journal of the Addictions, 20, 823–844.

Davis, R.C., Lurigio, A.J., & Rosenbaum, D.P. (1993). Drugs and the Community: Involving Community Residents in Combating the Sales of Illegal Drugs. Springfiled, IL: Charles C. Thomas.

Dear, M., Wilton, R., Gaber, S.L., & Takahashi, L.M. (1997). Seeing people differently: the sociospatial construction of disability. Environment and Planning: D Society and Space, 15, 455–480.

Dear, M., & Wolch, J. (1987). Landscapes of Despair: From Deinstitutionalization to Homelessness. Oxford: Polity Press.

Dear, M.J. (1988). The postmodern challenge: reconstructing human geography. Transactions of the Institute of British Geographers, 13, 262–274.

Dee, T.S. (2001). Alcohol abuse and economic conditions: evidence from repeated cross-sections of individual-level data. Health Economics, 10, 257–270.

Dehne, K., Khodakevich, L., Hamers, F., & Schwartlander, B. (1999). The HIV/AIDS epidemic in Eastern Europe: recent patterns and trends and the implications for policy-making. AIDS, 13, 741–749.

Dehne, K., & Kobyshcha, Y. (2000). The HIV Epidemic in Central and Eastern Europe Update 2000.Geneva: UNAIDS.

Delva, J., Mathiesen, S., & Kamata, A. (2001). Use of illegal drugs among mothers across racial/ethnic groups in the United States: a multi-level analysis of individual and community-level influences. Ethnicity and Disease, 11, 614–625.

DeMarco, R., & Johnsen, C. (2003). Taking action in communities: women living with HIV/AIDS lead the way. Journal of Community Health Nursing, 20(1), 51–62.

Deng, R., Li, J., Sringernyuang, L., & Zhang, K. (2007). Drug abuse, HIV/AIDS and stigmatisation in a Dai community in Yunnan, China. Social Science and Medicine, 64, 1560–1571.

Denton, N. (1994). Are African Americans still hypersegregated? In: R. Bullard, J. Grigsby, C. Lee, & J.R. Feagin, (eds.) Residential Apartheid: The American Legacy, vol. 2 (pp. 49–81) Los Angeles: CAAS Publications, pp. 244–245.

Deren, S., Shedlin, M., Hamilton, T., & Hagan, H. (2002). Impact of the September 11th attacks in New York City on drug users: a preliminary assessment. Journal of Urban Health, 79, 409–412.

Des Jarlais, D.C. (2000). Research, politics, and needle exchange. American Journal of Public Health, 90(79), 1392–1394.

Des Jarlais, D.C., & Friedman, S.R. (1998). Fifteen years of research on preventing HIV infection among injecting drug users: what we have learned, what we have not learned, what we have done, what we have not done. Public Health Reports, 113(Suppl 1), 182–188.

Des Jarlais, D.C., Marmor, M., Paone, D., Titus, S., Shi, Q., Perlis, T., & Friedman, S. (1996). HIV incidence among syringe exchange participants in New York City. The Lancet, 348, 987–991.

Des Jarlais, D.C., Paone, D., Friedman, S.R., Peyser, N., & Newman, R.G. (1995). Public health then and now: regulating controversial programs for unpopular people: methadone maintenance and syringe-exchange programs. American Journal of Public Health, 85(11), 1577–1584.

Des Jarlais, D.C., Perlis, T.E., & Settembrino, J.M. (2005). The use of electronic debit cards in longitudinal data collection with geographically mobile drug users. Drug and Alcohol Dependence, 77, 1–5.

Desimone, J., & Farrelly, M.C. (2003). Price and enforcement effects on cocaine and marijuana demand. Economic Inquiry, 41, 98–115.

Desmond, D.P., & Maddux, J.F. (1984). Mexican–American heroin addicts. The American Journal of Drug and Alcohol Abuse, 10(3), 317–346.

Devine, O.J., Louis, T.A., & Halloran, M.E. (1994). Empirical Bayes methods for stabilizing incidence rates before mapping. Epidemiology, 5, 622–630.

DFID (UK Department for International Development). (2006). Working Document Knowledge for Action in HIV/AIDS in the Russian Federation: Report of findings for consultation among partners. October 2006, available at http://www.dfid.gov.uk/pubs/files/russia-aids.doc.

Diez-Roux, A.V. (1998). Bringing context back into epidemiology: variables and fallacies in multilevel analysis. American Journal of Public Health, 88, 216–222.

Diez-Roux, A.V. (2000). Multilevel analysis in public health research. Annual Review of Public Health, 21, 171–192.

Diez Roux, A.V. (2001). Investigating neighborhood and area effects on health. American Journal of Public Health, 91(11), 1783–1789.

Diez-Roux, A.V. (2003). Residential environments and cardiovascular risk. Journal of Urban Health, 80(4), 569–589.

Diez Roux, A.V. (2004). Estimating neighborhood health effects: the challenges of causal inference in a complex world. Social Science and Medicine, 58(10), 1953–1960.

Diez-Roux, A.V., Kiefe, C.I., Jacobs, D.R. Jr, Haan, M., Jackson, S.A., Nieto, F.J., et al. (2001b). Area characteristics and individual-level socioeconomic position indicators in three population-based epidemiologic studies. Annals of Epidemiology, 11(6), 395–405.

Diez Roux, A.V., Merkin, S.S., Arnett, D., Chambless, L., Massing, M., Nieto, F.J., et al. (2001a). Neighborhood of residence and incidence of coronary heart disease. The New England Journal of Medicine, 345(2), 99–106.

Dijkhuis, H., Zwerling, C., Parrish, G., Bennett, T., & Kemper, H.C. (1994). Medical examiner data in injury surveillance: a comparison with death certificates. American Journal of Epidemiology, 139(6), 637–643.

Dijst, M., & Kwan, M.P. (2005). Accessibility and the quality of life: time-geographic perspectives. In: K. Donaghy, G. Rudinger, & S. Poppelreuter (eds.), Social Dimensions of Sustainable Transport: Transatlantic Perspectives (pp. 109–126). Aldershot: Ashgate.

DiNardo, J., & Lemieux, T. (1992). Alcohol, marijuana, and American youth: the unintended consequences of government regulation. National Bureau of Economic Research Working Paper No. 4212. Cambridge, MA.

Ditton, J., & Frischer, M. (2001). Computerized projection of future heroin epidemics: a necessity for the 21st century? Substance Use and Misuse, 36(1&2), 151–166.

Dixon, P. (2007). Managing acute heroin overdose. Emergency Nurse, 15(2), 30–35.

Donnelly, N., Poynton, S., Weatherburn, D., Bamford, E., & Nottage, J. (2006). Liquor outlet concentrations and alcohol-related problems. Alcohol Studies Bulletin, No. 8, NSW Bureau of Crime Statistics and Research, Sydney.

Doran, B.J., & Lees, B.G. (2005). Investigating the spatiotemporal links between disorder, crime, and the fear of crime. The Professional Geographer, 57(1), 1–12.

Dorling, D., Barford, A., & Newman, M. (2006). WORLDMAPPER: the world as you've never seen it before. IEEE Transactions on Visualization and Computer Graphics, 12, 757–764.

Downing, M., Vernon, K., Riess, T.H., Mulia, N., McKnight, C., Des Jarlais, D.C., & Edlin, B.R. (2005). Community acceptance and implementation of HIV prevention interventions for injection drug users. AIDS Education and Prevention, 17(1), 68–78.

Doyle, T.J., & Bryan, R.T. (2000). Infectious disease morbidity in the US region bordering Mexico, 1990–1998. The Journal of Infectious Diseases, 182(5), 1503–1510.

Drake, R.E., Xie, H., McHugo, G.J., & Shumway, M. (2004). Three-year outcomes of long-term patients with co-occurring bipolar and substance use disorders. Biological Psychiatry, 56, 749–756.

Drebing, C.E., Fleitas, R., Moore, A., Krebs, C., Van Ormer, A., Penk, W., Seibyl, C., & Rosenheck, R. (2002). Patterns in work functioning and vocational rehabilitation associated with coexisting psychiatric and substance abuse disorders. Rehabilitation Counseling Bulletin, 46, 5–13.

Drucker, E. (1990). Notes from the drug wars, through the eye of the needle III: In a dark season, one million points of light. Drugtext: the Internet Center for Substance Use Related Risk Reduction. Available at: http://www.drugtext.org/library/articles/902304.htm.

Drucker, E. (1999). Drug prohibition and public health: 25 years of evidence. Public Health Reports, 114, 14–29.

Drucker, E. (2002). Population impact of mass incarceration under New York's Rockefeller drug laws: an analysis of years of life lost. Journal of Urban Health, 79, 1–10.

Drucker, E., Lurie, P., Wodak, A., & Alcabes, P. (1998). Measuring harm reduction: the effects of needle and syringe exchange programs and methadone maintenance on the ecology of HIV. AIDS, 12(Suppl A), S217–S230.

Drummer, O.H. (2005). Recent trends in narcotic deaths. Therapeutic Drug Monitoring, 27(6), 738–740.

Du Bois, W.E.B. (1962). Black Reconstruction in American 1860–1880. New York: Meridian Books.

DUF. (1995). Annual Report on Adult and Juvenile Arrestees. Washington, DC: U. S. Department of Justice.

DUF. (1996). Annial Report on Adult and Juvenile Arrestees. Washington, DC: U. S. Department of Justice.

Duffy, M. (2004). Drug abuse patterns and trends in Louisiana. Proceedings of the Louisiana State Epidemiology Work Group Baton Rouge: Department of Health and Hospitals, Office of Addictive Disorders, State of Louisiana.

Dummer, T.J.B. (2002). GIS, mapping and health: a study of stillbirth rates in Cumbria. The Bulletin of the Society of University Cartographers, 36(2), 31–38.

Duncan, C., Jones, K., & Moon, G. (1993). Do places matter? A multi-level analysis of regional variations in health-related behaviour in Britain. Social Science and Medicine, 37, 725–733.

Duncan, C., Jones, K., & Moon, G. (1996). Health-related behaviour in context: a multilevel modelling approach. Social Science and Medicine, 42(6), 817–830.

Duncan, C., Jones, K., & Moon, G. (1998). Context, composition and heterogeneity: using multilevel models in health research. Social Science and Medicine, 4, 97–117.

Duncan, C., Jones, K., & Moon, G. (1999). Smoking and deprivation: are there neighbourhood effects? Social Science and Medicine, 48, 497–505.

Duncan, G.J., & Aber, J.L. (1997). Neighborhood models and measures. In: J. Brooks-Gunn, G.J. Duncan, & J.L. Aber (eds.), Neighborhood Poverty: Context and Consequences for Children (pp. 62–78). New York: Russell Sage Foundation.

Duncan, S.C., Duncan, T.E., & Strycker, L.A. (2002). A multilevel analysis of neighborhood context and youth alcohol and drug problems. Prevention Science, 3, 125–133.

Dunn, J., & Laranjeira, R. (1999). Transition in the route of cocaine administration – characteristics, direction, and associated variables. Addiction, 94, 813–824.

Dupere, V., & Perkins, D.D. (2007). Community types and mental health: a multilevel study of local environmental stress and coping. American Journal of Community Psychology, 39, 107–119.

Dyck, I. (1995). Hidden geographies: the hidden lifeworlds of women with multiple sclerosis. Social Science and Medicine, 40, 307–320.

Eberstadt, N. (2002). The Future of Aids. Foreign Affairs, November/December. Available at: http://www.foreignaffairs.org/20021101faessay9990-p10/nicholas-eberstadt/the-future-of-aids.html (accessed 21 July, 2005).

Eck, J. (1995). Drug markets and drug places: a case control study of the spatial structure of illicit drug dealing. Dissertation Abstracts International – The Humanities and Social Sciences, University of Maryland, 56(1).

Eck, J.E. (1995). A general model of the geography of illicit retail market places. In: J.E. Eck, & D. Weisburd (eds.), Crime and Place: Crime Prevention Studies, vol. 4 (pp. 67–94). Monsey, New York: Criminal Justice Press.

Eck, J.E., Spencer, C., James, C., Michael, L., & Ronald, W. (2005). Mapping Crime: Understanding the Hot Spots (NCJ 209393). Washington, DC: US Department of Justice.

Edwards, A., Elwyn, G., & Mulley, A. (2002). Explaining risks: turning numerical data into meaningful pictures. BMJ, 324, 827–830.

Elgethun, K., Fenske, R., Yost, M., & Palcisko, G. (2003). Time-location analysis for exposure assessment studies of children using a novel global positioning system instrument. Environmental Health Perspectives, 111(1), 115–122.

Ellaway, A., & MacIntyre, S. (1998). Does housing tenure predict health in the UK because it exposes people to different levels of housing related hazards in the home or its surroundings? Health and Place, 4, 141–150.

Elliot, J. (1999). An Unexpected Light: Travels in Afghanistan. New York: Picador.

Elliot, P., Cuzick, J., English, D., & Stern, R. (eds.) (1993). Geographical and Environmental Epidemiology: Methods for Small-Area Studies. Oxford, England: Oxford Medical Publications.

Elliot, P., Wakefield, J., Best, N., & Briggs, D. (eds.). (2000). Spatial Epidemiology: Methods and Applications. Oxford: Oxford University Press.

Elliott, D.S., Wilson, W.J., Huizinga, D., Sampson, R.J., Elliot, A., & Rankin, B. (1996). The effects of neighborhood disadvantage on adolescent development. Journal of Research in Crime and Delinquency, 33(4), 389–426.

Elliott, S., & Gillie, J. (1998). Moving experiences: a qualitative analysis of health and migration. Health and Place, 4, 293–311.

Emler, N. (1999). Moral character. In: V. Derlega, B. Winstead, & W. Jones (eds.), Personality: Contemporary Theory and Research. Chicago: Nelson-Hall 376–404.

English, P., Neutra, R., Scalf, R., Sullivan, M., Waller, L., & Zhu, L. (1999). Examining associations between childhood asthma and traffic flow using geographical information system. Environmental Health Perspectives, 107, 761–767.

Ennett, S.T., Flewelling, R.L., Lindrooth, R.C., & Norton, E.C. (1997). School and neighborhood characteristics associated with school rates of alcohol, cigarette, and marijuana use. Journal of Health and Social Behaviour, 38, 55–71.

Epstein, S. (1996). Impure Science: AIDS, Activism, and the Politics of Knowledge. Berkeley, CA: University of California Press.

Esbensen, F.-A., & Huizinga, D. (1990). Community structure and drug use: from a social disorganization perspective. Justice Quarterly, 7(4), 691–709.

Escobedo, L., & Ortiz, M. (2002). The relationship between liquor outlet density and injury and violence in New Mexico. Accident Analysis and Prevention, 34, 689–694.

ESRI. (2003a). ArcMap 8.3 [computer software]. Redlands, CA: ESRI.

ESRI. (2003b). ArcGIS StreetMap USA [DVD]. Redlands, CA: ESRI.

ESRI. (2005). ArcGIS. Redlands: ESRI.

ESRI. (2006). GIS and Emergency Management in Indian Ocean Earthquake/Tsunami Disaster. Report. No. ESRI White Paper, accessed 9 April, 2006. ESRI.

European Centre for the Epidemiological Monitoring of AIDS [ECEMA]. (2002). HIV/AIDS Surveillance in Europe. Available at: http://www.eurohiv.org/reports/report_68?pdf/report_eurohiv_68.pdf (accessed 20 July, 2005).

Evans, G.W. (2003). The built environment and mental health. Journal of Urban Health, 80(4), 536–555.

Ewart, C.K., & Suchday, S. (2002). Discovering how urban poverty and violence affect health: development and validation of a neighborhood stress index. Health Psychology, 21, 254–262.

Factor, S.H., Wu, Y., Monserrate, J., Edwards, V., Cuevas, Y., Del Vecchio, S., et al. (2002). Drug use frequency among street-recruited heroin and cocaine users in Harlem and the Bronx before and after September 11, 2001. Journal of Urban Health, 79, 404–408.

Fang, Y., Wang, Y., Shi, J., Liu, Z., & Lu, L. (2006). Recent trends in drug abuse in China. Acta Pharmacologic Sinica, 27(2), 140–144.

Faris, R.E., & Dunham, H.W. (1939). Mental Disorders in Urban Areas. Chicago: University of Chicago Press.

Farmer, P. (1992). AIDS and Accusation: Haiti and the Geography of Blame. Berkeley and Los Angeles, CA: University of California Press.

Farmer, P. (1999). Infections and inequalities: The modern plagues. Berkeley, CA: University of California Press.

Farrell, S.P., Blank, M., Koch, J.R., Munjas, B., & Clement, D.G. (1999). Predicting whether patients receive continuity of care after discharge from state hospitals: policy implications. Archives of Psychiatric Nursing, 13, 279–285.

Farrell, S.P., Koch, J.R., Blank, M., (1996). Rural and urban differences in continuity of care after state hospital discharge. Psychiatr. Serv. 47, 652–654.

Ferreira-Pinto, J., & Ramos, R.L. (1997). Migration of injecting drug users: an exploratory study and research setting agenda. Journal of Border Health, 2, 35–41.

Field, K., & Beale, L. (2004). Using GIS to model incidence, prevalence and spread of non-legal drug use. Transactions in GIS, 8(4), 423–439.

Field, K., Beale, L., Heatlie, H., & Frischer, M. (2001). Using a geographic information system to forecast the diffusion of drug misuse. Proceedings of the 21st International ESRI User Conference. San Diego, CA July 2001.

Fillingim, R.B., & Gear, R.W. (2004). Sex differences in opioid analgesia: clinical and experimental findings. European Journal of Pain, 8(5), 413–425.

Finan, K. (2005). Houston works to treat wave of evacuee addicts. Houston Chronicle, Houston, TX, Sept. 15, 2005, p.1/B.

Finkle, B.S., McCloskey, K.L., Kiplinger, G.F., & Bennett, I.F. (1976). A national assessment of propoxyphene in postmortem medicolegal investigation, 1972–1975. Journal of Forensic Sciences, 21(4), 706–741.

Foa, E.B., Stein, D.J., & McFarlane, A.C. (2006). Symptomatology and psychopathology of mental health problems after disaster. The Journal of Clinical Psychiatry, 67(Suppl 2), 15–25.

Foley, R., & Platzer, H. (2007). Place and provision: mapping mental health advocacy services in London. Social Science and Medicine, 64, 617–632.

Folsom, R.E., Shah, B., & Vaish, A. (1999). Substance abuse in states: a methodological report on model based estimates from the 1994–1996 National Household Surveys on Drug Abuse. Proceedings of the American Statistical Association, Survey Research Methods 1999, Section (pp. 371–375). Washington, DC: American Statistical Association.

Foner, E. (2002). Reconstruction: America's unfinished revolution 1863–1877. In: H.S. Commager, & R.B. Morris (eds.), The New American Nation Series. New York: Perennial Classics.

Ford, N., & Koetsawang, S. (1991). The socio-cultural context of the transmission of HIV in Thailand. Social Science and Medicine, 33, 405–414.

Fortney, J.C., Booth, B.M., Blow, F.C., Bunn, J.Y., & Cook, C.A.L. (1995). The effects of travel barriers and age on the utilization of alcoholism treatment aftercare. The American Journal of Drug and Alcohol Abuse, 21, 391–406.

Fortney, J.C., Lancaster, A.E., Owen, R.R., & Zhang, M. (1998). Geographic market areas for psychiatric and medical outpatient treatment. Journal of Behavioural Health Services and Research, 25, 108–116.

Fortney, J.C., Sullivan, G., Williams, K., Jackson, C., Morton, S.C., & Koegel, P. (2003). Measuring continuity of care for clients of public mental health systems. Health Services Research, 38, 1157–1175.

Fox, J. (1991). Regression diagnostics. In: M.S. Lewis-Beck (ed.), Quantitative Applications in the Social Sciences. Newbury Park, CA: Sage Publications.

Frabutt, J.M., & Gathings, M.J. (eds.). (2005). Research panel presentation at the North Carolina statewide Project Safe Neighborhoods Conference: The Power of Data to Drive the Strategies, Evaluating Programs and Strategic Interventions: Offender notification sessions across the Middle District. Winston-Salem, NC. Washington, DC: U.S. Department of Justice.

Frabutt, J.M., Gathings, M.J., Hunt, E.D., Loggins, T.J. (Eds). (2004). Report of the Center for Youth, Family, and Community Partnerships at the University of North Carolina at Greensboro. High Point West End Initiative: Project Description, Log, and Preliminary Impact Analysis. University of North Carolina at Greensboro, NC.

Francaviglia, R.V. (1978). The Mormon Landscape. New York: AMS Press.

Frank, R., Cerda, M., & Rendon, M. (2007). Barrios and burbs: residential context and health-risk behavior among Angeleno adolescents. Journal of Health and Social Behavior, Vol. 48, No. 3, September 2007, pp. 283–300 (18) American Sociological Association.

Freisthler, B., & Gruenewald, P.J. (2005). Social ecology and the invention of new regulatory strategies for preventing drug and alcohol problems. In: T. Stockwell, P.J. Gruenewald, J.W. Toumbourou, & W. Loxley (eds.), Preventing Harmful Substance Use: The Evidence Base for Policy and Practice (pp. 291–306). Chichester, West Sussex, England: John Wiley & Sons, Ltd.

Freisthler, B., Gruenewald, P.J., Johnson, F.W., Treno, A.J., & LaScala, E.A. (2005) An exploratory study examining the spatial dynamics of illicit drug availability and rates of drug use. Journal of Drug Education, 35, 15–27.

Freisthler, B., LaScala, E.A., Gruenewald, P.J., & Treno, A.J. (2005) An examination of drug activity: effects of neighborhood social organization on the development of drug distribution systems. Substance Use and Misuse, 40, 671–686.

Freisthler, B., Needell, B., & Gruenewald, P.J. (2005). Is the physical availability of alcohol and illicit drugs related to neighborhood rates of child maltreatment? Child Abuse and Neglect, 29, 1049–1060.

Friedman, J. (1977). Central business district and residential urban renewal: response to undesired change? Social Science Quarterly, 58, 45–49.

Friedman, J.J. (1977). Community action on water pollution. Human Ecology, 5, 329–353.

Friedmann, P.D., Lemon, S.C., Stein, M.D., Etheridge, R.M., & D'Aunno, T.A. (2001). Linkage to medical services in the drug abuse treatment outcome study. Medical Care, 39, 284–295.

Friedman, S.R. (1998). The political economy of drug-user scapegoating and the philosophy and politics of resistance. Drugs: Education, Prevention, and Policy, 5(1), 15–32.

Friedman, S.R. (2002). Sociopharmacology of drug use: initial thoughts. The International Journal of Drug Policy, 13, 341–347.

Friedman, S.R., Curtis, R., Neaigus, A., Jose, B., & Des Jarlais, D.C. (1999). Social Networks, Drug Injectors' Lives and HIV/AIDS. New York: Kluwer/Plenum.

Friedman, S.R., Lieb, S., Tempalski, B., et al. (2005). HIV among injection drug users in large US metropolitan areas, 1998. Journal of Urban Health, 82, 434–445.

Friedman, S.R., Perlis, T., & Des Jarlais, D.C. (2001). Laws prohibiting over-the-counter syringe sales to injection drug users: relations to population density, HIV prevalence, and HIV incidence. American Journal of Public Health, 91(5), 791–793.

Friedman, S.R., Perlis, T., Lynch, J., & Des Jarlais, D.C. (2000). Economic inequality, poverty, and laws against syringe access as predictors of metropolitan area rates of drug injection and HIV infection. Proceedings of the Global Research Network Meeting (pp. 147–149). Durban, South Africa.

Friedman, S.R., Quimby, E., Sufian, M., Abdul-Quader, A., & Des Jarlais, D.C. (1988). Racial aspects of the AIDS epidemic. California Sociologist, 11, 55–68.

Friedman, S.R., Sotheran, J.L., Abdul-Quader, A., et al. (1987). The AIDS epidemic among Blacks and Hispanics. The Milbank Quarterly, 65, 455–499.

Friedman, S.R., Tempalski, B., Brady, J.E., Friedman, J.J., Cooper, H.L.F., Flom, P.L., McGrath, M.M., Gostnell, K., & Des Jarlais, D.C. (2007). Predictors of the degree of drug treatment coverage for injection drug users in 94 metropolitan areas in the United States of America. The International Journal on Drug Policy, 18(6), 475–485.

Friedman, S.R., Tempalski, B., Cooper H., et al. (2004c). Estimating number of injecting drug users in metropolitan areas for structural analyses of community vulnerability and for assessing relative degrees of service provision for injecting drug users. Journal of Urban Health, 81, 377–400.

Friedman, S.R., Tempalski, B., Cooper, H., Keen, M., Friedman, R., Flom, P.L., & Des Jarlis, D.C. (2004a). Structural analysis for structural intervention: predictors of IDU population size and HIV prevalence among IDUs in U.S. cities. Oral Presentation at the XV International AIDS Conference 11–16 July 2004, Bangkok, Thailand.

Friedman, S.R., Tempalski, B., Cooper, H., Perlis, T., Keem, M., Friedman, R., & Flom, P.L. (2004b). Estimating numbers of IDUs in metropolitan areas for structural analyses of community vulnerability and for assessing relative degrees of service provision for IDUs. Journal of Urban Health, 8(3), 377–400.

Frye, M.A., & Salloum, I.M. (2006). Bipolar disorder and comorbid alcoholism: prevalence rates and treatment considerations. Bipolar Disorders, 8, 677–685.

Frye, V., Latka, M.H., Koblin, B., Halkitis, P.N., Putnam, S., Galea, S., & Vlahov, D. (2006). The urban environment and sexual risk behaviour among men who have sex with men. Journal of Urban Health: Bulletin of the New York Academy of Medicine, 83(2), 308–324.

Fuller, C., Borrell, L., Latkin, C.A., et al. (2005). Effects of race, neighborhood, and social network on age at initiation of injection drug use. American Journal of Public Health, 95, 689–695.

Fuller, C., Vlahov, D., Arria, A., Ompad, D., Garfein, R., & Strathdee, S.A. (2001). Factors associated with adolescent initiation of injection drug use. Public Health Reports, 116, 136–145.

Fullilove, M.T. (1993). Perceptions and misperceptions of race and drug use. JAMA, 269, 1034.

Fullilove, M.T. (2001). Root shock: the consequences of African American dispossession. Journal of Urban Health, 78, 72–80.

Fussell, E. (2005). Leaving New Orleans: Social Stratification, Networks, and Hurricane Evacuation. Available at: http://understandingkatrina.ssrc.org [On-line].

Galea, S., Ahern, J., & Karpati, A. (2005). A model of underlying socioeconomic vulnerability in human populations: evidence from variability in population health and implications for public health. Social Science and Medicine, 60(11), 2417–2430.

Galea, S., Ahern, J., Resnick, H., Kilpatrick, D., Bucuvalas, M., Gold, J., et al. (2002a). Psychological sequelae of the September 11 terrorist attacks in New York City. The New England Journal of Medicine, 346, 982–987.

Galea, S., Ahern, J., Tardiff, K., et al. (2003). Racial/Ethnic differences in overdose mortality trends in New York City, 1990–1998. Journal of Urban Health, 80, 201–211.

Galea, S., Nandi, A., & Vlahov, D. (2004). The social epidemiology of substance use. Epidemiologic Reviews, 26, 36–52.

Galea, S., Resnick, H., Ahern, J., Gold, J., Bucuvalas, M., Kilpatrick, D., et al. (2002). Posttraumatic stress disorder in Manhattan, New York City, after the September 11th terrorist attacks. Journal of Urban Health, 79, 340–353.

Galea, S., Rudenstine, S., & Vlahov, D. (2005b). Drug use, misuse, and the urban environment. Drug and Alcohol Review, 24, 127–136.

Galea, S., Vlahov, D., Ahern, J., Rudenstine, S., & Wallace, Z. (2005c). Urban built environment and depression: a multilevel analysis. Journal of Epidemiology and Community Health, 59(10), 822–827.

Galea, S., Worthington, N., Piper, T.M., Nandi, V.V., Curtis, M., & Rosenthal, D.M. (2006). Provision of naloxone to injection drug users as an overdose prevention strategy: early evidence from a pilot study in New York City. Addictive Behaviours, 31(5), 907–912.

Galobardes, B., Shaw, M., Lawler, A.L., Lynch, J.W., & Smith, G.D. (2006). Indicators of socioeconomic position (part 2). Journal of Epidemiology and Community Health, 60, 95–101.

Galster, G.C., & Mikelsons, M. (1995). The geography of metropolitan opportunity: a case study of neighborhood conditions confronting youth in Washington, DC. Housing Policy Debate, 6, 73–102.

Garcia, L.M., McGeary, K.A., Shultz, J.M., & McCoy, C.B. (1999). The impact of insurance status on drug abuse treatment completion. Journal of Health Care Finance, 26, 40–47.

Garcia, M.T. (1989). Mexican Americans: Leadership, Ideology, and Identity, 1930–1960. New Haven: Yale University Press.

Gaskin, S. (1970). Monday Night Class. Summertown, TN: The Farm.

Gatrell, A., & Bailey, T. (1996). Interactive spatial data analysis in medical geography. Social Science and Medicine, 42, 843–855.

GeoDa Software Manual (2004). GeoDa Center for Geospatial Analysis, Arizona State University, Tempe, AZ.

Georgia Vital Statistics Report. (2002). Office of Health Information and Policy, Division of Public Health, Georgia Department of Human Resources. Atlanta, GA.

Geronimus, A.T., & Bound, J. (1998). Use of census-bases aggregate variables to proxy for socioeconomic group: evidence from national samples. American Journal of Epidemiology, 148, 475–486.

Gesler, W. (1992). Therapeutic landscapes: medical issues in light of the new cultural geography. Social Science and Medicine, 34, 735–746.

Gesler, W., & Kearns, R. (2002). Culture/Place/Health. London: Routledge.

Gesler, W., & Meade, M. (1988). Locational and population factors in health-care-seeking behavior in Savannah, Georgia. Health Services Research, 23, 443–462.

Ghosh, M., & Rao, J.N.K. (1994). Small area estimation: an appraisal. Statistical Science, 9(1), 55–76.

Gibbons, F.X., Gerrard, M., Cleveland, M.J., Wills, T.A., & Brody, G. (2004). Perceived discrimination and substance use in African American parents and their children: a panel study. Journal of Personality and Social Psychology, 86, 517–529.

Gibbs, J.P. (ed.) (1982). Social Control: Views from the Social Sciences. Beverly Hills, CA: Sage Publications, Inc.

Gilchrist, V., et al. (2005). Physician activities during time out of the examination room. Annals of Family Medicine, 3, 494–499.

Gill, J.R., & Graham, S.M. (2002). Ten years of "body packers" in New York City: 50 deaths. Journal of Forensic Sciences, 47(4), 843–846.

Gillis, J.S., & Mubbashar, M.H. (1995). Risk factors for drug abuse in Pakistan: a replication. Psychological Reports, 76(1), 99–108.

Gilson, A.M., Ryan, K.M., Joranson, D.E., & Dahl, J.L. (2004). A reassessment of trends in the medical use and abuse of opioid analgesics and implications for diversion control: 1997–2002. Journal of Pain and Symptom Management, 28(2), 176–188.

Gladwell, M. (2002). The tipping point: how little things can make a big difference. New York: Back Bay Books.

Gladwin, H., & Peacock, W.G. (1997). Warning and evacuation: a night for hard houses. In: W.G. Peacock, B.H. Morrow, & H. Gladwin (eds.), Hurricane Andrew: Ethnicity, Gender, and the Sociology of Disasters (pp. 52–74). London, England: Routledge.

Gleghorn, A.A., Jones, T.S., Doherty, M.C., Celentano, D.D., & Vlahov, D. (1995). Acquisition and use of needles and syringes by injecting drug users in Baltimore, Maryland. Journal of Acquired Immune Deficiency Syndromes and Human Retrovirology, 10(1), 97–103.

Goel, R.K., & Morey, M.J. (1995). The interdependence of cigarette and liquor demand. Southern Economic Journal, 62, 451–459.

Goldgerg, J.F., et al. (1999). A history of substance abuse complicates remission from acute mania in bipolar disorder. Journal of Clinical Psychiatry, 60(11), 733–740.

Goldstein, P.J. (1998). Drugs, violence, and federal funding: a research odyssey. Substance Use and Misuse, 33, 1915–1936.

Golob, T.F. (2000). A simultaneous model of household activity and trip chain generation. Transportation Research B, 34, 355–376.

Gong, X., & Shao, C. (2001). The current status of STDs and the purposes and strategies of their prevention. In: C. Shao (ed.), Xingchuanbo Jibing de Yufang yu Kongzhi (pp. 1–7). Beijing: Science Publisher.

Goodchild, M.F., Janelle, D.G. (eds.), 2004. Spatially Integrated Social Science. Oxford: Oxford University Press.

Goodman, A.B., Siegel, C., Craig, T.J., & Lin, S.P. (1983). The relationship between socioeconomic class and prevalence of schizophrenia, alcoholism, and affective disorders treated by inpatient care in a suburban area. The American Journal of Psychiatry, 140, 166–170.

Gordon, D.A. (1995). Functional family therapy for delinquents. In: R.R. Ross, D.H. Antononwicz, & G.K. Dhaliwal (eds.), Going Straight: Effective Delinquency Prevention and Offender Rehabilitation. Ottawa, Ontario, Canada: AIR Training and Publications, pp. 244–245.

Gorman, D.M., Gruenewald, P.J., Hanlon, P.J., Mezic, I., Waller, L.A., Castillo-Chavez, C., Bradley, E., & Mezic, J. (2004). Implications of systems dynamic models and control theory for environmental approaches to the prevention of alcohol- and other drug use-related problems. Substance Use and Misuse, 39(10–12), 1713–1750.

Gorman, D.M., Speer, P.W., Gruenewald, P.J., & Labouvie, E.W. (2001). Spatial dynamics of alcohol availability, neighborhood structure and violent crime. Journal of Studies on Alcohol, 62, 628–636.

Gorman, D.M., Zhu, L., & Horel, S. (2005). Drug "Hot Spot", alcohol availability and violence. Drug and Alcohol Review, 24, 507–551.

Gottlieb, S., Arenberg, S., & Singh, R. (1994). Crime Analysis: From First Report to Final Arrest. Montclair, CA: Alpha Publishing, Co.

Grady, S.C. (2006). Racial disparities in low birthweight and the contribution of residential segregation: a multilevel analysis. Social Science and Medicine, 63(12), 3013–3029.

Graham, J. (2006). Katrina Survivors Getting Chance to Aid in Recovery: Harvard Recruiting 2,000 Victims to Help Disaster Relief Effort. Chicago: Chicago Tribune.

Grassly, N.C., Lowndes, C.M., Rhodes, T., Judd, A., Renton, A., & Garnett, G.P. (2003). Modelling emerging HIV epidemics: the role of injecting drug use and sexual transmission in the Russian Federation, China and India. The International Journal on Drug Policy, 14, 25–43.

Gray, J. (1995). Operating needle exchange programmes in the hills of Thailand. AIDS Care, 7(4), 489–499.

Gray, M. (1998). Drug Crazy: How We Got Into This Mess and How We Can Get Out. New York: Routledge.

Great Britain Royal Commission on Opium. (1894–1895). First Report of the Royal Commission on Opium: with Minutes of Evidences and Appendices. London: Printed for H.M. Stationary Office by Eyre and Spottiswoode.

Green, C.R., Ndao-Brumblay, S.K., West, B., & Washington, T. (2005). Differences in prescription opioid analgesic availability: comparing minority and white pharmacies across Michigan. The Journal of Pain, 6(10), 689–699.

Greenberg, G.A., Rosenheck, R.A., & Seibyl, C.L. (2002). Continuity of care and clinical effectiveness: outcomes following residential treatment for severe substance abuse. Medical Care, 40, 246–259.

Greene, J.D. (2002). Cities and Privatization: Prospects for the New Century. Upper Saddle River, NJ: Pearson Education.

Greenland, S. (2006). Bayesian perspectives for epidemiological research: I. Foundations and basic methods. International Journal of Epidemiology, 35, 765–775.

Grella, C.E., Joshi, V., & Hser, Y.I. (2000). Program variation in treatment outcomes among women in residential drug treatment. Evaluation Review, 24, 364–383.

Greve, H.R. (2002). An ecological theory of spatial evolution: local density dependence in Tokyo banking, 1894–1936. Social Forces, 80, 847–879.

Griffith, D.A., Wong, D.W.S., & Whitfield, T. (2003). Exploring relationships between the global and regional measures of spatial autocorrelation. Journal of Regional Science, 43(4), 683–710.

Groff, E., & La Vigne, N. (2001). Mapping an opportunity surface of residential burglary. Journal of Research in Crime and Delinquency, 38(3), 257–278.

Grossman, M., Chaloupka, F.J., & Sirtalan, I. (1998). An empirical analysis of alcohol addiction: results from the monitoring the future panels. Economic Inquiry, 36, 39–48.

Gruenewald, P. (2007). The spatial ecology of alcohol problems: niche theory and assortative drinking. Addiction, 102, 870–878.

Gruenewald, P.J., Freisthler, B., Remer, L., LaScala, E.A., & Treno, A. (2006). Ecological models of alcohol outlets and violent assaults: crime potentials and geospatial analysis. Addiction, 101(5), 666–677.

Gruenewald, P.J., Ponicki, W.R., & Holder, H.D. (1993). The relationship of outlet densities toalcohol consumption: a time series cross-sectional analysis. Alcoholism: Clinical and Experimental Research, 17, 38–47.

Gruenewald, P.J., & Remer, L. (2006). Changes in Outlet Densities Affect Violence Rates. Alcoholism: Clinical and Experimental Research, 30, 1184–1193.

Grund, J., Kaplan, C., & Adriaans, N. (1991). Featuring high-risk behavior: needle sharing in the Netherlands. American Journal of Public Health, 81(12), 1602–1607.

Guagliardo, M., Ronzio, C., Cheung, I., Chacko, E., & Joseph, J. (2004). Primary Care Physician Accessibility: An Urban Case Study of Pediatric Care. Health and Place, 10(3), 273–283.

Güereña-Burgueño, F., Benenson, A.S., Bucardo Amaya, J., Caudillo Carreno, A., & Curiel Figueroa, J.D. (1992). [Sexual behavior and drug abuse in homosexuals, prostitutes and prisoners in Tijuana, Mexico]. Revista Latinoamericana de Psicología, 24(1–2), 85–96.

Güereña-Burgueño, F., Benenson, A.S., & Sepulveda-Amor, J. (1991). HIV-1 prevalence in selected Tijuana sub-populations. American Journal of Public Health, 81(5), 623–625.

Guthrie, B.J., Young, A.M., Williams, D.R., Boyd, C.J., & Kinter, E.K. (2002). African American girls' smoking habits and day-to-day experiences with racial discrimination. Nursing Research, 51, 183–190.

Gwadz, M., Clatts, M., Yi, H., Leonard, N., Goldsamt, L., & Lankenau, S. (2006). Resilience among urban young men who have sex with men. Sexuality Research and Social Policy, 3(1), 13–21.

Gymiah-Brempong, K. (2006). Neighborhood income, alcohol availability, and crime rates. The Review of Black Political Economy, 3, 21–44.

Haenfler, R. (2006). Straight-Edge: Hardcore Punk, Clean-Living Youth, and Social Change. New Brunswick, NJ: Rutgers University Press.

Hagan, H., Thiede, H., & Des Jarlais, D.C. (2005). HIV/hepatitis C virus co-infection in drug users: risk behavior and prevention. AIDS, 19(Suppl 3), S199–S207.

Hagen, B.F., Armstrong-Esther, C., Quail, P., Williams, R.J., Norton, P., Le Navenec, C.L., et al. (2005). Neuroleptic and benzodiazepine use in long-term care in urban and rural Alberta: characteristics and results of an education intervention to ensure appropriate use. International Psychogeriatrics, 17(4), 631–652.

Hahn, J., Page-Shafer, K., Ford, J., Paciorek, A., Lum, P. (2008). Travelling young injection drug users at high risk for acquistion and transmission of viral infections. *Drug and Alcohol Dependence*, 93, 43–50.

Halloran, M.E. (1998). Concepts of infectious disease epidemiology. In: K.J. Rothman, & S. Greenland (eds.), Modern Epidemiology, 2nd Edition (pp. 529–554). New York: Lippincott Williams and Wilkins.

Hansten, M.L., Downey, L., Rosengren, D.B., & Donovan, D.M. (2000). Relationship between follow up rates and treatment outcomes in substance abuse research: more is better but when is enough enough? Addiction, 95, 1403–1418.

Harding, D.J. (2003). Counterfactual models of neighborhood effects: the effect of neighborhood poverty on high school dropout and teenage pregnancy. American Journal of Sociology, 109, 676–719.

Harlow, K. (1990). Patterns of rates of mortality from narcotics and cocaine overdose in Texas, 1976–1987. Public Health Reports, 105, 455–462.

Harocopos, A., & Hough, M. (2005). Drug dealing in open-air markets. Center for Problem Oriented Policing, Guide No.31. Available at: http://www.popcenter.org/Problems/problem-drug_open_markets.htm.

Harries, K.D. (1974). The Geography of Crime and Justice. New York: McGraw-Hill.

Harries, K.D. (1999). Mapping Crime: Principle and Practice (NCJRS Publication No. NCJ 178919). Washington, DC: National Institute of Justice.

Harries, K.D. (2002). Applications of geographic information systems in parole and probation. Perspective, Journal of the American Probation and Parole Association, 26, 26–31.

Harrison, J.P., & Sexton, C. (2004). The health care manager. Frederick, 23(3), 192.

Harrison, K., Vlahov, D., Jones, K., Charron, K., & Clements, M.L. (1995). Medical eligibility, comprehension of the consent process, and retention of injection drug users recruited for an HIV vaccine trial. Journal of Acquired Immune Deficiency Syndromes and Human Retrovirology, 10(3), 386–390.

Harrison, P., & Beck, A. (2005). Prisoners in 2004 (NCJRS Publication No. NCJ 210677). Washington, DC: U.S. Department of Justice, Bureau of Justice Statistics.

Harrison, P.M., & Beck, A.J. (2003). Prisoners in 2002 (Bulletin No. NCJ 200248). Bureau of Justice Statistics.

Hart, K., Kunitz, S., Sell, R., & Mukamel, D. (1998). Metropolitan governance, residential segregation, and mortality among African–Americans. American Journal of Public Health, 88, 434–438.

Hartwig, F., & Dearing, B.E. (1979). Exploratory data analysis. Beverly Hills: Sage Publications.

Harvey, D. (1996). On the history and present condition of geography: an historical materialist manifesto. In: J. Agnew, D.N. Livingstone, & A. Rogers (eds.), Human Geography: An Essential Anthology. Oxford: Blackwell Publishers, Ltd., pp. 95–107.

Harvey, E., Strathdee, S.A., Patrick, D.M., Ofner, M., Archibald, C.P., Eades, G., et al. (1998). A qualitative investigation into an HIV outbreak among injection drug users in Vancouver, British Columbia. AIDS Care, 10(3), 313–321.

Havens, J.R., Walker, R., & Leukefeld, C.G. (2007). Prevalence of opioid analgesic injection among rural nonmedical opioid analgesic users. Drug and Alcohol Dependence, 87(1), 98–102.

Hawkins, J.D., Catalano, R.F., & Miller, J.Y. (1992). Risk and protective factors for alcohol and other drug problems in adolescence and early adulthood: implications for substance abuse prevention. Psychological Bulletin, 112(1), 64–105.

Hawkins, J.D., Van Horn, M.L., & Arthur, M.W. (2004). Community variation in risk and protective factors and substance use outcomes. Prevention Science, 5(4), 213–220.

Haynes, R., Bentham, G., Lovett, A., & Gale, S. (1999). Effects of distances to hospital and GP surgery on hospital inpatient episodes controlling for needs and provision. Social Science and Medicine, 49, 425–433.

Health Insurance Portability and Accountability Act (HIPAA). (1996). Pub. L. No. 104–191, 119 Stat. 1936.

Heckathorn, D.D. (1997). Respondent-driven sampling: a new approach to the study of hidden populations. Social Problems, 44(2), 174–199.

Heckathorn, D. D. (2002). Respondent-driven sampling II: Deriving Valid Population Estimates from Chain referral Samples of Hidden Populations. Social Problems, 49, 11–34.

Heffernan, R., Mostashari, F., Das, D., Karpati, A., Kulldorff, M., & Weiss, D. (2004). Syndromic surveillance in public health practice, New York City. Emerging Infectious Diseases, 10, 858–864.

Hembree, C., Tracy, M., Markham Piper, T., Miller, J., Vlahov, D., Tardiff, K.J., Galea, S., & Ahern, J. (2005). The urban built environment and overdose mortality in New York City neighborhoods. Health and Place, 11(2), 147–156.

Henderson, G.L. (1991). Fentanyl-related deaths: demographics, circumstances, and toxicology of 112 cases. Journal of Forensic Sciences, 36(2), 422–433.

Herman, S.E., Frank, K.A., Mowbray, C.T., Ribisi, K.M., Davidson, W.S. II, Bootsmiller, B., Jordan, L., Greenfield, A.L., Loveland, D., & Luke, D.A., (2000). Longitudinal effects of integrated treatment on alcohol use for persons with serious mental illness and substance use disorders. The Journal of Behavioural Health Services and Research, 27, 286–304.

Herz, D.C. (2000). Drugs in the Heartland: Methamphetamine Use in Rural Nebraska. Research in Brief, National Institute of Justice, Rockville, MD.

Hickman, M., Cox, S., Harvey, J., et al. (1999). Estimating the prevalence of problem drug use in inner London: a discussion of three capture-recapture studies. Addiction, 94, 1653–1662.

High Point Police Department. (1999). High Point, NC: One city's success in reducing gun violence: Middle district of North Carolina. Unpublished workbook on building community participation and implementing the Violent Crime Task Force model.

High Point Police Department. (2000). One City's Success in Reducing Gun Violence. Award submission for the Herman Goldstein Award, Center for Problem-Oriented Policing. Available at: http://www.popcenter.org/library/goldstein/2000/00-09.PDF.

Hodgins, D.C., & El Guebaly, N. (1992). More data on the addiction severity index. Reliability and validity with the mentally ill substance abuser. The Journal of Nervous and Mental Disease, 180, 197–201.

Hodgson, N., Landsberg, L., Lehning, A., & Kleban, M. (2006). Palliative care services in Pennsylvania nursing homes. Journal of Palliative Medicine, 9(5), 1054–1058.

Hoffman, B. (2003). Health care reform and social movements in the United States. American Journal of Public Health, 93(1), 75–98.

Hoffmann, J.P. (2002). The community context of family structure and adolescent drug use. Journal of Marriage and the Family, 64(2), 314–330.

Hogan, D.P., & Kitagawa, E.M. (1985). The impact of social status, family structure, and neighborhood on the fertility of black adolescents. American Journal of Sociology, 90(4), 825–855.

Hohman, M., Oliver, R., & Wright, W. (2004). Methamphetamine abuse and manufacture: the child welfare response. Social Work, 49(3), 373–381.

Holder, H.D., Gruenewald, P.J., Ponicki, W.R., Treno, A.J., Grube, J.W., Saltz, R.F., Voas, R.B., Reynolds, R., Davis, J., Sanchez, L., Gaumont, G., & Roeper, P. (2000). Effect of community-based interventions on high-risk drinking and alcohol-related injuries. Journal of the American Medical Association, 284, 2341–2347.

Holmberg, S.D. (1996). The estimated prevalence and incidence of HIV in 96 large US metropolitan areas. American Journal of Public Health, 86, 642–654.

Holmberg, S.D., Conley, L.J., Luby, S.P., Cohn, S., Wong, L.C., & Vlahov, D. (1995). Recent infection with human immunodeficiency virus and possible rapid loss of CD4 T lymphocytes. Journal of Acquired Immune Deficiency Syndromes and Human Retrovirology, 9(3), 291–296.

Hong Kong Police. (2006). Crimes Statistics. Retrieved 29 November, 2006, from Hong Kong Police web site http://www.info.gov.hk/police/hkp-home/english/statistics/details.htm.

Hopfer, C.J., Mikulich, S.K., & Crowley, T.J. (2000). Heroin use among adolescents in treatment for substance use disorders. Journal of the American Academy of Child Adolescent Psychiatry, 39(10), 1316–1323.

Hoppe-Roberts, J.M., Lloyd, L.M., & Chyka, P.A. (2000). Poisoning mortality in the United States: comparison of national mortality statistics and poison control center reports. Annals of Emergency Medicine, 35(5), 440–448.

Howard, D.L., LaVeist, T.A., & McCaughrin, W.C. (1996). The effect of social environment on treatment outcomes in outpatient substance misuse treatment organizations: does race really matter? Substance Use and Misuse, 31(5), 617–638.

Hser, Y.I., Anglin, D., & Powers, K. (1993). A 24-year follow-up of California narcotics addicts. Archives of General Psychiatry, 50, 577–584.

Hser, Y.I., Evans, E., & Huang, D. (2004). Relationship between drug treatment services, retention, and outcomes. Psychiatric Services, 55, 767–774.

Hser, Y.I., Hoffman, V., Grella, C.E., & Anglin, M.D. (2001). A 33-year follow-up of narcotics addicts. Archives of General Psychiatry, 58, 503–508.

Hsing, Y. (1996). An analysis of arrests regarding illegal drugs: the determinants and policy implications. American Journal of Economics and Sociology, 55, 53–60.

Hu, Z., Liu, H., Li, X., Stanton, B., & Chen, X. (2006). HIV-related sexual behaviour among migrants and non-migrants in a rural area of China: role of rural-to-urban migration. Public Health, 120(4), 339–345.

Hubbard, R.L., Craddock, S.G., Flynn, P.M., Anderson, J., & Etheridge, R.M. (1997). Overview of 1-year follow-up outcomes in the Drug Abuse Treatment Outcomes Study (DATOS). Psychology of Addictive Behaviours, 11, 261–278.

Hughes, A.A., Bogdan, G.M., & Dart, R.C. (2007). Active surveillance of abused and misused prescription opioids using poison center data: a pilot study and descriptive comparison. Clinical Toxicology (Philadelphia, PA), 45(2), 144–151.

Hughes, P.H., & Crawford, G.A. (1974). The high drug use community: a natural laboratory for epidemiological experiments in addiction control. American Journal of Public Health, 64(Suppl 12), 11–15.

Human Rights Watch [HRW]. (2004). Lessons not learned: human rights abuses and HIV/AIDS in the Russian Federation. 16(5) D, New York: Human Rights Watch Publications.

Hunt, C.W. (1989). Migrant labor and sexually transmitted diseases: AIDS in Africa. Journal of Health and Social Behavior, 30, 353–373.

Hunt, D., Kuck, S., & Truitt, L. (2006). Methamphetamine Use: Lessons Learned. Cambridge, MA: Abt Associates.

Hunt, E.D. (1992). Upgrading site-catchment analyses with the use of GIS: investigating the settlement patterns of horticulturalists. World Archaeology, 24(2), 283–309.

Hunt, L., & Chambers C. (1976). The Heroin Epidemics: A Study of Heroin Use in the United States, 1965–1975. New York: Spectrum.

Hurley, S.F., Jolley, D.J., & Kaldor, J.M. (1997). Effectiveness of needle-exchange programs for prevention of HIV infection. The Lancet, 21, 1797–1800.

Hyde, J. (2005). From home to street: understanding young people's transitions into homelessness. Journal of Adolescence, 28, 171–183.

Imperial College and Medical Research Council. (2004). WinBUGS Version 1.4.1 [Computer Software]. London.

India Census Commissioner. (1891). Census of India. Calcutta, London: Printed at Eyre and Spottiswoode, 1892–1893.

India Census Commissioner. (1901). Census of India. Calcutta, India: Office of the Superintendent of Government Printing, 1903.

India Census Commissioner. (1911). Census of India. Calcutta, India: Office of the Superintendent of Government Printing, 1913.

India Census Commissioner. (1921). Census of India. Calcutta, India: Office of the Superintendent of Government Printing, 1921–24.

India Census Commissioner. (1931). Census of India. Delhi, India: Manager of Publications, 1932–37.

India Census Commissioner. (1941). Census of India. Delhi, India: Manager of Publications, 1941.

Indian Hemp Drugs Commission: Marijuana. (1894). Report of the Indian Hemp Drugs Commission. Simla: Government Central Printing Office.

Indian Hemp Drugs Commission: Marijuana. (1969). Report of the Indian Hemp Drugs Commission, 1893–1894. Silver Spring, MD: Thomas Jefferson Publishing Company, 1969. ("Hemp Report", Kaplan (1969). reprint in text).

INEGI. (2000a). Sistema para la consulta de informacion censal. SCINCE por colonias. XII Censo de Poblacion y Vivienda: Instituto Nacional de Estadística Geografía e Informática.

Institute of Medicine Committee on Understanding and Eliminating Racial and Ethnic Disparities in Health Care. (2003). In: B.D. Smedley, A.Y. Stith, A.R. Nelson (eds.), Unequal Treatment: Confronting Racial and Ethnic Disparities in Health Care. Washington, DC: The National Academies Press.

International Narcotics Control Board. (2006). Report of the International Narcotics Control Board for 2005. Retrieved from.

Irwin, K.L., Edlin, B., Faruque, S., et al. (1996). Crack cocaine smokers who turn to drug injection: characteristics, factors associated with injection, and implications for HIV transmission. The Multicenter Crack Cocaine and HIV Infection Study Team. Drug and Alcohol Dependence, 42, 85–92.

Jackson, J.B. (1984). Discovering the Vernacular Landscape. New Haven: Yale Univesity Press.

Jackson, J.E., Doescher, M.P., & Hart, L.G. (2006). Problem drinking: rural and urban trends in America, 1995/1997 to 2003. Preventive Medicine, 43(2), 122–124.

Jackson, S., Anderson, R., Johnson, N., & Sorlie, P. (2000). The relation of residential segregation to all-cause mortality: a study in black and white. American Journal of Public Health, 90, 615–617.

Jacobson, J. (1999). Policing Drug Hot-Spots. In: Barry Webb (ed.), Home Office Policing and Reducing Crime Unit Research, Development and Statistics Directorate.

Jacobson, J. (2004c). Place and attrition from substance abuse treatment. Journal of Drug Issues, 34, 23–49.

Jacobson, J.O. (2004a). Place and attrition from substance abuse treatment. Journal of Drug Issues, Winter 34, 23–50.

Jacobson, J.O. (2004b). The ecological context of substance abuse treatment outcomes: implications for NIMBY disputes and client placement decisions. Dissertation, The Rand Graduate School.

Jacobson, J.O. (2006). Do drug treatment facilities increase clients' exposure to potential neighborhood-level triggers for relapse? A small-area assessment of a large, public treatment system. Journal of Urban Health, 83(2), 150–161.

Jacobson, J.O., Robinson, P., & Bluthenthal, R.N. (2006). A multilevel decomposition approach to estimate the role of program location and neighborhood disadvantage in racial disparities in alcohol treatment completion. Social Science and Medicine.

James, K.E., Wagner, F.A., & Anthony, J.C. (2002). Regional variation in drug purchase opportunity among youths in the United States. Journal of Urban Health, 79(1), 104–112.

James, L., Matthews, I., & Nix, B. (2004). Spatial contouring of risk: a tool for environmental epidemiology. Epidemiology, 15(3), 287–292.

Jang, S.J., & Johnson, B.R. (2001). Neighborhood disorder, individual religiosity, and adolescent use of illicit drugs: a test of multilevel hypotheses. Criminology, 39(1), 109–144.

Jansen, K.L. (2001). Ketamine: Dreams and Realities. Sarasota, FL: Multidisciplinary Association for Psychedelic Studies.

Jauncey, M.E., Taylor, L.K., & Degenhardt, L.J. (2005). The definition of opioid-related deaths in Australia: implications for surveillance and policy. Drug and Alcohol Review, 24(5), 401–409.

Jayakody, A.A., Bhui, K.S., Head, J.A., Taylor, S.J.C., Booy, R., Klineberg, E., Clark, C., Stansfeld, S.A., Viner, R.M., & Haines, M.M. (2006). Illicit and traditional drug use among ethnic minority adolescents in East London. Public Health, 120(4), 329–338. Luo, W., & Wang, F. (2003). Measures of spatial accessibility to health care in a GIS environment: synthesis and a case study in the Chicago region. Environment and Planning B: Planning and Design, 30(6), 865–884.

Jencks, C., & Mayer, S.E. (1990). The social consequences of growing up in a poor neighborhood. In: L.E. Lynn, & M.G.H. McGeary (eds.), Inner-City Poverty in the United States (pp. 111–186). Washington, DC: National Academy Press.

Jenkins, J.L., O'Connor, R.E., & Cone, D.C. (2006). Differentiating large-scale surge versus daily surge. Academic Emergency Medicine, 13, 1169–1172.

Jenks, R.H., & Malecki, J.M. (2004). GIS – a proven tool for public health analysis. Journal of Environmental Health, 67(3), 32–34.

Jin, R.L., Shah, C.P., & Svoboda, T.J. (1997). The impact of unemployment on health: a review of the evidence. Journal of Public Health Policy, 18, 275–301.

Jochelson, K., Mothibeli, M., & Leger, J. (1991). Human immunodeficiency virus and migrant labor in South Africa. International Journal of Health Services, 21(1), 157–173.

Joe, G.W., Simpson, D.D., & Sells, S.B. (1994). Treatment process and relapse to opioid use during methadone maintenance. The American Journal of Drug and Alcohol Abuse, 20, 173–197.

Johnson, B.D., & Mulffler J. (1997). Sociocultural. In: J.H. Lowinson, P. Ruiz, R.B. Millman, & J.G. Langrod, (eds.), Substance Abuse: A Comprehensive Textbook, 3rd Edition (pp. 107–118). Baltimore: Williams and Wilkins.

Johnson, B.D., Williams, T., Dei, K.A., & Sanabria, H. (1990). Drug abuse in the inner city: impact on hard-drug users. In: M. Tonry, N. Morris, & N. I.O. Justice (eds.), Drugs and Crime, vol. 13 (pp. 9–30). Chicago and London: The University of Chicago Press.

Johnson, S., Scribner, R., Gruenewald, P.J., & Cohen, D. (2007). Geospatial methods for core group identification. Sexually Transmitted Infections, in press.

Johnston, L.D., O'Malley, P.M., Bachman, J.G., & Schulenberg, J.E. (2005). Monitoring the Future National Survey Results on Drug Use, 1876–2004: Volume I, Secondary School Students (NIH Publication No. 05-5727). Bethesda, MD: National Institute on Drug Abuse.

Johnston, R.J. (1991). A Question of Place: Exploring the Practices of Geography. Oxford, England & Cambridge, MA: Blackwell Publishing Ltd.

Joint United Nations Programme on HIV/AIDS (2000). Report on the Global AIDS Epidemic, June 2000. Geneva: Joint United Nations Programme on HIV/AIDS.

Jones, G., Davidson, J., Bisset, C., & Brettle, R.P. (1988). Mobility of injection drug users as a means of spread for the human immunodeficiency virus. ANSWER (AIDS News Supplement, CDC Weekly Report), A76, 1–4.

Jones-Webb, R., Snowden, L., Herd, D., Short, B., & Hannan, P. (1997). Alcohol-related problems among black, Hispanic, and white men: the contribution of neighborhood poverty. Journal of Studies on Alcohol, 58, 539–545.

Joseph, S., Yule, W., Williams, R., & Hodgkinson, P. (1993). Increased substance use in survivors of the Herald of Free Enterprise disaster. The British Journal of Medical Psychology, 66(Pt 2), 185–191.

Jospeh, A., & Phillips, D. (1984). Accessibility and Utilization: Geographical Perspectives on Health Care Delivery. New York: Harper and Row.

Joyner, L.M., Wright, J.D., & Devine, J.A. (1996). Reliability and validity of the addiction severity index among homeless substance misusers. Substance Use & Misuse, 31, 729–751.

Kalichman, S.C., Kathleen, J.A.K., Andrei, J.S., Alla, P.K., & Juliana, S. (2000). The emerging AIDS crisis in Russia: review of enabling factors and prevention needs. International Journal of STD & AIDS, 11, 71–75.

Kamel Boulos, M., Cai, Q., Padgett, J., & Rushton, G. (2005). Using software agents to preserve individual health data confidentiality in micro-scale geographical analyses. Journal of Biomedical Informatics, 39, 160–170.

Kaplan, G.A., Pamuk, E.R., Lynch, J.W., Cohen, R.D., & Balfour, J.L. (1996). Inequality in income and mortality in the United States: analysis of mortality and potential pathways. British Medical Journal, 312, 999–1003.

Karauppannan, J. (2005). Mapping and corrections: management of offenders with geographic information systems. Corrections Compendium, 30(1), 1–10.

Kato, K., et al. (1999). Genetic similarity of HIV type 1 subtype E in a recent outbreak among injecting drug users in northern Vietnam to strains in Guangxi Province of southern China. AIDS Research and Human Retroviruses, 15(13), 1157–1168.

Kawachi, I., & Berkman, L.F. (2003). Neighborhoods and Health. New York: Oxford University Press.

KCI. (n.d.). Methamphetamine frequently asked questions. Available at: http://www.kci.org/meth_info/faq_meth.htm. [On-line]

Kearns, R.A., & Joseph, A.E. (1993). Space in its place: developing the link in medical geography. Social Science and Medicine, 37(6), 711–717.

Keck, P.E. Jr, McElroy, S.L., Strakowski, S.M., et al. (1998). 12-month outcome of patients with bipolar disorder following hospitalization for a manic or mixed episode. The American Journal of Psychiatry 155, 646–652.

Kelen, G.D., & McCarthy, M.L. (2006). The science of surge. Academic Emergency Medicine, 13, 1089–1094.

Kelling, G.L., & Coles, C.M. (1996). Fixing Broken Windows. New York: The Free Press.

Kelly, M. (2000). Inequality and crime. The Review of Economics and Statistics, 82(4), 530–539.

Kelly, P.J., Blacksin, B., & Mason, E. (2001). Factors affecting substance abuse treatment completion for women. Issues in Mental Health Nursing, 22, 287–304.

Kendler, K.S., Thornton, L.M., & Gardner, C.O. (2000). Stressful life events and previous episodes in the etiology of major depression in women: an evaluation of the kindling hypothesis. The American Journal of Psychiatry, 157, 1243–1251.

Kennedy, B.P., Kawachi, I., & Prothrow-Suth, D. (1996a). Income distribution and mortality: cross sectional ecological study of the Robin Hood index in the United States. BMJ, 312, 1004–1007.

Kennedy, D.M. (1997). Pulling levers: chronic offenders, high-crime settings, and a theory of prevention. Valparaiso University Law Review, 31(2).

Kennedy, D.M. (1998). Pulling levers: getting deterrence right. National Institute of Justice Journal, 2(36), 238–249.

Kennedy, D.M., Anthony, A.B., & Anne M Piehl (2001). Reducing Gun Violence: The Boston Gun Project's Operation Ceasefire (NCJ 188741). Washington, DC: US Department of Justice.

Kennedy, D.M., Piehl, A.M., & Braga, A.A. (1996b). Youth violence in Boston: gun markets, serious youth offenders, and a use-reduction strategy. Law and Contemporary Problems, 59(1), 147–196.

Kessler, R.C., et al. (2005). Lifetime prevalence and age-of-onset distributions of DSM-IV disorders in the National Co-morbidity Survey Replication. Archives of General Psychiatry, 62, 593–602.

Kessler, R.C., Olfson, M., & Berglund, P.A. (1998). Patterns and predictors of treatment contact after first onset of psychiatric disorders. The American Journal of Psychiatry, 155, 62–69.

Kessler, R.C., Sonnega, A., Bromet, E., Hughes, M., & Nelson, C.B. (1995). Posttraumatic stress disorder in the National Comorbidity Survey. Archives of General Psychiatry, 52, 1048–1060.

Khanina, et al. (2000). A homogeneous subtype A HIV-1 strain caused new local outbreaks among injecting drug users in Russia. XIIIth International AIDS Conference, Durban, 2000.

Khantzian, E. (1997). The self-medication hypothesis of substance abuse disorders: a reconsideration and recent applications. Harvard Review of Psychiatry, 4, 231–244.

Khomdon Singh Lisam (1998). Personal communication with the author, Imphal, Manipur State.

King, G. (1997). The "race" concept in smoking: a review of the research on African Americans. Social Science and Medicine, 45, 1075–1087.

Kinzly, M. (2003). Personal communication. Research Associate, Department of Epidemiology and Public Health, Yale University, New Haven, CT.

Kipke, M., Montgomery, S.B., Simon, T., & Iverson, E. (1997). Homeless youth and their exposure to and involvement in violence. Journal of Adolescent Health, 20, 360–367.

Kipke, M., O'Connor, S., Palmer, R., & MacKenzie, R. (1995). Street youth in Los Angeles. Archives of Pediatrics & Adolescent Medicine, 149, 513–519.

KKTV. (2005). Meth – a social plague. June 6.

KKTV. (2007). Meth Lab Busts Decrease. Reported by Lauri Martin, February 24. Available at: http://www.kktv.com/home/headlines/6068256.html. [On-line]

Kleiman, A.R. (1993). Enforcement swamping: a positive-feedback mechanism in rates of illicit activity. Mathematical and computer modelling, 17(2).

Knight, J., Song, L., & Jia, H. (1999). Chinese rural migrants in urban enterprises: three perspectives. The Journal of Development Studies, 35(3), 73–104.

Knowles, G. (1999). Gambling, drugs, and sex: new drug trends and addictions in Honolulu, Hawaii, 1998. Sociological Practice, 1(1), 45–69.

Koester, S. (1994). Copping, running and paraphernalia laws: contextual and needle risk behaviors among drug users in Denver. Human Organization, 53, 287–295.

Korff, M.V., Koepsell, T., Curry, S., & Diehr, P. (1992). Multi-level analysis in epidemiologic research on health behaviors and outcomes. American Journal of Epidemiology, 135(10), 1077–1082.

Kornblum, W. (1993). Drug legalization and the minority poor. In: R. Bayer, & G.M. Oppenhiemer (eds.), Confronting Drug Policy (pp. 115–135). New York: Cambridge University Press.

Koshkina, E. (2001). Prevalence of drug addiction diseases in the Russian Federation in 2000 according to official statistics. Voprosy Narkologii [Addiction Problems], 3, 6171.

Koshkina, E., Koryakin, S., & Tsarev, S. (2002). A study of the 'latent' contingent of drug addicts by the methods of 'repetitive capture' in the town of Chapayevsk of Samara Region in 2000–2001 Voprosy Narkoloii (Addiction Problems) 4, 46–52.

Kottiri, B.J., Friedman, S.R., Neaigus, A., Curtis, R., & Des Jarlais, D.C. (2002). Risk networks and racial/ethnic differences in the prevalence of HIV infection among injection drug users. Journal of Acquired Immune Deficiency Syndromes, 30(1), 95–104.

Kramer, J. (2005). Drug Abuse and HIV/AIDS in Russia. Paper presented at the ICCEES Conference, 29 July, 2005.

Kraska, P.B, & Cubellis, L.J. (1997). Militarizing mayberry and beyond: making sense of American paramilitary policing. Justice Quarterly, 14(4), 607–629.

Krause, N. (1996). Neighborhood deterioration and self-rated health in later life. Psychology and Aging, 11, 1342–1352.

Krieger, N. (2001). A glossary for social epidemiology. Journal of Epidemiology and Community Health, 55, 693–700.

Krupitsky, E., Zvartau, E., Horton, N.J., Schoolwerth, K.R., Bryant, K., & Samet, J.H. (2004). The onset of HIV infection in the Leningrad region of Russia: a focus on drug and alcohol dependence. HIV Medicine, 5, 30–3.

Kruse, N., Behets, F.M., Vaovola, G., Burkhardt, G., Barivelo, T., Amida, X., & Dallabetta, G. (2003). Participatory mapping of sex trade and enumeration of sex workers using capture-recapture methodology in Diego-Suarez, Madagascar. Sexually Transmitted Diseases, 30(8), 664–670.

Ku, L., Sonenstein, F.L., & Pleck, J.H. (1993). Neighborhood, family, and work: influences on the premarital behaviors of adolescent males. Social Forces, 72(2), 479–503.

Kuldorff, M. (2005). SatScan: software for the spatial and space-time scan statistics. Information Management Services, Inc.,

Kulldorff, M. (1997). A spatial scan statistic. Communications in Statistics: Theory and Methods, 25, 1481–1496.

Kulldorff, M., Heffernan, R., Hartman, J., Assuncao, R., & Mostashari, F. (2005). A space-time permutation scan statistic for disease outbreak detection. PLoS Medicine, 2, e59.

Kulldorff, M., Song, C., Gregorio, D., Samociuk, H., & DeChello, L. (2006). Cancer map patterns: are they random or not? American Journal of Preventive Medicine, 30(Suppl 2), S37–S49.

Kumar, N., & Waylor, C. (2003). Proximity to alcohol-serving establishments and crime probabilities in Savannah, Georgia: a statistical and GIS analysis. Southeastern Geographer, 43(1), 125–142.

Kuntsche, E.N., & Kuendig, H. (2005). Do school surroundings matter? Alcohol outlet density, perception of adolescent drinking in public, and adolescent alcohol use. Addictive Behaviors, 30, 151–158.

Kwan, M.-P. (1999a). Gender, the home-work link, and space-time patterns of non-employment activities. Economic Geography 75, 370–394.

Kwan, M. (1999b). Gender and individual access to urban opportunities: a study using space-time measures. The Professional Geographer 51(2), 210–227.

La Nación. (1998) Costa Rica; September 12, 1998.

La Vigne, N.G., Cowan, J., & Brazzell, D. (2006). Mapping Prisoner Reentry: An Action Research Guidebook. Washington, DC: Urban Institute.

La Vigne, N.G., Naser, R.L., Brooks, L.E., & Castro, J.L. (2005). Examining the effect of incarceration and in-prison family contact on prisoners' family relationships. Journal of Contemporary Criminal Justice, 21(4), 1–22.

La Vigne, N.G., & Parthasarathy, B. (2005). Returning Home Illinois Policy Brief: Prisoner Reentry and Residential Mobility. Washington, DC: Urban Institute.

La Vigne, N.G., & Thomson, G.L. (2003). A Portrait of Prisoner Reentry in Ohio. Washington, DC: Urban Institute.

La Vigne, N.G., Visher, C., & Castro, J. (2004). Chicago Prisoners' Experiences Returning Home. Washington, DC: Urban Institute.

Lader, M. (2005). Medical Marijuana Act Passes State Legislature. The Brown Daily Herald, July 18.http://www.browndailyherald.com/media/paper472/news/2005/07/18/Metro/Medical. Marijuana.Act.Passes.State.Legislature-962839.shtml.

LaGrange, T.C. (1999). The impact of neighborhoods, schools, and malls on the spatial distribution of property damage. Journal of Crime and Delinquency, 36(4), 393–422.

Laguerre, M. (1994). The Informal City. New York: St. Martin's Press.

Lai, S., Chen, J., Celentano, D., Page, J.B., Lai, H., Yang, J., Liu, W., McCoy, C.B., & Yu, X. (2000). Adoption of injection practices in heroin users in Guangxi province, China. Journal of Psychoactive Drugs, 32, 285–292.

Lamont, A., Ukoumunne, O.C., Tyrer, P., Thornicroft, G., Patel, R., & Slaughter, J. (2000). The geographical mobility of severely mentally ill residents in London. Social Psychiatry and Psychiatric Epidemiology, 35, 164–169.

Landen, M.G., Castle, S., Nolte, K.B., Gonzales, M., Escobedo, L.G., Chatterjee, B.F., et al. (2003). Methodological issues in the surveillance of poisoning, illicit drug overdose, and heroin overdose deaths in New Mexico. American Journal of Epidemiology, 157(3), 273–278.

Landrine, H., & Klonoff, E.A. (2000). Racial discrimination and cigarette smoking among blacks: findings from two studies. Ethnicity & Disease, 10, 195–202.

Lang, L. (2000). GIS for health organizations. Redlands: ESRI Press.

Lang, M.A., & Belenko, S. (2000). Predicting retention in a residential drug treatment alternative to prison program. Journal of Substance Abuse Treatment, 19, 145–160.

Lankenau, S.E., & Clatts, M. (2002). Ketamine injection among high risk youth: preliminary findings from New York City. Journal of Drug Issues, 32, 893–905.

Lankenau, S.E., & Clatts, M. (2004). Drug injection practices among high-risk youths: the first shot of ketamine. Journal of Urban Health, 81, 232–248.

Lankenau, S.E., Clatts, M., Welle, D., Goldsamt, L., & Gwadz, M. (2005). Street careers: homelessness, drug use, and hustling among young men who have sex with men. International Journal of Drug Policy, 16, 10–18.

Lankenau, S., Sanders, B., Jackson Bloom, J., Hathazi, D. (forthcoming). Recruiting and retaining mobile young injection drug users in a longitudinal study. *Substance Use & Misuse*.

Lankenau, S.E., Sanders, B., Jackson Bloom, J., Hathazi, D., Alarcon, E., Tortu, S., & Clatts, M. (2007). First injection of ketamine among young injection drug users (IDUs) in three U.S. cities. Drug and Alcohol Dependence, 87, 183–193.

Lansky, A., Nakashima, A.K., Diaz, T., Fann, S.A., Conti, L., Herr, M., Smith, D., Karon, J., Jones, J.L., & Ward, J.W. (2000). Human immunodeficiency virus infection in rural areas and small cities of the southeast: contributions of migration and behavior. The Journal of Rural Health, 16, 20–30.

Larson, A., & Bammer, G. (1996). Why? Who? How? Estimating numbers of illicit drug users: lessons from a case study from the Australian Capital Territory. Australian and New Zealand Journal of Public Health, 20, 493–499.

Larson, A., Stevens, A., & Wardlaw, G. (1994). Indirect estimates of 'hidden' populations: capture–recapture methods to estimate the numbers of heroin users in the Australian Capital Territory. Social Science and Medicine, 39.

LaScala, E.A., Freisthler, B., & Gruenewald, P.J. (2005). Population ecologies of drug use, drinking and related problems. In: T. Stockwell, P.J. Gruenewald, J.W. Toumbourou, & W. Loxley (eds.), Prevention Harmful Substance Use: The Evidence Base for Policy and Practice (pp. 67–78). New York: John Wiley & Sons.

Latkin, C., Glass, G., & Duncan, T. (1998). Geographic information systems to assess spatial patterns of drug use, selection bias and attrition among a sample of injection drug users. Drug and Alcohol Dependence, 50, 167–175.

Latkin, C.A., Forman, V., Knowlton, A., & Sherman, S. (2003). Norms, social networks, and HIV-related risk behaviors among urban disadvantaged drug users. Social Science and Medicine, 56(3), 465–476.

Latkin, C.A., Mandell, W., Vlahov, D., Oziemkowska, M., & Celentano, D.D. (1996). The long-term outcome of a personal network-oriented HIV prevention intervention for injection drug users: the SAFE Study. American Journal of Community Psychology, 24(3), 341–364.

Latkin, C.A., Williams, C.T., Wang, J., & Curry, A. D. (2005). Neighborhood social disorder as a determinant of drug injection behaviors: a structural equation modeling approach. Health Psychology, 24, 96–100.

Lau, J.T.F., & Thomas, J. (2001). Risk behaviors of Hong Kong male residents travelling to mainland China: a potential bridge population for HIV infection. AIDS Care, 13, 71–81.

LaVeist, T. (1989). Linking residential segregation to the infant-mortality race disparity in U.S. cities. Sociology and Social Research, 73, 90–94.

LaVeist, T. (2002). Race, Ethnicity, and Health: A Public Health Reader. San Francisco: Jossey-Bass.

Law, J, & Haining, R. (2004). A Bayesian approach to modeling binary data: the case of high intensity crime areas. Geographical Analysis, 36, 197–216.

Lawson, A. (2001). Statistical Methods in Spatial Epidemiology. New York: John Wiley & Sons.

Lawson, A., Browne, W., Vidal Rodeiro, C. (2003). Disease Mapping with WinBUGS and ML-WiN. West Sussex: John Wiley and Sons Ltd.

Lee, B.A., Oropesa, R.S., & Kanan, J.W. (1994). Neighborhood context and residential mobility. Demography, 22, 199–222.

Lee, M.A., & Shlain, B. (1985). Acid Dreams: The CIA, LSD, and the Sixties Rebellion. New York: Grove Weidenfeld.

Lee, M.T., Reif, S., Ritter, G.A., Levine, H.J., & Horgan, C.M. (2004). Alcohol and Drug Services Study (ADSS) Phase II: Client Record Abstract Report (DHHS Publication No. SMA 04-3974, Analytic Series No. A-27). Rockville: Substance Abuse and Mental Health Services Administration.

Lee, P.I.E. (2001). ACAN in the 21st Century – A Continuing Challenge. Hong Kong: Narcotics Division, Security Bureau.

Lemann, N. (1991). The Promised Land: The Great Migration and How it Changed America. New York: Vintage Books.

Lettieri, D.J., Sayes, M., & Pearson, H.W. (1980). Theories on Drug Abuse: Selected Contemporary Perspectives (Research Monograph 30). Rockville: National Institute of Drug Abuse.

Leukefeld, C.G., Narevic, E., Hiller, M.L., Staton, M., Logan, T.K., Gillespie, W., et al. (2002). Alcohol and drug use among rural and urban incarcerated substance abusers. International Journal of Offender Therapy and Comparative Criminology, 46(6), 715–728.

Leventhal, T., & Brooks-Gunn, J. (2003). Moving to opportunity: an experimental study of neighborhood effects on mental health. American Journal of Public Health, 93(9), 1570–1582.

Levine, N. (2004). CrimeStat: A Spatial Statistics Program for the Analysis of Crime Incident Locations. Washington, DC: National Institute of Justice.

Li, X., Stanton, B., Fang, X., Lin, D., Mao, R., Wang, J., Cottrell, L., & Harris, C. (2004). HIV/STD risk behaviors and perceptions among rural-to-urban migrants in China. AIDS Education and Prevention, 16(6), 538–556.

Liang, Z., & Ma, Z. (2004). China's floating population: new evidence from the 2000 census. Population and Development Review, 30(3), 467–488.

Lieb, S., Friedman, S.R., Zeni, M.B., et al. (2004). An HIV prevalence-based model for estimating risk-populations of injection drug users and men who have sex with men. Journal of Urban Health, 81.

Lieberson, S., & Carter, D.K. (1982). A model for inferring the voluntary and involuntary causes of residential segregation. Demography, 19, 511–526.

Lilienfeld, D.E., & Stolley, P.D. (1994). Foundations of Epidemiology, 3rd Edition. New York: Oxford University press.

Lillie-Blanton, M., Anthony, J., Schuster, C.R. (1993). Probing the meaning of racial/ethnic group comparisons in crack cocaine smoking. JAMA, 269, 993–997.

Lin, S.J. (2004). Access to community pharmacies by the elderly in Illinois: a geographic information systems analysis. Journal of Medical Systems, 28, 301–309.

Lipton, R., Gorman, D.M., Wieczorek, W.F., & Gruenewald, P. (2003). The application of spatial analysis to the public health understanding of alcohol and alcohol-related problems. In: O.A. Khan, & R. Skinner (eds.), Geographic Information Systems and Health Applications (pp. 57–79). Hershey, PA: Idea Group Publishing.

Lipton, R.I., & Gruenewald, P.J. (2003). The spatial dynamics of violence and alcohol outlets. Journal of Studies on Alcohol, 63(2), 187–195.

Liu, J.L., Liu, J.T., Hammitt, J.K., & Chou, S.Y. (1999). The price elasticity of opium in Taiwan, 1914–1942. Journal of Health Economics, 18, 795–810.

Lix, L., DeVerteuil, G., Walker, J., Robinson, J., Hinds, A., & Roos, L. (2007). Residential mobility of individuals with diagnosed schizophrenia. Social Psychiatry and Psychiatric Epidemiology, 42(3), 221–228.

Long, E.F., Brandeau, M.L., Galvin, C.M., Vinichenko, T., Tole, S.P., Schwartz, A., et al. (2006). Effectiveness and cost-effectiveness of strategies to expand antiretroviral therapy in St. Petersburg, Russia. AIDS, 20, 2207–2215.

Longley, P.A., Goodchild, M.F., Maguire, D.J., & Rhind, D.W. (2001). Geographic Information Systems and Science. Chichester: John Wiley.

Low, N., Arauz, R., Gorter, A., Guevara, M., González, A., Sequeira, G., Morgan, N., Rocha, R., Serna, R., González, A., & Davey S.G. (1992). Conocimientos Acerca del SIDA de la Población Adulta de Managua. Boletín de la Oficina Sanitaria Panamericana, 112, 319–326. Hamilton, R. (2004, November). IDB America. Internet Page.

Lowndes, C.M., Alary, M., & Platt, L.,(2003). Injection drug use, commercial sex work, and the HIV/STI epidemic in the Russian Federation. Sexually Transmitted Diseases, 30, 46–48.

Lu, M. (1998). Analyzing migration decisionmaking: relationships between residential satisfaction, mobility intentions, and moving behavior. Environment & Planning A, 30, 1473–1495.

Lu, X., & Pas, E.I. (1999). Socio-demographics, activity participation and travel behavior. Transportation Research Part A: Policy and Practice, 33, 1–18.

Lukashov, V., et al. (2000). Molecular characterization of a growing HIV-1 epidemic among IDUs in the former Soviet Union. XIIIth International AIDS Conference, Durban, 2000.

Luke, D.A. (2005). Getting the big picture in community science: methods that capture context. American Journal of Community Psychology, 35, 185–200.

Lupton, R., Wilson, A., May, T., Warburton, H., & Turnbull, P.J. (2002). A Rock and a Hard Place: Drug Markets in Deprived Neighborhoods. London: CASE, London School of Economics and the Criminal Policy Research Unit, South Bank University Home Office Research, Development and Statistics Directorate.

Lurie, M.N., Williams, B.G., Suma, K., et al. (2003). The impact of migration on HIV-1 transmission in South Africa: a study of migrant and nonmigrant men and their partners. Sexually Transmitted Diseases, 30, 149–156.

Lurie, P., & Drucker, E. (1997). An opportunity lost: HIV infections associated with lack of a national needle-exchange programme in the U.S.A. Lancet, 349(9052), 604–608.

Lynch, J.W., Kaplan, G.A, Pamuk, E.R., Cohen, R.D., et al. (1998). Income inequality and mortality in metropolitan areas of the United States. American Journal of Public Health, 88.

Lyson, T., Torres, R., & Welsh, R. (2001). Scale of agricultural production, civic engagement, and community welfare. Social Forces, 80, 311–327.

Ma, L.J., & Xiang, B. (1998). Native place, migration and the emergence of peasant enclaves in Beijing. The China Quarterly, 155, 546–581.

MacDonald, M., Law, M., Kaldor, J., Hales, J., & Dore, G. (2003). Effectiveness of needle and syringe programmes for preventing HIV transmission. International Journal of Drug Policy, 14, 351–352.

Macdonald, S., Cherpitel, C.J., Borges, G., DeSouza, A., Giesbrecht, N., & Stockwell, T. (2005). The criteria for causation of alcohol in violent injuries based on emergency room data from six countries. Addictive Behaviors, 30, 103–113.

MacDonald, Z., & Pudney, S. (2000). Illicit drug use, unemployment, and occupational attainment. Journal of Health Economics, 19(6), 1089–1115.

MacDonald, Z., & Pudney, S. (2001). Illicit drug use and labour market achievement: evidence from the UK. Applied Economics, 33(13), 1655–1668.

Macintyre, S., Ellaway, A., & Cummins, S. (2002). Place effects on health: how can we conceptualise, operationalise and measure them? Social Science and Medicine, 55, 125–139.

Mack, K.A. (2004). Fatal and nonfatal unintentional injuries in adult women, United States. Journal of Women's Health, 13(7), 754–763.

Mackenbach, J.P., & Bakker, M.J., for the European Network on Interventions and Policies to Reduce Inequalities in health. Lancet, 362, 1409–1414.

Magis-Rodríguez, C., Marques, L.F., & Touze, G. (2002). HIV and injection drug use in Latin America. AIDS, 16(Suppl 3), S34–S41.

Maher, L., & Dixon, D. (1999). Policing and public health: law enforcement and harm minimization in a street-level drug market. The British Journal of Criminology, 39, 488–512.

Malczewski, J., & Poetz, A. (2005). Residential burglaries and neighborhood socioeconomic context in London, Ontario: global and local regression analysis. The Professional Geographer, 57(4), 516–529.

Malm, A.E., & Tita, G.E. (2006). A spatial analysis of green teams: a tactical response to marijuana production in British Columbia. Policy Sciences, 39(4), 361–377.

Mankowski, E.S., Humphreys, K., & Moos, R.H. (2001). Individual and contextual predictors of involvement in twelve-step self-help groups after substance abuse treatment. American Journal of Community Psychology, 29, 519–536.

Manoff, S.B., Vlahov, D., Herskowitz, A., Solomon, L., Munoz, A., Cohn, S., et al. (1996). Human immunodeficiency virus infection and infective endocarditis among injecting drug users. Epidemiology, 7(6), 566–570.

Mansvelt, W.M.F., & Creutzberg P. (1978). Changing Economy in Indonesia: Rice Prices, vol. 4. The Hague: Martinus Nijhoff.

Marcus, S.C., Fortney, J.C., Olfson, M., & Ryan, N.D. (1997). Travel distance to outpatient treatment for depression. Psychiatric Services, 48, 1005.

Marshall, P.L., Singer, M., & Clatts, M.C. (eds.) (1999). Integrating Cultural, Observational, and Epidemiological Approaches in HIV/AIDS. NIH Publication No. 99-4565. Rockville: National Institute on Drug Abuse.

Marshall, R. (1991). A review of methods for the statistical analysis of spatial patterns of disease. Journal of the Royal Statistical Society. Series A (Statistics in Society), 154(Pt 3), 421–441.

Marston, S.A. (2003). Mobilizing geography: locating space in social movement theory. Mobilization: An International Journal, 8(2), 227–231.

Martin, D.G. (2003). 'Place-framing' as place-making: constituting a neighborhood for organizing and activism. Annals of the Association of American Geographers, 93(3), 730–750.

Martin, D.G., & Miller, B. (2003). Space and contentious politics. Mobilization: An International Journal, 8(2), 143–156.

Martin, J.K., Tuch, S.A., & Roman, P.M. (2003). Problem drinking patterns among African-Americans: the impacts of reports of discrimination, perceptions of prejudice, and "risky" coping strategies. Journal of Health and Social Behavior, 44, 408–425.

Martinez, R., Jr, Lee, M.T., & Nielsen, A.L. (2004). Segmented assimilation, local context and determinants of drug violence in Miami and San Diego: does ethnicity and immigration matter? The International Migration Review, 38(1), 131–157.

Mashkilleyson, N., & Leinikki, P. (1999). Evolution of the HIV epidemic in Kaliningrad, Russia. Journal of Clinical Virology, 12, 37–42.

Mason, M., Cheung, I., & Walker, L. (2004a). Substance use, social networks and the geography of risk and protection of urban adolescents. Substance Use & Misuse, 39(10–12), 1–27.

Mason, M., Cheung, I., & Walker, L. (2004b). The social ecology of urban adolescent substance use: a case study utilizing geographic information systems. The Journal of Primary Prevention, 25(2), 271–282.

Massey, D. (1997). A global sense of place. In: T. Cresswell (ed.), Place a Short Introduction (pp. 63–70). Oxford: Blackwell Publishing.

Massey, D., & Denton, N. (1988). The dimensions of residential segregation. Social Forces, 67, 281–315.

Massey, D., & Denton, N. (1993). American Apartheid: Segregation and the Making of the American Underclass. Cambridge, MA: Harvard University Press.

Massey, D., & Denton, N. (1998). The elusive quest for the perfect index of concentration: reply to Egan, Anderton, and Weber. Social Forces, 76, 1123–1133.

Massey, D., & Mullen, B.P. (1984). Process of Hispanic and Black spatial assimilation. American Journal of Sociology, 89, 836–873.

Massey, D., White, M.J., & Phua, V.-C. (1996). The dimensions of segregation revisited. Sociological Methods & Research, 25, 172–206.

Massey, D.S. (1996). The age of extremes: concentrated affluence and poverty in the twenty-first century. Demography, 33(4), 395–412.

Massey, D.S., Arango, J., Hugo, G., Kouaouci, A., & Pellegrino, A. (1994). An evaluation of international migration theory: the North American case. Population and Development Review, 20(4), 699–751.

Massey, D.S., & Denton, N.A. (1992). American Apartheid: Segregation and the Making of the Underclass. Cambridge: Harvard University Press.

Matthews, S.A., Detwiler, J.E., & Burton, L.M. (2006). Viewing people and places: coupling geographic information analysis and ethnographic research. Cartographica.

Maxwell, J., & Spence, R. (2005). Profiles of club drug users in treatment. Substance Use & Misuse, 40(9–10), 1409–1426.

Maxwell, J.C. (1999). Heroin Addicts in Texas: The Nature and Size of a Hidden Population. Austin, TX: Texas Commission on Alcohol and Drug Abuse.

Maxwell, J.C. (2006). Substance Abuse Trends in Texas, January 2006. Austin: The Gulf Coast Addiction Technology Transfer Center (GCATTC), The Center for Social Work Research, School of Social Work, The University of Texas at Austin.

Maxwell, J.C., Podus, D., & Walsh, D. (2006a). Katrina and Rita – Who Got Treated? Scottsdale, AZ: College on Problems of Drug Dependence.

Maxwell, J.C., Pullum, T.W., & Tannert, K. (2005). Deaths of clients in methadone treatment in Texas: 1994–2002. Drug and Alcohol Dependence, 78, 73–81.

Maxwell, S., Bigg, D., Stanczykiewicz, K., & Carlberg-Racich, S. (2006b). Prescribing naloxone to actively injecting heroin users: a program to reduce heroin overdose deaths. Journal of Addictive Diseases, 25(3), 89–96.

Mazerolle, L.G., Kadleck, C., & Roehl, J. (1998). Controlling drug and disorder problems: the role of place manager. Criminology, 36, 371–403.

McAuliffe, W.E., & Dunn, R. (2004). Substance abuse treatment needs and access in the USA: interstate variations. Addiction, 99, 999–1014.

McAuliffe, W.E., Woodworth, R., Zhang, C., & Dunn, R. (2002). Identifying substance abuse treatment gaps in substrate areas. Journal of Substance Abuse Treatment, 23(33), 199–208.

McCague, D. (2004). Personal communication. Director, Health Care for the Homeless, Albuquerque, New Mexico.

McCarthy, J.D., Wolfson, M., Baker, D.P., & Mosakowski, E. (1988). The founding of social movement organizations: local citizens' groups opposing drunken driving. In: G.R. Carroll (ed.), Ecological Models of Organization. Cambridge, MA: Ballinger.

McCarthy, M.L., Aronsky, D., & Kelen, G.D. (2006). The measurement of daily surge and its relevance to disaster preparedness. Academic Emergency Medicine, 13, 1138–1141.

McCarty, D., Rieckmann, T., Green, C., Gallon, S., & Knudsen, J. (2004). Training rural practitioners to use buprenorphine: using The Change Book to facilitate technology transfer. Journal of Substance Abuse Treatment, 26(3), 203–208.

McDuff, J., & Ford, W.E. (2005). A Report on the Post-September 11 State Disaster Relief Grant Program of SAMHSA's Center for Substance Abuse Treatment (DHHS Publication No. SMA

3993). Rockville, MD: Center for Substance Abuse Treatment, Substance Abuse and Mental Health Services Administration.

McEwan, R.T., McCallum, A., Bhopal, R.S., & Madhok, R. (1992). Sex and the risk of HIV infection: the role of alcohol. Addiction, 87, 577–584.

McLafferty, S. (2003). GIS and health care. Annual Review of Public Health, 24, 25–42.

McLafferty, S., & Grady, S. (2005). Immigration and geographic access to prenatal clinics in Brooklyn, NY: a geographic information systems analysis. American Journal of Public Health, 95, 638–640.

McLellan, A.T. (2002). Have we evaluated addiction treatment correctly? Implications from a chronic care perspective. Addiction, 97, 249–252.

McLellan, A.T., Kushner, H., Metzger, D., Peters, R., Smith, I., Grissom, G., et al. (1992). The fifth edition of the Addiction Severity Index. Journal of Substance Abuse Treatment, 9, 199–213.

McLellan, A.T., Lewis, D.C., O'Brien, C.P., & Kleber, H.D. (2000). Drug dependence, a chronic medical illness: implications for treatment, insurance, and outcomes evaluation. JAMA, 284, 1689–1695.

McLellan, A.T., Luborsky, L., Cacciola, J., Griffith, J., Evans, F., Barr, H.L., et al. (1985). New data from the Addiction Severity Index. Reliability and validity in three centers. The Journal of Nervous and Mental Disease, 173, 412–423.

McLellan, A.T., Luborsky, L., Woody, G.E., & O'Brien, C.P. (1980). An improved diagnostic evaluation instrument for substance abuse patients. The Addiction Severity Index. The Journal of Nervous and Mental Disease, 168, 26–33.

McMurry, L.O. (1998). To Keep the Waters Troubled: The Life of Ida B. Wells. New York: Oxford University Press.

McNicol, E., Horowicz-Mehler, N., Fisk, R.A., Bennett, K., Gialeli-Goudas, M., Chew, P.W., et al. (2003). Management of opioid side effects in cancer-related and chronic noncancer pain: a systematic review. The Journal of Pain, 4(5), 231–256.

McNulty, T.L., & Bellair, P.E. (2003). Explaining racial and ethnic differences in serious adolescent violent behavior. Criminology, 41(3), 701–730.

MCP (2002). Commitment and Action Against AIDS, Tuberculosis and Malaria. Proposal Submitted by the Country Coordinating Mechanism (MCP) to the World Fund Against AIDS, Tuberculosis and Malaria. Managua, Nicaragua. Retrieved from www.theglobalfund.org/en/.

McQuie, H. (2003). Campaign Coordinator to legalize the medical use of marijuana.

Meade, M.S., & Earickson, R.J. (2000). Medical Geography, 2nd Edition. New York: Guilford Press.

Mears, D.P., & Bhati, A.S. (2006). No community is an island: the effects of resource deprivation on urban violence in spatially and socially proximate communities. Criminology, 44, 509–548.

Medina-Mora, M.E., & Rojas Guiot, E. (2003). Demand of drugs: Mexico in the international perspective. Salud Mental, 26(2), 1–11.

Mehren, E. (2006). Study: Katrina evacuees having trouble coping. *Seattle Times*. Retrieved May 7, 2008, from http://www.seattletimes.com.

Mellor, J.M., & Milyo, J. (2003). Is exposure to income inequality a public health concern? Lagged effects of income inequality on individual and population health. Health Services Research, 1, 137–151.

Meredith, T., Speir, J., Johnson, S., & Hull, H. (2003). Enhancing parole decision-making through the automation of risk assessment. Atlanta, GA: Applied Research Services, Inc.

Metraux, S., & Culhane, D.P. (2004). Homeless shelter use and reincarceration following prison release: assessing the risk. Criminology & Public Policy, 3, 201–222.

Mikolaenko, I., Robinson, C.A., Jr, & Davis, G.G. (2002). A review of methadone deaths in Jefferson County, Alabama. The American Journal of Forensic Medicine and Pathology, 23(3), 299–304.

Mikuriya, T.H., & Aldrich, M. (1998). Cannabis 1988 old drug, new dangers, the potency debate. Journal of Psychoactive Drugs, 20, 1.

Milburn, N.G., Rotheram-Borus, M.J., Rice, E., Mallet, S., & Rosenthal, D. (2006). Cross-national variations in behavioral profiles among homeless youth. American journal of community psychology, 37(1/2), 63–76.

Milillo, T.M., & Gardella, J. (2006). Spatial statistics and interpolation methods for TOF SIMS imaging. Applied Surface Science, 252, 6883–6890.

Millar, A.B., & Gruenewald, P.J. (1997). Use of spatial models for community program evaluation of changes in alcohol outlet distribution. Addiction, 92(6s1), 273–284.

Miller, B.A. (2000). Geography and social movements: comparing antinuclear activism in the Boston area. Minneapolis, MN: University of Minnesota Press.

Miller, C., Firestone, M., Ramos, R., Burris, S., Ramos, M.E., Case, P., et al. (2008). Injecting drug users' experiences of policing practices in two Mexican-U.S. border cities: public health perspectives. International Journal of Drug Policy 19(4), 324–31.

Millin, M.G., Jenkins, J.L., & Kirsch, T. (2006). A comparative analysis of two external health care disaster responses following Hurricane Katrina. Prehospital Emergency Care, 10, 451–456.

Mills, J. (2003). Cannabis Britannica: Empire, Trade, and Prohibition, 1800–1928. Oxford: Oxford University Press.

Mills, J. (2005). Cannabis in the commons: colonial networks, missionary politics, and the origins of the Indian Hemp Drugs Commission 1893-4. Journal of Colonialism and Colonial History, 6, 1.

Minehan, T. (1934). Boy and Girl Tramps of America. New York: Farrar and Reinhart.

Minino, A.M., Anderson, R.N., Fingerhut, L.A., Boudreault, M.A., & Warner, M. (2006). Deaths: injuries, 2002. National Vital Statistics Reports, 54(10), 1–124.

Ministry of Health, China; Joint United Nations Programme on HIV/AIDS & World Health organization (2006). 2005 Update on the HIV/AIDS Epidemic and Response in China. Beijing: China CDC.

MINSA (2000). Nicaraguan Demographic and Health Survey 1998 (ENDESA-98). Managua: MINSA/OPS/OMS.

MINSA (2001). Nicaraguan Demographic and Health Survey 2001. ENDESA, 2001. Nicaragua: MINSA.

MINSA (2003). Estudio Multicéntrico Centroamericano de Prevalencia de VIH/ITS y Comportamientos. Nicaragua: Division General de Vigilancia para la Salud, Ministerio de Salud.

MINSA (2004). National Program of STDs and HIV/AIDS. Nicaragua: Divisi6n General de Vigilancia para la Salud, Ministerio de Salud.

Miron, J.H. (2005). The Budgetary Implications of Marijuana Prohibition. Mimeo, Cambridge: Harvard University.

Mitchell, A. (2005). The ESRI Guide to GIS Analysis: Spatial Measurements and Statistics. Redlands, CA: ESRI Press.

Mitchell, O., MacKenzie, D.L., & Wilson, D.B. (2006). Incarceration-based drug treatment. In: B.C. Welsh, & D.P. Farrington (eds.), Preventing Crime: What Works for Children, Offenders, Victims, and Places (pp. 103–116). New York: Springer.

Mitchell, R., Dorling, D., & Shaw, M. (2002). Population production and modelling mortality – an application of geographic information systems in health inequalities research. Health & Place, 8, 15–24.

Mobley, L., Root, E., Anselin, L., Lozano, N., & Koschinsky, J. (2006). Spatial analysis of elderly access to primary care services. International Journal of Health Geographics. Retrieved May 2, 2008, from http://www.ijhealthgeographics.com/contents/5/1/19.

Moffitt, T. (1993). Adolescence-limited and life-course-persistent antisocial behavior: a developmental taxonomy. Psychological Review, 100, 674–701.

Mohan, J. (2002). Planning, Markets and Hospitals. New York: Routledge.

Monitoring the AIDS Pandemic (2001). The Status and Trends of HIV/AIDS/STI Epidemics in Asia and the Pacific, Report Presented at the Sixth International Congress on AIDS in Asia and the Pacific, Melbourne, 4 October 2001.

Montana Meth Project (2006). Mission. Available at: http://www.montanameth.org/about_mission.aspx (accessed 10 June, 2006) [On-line].

Montejano, D. (1987). Anglos and Mexicans in the Making of Texas 1836–1986. Austin, TX: The University of Texas Press.

Montiel-Hernandez, A.M., Muniz, M., Baez-Villasenor, J., & del Rio, C.C. (1996). HIV/AIDS in the Mexico-US border: specific risk factors. International Conference on AIDS, 11(2), 483 (abstract no. Pub.D.1293).

Moon, G., Quarendon, G., Barnard, S., Twigg, L., & Blyth, B. (2007). Fat nation: deciphering the distinctive geographies of obesity in England. Social Science and Medicine, 65, 20–31.

Mooney, C., Zwanziger, J., Phibbs, C.S., & Schmitt, S. (2000). Is travel distance a barrier to veterans' use of VA hospitals for medical surgical care? Social Science and Medicine, 50, 1743–1755.

Moore, J.W. (1978). Homeboys: Gangs, Drugs, and Prison in the Barrios of Los Angeles. Philadelphia, PA: Temple University Press.

Morales, S., Lozada, R., Magis, C., & Saavedra, J.A. (2004). Monitoreo and Evaluation of HIV Prevention Interventions in Injectable Drug Users in Mexico. Paper Presented at the XV International Conference on AIDS, Bangkok, Thailand.

Moran, D. (2005). The geography of HIV/AIDS in Russia: risk and vulnerability in transition. Eurasian Geography and Economics, 7, 525–551.

Moran, D. (2007). HIV/AIDS in Russia: local and global perspectives. In: M. Flynn, R. Kay, & J.D. Oldfield (eds.), Trans-National Issues, Local Concerns and Meanings of Post-Socialism: Insights from Russia, Central Eastern Europe and Beyond. Lanham: University Press of America, in press.

Moran, D., & Jordaan, J.A. (2007). HIV/AIDS in Russia: determinants of regional prevalence. International Journal of Health Geographics, 6, 22.

Moran, P.A.P. (1950). Notes on continuous stochastic phenomena. Biometrika, 37, 17–23.

Morenoff, J.D. (2003). Neighborhood mechanisms and the spatial dynamics of birth weight. American Journal of Sociology, 108, 976–1018.

Morenoff, J.D., & Lynch, J. (2004). What makes a place healthy? Neighborhood influences on racial/ethnic disparities in health over the life course. In: N.B. Anderson, R.A. Bulatao, & B. Cohen (eds) Critical Perspectives on Racial and Ethnic Differences in Health in Late Life (pp. 406–449). Washington, DC: National Academies Press.

Morenoff, J.D., Sampson, R.J., & Raudenbush, S.W. (2001). Neighborhood inequality, collective efficacy, and the spatial dynamics of urban violence. Criminology, 39, 517–559.

Motomura, K., et al. (2000). Emergence of new forms of human immunodeficiency virus type 1 intersubtype recombinants in central Myanmar. AIDS Research and Human Retroviruses, 16(17), 1831–1843.

Movaghar, A.R., Goodarzi, R.R., Izadian, E., Mohammadi, M.R., Hosseini, M., & Vazirian, M. (2005). The impact of Bam earthquake on substance users in the first 2 weeks: a rapid assessment. Journal of Urban Health, 82, 370–377.

Mowbray, C.T., Jordan, L.C., Ribisi, K.M., Kewalramani, A., Luke, D., Herman, S., & Bybee, D. (1999). Analysis of post-discharge change in a dual diagnosis population. Health & Social Work, 24, 91–103.

Mueller, M.R., Shah, N.G., & Landen, M.G. (2006). Unintentional prescription drug overdose deaths in New Mexico, 1994–2003. American Journal of Preventive Medicine, 30(5), 423–429.

Mullner, R.M., Chung, K., Croke, K.G., & Mensah, E.K. (2004). Geographic information systems in public health and medicine. Journal of Medical Systems, 28(3), 215–221.

MultiQuest, I. (2006). 2006 Public Opinion Survey. Available at: http://www.dps.state.la.us/tiger/2006PublicOpinionSurvey.pdf [On-line].

Mumola, C., & Karberg, J. (2006). Drug Use and Dependence, State and Federal Prisoners, 2004 (NCJRS Publication No. NCJ 213530). Washington, DC: U.S. Department of Justice, Bureau of Justice Statistics.

Muntaner, C., Eaton, W.W., Diala, C., Kessler. R.C., & Sorlie, P.D. (1998). Social class, assets, organizational control and the prevalence of common groups of psychiatric disorders. Social Science and Medicine, 47(12), 2043–2053.

Musshoff, F., Padosch, S.A., & Madea, B. (2005). Death during patient-controlled analgesia: piritramide overdose and tissue distribution of the drug. Forensic Science International, 154(2–3), 247–251.

Myanmar Ministry of Health (2000). National HIV/AIDS Sentinel Surveillance Reports, 1998, 1999. Yangon: Myanmar Ministry of Health.

Nandi, A., Galea, S., Ahern, J., Bucciarelli, A., Vlahov, D., & Tardiff, K. (2006). What explains the association between neighborhood-level income inequality and the risk of fatal overdose in New York City? Social Science and Medicine, 63(3), 662–674.

National Association for the Advancement of Colored People (2005). Legal Affairs History: New York: National Association for the Advancement of Colored People.

National Bureau of Economic Research (2006). Vital Statistics NCHS's Multiple Cause of Death Data, 1959–2003. Retrieved June, 2006, from http://www.nber.org/data/multicause.html. [On-line]

National Center for Health Statistics (2006). The Health Literacy of America's Adults: Results from the 2003 National Assessment of Adult Literacy (NCES 2006–483). Washington, DC: US Department of Education.

National Drug Intelligence Center (NDIC) (2003). Colorado Drug Threat Assessment: Methamphetamine. Available at: http://www.usdoj.gov/ndic/pubs4/4300/meth.htm. [On-line]

National Institute of Drug Abuse (2004). Lessons from Prevention Research. Washington, DC: National Institute on Drug Abuse (NIDA), U.S. Department of Health and Human Services.

National Institute of Drug Abuse (2007). Drugs, Brains, and Behavior: The Science of Addiction. Washington, DC: National Institute on Drug Abuse (NIDA), U.S. Department of Health and Human Services.

National Institute of Justice (1996). 1995 drug use Forecasting: Annual Report on Adult and Juvenile Arrestees. Rockville, MD: National Institute of Justice, Office of Justice Programs, U.S. Department of Justice.

National Institute on Drug Abuse (2001). Strategic Plan on Reducing Health Disparities, vol. 2004. Washington, DC: National Institute on Drug Abuse.

National Institute on Drug Abuse (2004). 2004 National Survey on Drug Use and Health. Washington, DC: National Institute on Drug Abuse (NIDA), U.S. Department of Health and Human Services.

National Mental Health Association (2006). Hurricane Katrina Anniversary: Mental Health Needs Continue Unmet. National Mental Health Association News Release Database. Available at: www.nmha.org. [On-line]

Neaigus, A., Hagen, D.L., Friedman, S.R., Miller, M., & Des Jarlais, D.C. (1998). Transitions to Injecting Drug Use Among Non-Injecting Heroin Users. Paper Presented at the 126th American Public Health Association Meeting, Washington, DC.

Neaigus, A., Miller, M., Friedman, S.R., et al. (2001). Potential risk factors for the transition to injecting among non-injection heroin users: a comparison of former injectors and never injectors. Addiction, 96, 847–860.

Nelson, A.L., Bromley, R.D.F., & Thomas, C.J. (2001). Identifying micro and temporal patterns of violent crime and disorder in the British city center. Applied Geography, 21, 249–274.

Nelson, E.A., Maruish, M.E., & Alexander, J. (2000). Effects of discharge planning and compliance with outpatient appointments on readmission rates. Psychiatric Services, 51, 885–889.

Ness, M. (2005). North Dakota Bureau of Criminal Investigation. July 28, 2005.

Neubuerger, O.W., Miller, S.I., Schmitz, R.E., Matarazzo, J.D., Pratt, H., & Hasha, N. (1982). Replicable abstinence rates in an alcoholism treatment program. JAMA, 248(8), 960–963.

Nielsen, A.L., & Martinez, R. Jr (2003). Reassessing the alcohol-violence linkage: results from a multiethnic city. Justice Quarterly, 20, 445–469.

Nielsen, A.L., Martinez, R. Jr, & Lee, M.T. (2005). Alcohol, ethnicity, and violence: the role of alcohol availability for latino and black aggravated assaults and robberies. The Sociological Quarterly, 46, 479–502.

Nisbet, C.T., & Vakil F. (1972). Some estimates of price and expenditure elasticities of demand for marijuana among U.C.L.A. students. The Review of Economics and Statistics, 54, 473–475.

Noble, M, Treadwell, J.R., Tregear, S.J., Coates, V.H., Wiffen, P.J., Akafomo, C, Schoelles, K.M. (2007). Long-Term Opioid Management for Chronic Noncancer Pain. Cochrane Database of Systematic Reviews: Protocols 2007 Issue 3. Chichester, UK: John Wiley & Sons, Ltd.

Noriega-Minichiello, S., Magis, C., Uribe, P., Anaya, L., & Bertozzi, S. (2002). The Mexican HIV/AIDS surveillance system: 1986–2001. AIDS, 16(Suppl 3), S13–S17.

Norris, K. (1993). Dakota: A Spiritual Geography. New York: Ticknor and Fields.

Norris, F. (2005). Range, Magnitude, and Duration of the Effects of Disasters on Mental Health: Review Update 2005. Available at: http://redmh.org/research/general/REDMH_effects.pdf [On-line].

North, C.S., Nixon, S.J., Shariat, S., Mallonee, S., McMillen, J.C., Spitznagel, E.L., et al. (1999). Psychiatric disorders among survivors of the Oklahoma City bombing. JAMA, 282, 755–762.

Northridge, M.E., Sclar, E.D., Biswas, P. (2003). Sifting out the connections between the built environment and health: a conceptual framework for navigating pathways and planning healthy cities. Journal of Urban Health, 80(4), 556–568.

Novick, D.M., Haverkos, H.W., & Teller, D.W. (1997). The medically ill substance abuser. In: J.H. Lowinson, P. Ruiz, R.B. Millman, & J.G. Landgrod (eds.), Substance Abuse: A Comprehensive Textbook, 3rd Edition (pp. 534–550). Baltimore: Williams & Wilkins.

Novick, D.M., Joseph, H., Richman, B.L., et al. (1989). Medical maintenance: a new model for continuing treatment of socially rehabilitated methadone maintenance patients. In: L.S. Harris (ed.), Problems of Drug Dependence, 1988. Rockville, MD: NIDA, 168–176.

Nurco, D.N. (1972). An ecological analysis of heroin addicts in Baltimore. The International Journal of the Addictions, 7, 341–353.

Nurco, D., Shaffer, J., & Cisin, I. (1984). An ecological analysis of the interrelationships among drug abuse and other indices of social pathology. The International Journal of the Addictions, 19, 441–451.

Oakes, J.M. (2004). The (mis)estimation of neighborhood effects: causal inference for a practicable social epidemiology. Social Science and Medicine, 58(10), 1929–1952.

O'Campo et al. (2005). Uncovering neighborhood influences in intimate partner violence using concept mapping. Journal of Epidemiology and Community Health, 59(3), 603–308.

O'Campo, P., Xue, X., Wang, M.C., & Caughy, M. (1997). Neighborhood risk factors for low birthweight in Baltimore: a multilevel analysis. American Journal of Public Health, 87, 1113–1118.

Ochoa-Diaz Lopez, H., Sanchez-Perez, H.J., & Martinez-Guzman, L.A. (1996). Use of an index of social welfare for health planning at a municipal level. Salud pública de México, 38(4), 257–267.

O'Connell, M., & Fleury, J.K. (1999). Mapping for corrections. Paper Presented at the 19th Annual ESRI International User Conference, San Diego, CA, July 26–30.

O'Donnell, J., & Jones, J. (1970). Diffusion of the intravenous technique among narcotic addicts. In: J. Ball, & C. Chambers (eds.), The Epidemiology of Opiate Addiction in the United States (pp. 147–164). Springfield: Charles C Thomas Publisher.

Office of Management and Budget (2000). Standards for defining metropolitan and micropolitan statistical areas. Federal Register, 65, 82228–82238.

Office of Management and Budget (2004). Update of Statistical Area Definitions and Additional Guidance on Their Uses. OMB Bulletin, No. 04-03. Washington, DC: Office of Management and Budget.

Office of National Drug Control Policy (ONDCP) (2001). The Economic Costs of Drug Abuse in the United States, 1992–1998 (NCJ-190636). Washington, DC: Office of National Drug Control Policy.

Office of National Drug Control Policy (ONDCP) (2005). Methamphetamine. Available at: http://www.whitehousedrugpolicy.gov. [On-line]

Oficina de la presidencia de la republica (2002). Prevención y Redes Sociales en Contra de las Adicciones. Retrieved 14 March, 2005, from http://www.presidencia.gob.mx/?P=2&Orden=Leer&Tipo=PP&Art=3272. [On-line]

Oliver, M.N., Mathews, K., Siadaty, M., Hauck, F., & Pickle, L. (2005). Geographic bias related to geocoding in epidemiologic studies. International Journal of Health Geographics. Retrieved May 2, 2008, from http://www.ijhealthgeographics.com/content/4/1/29.

Oliver, P., & Keen, J. (2003). Concomitant drugs of misuse and drug using behaviours associated with fatal opiate-related poisonings in Sheffield, UK, 1997–2000. Addiction, 98(2), 191–197.

O'Malley, P.M., Johnston, L.D., & Bachman, J.G. (1991). Quantitative and qualitative changes in cocaine use among American high school seniors, college students, and young adults. NIDA Research Monograph, 110, 19–43.

Openshaw, S. (1984). The modifiable areal unit problem'. Concepts and Techniques in Modern Geography, 38, 41.

Organista, K.C., & Organista, P.B. (1997). Migrant laborers and AIDS in the United States: a review of the literature. AIDS Education and Prevention, 9(1), 83–93.

Osypuk, T., & Galea, S., (eds.) (2007). What level macro? Choosing appropriate levels to assess how place influences population health. In: S. Galea (ed.), Macrosocial Determinants of Population Health. Springer.

Ozcan, Y.A., Shukla, R.K., & Tyler, L.H. (1997). Evaluation of effectiveness of community mental health care: the need fulfillment perspective. Organization Science, 8(2), 176–191.

Pacula, R.L. (1998). Does increasing the beer tax reduce marijuana consumption? Journal of Health Economics, 17, 557–585.

Palka, E. (1999). Accessible wilderness as a therapeutic landscape: experiencing the nature of Denali National Park, Alaska. In: A. Williams (ed.), Therapeutic Landscapes: The Dynamic Between Place and Wellness (pp. 123–152). Lanham, MD: University Press of America.

Panda S., et al. (2001). Interface between drug use and sex work in Manipur. The National Medical Journal of India, 14(4), 209–211.

Pandey, G. (2005). Bhaisajya Ratnavali: Susamskrta Sambarddhita Pariskrtyabhinavya Parisista Sahita, vol. 1. Varanasi: Chaukhamba Sanskrit Series Office.

Paoli, L. (2002). The price of freedom: illegal drug markets and policies in post-Soviet Russia. The Annals of the American Academy of Political and Social Science, 582, 167–180.

Paone, D., Clark, J., Shi, Q., Purchase, D., & Des Jarlais, D.C. (1999). Syringe exchange in the United States, 1996: a national profile. American Journal of Public Health, 89, 43–46.

Parfitt, T. (2003). Drug addiction and HIV infection on rise in Tajikistan. Lancet, 362(9391), 1206.

Parr, H. (1997). Mental health, public space, and the city: questions of individual and collective access. Environment and Planning. D, Society and Space 15(4), 435–454.

Patterson, K.D. & Pyle G.F. (1991). The geography and mortality of the 1918 influenza pandemic. Bulletin of the History of Medicine, 65, 4–21.

Paulozzi, L.J. (2006). Opioid analgesic involvement in drug abuse deaths in American metropolitan areas. American Journal of Public Health, 96(10), 1755–1757.

Paulozzi, L.J., Budnitz, D.S., & Xi, Y. (2006). Increasing deaths from opioid analgesics in the United States. Pharmacoepidemiology and Drug Safety, 15(9), 618–627.

Pauly, G.A., McEwen, J.T., & Finch, S. (1967). Computer mapping – a new technique in crime analysis. In: S.A. Yefsky (ed.), Law Enforcement Science and Technology. New York: Thompson Book Company, 739–748.

PDFA. (2005). Meth Frequently Asked Questions. Available at: http://www.drugfree.org/Portal/DrugIssue/MethResources/meth_faq.html. [On-line]

Peele, S.M. (1989). The Diseasing of America: Addiction Treatment Out of Control. Lexington, MA: D.C. Heath.

Pendyala, R.M., Goulias, K.G., & Kitamura, R. (1991). Impact of telecommuting on spatial and temporal patterns of household travel. Transportation, 18, 411–432.

Penrod, J., Preston, D.B., Cain, R.E., & Starks, M.T. (2003). A discussion of chain referral as a method for sampling hard-to-reach populations. Journal of Transcultural Nursing, 14, 100–107.

Perlis, R. Ostacher, M., & Patel J.K. (2006). Predictors of recurrence in bipolar disorder: primary outcomes from the systematic treatment enhancement program for bipolar disorder (STEP-BD). The American Journal of Psychiatry, 163, 217–224.

Perlis, T., Torrico, C., & Settembrino, J. (2002). Travel patterns among young urban nomad injecting drug users in the United States. Paper Presented at the XIV International Conference on AIDS, Barcelona, Spain.

Perry, M., Dulio, A., Artiga, S., Shartzer, A., & Rousseau, D. (2006). Voices of the Storm: Health Experiences of Low-Income Katrina Survivors. Menlo Park, CA: The Henry J. Kaiser Family Foundation.

Petchey, R., Williams, J., Farnsworth, B., & Starkey, K. (1998). A tale of two (low prevalence) cities: social movement organizations and the local policy response to HIV/AIDS. Social Science and Medicine, 47(9), 1197–1208.

Peters, R.J. Jr, Williams, M., Ross, M.W., Atkinson, J., & Yacoubian, G.S. Jr (2007). Codeine cough syrup use among African-American crack cocaine users. Journal of Psychoactive Drugs, 39(1), 97–102.

Peterson, R., & Krivo, L. (1993). Racial segregation and black urban homicide. Social Forces, 71, 1001–1026.

Peterson, R.D., & Krivo, L.J. (2005). Macro-structural analyses of race, ethnicity and violent crime: recent lessons and new directions for research. Annual Review of Sociology, 31, 331–356.

Peterson, W. (1958). A general typology of migration. American Sociological Review, 23, 256–265.

Petronis, K.R., & Anthony, J.C. (2003). A different kind of contextual effect: geographical clustering of cocaine incidence in the USA. Journal of Epidemiology and Community Health, 57, 893–900.

Pickett, K., & Pearl, M. (2001). Multilevel analyses of neighbourhood socioeconomic context and health out-comes: a critical review. Journal of Epidemiology and Community Health, 55, 111–122.

Pile, S., & Keith, M. (eds.) (1997). Geographies of Resistance. London and New York: Routledge.

Pilkington, H. (2005). Full Research Report, "Everyday" but Not "Normal": Drug Use and Youth Cultural Practice in Russia. Retrieved 26 June, 2007 from www.esrcsocietytoday.ac.uk. [On-line]

Pilkington, H. (2006). 'For us it is normal': exploring the 'recreational' use of heroin within Russian youth cultural practice. Journal of Communist Studies and Transition Politics, 22, 24–53 (1352–3279).

Piper, T.M., Rudenstine, S., Stancliff, S., Sherman, S., Nandi, V., Clear, A., et al. (2007). Overdose prevention for injection drug users: lessons learned from naloxone training and distribution programs in New York City. Harm Reduction Journal, 4, 3.

Piyasirisilp, S., et al. (2000) A recent outbreak of HIV-1 infection in southern China was initiated by two highly homogeneous, geographically separated strains, circulating recombinant form AE and a novel BC recombinant. Journal of Virology, 74(23), 11286–11295.

Platt, L., Hickman, M., Rhodes, T., Mikhailova, L., Karavashkin, V., Vlasov, A., et al. (2004). The prevalence of injecting drug use in a Russian city: implications for harm reduction and coverage. Addiction, 99, 1430–1438.

Polednak, A. (1996). Trends in US urban black infant mortality, by degree of racial segregation. American Journal of Public Health, 86, 723–726.

Police Employee (LEOKA) (2002) Data. U.S. Dept. of Justice, Federal Bureau of Investigations. Uniform Crime Reporting Program, 1990–2000 [Computed file]. Compiled by the U.S. Dept. of Justice, Federal Bureau of Investigation. ICPSR ed. Ann Arbor, MI: Inter-University Consortium for Political and Social Research [producer and distributor], 2002. Ann Arbor, Michigan.

Pollack, C.E., Cubbin, C., Ahn, D., & Winkleby, M. (2005). Neighbourhood deprivation and alcohol consumption: does the availability of alcohol play a role? International Journal of Epidemiology, 34, 772–780.

Population Figures (2006). 2005 Population Estimates for cities in Nicaragua. Retrieved 1 December, 2006, from http://www.mongabay.com/igapo/2005_world_city_populations/Nicaragua.html. [On-line]

Porter, D. (1999). Health, Civilization and the State: A History of Public Health from Ancient to Modern Times. London and New York: Routledge.

Poser, W., Poser, S., & Eva-Condemarin, P. (1992). Mortality in patients with dependence on prescription drugs. Drug Alcohol Depend, 30(1), 49–57.

Poshyachinda, V. (1993). Drug injecting and HIV infection among the population of drug abusers in Asia. Bulletin on Narcotics, 45(1), 77–90.

Post, R.M., Rubinow, D.R., & Ballinger, J.C. (1986). Conditioning and sensitization in the longitudinal course of affective illness. The British Journal of Psychiatry, 149, 191–201.

Poulin, C., Stein, J., & Butt, J. (1998). Surveillance of drug overdose deaths using medical examiner data. Chronic Diseases in Canada, 19(4), 177–182.

Poulsen, E., & Kennedy, L. (2004). Using dasymetric mapping for spatially aggregated crime data. Journal of Quantitative Criminology, 20(3), 243–262.

Poundstone, K.E., Strathdee, S.A., & Celentano, D.D. (2004). The social epidemiology of human immunodeficiency virus/acquired immunodeficiency syndrome. Epidemiologic Reviews, 26, 22–35.

Prendergast, M.L., Podus, D., & Chang, E. (2000). Program factors and treatment outcomes in drug dependence treatment: an examination using meta-analysis. Substance Use & Misuse, 35, 1931–1965.

Prince, R., Kearns, R., & Craig, D. (2006). Governmentality, discourse and space in the New Zealand health care system, 1991–2003. Health & Place, 12, 253–266.

Prue, D.M., Keane, T.M., Cornell, J.E., & Foy, D.W. (1979). An analysis of distance variables that affect aftercare attendance. Community Mental Health Journal, 15, 149–154.

Purchase, D. (2003). Personal communication. Director, Point Defiance AIDS Projects, and the North American Syringe Exchange Network. Tacoma, WA.

Putnam, R. (2000). Bowling Alone: The Collapse and Revival of American Community. New York: Simon & Schuster.

Putnam, R.D. (2000). Bowling Alone. New York: Simon & Schuster.

Quan, V.M., Chung, A., & Abdul-Quader, A.S. (1998). The feasibility of a syringe-needle-exchange program in Vietnam. Substance Use & Misuse, 33(5), 1055–1067.

Quan, V.M., et al. (2000). HIV in Vietnam: the evolving epidemic and the prevention response. Journal of Acquired Immune Deficiency Syndromes, 25(4), 360–369.

Quillian, L. (1999). Migration patterns and the growth of high-poverty neighborhoods 1970–1990. American Journal of Sociology, 105(1), 1–37.

Quintero, G.A., & Estrada, A.L. (1998). Cultural models of masculinity and drug use: "machismo," heroin, and street survival on the US-Mexico border. Contemporary Drug Problems, 25, 147–168.

Rachlis, B., Brouwer, K.C., Mills, E.J., Hayes, M., Kerr, T., & Hogg, R.S. (2007). Migration and transmission of blood-borne infections among injection drug users: understanding the epidemiologic bridge. Drug and Alcohol Dependence, 90(2–3), 107–119.

Ramos, R. (1995). An Ethnographic Study of Heroin Abuse by Mexican American in San Antonio, Texas. Austin, TX: Texas Commission on Alcohol and Drug Abuse.

Ramos, R.L. (1990). Migratory patterns and the spread of HIV across the Mexico-United States Border. Salud Fronteriza, 1(3), 23–32.

Ramos, R.L. (2000). De la investigación a la acción: la experiencia del programa Compañeros, AC. In: P. Uribe, & C. Magis-Rodríguez (eds.), La Respuesta Mexicana al SIDA: Mejores Prácticas (pp. 129–130). México: Ángulos del SIDA 2000. Consejo Nacional para la Prevencion y Control del SIDA.

Rannazzisi, J.T. (2007). The DEA's balancing act to ensure public health and safety. Clinical Pharmacology and Therapeutics, 81(6), 805–806.

Ratcliffe, J.H. (2002). The hotspot matrix: a framework for the spatio-temporal targeting of crime reduction. Police Practice and Research, 5(1), 5–23.

Ratcliffe, J.H. (2005). Detecting spatial movement of intra-region crime patterns over time. Journal of Quantitative Criminology, 21(1), 103–123.

Ratcliffe, J.H., & McCullagh, M.J. (1999). Hotbeds of crime and the search for spatial accuracy. Journal of Geographical Systems, 1(4):385–398.

Rawson, R.A., et al. (2004). A multi-site comparison of psychosocial approaches for the treatment of methamphetamine dependence. Addiction, 99(6), 708–717.

Redlinger, L.J., & Michel, J.B. (1970). Ecological variations in heroin abuse. The Sociological Quarterly, 11, 219–229.

Reibel, M., & Bufalino, M.E. (2005). Street-weighted interpolation techniques for demographic count estimation in incompatible zone systems. Environment & Planning A, 37, 127–139.

Reid, R.J., Hughey, J., & Peterson, N.A. (2003). Generalizing the alcohol outlet-assaultive violence link: evidence from a U.S. midwestern city. Substance Use & Misuse, 38, 1971–1982.

Reinertson-Sand, M. (2006). Methamphetamine Frequently Asked Questions. Rural Assistance Center. Available at: http://www.raconline.org/info_guides/meth/methfaq.php. [On-line]

Reisig, M.D., Bales, W.D., Hay, C., & Wang, X. (2007). The effect of racial inequality on black male recidivism. Justice Quarterly, 24(3), 408–434.

Rengert, G.F., Ratcliffe, J.H., & Chakravorty, S. (2005). Policing Illegal Drug Markets: Geographic Approaches to Crime Reduction. Monsey, NY: Criminal Justice Press.

Rhodes, T., Ball, A., Stimson, G.V., Kobyshcha, Y., Fitch, C., Pokrovsky, V., et al. (1999). HIV infection associated with drug injecting in the newly independent states, eastern Europe: the social and economic context of epidemics. Addiction, 94, 1323–1336.

Rhodes, T., Lowndes, C., Judd, A., Mikhailova, L.A., Sarang, A., Rylkov, A., et al. (2002). Explosive spread and high prevalence of HIV infection among injecting drug users in Togliatti city, Russia. AIDS, 16, F25–F31.

Rhodes, T., Mikhailova, L., Sarang, A., Lowndes, C.M., Rylkov, A., Khutorskoy, M., et al. (2003). Situational factors influencing drug injecting, risk reduction and syringe exchange in Togliatti city, Russian Federation: a qualitative study of micro risk environment. Social Science and Medicine 57, 39–54.

Rhodes, T., Platt, L., Sarang, A., Mikhailova, L., & Monaghan, G. (2006). Street policing, injecting drug use and harm reduction in a Russian city: a qualitative study of police perspectives. Journal of Urban Health, 83, 911–25.

Rhodes, T., Singer, M., Bourgois, P., Friedman, S., & Strathdee, S. (2005). The social structural production of HIV risk among injecting drug users. Social Science and Medicine, 61, 1026–1044.

Rhodes, T., Singer, M., Bourgois, P., Friedman, S.R., & Strathdee, S. (2005). The social structural production of HIV risk among injecting drug users. Social Science and Medicine, 61(5), 1026–1044.

Richards, J.F. (2002). Opium and the British Indian Empire: The Royal Commission of 1895. Modern Asian Studies, 36(2), 375–420.

Richards, T.B., Croner, C.M., Rushton, G., Brown, C.K., & Fowler, L. (1999). Geographic information systems and public health: mapping the future. Public Health Reports 114(4), 359.

Richardson, D. (2006). GIS & T: Transforming science and society. In: D. DiBiase, et al. (eds.) (UCGIS Education Committee), Geographic Information Science and Technology: Body of Knowledge. Washington, DC: Association of American Geographers.

Richardson, D., & Solis, P. (2004). Confronted by insurmountable opportunities: geography in society at the AAG's centennial. The Professional Geographer, 56(1), 4–11.

Richardson, S., Abellan, J.J., & Best, N. (2006). Bayesian spatio-temporal analysis of joint patterns of male and female lung cancer risks in Yorkshire (UK). Statistical Methods in Medical Research, 15, 385–407.

Richman, J.A., Flaherty, J.A., & Rospenda, K.M. (1996). Perceived workplace harassment experiences and problem drinking among physicians: broadening the stress/alienation perspective. Addiction, 91, 391–403.

Ricketts, T., Randolph, R., Howard, H., Pathman, D., & Carey, T. (2001). Hospitalization rates as indicators of access to primary care. Health & Place, 7, 27–38.

Riner, M.E., Cunningham, C., & Johnson, A. (2004). Public health education and practice using geographic information system technology. Public Health Nursing, 21(1), 57–65.

Robert Wood Johnson Foundation (2001). Substance Abuse: The Nation's Number One Health Problem. Key Indicators for Policy. Princeton: Robert Wood Johnson Foundation.

Roberts, A. (2000). Psychiatric comorbidity in white and African-American illicit substance abusers: evidence for differential etiology. Clinical Psychology Review, 20, 667–677.

Roberts, K. (1997). China's 'tidal wave' of migrant labor: what can we learn from Mexican undocumented migration to the United States? The International Migration Review, 31(2), 249–293.

Robinson, J.B, & Rengert, G.F. (2006). Illegal drug markets: the geographic perspective and crime propensity. Western Criminology Review, 7(1), 20–32. Available at: http://wcr.sonoma.edu/v07n1/20/sexoffender.html [On-line].

Rockwell, R., Des Jarlais, D.C., Friedman, S.R., Perlis, T.E., & Paone, D. (1999). Geographic proximity, policy and utilization of syringe exchange programmes. AIDS Care, 11(4), 437–442.

Romley, J.A., Cohen, D., Ringel, J., & Sturm, R. (2006). Alcohol and environmental justice: the density of liquor stores and bars in urban neighborhoods in the United States. Journal of Studies on Alcohol and Drugs, 1, 48–55.

Rose, D.R., & Clear, T.R. (1998). Incarceration, social capital and crime: implications for social disorganization theory. Criminology, 36, 441–479.

Ross, C.E., & Jang, S.J. (2000). Neighborhood disorder, fear, and mistrust: the buffering role of social ties with neighbors. American Journal of Community Psychology, 28, 401–420.

Rothenberg, R., Malone, S., Muth, S., Potterrat, J., & Woodhouse, D. (2005). Social and geographic distance in HIV risk. Sexually Transmitted Diseases, 32(8), 506–512.

Rothenberg, R.B., Potterat, J.J., Woodhouse, D.E., Muth, S.Q., Darrow, W.W., & Klovdahl, A.S. (1998). Social network dynamics and HIV transmission. AIDS, 12(12), 1529–1536.

Rothman, R.E., Hsu, E.B., Kahn, C.A., & Kelen, G.D. (2006). Research priorities for surge capacity. Academic Emergency Medicine, 13, 1160–1168.

Roy, E., Haley, N., Leclerc, P., Cedras, L., & Boivin, J. (2002). Drug injection among street youth: the first time. Addiction, 97, 1003–1009.

Roy, E., Haley, N., Leclerc, P., Lemire, N., Boivin, J-F., Frappier, J.-Y., & Claessens, C. (2000). Prevalence of HIV infection and risk behaviors among Montreal street youth. International Journal of STD & AIDS, 11, 241–247.

Roy, E., Haley, N., Leclerc, P., Sochanski, B., Boudrenau, J.F., & Boivin, J.F. (2004). Mortality in a cohort of street youth in Montreal. JAMA, 292(5), 569–574.

Royce, R., Sena, A., Cates, W. Jr, & Cohen, M. (1997). Sexual transmission of HIV. The New England Journal of Medicine, 336, 1072–1078.

Rudowitz, R., Rowland, D., & Shartzer, A. (2006). Health care in New Orleans before and after Hurricane Katrina. Health Affairs, 25, 393–406.

Ruhm, C.J. (2000). Are recessions good for your health? The Quarterly Journal of Economics, 115, 617–650.

Rush, J.R. (1990). Opium to Java: Revenue Farming and Chinese Enterprise in Colonial Indonesia, 1860–1910. Ithaca, NY: Cornell University Press.

Rushton, G., & Lolonic, P. (1996). Exploratory spatial analysis of birth defect rates in an urban population. Statistics in Medicine, 15, 717–726.

Russell, D., & Cutrona, C. (1988). Development and Evolution of the UCLA Loneliness Scale. Iowa City: Center for Health Services Research, College of Medicine, University of Iowa.

Russian Federation Ministry of Health (2002). National HIV/AIDS Sentinel Surveillance Reports, 2001. Moscow: Russian Federation Ministry of Health.

Saarento, O., Oiesvold, T., Sytema, S., Gostas, G., Kastrup, M., Lonnerberg, O., Muus, S., Sandlund, M., & Hansson, L. (1998). The Nordic comparative study on sectorized psychiatry: continuity of care related to characteristics of the psychiatric services and the patients. Social Psychiatry and Psychiatric Epidemiology, 33, 521–527.

Sajatovic, M., Bauer, MS., Kilbourne, A.M. at al. (2006). Self-reported medication treatment adherence among veterans with bipolar disorder. Psychiatric Services, 57, 56–62.

SAMHSA (2006). Retrieved May 2, 2008, from (http://www.drugabusestatistics.samhsa.gov/sub-State2k6/alcDrugs.htm#Tab).

Sampson, R.J. (2003). Neighborhood-level context and health: lessons from sociology. In: I. Kawachi, & L. Berkman (eds.), Neighborhoods and Health (pp. 132–146). New York: Oxford University Press.

Sampson, R.J., & Bean, L. (2006). Cultural mechanisms and killing fields: a revised theory of community-level racial inequality. In: R.D. Peterson, L.J. Krivo, & J. Hagan (eds.), The Many Colors of Crime: Inequalities of Race, Ethnicity, and Crime in America (pp. 8–36). New York: New York University Press.

Sampson, R.J., Morenoff, J.D., & Gannon-Rowley, T. (2002). Assessing neighborhood effects: social processes and new directions in research. Annual Review of Sociology, 28, 443–478.

Sampson, R.J., Morenoff, J.D., & Raudenbush, S.W. (2005). Social anatomy of racial and ethnic disparities in violence. American Journal of Public Health, 95, 224–232.

Sampson, R.J., Raudenbush, S.W., & Earls, F.E. (1997). Neighborhoods and violent crime: a multilevel study of collective efficacy. Science, 277, 918–924.

Sanchez, J., Shedlin, M., & Araica, A. (2000). Needs & Resource Assessment: STIs/HIV in Nicaragua. Washington, DC: USAID

Sanders, B. (2006). Young people, clubs and drugs. In: B. Sanders (ed.), Drugs, Clubs and Young People: Sociological and Public Health Perspectives. Aldershot, UK: Ashgate.

Sanders, B., Lankenau, S., Jackson Bloom, J., Hathazi, D. (2008). Multiple drug use and polydrug use amongst homeless traveling youths. Journal of Ethnicity in Substance Abuse, 17, 23–40.

San Francisco AIDS Foundation (2007). Frequently Asked Questions About Crystal Methamphetamine. Available at: http://www.aidshotline.org/crm/asp/refer/faq_meth.asp. [On-line]

Santamaría, B.M. (1997). Census of Boys, Girls and Adolescents in the Street and Public Spaces. Managua, Nicaragua: FONIF, UNICEF.

SAS Institute (2004). I: SAS/STAT 9.1 User's Guide. Cary: SAS Institute.

Sauer, C.O. (1925). The Morphology of Landscape. University of California Publications in Geography, 2, 19–54.

Saunders-Phillips, K. (2002). Factors influencing HIV/AIDS in women of color. Public Health Reports, 117, S151–S156.

Savitch, H.V. (2003). How suburban sprawl shapes human well-being. Journal of Urban Health, 80(4), 590–607.

Saxe, L., Kadushin, C., Beveridge, A., Livert, D., Tighe, E., Rindskopf, D., Ford, J., & Brodsky, A. (2001). The visibility of illicit drugs: implications for community-based drug control strategies. American Journal of Public Health, 91, 1987–1994.

Saxena, S., & Mokhtarian, P.L. (1997). The impact of telecommuting on the activity spaces of participants. Geographical Analysis, 29, 124–144.

Schabner, D. (2005). Do Meth Addicts Want Your Identity? ABC News, June 2. Available at: http://abcnews.go.com/Business/FinancialSecurity/story?id=793760&page=1. [On-line]

Schilling, R.F., Monterroso, E., Fontdeveila, J., Fernando, D., & El-Bassel, N. (2004). Proximity to needle exchange programs and HIV-related risk behavior among injection drug users in Harlem. Evaluation and Program Planning, 27(1), 25–33.

Schmitt, S.K., Phibbs, C.S., & Piette J.D. (2003). The influence of distance on utilization of outpatient mental health aftercare following inpatient substance abuse treatment. Addictive Behaviors, 28, 1183–1192.

Schoeneberger, M.L., Leukefeld, C.G., Hiller, M.L., & Godlaski, T. (2006). Substance abuse among rural and very rural drug users at treatment entry. The American Journal of Drug and Alcohol Abuse, 32(1), 87–110.

Schulte Gary, S.L., Aultman-Hall, L., McCourt, M., & Stamatiadis, N. (2003). Consideration of driver home county prohibition and alcohol-related vehicle crashes. Accident; Analysis and Prevention, 35, 641–648.

Schultz, A., Williams, D.R., Israel, B., et al. (2000). Unfair treatment, neighborhood effects, and mental health in the Detroit metropolitan area. Journal of Health and Social Behavior, 41, 314–332.

Schumacher, J.E., Mennemeyer, S.T., Milby, J.B., Wallace, D., & Nolan, K. (2002). Costs and effectiveness of substance abuse treatments for homeless persons. The Journal of Mental Health Policy and Economics, 5(1), 33–42.

Schuster, M.A., Stein, B.D., Jaycox, L., Collins, R.L., Marshall, G.N., Elliott, M.N., et al. (2001). A national survey of stress reactions after the September 11, 2001, terrorist attacks. The New England Journal of Medicine, 345, 1507–1512.

Schuurman, N., Bell, N., Dunn, J., & Oliver, M. (2007). Deprivation indices, population health and geography: an evaluation of the spatial effectiveness of indices at multiple scales. Journal of Urban Health, 84(4), 591–603.

Schwartz, S., & Diez-Roux, A.V. (2001). Commentary: causes of incidence and causes of cases – a Durkheimian perspective on Rose. International Journal of Epidemiology, 30(3), 435–439.

Scott, M.S. (2000). Clandestine Drug Labs: Chemical Time Bombs. Washington, DC: Office of Community Oriented Policing Services, GPO.

Scott, M.S. (2002). The Benefits and Consequences of Police Crackdowns. Problem-Oriented Guides for Police, Response Guides Series No. 1. Washington, DC: Office of Community Oriented Policing Services, U.S. Department of Justice 1-932582-40-1

Scott, W.R. (2003). Organizations: Rational, Natural, and Open Systems, 5th Edition. Thousand Oaks, CA: Sage.

Scribner, R.A., Cohen, D.A., & Fisher, W. (2000). Evidence of a structural effect for alcohol outlet density: a multilevel analysis. Alcoholism, Clinical and Experimental Research, 24, 188–195.

Scribner, R.A., MacKinnon, D.P., & Dwyer, J.H. (1994). Alcohol outlet density and motor vehicle crashes in Los Angeles County cities. Journal of Studies on Alcohol, 55, 447–453.

Scribner, R.A., MacKinnon, D.P., & Dwyer, J.H. (1995). The risk of assaultive violence and alcohol availability in Los Angeles County. American Journal of Public Health, 85(3), 335–340.

Seage, G., et al. (2001). Are US populations appropriate for trials of HIV vaccine? American Journal of Epidemiology, 153(7), 619–627.

Seeman, B.T. (2005). Hurricane Victims With Mental Problems Parted From Health Care. Washington, DC: Newshouse News Service.

Selik, R.M., Castro, K.G., & Pappaioanou, M. (1988). Racial/ethnic differences in the risk of AIDS in the United States. American Journal of Public Health, 78, 1539–1545.

Selik, R.M., Castro, K.G., Pappaioanou, M., & Buehler, J.W. (1989). Birthplace and the risk of AIDS among Hispanics in the United States. American Journal of Public Health, 79, 836–839.

Shao, Y., et al. (1998). Genetic recombination of HIV-1 strains identified in China. XIIth International AIDS Conference, Geneva.

Shaw, C.R., & McKay, H. (1942). Juvenile Delinquency in Urban Areas. Chicago: University of Chicago Press.

Shedlin, M., Arauz, R., Ortels, P., Aburto, M., & Norori, D. (Fundacion Nimehuatzin) (2006). Drugs and AIDS: National Alert. Envio (Journal of the Central American University, Managua, Nicaragua) 292.

Sheehan, T., DeChello, L., Kulldorff, M., Gregorio, D. Gershman, S., & Mroszczyk, M. (2004). The geographic distribution of breast cancer incidence in Massachusetts 1988 to 1997, adjusted for covariates. International Journal of Health Geographics, 3, 17.

Shelley, L. (2006). The drug trade in contemporary Russia. China and Eurasia Forum Quarterly, 4, 15–20.

Shepard, E.M, & Blackley, P.R. (2005). Drug enforcement and crime: recent evidence from New York State. Social Science Quarterly, 86(2), 323.

Sherman, J., Spencer, J., Preisser, J., & Gesler, W. (2005). A suite of methods for representing activity space in a health care accessibility study. International Journal of Health Geographics, 4:24.

Sibley, D. (1995). Geographies of Exclusion. London: Routledge.

Silver, E., Mulvey, E.P., & Swanson, J.W. (2002). Neighborhood structural characteristics and mental disorder: Faris and Dunham revisited. Social Science and Medicine, 55(8), 1457–1470.

Simons, R.L., Johnson, C., Beaman, J., Conger, R.D., & Whitbeck, L.B. (1996). Parents and peer group as mediators of the effect of community structure on adolescent problem behavior. American Journal of Community Psychology, 24(1), 1573–2770.

Simpson, D.D. (1981). Treatment for drug abuse: follow-up outcomes and length of time spent. Archives of General Psychiatry 38, 875–880.

Simpson, D.D., Joe, G.W., Rowan-Szal, G.A., & Greener, J.M. (1997). Drug abuse process components that improve retention. Journal of Substance Abuse Treatment 14, 565–572.

Simpson, L. (2002). Geography conversion tables: a framework for conversion of data between geographical units. International Journal of Population Geography, 8, 69–82.

Sims, S.A., Snow, L.A., & Porucznik, C.A. (2007). Surveillance of methadone-related adverse drug events using multiple public health data sources. Journal of Biomedical Informatics, 40(4), 382–389.

Sinatra, R. (2006). Opioid analgesics in primary care: challenges and new advances in the management of noncancer pain. Journal of the American Board of Family Medicine, 19(2), 165–177.

Sindelar, J.L., & Fiellin, D. (2001). Innovations in treatment for drug abuse. Annual Review of Public Health, 22, 249–72.

Singer, M. (1998). Forging a political economy of AIDS. In: M. Singer (ed.), The Political Economy of AIDS. Amityville: Baywood Publishing Company, Inc.

Sitt, B.G., & Giacopassi, D.J. (1992). Alcohol availability and alcohol-related crime. Criminal Justice Review, 17, 268–279.

Situ, Y., & Liu, W. (1996). Transient population, crime, and solution: the Chinese experience. International Journal of Offender Therapy and Comparative Criminology, 40(4), 293–299.

Skeldon, R. (2000). Population Mobility and HIV Vulnerability in South East Asia: An Assessment and Analysis. Bangkok: UNDP.

Skogan, W.G. (1990). Disorder and Decline: Crime and the Spiral of Urban Decay. New York: Free Press.

Smith, C.J., & Yang, X. (2005). Examining the connection between temporary migration and the spread of STDs and HIV/AIDS in China. The China Review, 5(1), 109–137.

Smith, D.M. (1994). Geography and Social Justice. Oxford & Cambridge, MA: Blackwell Publishers Ltd.

Snedker, K.A., Herting, J.R., & Walton, E., (under review) Contextual effects and adolescent substance use: exploring the role of neighborhoods.

Soja, E.W. (1980). The Socio-spatial dialectic. Annals of the Association of American Geographers, 70(2), 207–225.

Soja, E.W. (1997). Reassertions: towards a spatialized ontology. In: J. Agnew, D.N. Livingstone, & A. Rogers (eds.), Human Geography: An Essential Anthology (pp. 623–635). Oxford & Cambridge, MA: Blackwell Publishers Ltd.

Solinger, D.J. (1999). Contesting Citizenship in Urban China: Peasant Migrants, the State, and the Logic of the Market. Berkeley: University of California, Berkeley.

Sorensen, J.L., & Copeland, A.L. (2000). Drug abuse treatment as an HIV prevention strategy: a review. Drug and Alcohol Dependence, 59(1), 17–31.

Sorenson, S.B., Shen, H., & Kraus, J.F. (1997). Undetermined manner of death. A comparison with unintentional injury, suicide, and homicide death. Evaluation Review, 21(1), 43–57.

Sorg, M., & Greenwald, M. (2002). Maine Drug-Related Mortality Patterns: 1997–2002. Retrieved 9 October, 2006, from http://mainegov-images.informe.org/ag/pr/drugreport.pdf. [On-line]

Sosin, M.R., & Grossman, S.F. (2003). The individual and beyond: a socio-rational choice model service participation among homeless adults with substance abuse problems. Substance Use & Misuse, 38, 503–549.

Soskolne, V., & Shtarkshall, R.A. (2002). Migration and HIV prevention programmes: linking structural factors, culture, and individual behaviour – an Israeli experience. Social Science and Medicine, 55, 1297–1307.

Speer, P.W, Gorman, D.M., Labouvie, E.W., & Ontkush, M.J. (1998). Violent crime and alcohol availability: relationships in an urban community. Journal of Public Health Policy, 19, 303–318.

Sperry, K. (1988). An epidemic of intravenous narcoticism deaths associated with the resurgence of black tar heroin. Journal of Forensic Sciences, 33(5), 1156–1162.

Spiegelhalter, D., Thomas, Q., Best, N., & Lunn, D. (2003). WinBUGS User Manual, Version 1.4. Cambridge, UK: Medical Research Council Biostatistics Unit.

Sporer, K.A., & Kral, A.H. (2007). Prescription naloxone: a novel approach to heroin overdose prevention. Annals of Emergency Medicine, 49(2), 172–177.

Squires, N.F., Beeching, N.J., Schlecht, B.J., & Ruben, S.M. (1995). An estimate of the prevalence of drug misuse in Liverpool and a spatial analysis of known addiction. Journal of Public Health Medicine, 17, 103–109.

SSA (1998). Encuesta Nacional de Adicciones. Datos Epidemiologicos. Retrieved 20 September, 2006, from http://www.salud.gob.mx/unidades/conadic/epidem.htm. [On-line]

SSA (2002). Sistema de Vigilancia Epidemiológica de las Adicciones (SISVEA). Informe 2002. Retrieved 7 January, 2005, from http://www.epi.org.mx./sis/descrip.htm. [On-line]

Stahler, G.J., Cohen, E., & Shipley, T.E. (1993). Why clients drop out of treatment: ethnographic perspectives on treatment attrition among homeless male "crack" cocaine users. Contemporary Drug Problems, 20(4), 651–680.

Star, J., & Estes, J. (1990). Geographic Information Systems: An Introduction. Englewood Cliffs, NJ: Prentice Hall.

StataCorp. (2001). Stata Reference Manual, Release 7. College Station, TX: Stata Press.

State of New Jersey Department of Health and Senior Services (NJDHSS) (2002). Division of AIDS, HIV/AIDS Epidemiologic Profile for the State of New Jersey. Trenton, NJ.

Stephens, B.G., Jentzen, J.M., Karch, S., Wetli, C.V., & Mash, D.C. (2004). National Association of Medical Examiners position paper on the certification of cocaine-related deaths. The American Journal of Forensic Medicine and Pathology, 25(1), 11–13.

Sterk-Elifson, C, & Elifson, K.W. (1992). Someone to count on: homeless, male drug users and their friendship relations. Urban Anthropology, 21, 235–251.

Stimson, G.V., Des Jarlais, D., Ball, A.L., & Organization., W.H. (1998). Drug injecting and HIV infection: global dimensions and local responses. London; Bristol, PA.: UCL Press.

St. John, C. Edwards, M., & Wenk, D. (1995). Racial differences in intraurban residential mobility. Urban Affairs Review, 30(5), 709–729.

Stockwell, T., & Gruenewald, P.J. (2004). Controls on the physical availability of alcohol. In: N. Heather, & T. Stockwell (eds.), The Essential Handbook of Treatment and Prevention of Alcohol Problems (pp. 213–233). New York: John Wiley & Sons, Ltd.

Stockwell, T., Gruenewald, P.J., Toumbourou, J.W., & Loxley, W. (eds.) (2005). Preventing Harmful Substance Use: The Evidence Base for Policy and Practice (pp. 225–236). Chichester, West Sussex, England: John Wiley & Sons, Ltd.

Stoller, N.E. (1998). Lessons from the Damned: Queers, Whores, and Junkies Response to AIDS. New York, NY: Routledge.

Storr, C.L., Chen, C.Y., & Anthony, J.C. (2004). Unequal opportunity: neighborhood disadvantage and the chance to buy illegal drugs. Journal of Epidemiology and Community Health, 58, 231–237.

Strakowski, S.M., et al. (2000). The impact of substance abuse on the course of bipolar disorder. Biological Psychiatry, 48, 477–485.

Strang, J., Kelleher, M., Best, D., Mayet, S., & Manning, V. (2006). Emergency naloxone for heroin overdose. BMJ, 333(7569), 614–615.

Strathdee, S., Celentano, D., Shah, N., Lyles, C., Stambolis, V.S., Macalino, G., Nelson, K., & Vlahov, D. (1999). Needle exchange attendance and health care utilization promote entry into detoxification. Journal of Urban Health, 76, 448–460.

Strathdee, S.A., Fraga, W.D., Case, P., Firestone, M., Brouwer, K.C., Perez, S.G., et al. (2005). Vivo para consumirla y la consumo para vivir [I live to inject and inject to live]: high-risk injection behaviors in Tijuana, Mexico. Journal of Urban Health, 3_Suppl_4(82), iv58–iv73.

Strathdee, S.A., Ricketts, E.P., Huettner, S., Cornelius, L., Bishai, D., Havens, J. R., et al. (2006). Facilitating entry into drug treatment among injection drug users referred from a needle exchange program: results from a community-based behavioral intervention trial. Drug and Alcohol Dependence, 83(3), 225–232.

Strathdee, S.A., Zafar, T., Brahmbhatt, H., Baksh, A., & ul Hassan, S. (2003). Rise in needle sharing among injection drug users in Pakistan during the Afghanistan war. Drug and Alcohol Dependence, 71(1), 17–24.

Subramanian, S., Acevedo-Garcia, D., & Osypuk, T.L. (2005). Racial residential segregation and geographic hetergeneity in black/white disparity in poor self-rated health in the US: a multilevel statistical analysis. Social Science and Medicine, 60, 1667–1679.

Subramanian, S.V., Chen, J.T., Rehkopf, D.H., Waterman, P.D., & Krieger, N. (2005). Neighborhood disparities in context: multilevel analysis of census tract variations in black excess mortality and poverty. American Journal of Public Health, 95, 260–265

Substance Abuse and Mental Health Administration (2004). 2004 National Survey on Drug Use and Health. Rockville: NSDUH.

Substance Abuse and Mental Health Services Administration. (1992). National Household Survey on Drug Abuse: Main Findings 1992. 93. Rockville, MD: SAMSHA, Office of Applied Studies.

Substance Abuse and Mental Health Services Administration (2003). Emergency Department Trends from the Drug Abuse Warning Network, Final Estimates 1995–2002. Rockville, MD: Substance Abuse and Mental Health Services Administration.

Substance Abuse and Mental Health Services Administration (2004). Treatment Episode Data Set (TEDS), vol. 2004. Rockville: Substance Abuse and Mental Health Services Administration.

Substance Abuse and Mental Health Services Administration (SAMHSA) (2006). Results from the 2005 National Survey on Drug Use and Health: National Findings. Office of Applied Studies, NSDUH Series H-30, DHHS Publication No. SMA 06-4194. Rockville, MD: U.S. Department of Health and Human Services.

Sullivan, B. (2004). The Meth Connection to Identity Theft: Drug Addiction Plays a Part in Many Crime Rings, Cops Say. MSNBC. Retrieved 10 March, 2004, from http://www.msnbc.msn.com/id/4460349/. [On-line]

Sullivan, L.E., Metzger, D.S., Fudala, P.J., & Fiellin, D.A. (2005). Decreasing international HIV transmission: the role of expanding access to opioid agonist therapies for injection drug users. Addiction, 100, 150–158.

Sumartojo, E. (2000). Structural factors in HIV prevention: concepts, examples, and implications for research. AIDS, 14(Suppl 1), S3–S10.

Swartz, J. (2005). Meth addicts hack into identity theft. USA Today. September 30, p.1B.

Swendsen, J., & Merikangas, K. (2000). The comorbidity of depression and substance use disorders. Clinical Psychology Review, 20(2), 173–189.

Szreter, S. (1988). The importance of social intervention in Britain's mortality decline c. 1850–1914: a re-interpretation of the role of public health. Social History of Medicine, 1, 1–37.

Takahashi, L.M. (1998). Homelessness, AIDS, and Stigmatization. The NIMBY Syndrome in the United States at the End of the Twentieth Century. Oxford: Clarendon Press.

Tang, Y., Zhao, D., Zhao, C., & Cubells, J.F. (2006). Opiate addition in China: current situation and treatment. Addiction, 101, 657–665.

Tardiff, K., Gross, E., & Wu, J. (1989). Analysis of cocaine-positive fatalities. Journal of Forensic Sciences, 34, 53–63.

Taussig, J.A., Weinstein, B., Burris, S., & Jones, T.S. (2000). Syringe laws and pharmacy regulations are structural constraints on HIV prevention in the US. AIDS, 14(S1), S47–S51.

Taylor, J., & Jackson, B. (1990). Factors affecting alcohol consumption in black women. Part II. The International Journal of the Addictions, 25, 1415–1427.

Teen Challenge of the Dakotas. (2006). Mainpage. http://teenchallengeusa.com/brookings/(accessed 11 October, 2006).

Tempalski, B. (2005). The uneven geography of syringe exchange programs in the United States: Need, politics, and place. Dissertation, Department of Geography, University of Washington.

Tempalski, B. (2007). Placing the dynamics of syringe exchange programs in the United States. Health and Place, 13(2), 417–431.

Tempalski, B., Des Jarlais, D.C., Friedman, S.R., & Burris, S. (2003a). Spatial variations of legality, funding and services among U.S. SEPs. Published Abstract, The14th International Conference on the Reduction of Drug Related Harm. 6–10 April 2003. Chiang Mai, Thailand.

Tempalski, B., Friedman, S.R., Des Jarlais, D.C., McKnight, C., Keem, M., & Friedman, R. (2003b). What predicts which metropolitan areas in the USA have syringe exchange programs? International Journal on Drug Policy, Special Issue on Syringe Exchange and Access, 14(5–6), 417–424.

Texas Department of Health Services. (2006). Substance abuse treatment center admissions 9/2005-6/2006. Unpublished raw data.

Thai Working Group on HIV/AIDS Projections. (2001). Projections of HIV/AIDS in Thailand, 2000–2020. Bangkok, Thailand: Ministry of Public Health, AIDS Division.

Thailand Ministry of Health. (2001). National HIV/AIDS Sentinel Surveillance Reports, 2000. Bangkok: Thailand Ministry of Health.

Thaung, B., Gyee, K.M., & Kywe, B. (1996). Rapid assessment study of drug abuse in Myanmar. Ninth International Conference on AIDS, Vancouver.

The Global Fund to Fight AIDS, Tuberculosis and Malaria. (2006). Nicaragua and the Global Fund. Retrieved 30 November, 2006, from http://www.theglobalfund.org/programs/ grantdetails.aspx?compid=179&grantid=195&lang=en&CountryId=NIC.

Thematic Catrography and Geographic Visualization, 2nd Edition. Upper Saddle River, NJ: Pearson Prentice Hall.

Theobald, D.M. (2005). GIS Concepts and ArcGIS Methods, 2nd Edition. Fort Collins, CO: Conservation Planning Technologies.

Thies, C., & Register, C. (1993). Decriminalization of marijuana and the demand for alcohol, marijuana, and cocaine. The Social Science Journal, 30, 385–399.

Thornton, S. (1995). Club Cultures: Music, Media and Subcultural Capital. Middletown, CT: Wesleyan University Press.

Tizon, T.A., & Smith, D. (2005). Evacuees of Hurricane Katrina Resettle Along a Racial Divide. Los Angeles, CA: Los Angeles Times.

Tobler, W. (1970). A computer movie simulating urban growth in the Detroit region. Economic Geography, 46(2), 234–240.

Toriello, P.J., Morse, P.A., Morse, E.V., Kissinger, P., & Pedersen-Wasson, E. (2007). The resuscitation of a New Orleans substance abuse treatment agency after Hurricane Katrina. Journal of Health Care for the Poor and Underserved, 18, 482–486.

Treno, A.J., Cooper, K., & Roeper, P. (1994). Estimating alcohol involvement in trauma patients: search for a surrogate. Alcoholism: Clinical and Experimental Research, 18(6), 1306–1311.

Treno, A.J., Grube, J.W., & Martin, S.E. (2003). Alcohol availability as a predictor of youth drinking and driving: a hierarchical analysis of survey and archival data. Alcoholism: Clinical and Experimental Research, 27, 835–840.

Treno, A.J., Gruenewald, P.J., & Johnson, F.W. (2001). Alcohol availability and injury: the role of local outlet densities. Alcoholism: Clinical and Experimental Research, 25, 1467–1471.

Treno, A.J., Gruenewald, P.J., Remer, L.G., Johnson, F., & LaScala, E.A. (2008). Social selection or social influence: Examining multi-level relationships between bars, hostility and aggression, Addiction, 103(1), 66–77.

Trillo, A. (2002). Adictos: enfermos pero no criminales. Border Reflections. Retrieved 17 February, 2006, from http://www.pciborderregion.com/newsletter/1700_SP/1700_1_EN.htm.

Trocki, C. (1990). Opium and Empire. Chinese Society in Colonial Singapore, 1800–1910. Ithaca: Cornell University Press.

Trooskin, S., Hadler, J., St. Luois, T., & Navarro, V. (2005). Geospatial analysis of hepatitis C in Connecticut: a novel application of a new public health tool. Public Health, 119(11), 1042–1047.

Troyer, R., Clark, J., & Rojek, D. (1989). Social Control in the People's Republic of China. New York: Praeger.

Tucker, J.A., Vuchinich, R.E., & Gladsjo, J.A. (1990). Environmental infuences on relapse in substance use disorders. The International Journal of The Addictions, 25, 1017–1050.

Turk, H. (1977). Organizations in Modern Life. Washington, D.C.: Jossey-Bass.

Turner, C., Ku, L., Rogers, S., Lindberg, L., Pleck, J., & Sonenstein, F. (1998). Adolescent sexual behavior, drug use, and violence: increased reporting with computer survey technology. Science, 280, 867–873.

Umbricht-Schneiter, A., Ginn, D.H., Pabst, K.M., & Bigelow, G.E. (1994). Providing medical care to methadone clinic patients: referral vs. on-site care. American Journal of Public Health, 84, 207–210.

UNAIDS. (2001). Population Mobility and AIDS. Geneva: UNAIDS.

UNAIDS. (2004). 2004 Report on the Global AIDS Epidemic. Geneva, Switzerland: UNAIDS.

UNAIDS/WHO. (2003). AIDS Epidemic Update 2003. Joint United Nations Programme on HIV/AIDS. Geneva, Switzerland: World Health Organization.

UNDP. (2000). El Desarrollo Humano en Nicaragua 2000: Equidad para superar la vulnerabilidad. Programa de las Naciones Unidas para el Desarrollo, 1st Edition, Managua, Nicaragua: United Nations Development Programme.

UNDP (United Nations Development Programme). (2004). Reversing the Epidemic: Facts and Policy Options. Bratislava, Slovakia: UNDP.

UNICEF. (2004). Retrieved from http://www.unicef.org/infobycountry/nicaragua_statistics.html "Basic Indicators" (2004 data).

Uniform Crime Reports County Data. (n.d.) Retrieved 25 February, 2006, from http://www.icpsr.umich.edu. Note: (n.d.) = no date.

United Nations. (2006). Kofi Annan: Message on the international day against drug abuse and illicit trafficking. Retrieved from http://www.unodc.org/pdf/26june06/26June2006_SG_statement.pdf.

United Nations Office on Drugs and Crime (UNODC). (2000). The Seventh United Nations Survey on Crime Trends and the Operations of Criminal Justice Systems (1998–2000). New York: UNODC.

United States Department of Agriculture. (2004). Final 2004 County Typology Codes. Measuring Rurality. Retrieved 11 October, 2006, from http://www.ers.usda.gov/Data/TypologyCodes/.

United States-Mexico Border Health Commission. (2005). The U.S. – Mexico Border Region. Retrieved 17 October, 2006, from http://www.borderhealth.org/border_region.php.

UN Regional Task Force on Drug Use and HIV Vulnerability. (2002). Hong Kong: methadone treatment programme. In: Preventing HIV/AIDS among Drug Users – Case Studies from Asia (pp. 57–70). Bangkok: UNODC.

Unwin, D.J. (1996). GIS, spatial analysis and spatial statistics'. Progress in Human Geography, 20(4), 540–441.

U.S. Agency for International Development. (2004). Health Profile: Nicaragua. Retrieved 28 November, 2006, from www.usaid.gov.

U.S. Bureau of the Census. (1998). State and Metropolitan Area Data Book, 1997–1998. Washington, D.C.: U.S. Bureau of the Census.

U.S. Bureau of the Census. (2000). Quick Tables: QT-PL. Race, Hispanic or Latino, and Age: 2000 for San Antonio city, Texas. Retrieved 16 July, 2003, from http://factfinder.census.gov.

U.S. Bureau of the Census. (2001, 2002). Annual Demographic Survey March Supplement: Table 2. Age, Sex, Household Relationship, Race and Hispanic Origin by Ratio of Income to Poverty Level: 2001. Retrieved 17 July, 2003, from http://ferret.bls.census.gov/macro/032002/pov/new02_001.htm.

U.S. Bureau of the Census. (2002). Poverty in the United States: 2001, Current Population Reports. Washington, DC: U.S. Government Printing Office.

U.S. Bureau of the Census. (2005). Historical Metropolitan Area Definitions. Suitland, MD: U.S. Bureau of the Census.

U.S. Census. (2001). 2000 Summary File 1: Census of Population and Housing. U.S. Department of Commerce, Economics and Statistics Administration U.S. Census Bureau, Washington, DC DVD issued September 2001 (V1-D00-S1AS-08-US1).

U.S. Census Bureau. Census (2003). 2000 American Factfinder Available at http://factfinder.census.gov/home/saff/main.html?_lang=en/.

U.S. Census Bureau: United States Census (1992). Finance Statistics http://www.icpsr.umich.edu/cgi-bin/bob/archive2?study=4420&path=ICPSR&docsonly=yes.

U.S. Department of Health and Human Services. (2000). AIDS Public Information Data Set (pp. 1–52). Washington, D.C.: U.S. Department of Health and Human Services.

US Department of Health and Human Services. (2005). Hurricane Katrina. Press Briefing on Mental Health, 9-7-2005. Retrieved July 7, 2006 from http://www.hhs.gov/disasters/emergency/naturaldisasters/hurricanes/katrina/pressbriefingtranscript0907.html.

U.S. Department of Health and Human Services, Substance Abuse and Mental Health Services Administration, Office of Applied Studies. (2001). The DASIS Report. Substance Abuse and Mental Health Services Administration. Available at: http:// www.dasis.samhsa.gov.99ufds/Resource-556/t35.htm.

U.S. Department of Health and Human Services, Substance Abuse and Mental Health Services Administration, Office of Applied Studies. (2004). National Survey of Substance Abuse Treatment Services (N-SSATS) (2004): [United States] [computer file]. icpsr04256-v2. Arlington, VA: Synectics for Management Decisions, Inc. [producer], 2004. Ann Arbor, MI: Inter-university Consortium for Political and Social Research [distributor], 2006-08-01.

U.S. Department of Justice. (1999). Mapping Crime: Principle and Practice. (U.S. DOJ Publication No.178919). Washington, D.C.

U.S. Department of Justice. (2001). Reducing Gun Violence: The Boston Gun Project's Operation Ceasefire (NCJ 188741). Washington, D.C.: U.S. Department of Justice

U.S. Department of Justice. National Drug Intelligence Center. (2002). North Dakota: Drug Threat Assessment. Available at: http://www.usdoj.gov/ndic/pubs1/1052/1052p.pdf (last accessed 3 October, 2005).

U.S. Department of Labor, Bureau of Labor Statistics. (2001). Local Area Unemployment, Housing and Household Economic Statistics Division, Small Area Estimates Branch [Computer file]. Washington, DC: U.S. Department of Labor, Bureau of Labor Statistics [producer]. Ann Arbor, MI: Inter-university Consortium for Political and Social Research [distributor].

US Department of State, Bureau of International Narcotics and Law Enforcement Affairs (2002). International Narcotics Control Strategy Report. Washington, D.C.: U.S. Department of State.

U.S. Drug Enforcement Administration (DEA). (n.d.). Amphetamines/Methamphetamine. Retrieved on May 2, 2008, from http://counties.cce.cornell.edu/monroe/ag/Amphetamines-DEA1.htm. [On-line]

U.S. General Accounting Office. (1993). Needle Exchange Programs: Research Suggests Promise as an AIDS Prevention Strategy. Washington, D.C.: U.S. General Accounting Office.

U.S. General Services Administration. (2006). Expansion Project Overview [Electronic Version]. Available at: http://www.gsa.gov/Portal/gsa/ep/channelView.do?pageTypeId=8199&channelPage=%252Fep%252Fchannel%252FgsaOverview.jsp&channelId=-15895.

Uys, E.L. (1999). Riding the Rails: Teenagers on the Move during the Great Depression. New York: TV Books.

Vaillant, G.E. (1983). Natural history of male alcoholism V: is alcoholism the cart or the horse to sociopathy? British Journal of Addiction, 78, 317–326.

Valdez, A. (2005). Mexican American youth and adult prison gangs in a changing heroin market. Journal of Drug Issues, 35(4), 843–867.

Valdez, A., Kaplan, C.D., & Cepeda, A. (2000). The process of paradoxical autonomy and survival in the heroin careers of Mexican American women. Contemporary Drug Problems, 27(Spring 2000), 189–212.

Valdez, A., & Sifaneck, S.J. (2004). Getting high and getting by: dimensions of drug selling behaviors among American Mexican gang members in South Texas. Journal of Research in Crime and Delinquency, 41(1), 82–105.

Van Gundy, K. (2006). Substance Abuse in Rural and Small Town America. Reports on Rural America. Durham, NH: Carsey Institute.

Van Ours, J.C. (1995). The price elasticity of hard drugs: the case of opium in the Dutch East Indies, 1923–1938. The Journal of Political Economy, 103, 261–279.

Van Wichen, H., Largaespada, M.J., Ormel, H., & Montesdeoca, E.A. (1995). Es tiempo de Actuar. La Situación de VIH/SIDA en Nicaragua (86). Managua, Nicaragua: Servicio Holandés de Cooperación al Desarrollo, UNFPA.

Vaslamatzis, V., Kontaxakis, M., Markidis, M., & Katsouyanni, K. (1987). Social and resource factors related to the utilization of emergency psychiatric services in the Athens area. Acta Psychiatrica Scandinavica, 75, 95–98.

Veen, J. (1997). The BJA firearms trafficking program: demonstrating effective strategies to control violent crime. NCJ 166818. BJA Bulletin (November). Washington, DC: U.S. Department of Justice.

Ventafridda, V., Saita, L., Ripamonti, C., & De Conno, F. (1985). WHO guidelines for the use of analgesics in cancer pain. International Journal of Tissue Reaction, 7(1), 93–96.

Vermeiren, R., Schwarb-Stone, M., Deboutte, D., Leckman, P.E., & Ruchkin, V. (2003). Violence exposure and substance use in adolescents: findings from three countries. Pediatrics, 111, 535–540.

Vest, J.R., & Valadez, A.M. (2006). Health conditions and risk factors of sheltered persons displaced by Hurricane Katrina. Prehospital and Disaster Medicine, 21, 55–58.

Viani, R.M., Araneta, M.R., Ruiz-Calderon, J., Hubbard, P., Lopez, G., Chacon-Cruz, E., et al. (2006). Perinatal HIV counseling and rapid testing in Tijuana, Baja California, Mexico: seroprevalence and correlates of HIV infection. Journal of Acquired Immune Deficiency Syndromes, 41(1), 87–92.

Vivian, J., Saleheen, H., Singer, M., Navarro, J., & Mirhej, G. (2005). Under the counter: the diffusion of narcotic analgesics to the inner city street. Journal of Ethnicity in Substance Abuse, 4(2), 97–114.

Vlahov, D., Des Jarlais, D.C., Goosby, E., Hollinger, P.C., Lurie, P.G., Shiver, M.D., & Strathdee, S. (2001). Needle exchange programs for the prevention of human immuno-deficiency virus infection: epidemiology and policy. American Journal of Epidemiology, 154(12), S70–S77.

Vlahov, D., Galea, S., Resnick, H., Ahern, J., Boscarino, J.A., Bucuvalas, M., et al. (2002). Increased use of cigarettes, alcohol, and marijuana among Manhattan, New York, residents after the September 11th terrorist attacks. American Journal of Epidemiology, 155, 988–996.

Volkow, N.D. (2005). The Effects of Stress on Drug Abuse. National Institute on Drug Abuse. Available at: http://www.drugabuse.gov/Drugpages/Stress.html. [On-line]

Voltz, D. (2000). Mapping illicit drug markets. Paper Presented at "Crime Mapping: Adding Value to Crime Prevention and Control" Conference. Australian Mineral Foundation, Adelaide, 21–22 September. Retrieved 7 November, 2005, from http://www.aic.gov.au/conferences/mapping/voltz.pdf.

Vornick, L.A., & Brown, S.E. (2006). Management of co-morbid bipolar disorder and substance abuse. The Journal of Clinical Psychiatry, 67(Suppl 7), 24–30.

Vyrostek, S.B., Annest, J.L., & Ryan, G.W. (2004). Surveillance for fatal and nonfatal injuries – United States, 2001. MMWR Surveillance Summaries, 53(7), 1–57.

Wakefield, S., Elliott, S., Cole, D., & Eyles, J. (2001). Environmental risk and (re)action: air quality, health, and civic involvement in an urban industrial neighbourhood. Health and Place, 7, 163–177.

Walberg, P., McKee, M., Shkolnikov, V., Chenet, L., & Leon, D.A. (1998). Economic change, crime and mortality crisis in Russia: regional analysis. British Medical Journal, 317, 312–318.

Wallace, D., & Wallace, R. (1998). A plague on Your Houses. New York: Verso.

Wallace, R., Huang, Y., Gould, P., & Wallace, D. (1997). The hierarchical diffusion of AIDS and violent crime among U.S. metropolitan regions: inner-city decay, stochastic resonance and reversal of the mortality transition. Social Science and Medicine, 44(7), 935–947.

Wallace, R., Wallace, D., Andrews, H., Fullilove, R., & Fullilove, M.T. (1995). The spatiotemporal dynamics of AIDS and TB in the New York metropolitan region from a sociogeographic perspective: Understanding the linkages of central city and suburbs. Environment and Planning A, 27, 1085–1108.

Waller, L.A., Carlin, B.P., Xia, H., & Gelfand, A.E. (1997). Hierarchical spatio-temporal mapping of disease rates. Journal of the American Statistical Association, 92, 607–617.

Waller, L.A., & Gotway, C.A. (2004). Applied spatial statistics for public health data. New York: John Wiley & Sons.

Waller, L.A., Zhu, L., Gotway, C.A., Gorman, D.M., & Gruenewald, P.J. (2007). Quantifying geographic variations in associations between alcohol distribution and violence: a comparison of geographically weighted regression and spatially varying coefficient models. Stochastic Environmental Research and Risk Assessment, 21(5), 573–588.

Wang, F., & Luo, W. (2005). Assessing spatial and nonspatial factors in healthcare access in Illinois: towards an integrated approach to defining health professional shortage areas. Health and Place, 11, 131–146.

Wang, F., Zuo, X., & Ruan, D. (2002). Rural migrants in Shanghai: living under the shadow of socialism. The International Migration Review, 36(2), 520–645.

Wang, J., Siegal, H.A., Falck, R.S., & Carlson, R.G. (1998). Needle transfer among injection drug users: a multilevel analysis. The American Journal of Drug and Alcohol Abuse, 24, 225–237.

Wang, P.S., et al. (2005). Failure and delay in initial treatment contact after first onset of mental disorders in the National Comorbidity Survey replication. Archives of General Psychiatry, 62, 603–613.

Wang, P.S., Demler, O., Olfson, M., Pincus, H.A., Wells, K.B., & Kessler, R.C. (2006). Changing profiles of service sectors used for mental health care in the United States. The American Journal of Psychiatry, 163(7), 1187–1198.

Warner-Smith, M., Darke, S., Lynskey, M., & Hall, W. (2001). Heroin overdose: causes and consequences. Addiction, 96(8), 1113–1125.

Washington Coalition of Sexual Assault Programs. (2004). Research & Advocacy Digest, vol. 7. Olympia, WA: Anderson, J.

Watters, J.K., & Biernacki, P. (1989). Targeted sampling: options for the study of hidden populations. Social Problems, 36, 416–430.

Weber, J., & Kwan, M.P. (2002). Bringing time back in: a study on the influence of travel time variations and facility opening hours on individual accessibility. The Professional Geographer, 54(2), 226–240.

Weeks, J.R., Hill, A.G., Getis, A., & Stow, D. (2006). Ethnic residential patterns as predictors of intra-urban child mortality inequality in Accra, Ghana. Urban Geography, 7(7), 526–548.

Weich, S., Burton, E., Blanchard, M., Prince, M., Sproston, K., & Erens, B. (2001). Measuring the built environment: validity of a site survey instrument for use in urban settings. Health and Place, 7, 283–292.

Weinberg, N., & Glantz, M. (1999). Child psychopathology risk factors for drug abuse: overview. Journal of Clinical Child Psychology, 28, 290–297.

Weisburd, D., & Green, L. (1995). Policing drug hot spots: the Jersey City drug market analysis experiment. Justice Quarterly, 12(4), 711–736.

Weisburd, D., & Mazerolle, L.G. (2000). Crime and disorder in drug hot spots: implications for theory and practice in policing. Police Quarterly, 3(3), 331–349.

Weiss, L., Fabri, A., McCoy, K., Coffin, P., Netherland, J., & Finkelstein, R. (2002). A vulnerable population in a time of crisis: drug users and the attacks on the World Trade Center. Journal of Urban Health, 79, 392–403.

Weiss, R.A., & McMichael, A.J. (2004). Social and environmental risk factors in the emergence of infectious diseases. Nature Medicine Supplement, 10(12), S70–S76.

Weiss, R.D., et al. (2005). Does recovery from substance use disorder matter in patients with bipolar disorder? The Journal of Clinical Psychiatry, 66, 730–735.

Weitzman, E.R., Folkman, A., Folkman, M.P., & Wechsler, H. (2003). The relationship of alcohol outlet density to heavy and frequent drinking and drinking-related problems among college students at eight universities. Health and Place, 9(1), 1–6.

Weniger, B., et al. (1991). The epidemiology of HIV infection and AIDS in Thailand. AIDS, 5(Suppl 2), 571–585.
Weniger, B.G., & Berkley, S. (1996). The evolving HIV/AIDS pandemic. In: J.M. Mann, & D. Tarantola (eds.), AIDS in the World II: Global Dimensions, Social Roots, and Responses (pp. 57–70). New York: Oxford University Press.
Wennberg, J., & Gittelsohn, A. (1973). Small area variations in health care delivery. Science, 182, 1102–1108.
Westermeyer, J. (1976). The pro-heroin effects of anti-opium laws in Asia. Archives of General Psychiatry, 33(9), 1135–1139.
Whitbeck, L.B., Chen, X., Hoyt, D.R., Tyler, K.A., & Johnson, K.D. (2004). Mental disorder, subsistence strategies, and victimization among gay, lesbian, and bisexual homeless and runaway adolescents. Journal of Sex Research, 41(4), 329–342.
White, J.M., & Irvine, R.J. (1999). Mechanisms of fatal opioid overdose. Addiction, 94(7), 961–972.
White, W.L. (1998). Slaying the Dragon: The History of Addiction Treatment and Recovery in America. Illinois: Chestnut Health Systems/Lighthouse Institute.
Wieczorek, W.F., & Hanson, C. (1997). New modeling methods: geographic information systems and spatial analysis. Alcohol Health and Research World, 21(4), 331–339.
Wiffen, P.J., Edwards, J., Barden, J., & McQuay, H.J.M. (2003). Oral Morphine for Cancer Pain. Cochrane Database of Systematic Reviews: Reviews 2003 Issue 4. Chichester, UK: John Wiley & Sons.
Willaims, D.R., & Collins, C. (2001). Racial residential segregation: a fundamental cause of racial disparities in health. Public Health Reports, 116, 404–416.
Williams, A. (1998). Therapeutic landscapes in holistic medicine. Social Science and Medicine, 46, 1193–1203.
Williams, A. (1999). Therapeutic Landscapes: The Dynamic between Place and Wellness. Lanham, MD: University Press of America.
Williams, A. (2002). Changing landscapes of care: employing the concept of therapeutic landscapes as a framework in examining home space. Social Science and Medicine, 55, 141–154.
Williams, D.R., Neighbors, H.W., & Jackson, J.S. (1993). Racial/ethnic discrimination and health: findings from community studies. American Journal of Public Health, 93(2), 200–208.
Williams, E.D. (2005). Hurricanes Katrina & Rita: Addressing the Victims' Mental Health and Substance Abuse Treatment Needs: CRS Report for Congress Washington, D.C.: Congressional Research Service: The Library of Congress.
Williams, J. (2004). The effects of price and policy on marijuana use: what can be learned from the Australian Experience? Health Economics, 13, 123–137.
Williams, M., McCoy, H.V., Brown, A., Saunders, L., Freeman, R., & Chen, D. (2001). An evaluation of a brief HIV risk reduction intervention using empirically derived drug use and sexual risk indices. AIDS and Behavior, 5(1), 31–43.
Wilson, R., & Dufour, M. (eds.) (2000). Epidemiology of Alcohol Problems in Small Geographic Areas. NIAAA Research Monograph No. 36. NIH Publication No. 00–4357. Bethesda, MD: National Institute on Alcohol Abuse and Alcoholism.
Wilson, W.J. (1987). The Truly Disadvantaged: The Inner City, the Underclass, and Public Policy. Chicago, IL: University of Chicago Press.
Wilson, W.J. (1996). When Work Disappears: The World of the New Urban Poor. New York: Vintage Books.
Wilton, R. (1996). Diminished worlds? The geography of everyday life with HIV/AIDS. Health and Place, 2, 69–83.
Wilton, R. (2000). Grounding hierarchies of acceptance: the social construction of disability in NIMBY conflicts. Urban Geography, 215, 86–608.
Wilton, R., & Deverteuil, G. (2006). Spaces of sobriety/sites of power: examining social model alcohol recovery programs as therapeutic landscapes. Social Science and Medicine, 63(3), 649–661.

Wingo, L.K. (2001). Substance abuse in African American women. Journal of Cultural Diversity, 8, 21–25.

Winterfield, L., & Castro, J. (2005). Matching drug treatment to those in need: an analysis of correctional service delivery in Illinois and Ohio. Justice Research and Policy, 7(2), 29–55.

Wolch, J., Rahimian, A., & Koegel, P. (1993). Daily and periodic mobility patterns of the urban homeless. The Professional Geographer, 45(2), 159–169.

Wolfe, T. (1967). The Electric Kool-Acid Test. New York: Bantam.

Wong, D.W.S., & Lee, J. (2005). Statistical Analysis of Geographic Information with ArcView GIS and ArcGIS. Hoboken, NJ: Wiley.

Wong, K.H., Lee, S.S., Lim, W.L., & Low, H.K. (2003). Adherence to methadone is associated with a lower level of HIV-related risk behaviours in drug users. Journal of Substance Abuse Treatment, 24, 233–239.

Wong, Y.-L.I. (2002). Tracking change in psychological distress among homeless adults: an examination of the effect of housing status. Health and Social Work, 27(4), 262–273.

Wood, E., Chan, K., Montaner, J.S.G., et al. (2000). The end of the line: has rapid transit contributed to the spatial diffusion of HIV in one of Canada's largest metropolitan areas? Social Science and Medicine, 51, 741–748.

Wood, E., Spittali, P.M., Small, W., Kerr, T., Li, K., Hogg, R.S., et al. (2004). Displacement of Canada's largest public illicit drug market in response to a police crackdown. CMAJ, 170(10), 1551–1556.

World Health Organization. (2005). Summary Country Profile for HIV/AIDS Treatment Scale-up. Retrieved 29 November, 2006, from http://www.who.int/hiv/HIVCP_NIC.pdf.

World Health Organization (WHO). (1992). International Statistical Classification of Diseases and Related Health Problems, 10th Revision. Geneva, Switzerland: WHO.

World Health Organization (WHO). (2002). World report on violence and health 2002. Available at: http://www3.who.int/whosis/mort/table1.cfm?path-whosis. Accessed 31 July, 2007.

World Health Organization (WHO). (2000). WHO Report on Global Surveillance of Epidemic-Prone Infectious Diseases. Geneva, Switzerland: WHO.

Wright, D., & Gfroerer, J. (1997). Ratio estimation of hardcore drug users. Journal of Official Statistics, 13, 401–416.

Wu, Z., et al. (1996). Risk factors for intravenous drug use and sharing equipment among young male drug users in Longchuan County, south-west China. AIDS, 10(9), 1017–1024.

Wylie, J.L., Shah, L., & Jolly, A. (2007). Incorporating geographic settings into a social network analysis of injection drug use and bloodborne pathogen prevalence. Health and Place, 13(3), 617–628.

Yablonsky, L. (1968). The Hippie Trip. New York: Pegasus.

Yamada, I., & Thill, J. (2007). Local indicators of network-constrained clusters in spatial point patterns. Geographical Analysis, 39, 268–292.

Yang, D., Goerge, R., & Mullner, R. (2006). Comparing GIS-based methods of measuring spatial accessibility to health services. Journal of Medical Systems, 30, 23–32.

Yang, X. (2005). Does where we live matter? Community characteristics and HIV/STD prevalence in southwestern China. International Journal of STD and AIDS, 16(1), 31–37.

Yang, X. (2006). Temporary migration and HIV risk behaviors in China. Environment and Planning A, 38(8), 1527–1543.

Yang, X., Derlega, V., & Luo, H. (2007). Migration, behavior change, and HIV/STD risks in China. AIDS Care, 19(2), 282–288.

Yen, I., Ragland, D., Greiner, B., & Fisher, J. (1999). Racial discrimination and alcohol-related behavior in urban transit operators: findings from the San Francisco Muni Health and Safety Study. Public Health Reports, 114, 448–458.

Yin, Z., Valdez, A., Mata, A.G., & Kaplan, C.D. (1996). Developing a field-intensive methodology for generating a randomized sample for gang research. Free Inquiry-Special Issue: Gang, Drugs and Violence, 24(2), 195–204.

Young, R. (1999). Prioritising family health needs: a time-space analysis of women's health-related behaviors. Social Science and Medicine, 48(6), 797–813.

Yu, X.F., et al. (1999). Emerging HIV infections with distinct subtypes of HIV-1 infection among injection drug users from geographically separate locations in Guangxi Province, China. Journal of Acquired Immune Deficiency Syndromes, 22(2), 180–188.

Zhang, C. et al. (2002). High prevalence of HIV-1 and hepatitis C virus coinfection among injection drug users in the southeastern region of Yunnan, China. Journal of Acquired Immune Deficiency Syndromes, 29(2), 191–196.

Zhang, L. (2001). Strangers in the city: reconfigurations of space, power, and social networks within China's floating population. Stanford: Stanford University Press.

Zheng, X. (2001). Epidemiologic monitoring of AIDS in China. Zhongguo Xingbing Aizhibing Fangzhi, zhengkan, 1–3.

Zhou, Y., & Li, X. (1999). Demographic characteristics and illegal drug use patterns among attendees of drug cessation programs in China. Substance Use and Misuse, 34, 907–920.

Zhu, L., Gorman, D.M., & Horel, S. (2006). Hierarchical Bayesian spatial models for alcohol availability, drug "hot spots" and violent crime. International Journal of Health Geographics, 5, 54.

Index

Contributors

Mariana Aburto
Mariana Aburto, MPH, Nimehuatzin Foundation, is a clinical psychologist and university professor with experience in communications.

Matt Adams
Matthew Adams received his Masters Degree from Boston University and is presently working in Robert K. Futterman & Associates in New York as a GIS practitioner.

Erica Alarcon
Erica Alarcon, BA, has applied ethnographic methods to recruit and interview young injection drug users living in New Orleans. Currently, she provides HIV prevention and medical supportive services to individuals living with HIV in Louisiana.

Rita Arauz
Rita Arauz, MA, MPH, a psychologist, is an internationally recognized HIV/AIDS service and policy expert. She is the Founding President of the Nimehuatzin Foundation, the first Nicaraguan non-governmental organization dedicated exclusively to respond to AIDS.

Matthew Archibald
Matthew Archibald is an Assistant Professor of Sociology at Emory University. He studies healthcare organizations and social movements. His recent book, The Evolution of Self-Help, published by Palgrave Macmillan, investigates the sociopolitical and economic conditions shaping the self-help movement. marchib@emory.edu

Aniruddha Banerjee
Aniruddha Banerjee, MSURP, PhD, is an assistant professor of Geography at Indiana University Purdue University at Indianapolis and an affiliated Research Scientist with the Prevention Research Center, Berkeley. He has applied combinatorial optimization methods using both integer programming and graph theoretic approaches to analyze health information. He is currently working on simulating hospital infections by simultaneously applying micro-scale, meso-scale,

and macro-scale ecological models. He was awarded the Jacques May Thesis Prize by the Medical Geography Specialty Group in 2005. rbanerje@iupui.edu

Chris Beyrer
Chris Beyrer is a director of the Johns Hopkins Fogarty AIDS International Training and Research Program, a director of the Johns Hopkins Center for Public Health & Human Rights, a Senior Scientific Liaison of the HIV Vaccine Trials Network, and an Associate Director of the Center for Global Health. cbeyrer@jhsph.edu

Jennifer Jackson Bloom
Jennifer Jackson Bloom, MPH, a biostatistician, provides data management and qualitative and quantitative analytical expertise for several studies of high-risk youth populations in Los Angeles. Her research interests include behavioral health risks, modeling longitudinal change, the application of geography to research on high-risk youth, and mixed methods research.

Joanne Brady
Joanne Brady is a Principal Research Associate at NDRI and works as a data analyst on the Community Vulnerability and Response to IDU-related HIV project. She also has conducted data analyses on research projects investigating the individual and societal costs of cancer, the effects of different drug dosing regimens, and the relationship between douching, pregnancy, and pelvic inflammatory disease. Ms. Brady has a BS in Biology from Tufts University and a Master's in Epidemiology from Harvard School of Public Health

Kimberly Brouwer
Kimberly C. Brouwer, Assistant Professor at the University of California San Diego School of Medicine, pursues research in the spatial and molecular epidemiology of infectious diseases. She is currently collaborating with Mexican officials in studies of HIV/AIDS and injection drug use in cities along the Mexico/US border, in an effort to inform public health interventions. As principal investigator of a five-year grant from NIDA to explore social and environmental factors affecting disease transmission and risk behaviors in Tijuana, Mexico, her group is building a geographic information system model to measure disease clustering and recommend better distribution of health services. kbrouwer@ucsd.edu

Christopher Browning
Christopher R. Browning is an Associate Professor of Sociology at the Ohio State University. His current research focuses on the causes and consequences of community social organization; the neighborhood context of crime, risk behavior, and health; the long-term effects of maltreatment during childhood; and multilevel statistical models. He is a co-investigator on the Children's Exposure to Violence over Space and Time project funded by the National Institute of Child Health and Human Development.

John S. Brownstein
John S. Brownstein is an Instructor of Pediatrics at Harvard Medical School, and has joint appointments in the Children's Hospital Boston Informatics Program at

the Harvard-MIT Division of Health Sciences and Technology and the Division of Emergency Medicine. Dr. Brownstein was trained as an epidemiologist at Yale University where he received his PhD. He works on novel statistical modeling and medical informatics approaches for accelerating the translation of public health surveillance research into practice.

Angela Bucciarelli

Angela Bucciarelli, MPH, is a statistical/spatial analyst at the Division of Epidemiology, World Trade Center Health Registry, New York City Department of Health and Mental Hygiene, New York, NY.

Jessica Burnum

Jessica Burnum is a geographic information services technician with the North Carolina section of the American Society of Civil Engineers. She holds an MA in geography from Kansas State University and a BA in geography and environmental studies from the University of Colorado at Colorado Springs.

Lori Burrington

Lori Burrington is a PhD candidate in the Department of Sociology at the Ohio State University. Her research focuses on neighborhood and community effects on adolescent delinquency and problem behavior.

Meagan Cahill

Meagan Cahill is a Research Associate at the Urban Institute's Justice Policy Center, where she is a project manager for the multi-year, multi-site evaluation of the federally funded Gang Reduction Program. In addition, she is currently investigating the displacement of crime from public housing that is undergoing redevelopment. Cahill holds a PhD and a Master's degree in geography from the University of Arizona and a BA in geography from Mary Washington College. mcahill@ui.urban.org

Catherine Calder

Catherine Calder is Assistant Professor of Statistics at the Ohio State University. Her research interests include spatial statistics and Bayesian statistical methodology.

Alice Cepeda

Dr. Alice Cepeda is currently Assistant Professor in the Department of Sociology at the University of Houston (UH) where she is also Senior Researcher at the Office for Drug and Social Policy Research (ODSPR). She received her PhD in sociology from the City University of New York, Graduate Center. Her experience and research interest has been in substance use, crime, violence, and urban health issues among the Mexican-origin (Mexican Americans and Mexicans) population.

Pascual Ortells Chabrera

Pascual Ortells Chabrera, Technical Director, Nimehuatzin Foundation, is a sociologist with extensive experience in research, education, and communications.

Sanjoy Chakravorty

Sanjoy Chakravorty is professor and chair of geography and urban studies at Temple University in Philadelphia. He is the author of Made in India (2007, Oxford University Press, with S. Lall) and Fragments of Inequality (2005, Routledge) and more than three dozen journal papers, book chapters, and reports. His research has been funded by the US National Science Foundation, the US National Institute of Justice, the American Institute of Indian Studies, and the World Bank.

Siddharth Chandra

Siddharth Chandra is Associate Professor at the Graduate School of Public and International Affairs at the University of Pittsburgh. He holds a PhD from Cornell University. His research interests include health economics and policy, economic history, and the economics of addiction. schandra@pitt.edu

Ivan Cheung

Ivan Cheung is a research scientist at the Insurance Institute for Highway Safety. Previously, he held the positions of Director of GIScience Programs at the Association of American Geographers and Assistant Professor of Geography at the George Washington University. His applied research projects integrate geographic information systems, remote sensing technologies, and geographic analytical techniques in assessing access to health services, epidemiology, environmental health, immigrant settlement pattern, and urban climatology.

Michael C. Clatts

Michael Clatts, PhD, a medical anthropologist, has conducted a number of ethno-epidemiological studies of HIV risk, largely among vulnerable out-of-treatment populations, including drug users, commercial sex workers, homeless youth, and men who have sex with men. Recently, Dr. Clatts has extended the use of an ethno-epidemiological model to understanding HIV risk and protective factors among vulnerable youth and young adult populations in Vietnam, China, and India

Hannah Cooper

Dr. Hannah Cooper is an Assistant Professor at the Rollins School of Public Health at Emory University. Dr. Cooper's research concerns the structural determinants of drug use and drug-related harms, including HIV, both in the general population and within particular racial/ethnic groups. hannah.cooper@sph.emory.edu

Nabarun Dasgupta

Nabarun Dasgupta is an epidemiologist specializing in the medical and non-medical use of opioids and their consequences. He has extensive field experience domestically and internationally with substance use and harm reduction. Nabarun is currently a student at the University of North Carolina School of Public Health, and his dissertation research is on surveillance for non-fatal opioid poisonings. nabarund@unc.edu

Don C. Des Jarlais

Don C. Des Jarlais is the Director of Research of the Baron Edmond de Rothschild Chemical Dependency Institute at Beth Israel Medical Center, a Professor of Epidemiology and Population Health at the Albert Einstein College of Medicine, and a Senior Research Fellow at the National Development and Research Institutes. He has served as a consultant to the World Health Organization, the National Institutes of Health, and the Centers for Disease Control, and as a Commissioner on the US National Commission on AIDS.

Charlie DiMaggio

Charles DiMaggio, PhD, MPH, PA-C, is Assistant Professor of Clinical Epidemiology, Department of Epidemiology, Mailman School of Public Health, Columbia University, New York, NY. cjd11@columbia.edu

Alex Feidler

Alex Feidler has a MA in geography from the University of North Dakota and a bachelor's in geography from James Madison University. His main research interests lie in crime geography, law enforcement, and geospatial intelligence

Peter L. Flom

Peter L. Flom, PhD, is a statistician at NDRI, at Brainscope, Inc., and is also an independent consultant. He received his PhD in psychometrics from Fordham University in 1999, and has worked with Dr. Friedman and his team on various projects for the past 10 years.

James M. Frabutt

Dr. Jim M. Frabutt, PhD, is a faculty member in the Institute for Educational Initiatives and a Concurrent Associate Professor in the Department of Psychology at the University of Notre Dame. He previously served as Deputy Director of the Center for Youth, Family, and Community Partnerships at the University of North Carolina at Greensboro. He has directed a local collaborative, community–university partnership featuring the involvement of schools, parents, law enforcement, mental health practitioners, the faith community, and juvenile justice professionals. He has a bachelor's degree in psychology and masters and doctoral degrees in human development and family studies.

Bridget Freisthler

Bridget Freisthler, PhD, is an assistant professor of social welfare at the University of California, Los Angeles, and an affiliated Research Scientist with the Prevention Research Center. Her research focuses on understanding how social problems vary across geographic areas, and identifying those areas in a community that are at risk for developing or already experiencing high levels of social problems based on a growing understanding of neighborhood ecologies.

Risa Friedman

Risa Friedman is an Education Coordinator and Instructor at the Mountain Plains AIDS Education and Training Center in Denver, Colorado. Her research interests include evaluation of HIV prevention and education programs and cultural competency in the delivery of HIV services.

Samuel R. Friedman
Samuel R. Friedman, PhD, is Senior Research Fellow and the Director of the Theoretical Synthesis Core in the Center for Drug Use and HIV Research at NDRI, and is an author of over 350 publications on HIV, STI, and drug use epidemiology and prevention, as well as being a published poet. His current research includes how some long-term drug injectors remain uninfected with HIV and HCV; socioeconomic and policy predictors of the extent of injection drug use, HIV epidemics, and HIV prevention efforts in US metropolitan areas; and how war and sociopolitical transitions affect HIV. He is Associate Editor for Social Science of the International Journal of Drug Policy and has recently been on the editorial boards of AIDS, JAIDS, AIDScience, AIDS Education and Prevention, Drug and Alcohol Professional, and Harm Reduction Journal. friedman@ndri.org

Michele Jonsson
Dr. Jonsson Funk is a Research Assistant Professor in the Department of Epidemiology at the University of North Carolina at Chapel Hill. Her research focuses on problems in pharmacoepidemiology and epidemiologic methods, including propensity score analysis and doubly-robust estimation of treatment effects.

Sandro Galea
Sandro Galea, MD, DrPH, is Associate Professor of Epidemiology, Center for Social Epidemiology and Population Health Department of Epidemiology, University of Michigan, and an Adjunct Associate Professor of Epidemiology, Mailman School of Public Health, Columbia University.

Sucharita Gopal
Sucharita Gopal is a professor of geography and environment at Boston University. Her areas of research include neural network applications in remote sensing and GIS domains, spatial analysis for environmental and public health applications, spatial data mining through visualization and knowledge engineering, and multi-scale spatial modeling. She has published widely on these topics in multi-disciplinary journals. suchi@bu.edu

Karla Gostnell
Karla Gostnell, MPH, is a senior research associate at the Center for Drug Use and HIV Research at NDRI. Her professional interests include cross-cultural health issues and the socio-spatial dimensions of public health. She majored in geography at the University of Washington, and received her Masters in Public Health in Sociomedical Sciences at the Mailman School of Public Health at Columbia University in 2005.

Traci Green
Traci Craig Green, MSc, MPhil, is a PhD candidate in the Division of Chronic Disease Epidemiology at the Yale School of Public Health, a Pre-Doctoral Fellow

at the Yale Center For Interdisciplinary Research on AIDS, and a recipient of the National Institute on Drug Abuse Kirschstein National Research Service Award Pre-Doctoral Fellowship. Her research interests pertain to HIV/AIDS epidemiology and the social determinants of health with a focus on injecting drug users, prescription opioid abuse, women who use drugs, and structural interventions. traci.c.green@yale.edu

Paul J. Gruenewald

Dr. Paul J. Gruenewald is Scientific Director and Senior Research Scientist at Prevention Research Center, Pacific Institute for Research and Evaluation, Berkeley, California. His research focuses upon studies of the social, economic and physical availability of alcohol, alcohol use and alcohol-related problems, mathematical and statistical models of alcohol use and abuse, the development of methods for community-based evaluations of preventive interventions, and the environmental prevention of violence. Dr. Gruenewald also directs the Spatial Systems Group, a coordinating center for research using Geographic Information Systems, Spatial Statistical Systems, and Spatial Dynamic Models. He has been a Principal or Co-investigator on 20 funded research projects and is a MERIT Award researcher with the National Institute on Alcohol Abuse and Alcoholism.

Dodi Hathazi

Dodi Hathazi, BS, has applied ethnographic methods to recruit, track, and interview a cohort of young injection drug users in Los Angeles. Her research interests include pregnancy and reproductive health issues as well as healthcare utilization among homeless youth.

Jerald R. Herting

Jerald R. Herting is a research faculty member in the University of Washington's Department of Sociology and School of Nursing. He has been engaged in NIH and CDC funded research on adolescent physical and mental health and the role of health on the transition to young adulthood. Of particular interest is the role of context on adolescent health and social behaviors.

Eleazer Hunt

Dr. Eleazer D. Hunt, PhD, has been working in the area of GIS since 1981. Prior to coming to High Point, he worked for ESRI as a Project Manager in the area of Public Safety – leading the development of a National Institute of Justice funded crime analysis extension for ArcView, and consulting with law enforcement agencies around the world. He received his PhD in anthropology from State University of New York at Buffalo. Lee.hunt@highpointnc.gov

David M. Kennedy

David M. Kennedy joined the faculty of John Jay College of Criminal Justice. From 1991 to 2004, David was a senior researcher at the Program in Criminal Justice Policy and Management, Kennedy School of Government, Harvard University. He is the co-author of a seminal work on community policing, 'Beyond 911: A New Era for Policing', and numerous articles on police management, illicit drug markets, illicit firearms markets, youth violence, and deterrence theory. He directed the Boston Gun Project, a groundbreaking problem-solving policing exercise aimed at serious youth violence.

Lauren J. Krivo
Lauren J. Krivo is a Professor of Sociology at the Ohio State University. Her current research focuses on explaining race-ethnic differences in neighborhood crime, and the interrelationships among residential segregation, socioeconomic disadvantage, and crime. With Ruth D. Peterson and John Hagan, she is the co-editor of a recent book The Many Colors of Crime published by New York University Press (2006).

Mei-Po Kwan
Mei-Po Kwan is a Distinguished Professor of Social and Behavioral Sciences and Professor of Geography at the Ohio State University, and she is an Adjunct Professor at the Department of Epidemiology and Biostatistics in the School of Medicine of Case Western Reserve University. She is Editor of the Annals of the Association of American Geographers and section editor of the International Encyclopedia of Human Geography (10 volumes, Elsevier). Her research interests include GIS and public health, geography of crime, human activity–travel behavior, gender/ethnic issues in job access and urban travel, and information and communication technologies. kwan.8@osu.edu

Steve E. Lankenau
Stephen E. Lankenau, PhD, a sociologist, applies qualitative and quantitative methodologies to the study of high-risk, out-of-treatment populations, including homeless youth, injection drug users, and others involved in the informal street economy. Currently, Dr. Lankenau is a principal investigator on a four-year study examining prescription drug misuse among high-risk youth in New York and Los Angeles. slankenau@chla.usc.edu

Elizabeth LaScala
Dr. Elizabeth A. LaScala has served for over a decade as a Consulting Research Scientist at Prevention Research Center in Berkeley, California. She earned her MPH from University of Michigan and her PhD in Public Health from University of California, Los Angeles, with a specialty in the Behavioral Sciences and Health Education. At Prevention Research Center, her research includes ecological studies of alcohol-related problems, such as violence, accidents and injuries and, most recently, binge drinking on college campuses. Dr. LaScala has extended her interests to include the ecology of illegal drug markets.

Nancy LaVigne
Nancy LaVigne is a Senior Research Associate at the Urban Institute's Justice Policy Center, where she directs several projects involving the spatial analysis of crime and corrections data. Prior to her current position, LaVigne was Founding Director of the National Institute of Justice's Crime Mapping Research Center. LaVigne holds a PhD from Rutgers University, a master's degree from the LBJ School of Public Affairs at the University of Texas, and a BA from Smith College.

Spencer Lieb

Spencer Lieb, MPH, is Senior Epidemiologist with the Florida Department of Health, Bureau of HIV/AIDS. His work has focused on AIDS epidemiology since 1981 in Florida. He has authored or co-authored numerous papers on HIV/AIDS- and IDU-related issues, including several in collaboration with Dr. Sam Friedman and colleagues at NDRI. His main interest is in utilizing routinely collected HIV/AIDS surveillance data in non-routine ways to address a number of public health problems.

Shui Shan Lee

Shui Shan Lee is Professor of Infectious Diseases at the Stanley Ho Centre for Emerging Infectious Diseases, The Chinese University of Hong Kong. His main area of research interest is infectious disease epidemiology. sslee@cuhk.edu.hk

Remedios Lozarda

Remedios Lozarda, MD, is the HIV/STD Coordinator for the Mexican state of Baja California and runs Tijuana's Municipal AIDS Committee (Patronato Pro-Comusida, Tijuana A.C.). In 1991, Dr. Lozarda founded the Center for Integration and Recovery of Drug and Alcohol Addicts (CIRAD, A.C.), and she has also been an active collaborator on a number of projects involving HIV-positive marginalized populations. In 2007, Dr. Lozarda was recognized by the US–México Border Health Commission as a Border Hero for her work to improve public health in the region.

Max Lu

Max Lu is an Associate Professor of Geography at Kansas State University. He received his BS from Zhejiang University in 1983, his MS from the Chinese Academy of Sciences in 1986, and his PhD from Indiana University, Bloomington, in 1996. His current research interests include population migration and distribution, population and environment, and spatial patterns of health outcomes and crime. maxlu@ksu.edu

Silvana Mazzella

Mrs. Mazzella is currently the Local Health Systems Planner for the Division of Maternal, Child, and Family Health of the Philadelphia Department of Public Health, and has worked with the Health Department since 2000. Mrs. Mazzella has an extensive background in community organizing, community development, and community needs assessments, and prior to her work with the city of Philadelphia, she ran a city and statewide chapter providing recruitment consulting services to non-profits. Mrs. Mazzella has a Bachelor's Degree in Social Work and a Master's Degree in Geography and Urban Studies from Temple University.

Sara McLafferty

Sara McLafferty is a Professor of Geography at the University of Illinois at Urbana-Champaign. Her research investigates place-based disparities in health and access to health care for women, immigrants and racial/ethnic minorities in US cities, and the use of GIS and spatial analysis methods in examining health disparities. Her books include GIS and Public Health (with Ellen Cromley) and Geographies of Women's Health (with Isabel Dyck and Nancy Lewis). smclaff@uiuc.edu

Jeremy Mennis

Jeremy Mennis (PhD in Geography, Pennsylvania State University) is an Associate Professor in the Department of Geography and Urban Studies at Temple University. He is currently involved in a number of research projects which incorporate modeling neighborhood effects on individual behavior outcomes in the context of mental health, adolescent substance abuse, and juvenile delinquency.

Dominique Moran

Dominique Moran is lecturer in the geographies of transitional economies at the School of Geography, Earth and Environmental Sciences at the University of Birmingham, and an associate member of the Centre for Russian and East European Studies at that university. She is currently working on an ESRC-funded project exploring the geography of contemporary women's imprisonment in post-Soviet Russia, and has ongoing research interests in HIV/AIDS, specifically in examining the relationship between HIV prevalence and the circumstances of post-socialist transformation. d.moran@bham.ac.uk

Danilo Norori

Danilo Norori is a social worker, a musician, and a songwriter involved in treatment counseling and advocacy. His last two records are dedicated to drug use prevention: 'Cantarapia'.

Phoebe T.T. Pang

Phoebe T.T. Pang is a research assistant at the Stanley Ho Centre for Emerging Infectious Diseases, The Chinese University of Hong Kong. Her research interest is the application of geographic information systems in public health.

Ruth D. Peterson

Ruth D. Peterson is a Distinguished Professor of Social and Behavioral Sciences, Professor of Sociology, and Director of the Criminal Justice Research Center at the Ohio State University. Her research focuses on community conditions and crime, racial and ethnic inequality in patterns of crime, and the consequences of criminal justice policies for racially and ethnically distinct communities. She is Co-principal Investigator with Lauren J. Krivo on a project 'Race/Ethnicity, Crime, and Criminal Justice: Diverse Research and Participation in the Academy' funded by the National Science Foundation.

Cynthia Pope

Cynthia Pope, PhD, is an Associate Professor in the Department of Geography at Central Connecticut State University and a Visiting Lecturer at the Yale School of Public Health. Her publications, teaching, and research focus on gendered dimensions of HIV/AIDS risk as a product of globalization and nation building in Latin America, the Caribbean, and East Africa.

Lillian G. Remer

Lillian G. Remer is an Associate Research Scientist at the Prevention Research Center of the Pacific Institute for Research and Evaluation and a founding member of the Spatial Systems Group. Her primary expertise is in the area of database design and development including Geographic Information Systems and in geospatial analysis.

George F. Rengert

George F. Rengert is Professor of Criminal Justice at Temple University in Philadelphia. His area of specialty is the spatial and temporal behavior of criminals. He has written or edited seven books on the geographic aspects of crime.

Douglas Richardson

Douglas Richardson is the Executive Director of the Association of American Geographers (AAG). Prior to joining the AAG, Dr. Richardson had founded and was the President of GeoResearch, Inc., which developed and patented the first real-time interactive GPS/GIS technology and pioneered its embodiment in revolutionary new tools for real-time mapping and modeling, and for a wide range of interactive GPS/GIS applications in government, business, and society. He sold the company and its key patents in 1998. drichardson@aag.org

Kevin Romig

Kevin Romig is an Assistant Professor of Geography at Texas State University-San Marcos. His main research interests are in urbanization, community and citizenship issues, and geographic dimensions of popular culture. This research was supported by the University of North Dakota and largely conducted while Dr. Romig had an appointment at that institution. Kevin_Romig@txstate.edu

Bill Sanders

Bill Sanders, PhD, a sociologist, studies health risk behaviors, such as substance abuse, and violence and unsafe sexual practices among high-risk youth, such as injection drug users, the homeless, and gang-identified youth. Currently, Dr. Sanders is a principal investigator on a study focused on gang youth in Los Angeles, and had published Drugs, Clubs and Young People: Sociological and Public Health Perspectives in 2006.

Tom J. Scholten

Tom J. Scholten received his bachelor's degree from Grand Valley State University and his master's degree in geography from the University of North Carolina at Greensboro. He has worked as a GIS consultant for law enforcement, focusing on crime analysis and mapping community demographic conditions.

Michele G. Shedlin

Michele G. Shedlin, PhD, is the Leavell Professor of Nursing and Health Sciences; Co-Director, Hispanic Health Disparities Research Center; and Professor, Department of Sociology and Anthropology at the College of Health Sciences, University of Texas at El Paso. She is a medical anthropologist, currently involved in teaching and research on the risk factors for drug use and HIV/AIDS among mobile and immigrant Hispanic populations. mshedlin@utep.edu

Karen A. Snedker
Karen A. Snedker is an Assistant Professor of Sociology at Seattle Pacific University and a Research Scientist at the University of Washington's School of Nursing. Her interests include crime and community, fear of crime and risk assessment, and adolescent health and delinquency. Currently, she is working on several NIH grants to investigate the influence of neighborhood context on health and risky behaviors among youth. snedker@u.washington.edu

Ralph Spiga
Ralph Spiga, PhD, is Associate Professor of Psychiatry and Director of Research in the Department of Psychiatry, Temple University of Medical School. His research focus is substance abuse and dependence and comorbid mental health and substance use disorders.

Gerald Stahler
Gerald Stahler (MA in Geography, University of Toronto; PhD in Clinical Psychology, Temple University) is an Associate Professor in the Department of Geography and Urban Studies at Temple University. His research interests have focused on developing and evaluating community-based interventions for homeless substance abusers and, more recently, on examining the neighborhood influences on continuity of care, relapse, and recidivism for substance abusers. Formerly the Associate Vice-Provost for Research and Graduate Studies, his research has been funded by the National Science Foundation (NSF), the National Institute on Drug Abuse (NIDA), the National Institute on Alcohol Abuse and Alcoholism (NIAAA), the Center for Substance Abuse Treatment (CSAT), and the Pennsylvania Department of Health. Jstahler@temple.edu

Steffanie A. Strathdee
Steffanie A. Strathdee, a professor, is an infectious disease epidemiologist who has spent the last two decades focusing on underserved, marginalized populations in developed and developing countries. Since January, 2004, she has been the Harold Simon Chair and Chief of the Division of International Health and Cross Cultural Medicine in the Family and Preventive Medicine Department at the University of California, San Diego. Her recent work has focused on the prevention of blood-borne infections and barriers to care among injection drug-using populations, including leading a NIDA funded project (Proyecto El Cuete) to characterize the epidemiology of HIV and tuberculosis among injecting drug users in Tijuana, Mexico.

Marty A. Sumner
Major Marty A. Sumner is a twenty-three-year veteran with the High Point, North Carolina Police Department, and a graduate of John Wesley College. Major Sumner is the Chief of Staff currently responsible for the Major Crime Deterrence and Prevention Division. Major Sumner is a leader in the Department's violence

reduction strategies; he has worked closely with community members to insure the success of violence- and drug-reduction efforts.

Aaron Swoboda
Aaron Swoboda is an Assistant Professor with the Graduate School of Public and International Affairs at the University of Pittsburgh. He holds a PhD from the University of California, Berkeley. His research interests include land-use, spatial econometrics, and environmental economics.

Kenneth J. Tardiff
Kenneth J. Tardiff, MD, MPH, is Professor of Psychiatry, Department of Public Health, Weill Medical College of Cornell University, New York, NY.

Barbara Tempalski
Barbara Tempalski, PhD, MPH, received her doctorate degree in geography at the University of Washington in Seattle. She works as project director for the Community Vulnerability and Response to IDU-related HIV study at the Center for Drug Use and HIV Research in New York City. Her research examines the geographic variability in injection drug use-related HIV transmission and program response. tempalski@ndri.org

Yonette F. Thomas
Yonette F. Thomas, PhD, is the Chief of the Epidemiology Research Branch in the Division of Epidemiology, Services, and Prevention Research at the National Institute on Drug Abuse (NIDA) of the National Institutes of Health. Dr. Thomas is a medical sociologist/social epidemiologist with postgraduate training in demography and epidemiology. ythomas@mail.nih.gov

Stephanie Tortu
Stephanie Tortu, PhD, a social psychologist, has studied the relationship between social context and women's risk for HIV/AIDS as well as social and behavioral factors impacting risk for HIV/AIDS and hepatitis C among drug-involved populations. Dr. Tortu has taught at several colleges and universities, presented at national and international conferences, edited two books, and authored many scientific publications.

Andrew J. Treno
Andrew J. Treno received his PhD in sociology from the University of California at Berkeley in 1983. He has been at the Prevention Research Center for over 17 years. His primary area of interest is alcohol problem prevention and evaluation. He has published in the areas of community mobilization, alcohol involvement in injury, and evaluation of environmental approaches to the prevention of alcohol-related problems. He recently served as Project Director for the Sacramento Neighborhood Alcohol Prevention Project (SNAPP). In this capacity, he has given numerous community presentations on the project and has successfully obtained CSAP funding in support of project interventions. He has also worked to adapt many of the evaluation instruments from the Community Trials project to the SNAPP project.

Avelardo Valdez
Dr. Avelardo Valdez is a professor at the Graduate College of Social Work, University of Houston, and Director of the Office for Drug and Social Policy Research. His research and publications focus on the health and social consequences of drugs, violence, gangs, high risk sexual behaviors, and drug markets in South Texas and the US/Mexico border. avaldez2@uh.edu

Mark Vanelli
Mark Vanelli, MD, MBA, is a faculty member of the Levinson Institute and Harvard Medical School. He is a practicing psychiatrist and conducts CME classes at Harvard Medical School's Division of Continuing Education. He obtained his BA from Harvard University, Master's in Public Health at Brown University School of Medicine, and Psychiatry Residency at Harvard Medical School. He has published and conducted research in healthcare.

David Vlahov
David Vlahov, PhD, is director of the Center for Urban Epidemiologic Studies, New York Academy of Medicine, and Professor of Clinical Epidemiology, Mailman School of Public Health, Columbia University, New York, NY

John R. Weeks
John R. Weeks, PhD, is Professor of Geography and Director of the International Population Center at San Diego State University (SDSU). His research focuses on geodemographic analyses of the US–Mexico border region, and on spatial analyses of health outcomes in cities of developing countries. He is also the author of a bestselling textbook in the field of population studies.

Xiushi Yang
Xiushi Yang, PhD, is Professor of Sociology and Criminal Justice, Old Dominion University. His research focuses on migration, HIV risk drug using and sexual behaviors, and behavioral intervention of HIV risk behaviors. xyang@odu.edu